106 Springer Series in Solid-State Sciences

Edited by Peter Fulde

Springer Series in Solid-State Sciences

Editors: M. Cardona P. Fulde K. von Klitzing H.-J. Queisser

Managing Editor: H. K. V. Lotsch Volumes 1–89 are listed at the end of the book

90 **Earlier and Recent Aspects of Superconductivity** Editors: J. G. Bednorz and K. A. Müller

91 **Electronic Properties of Conjugated Polymers III** Basic Models and Applications
Editors: H. Kuzmany, M. Mehring, and S. Roth

92 **Physics and Engineering Applications of Magnetism** Editors: Y. Ishikawa and N. Miura

93 **Quasicrystals** Editors: T. Fujiwara and T. Ogawa

94 **Electronic Conduction in Oxides** By N. Tsuda, K. Nasu, A. Yanase, and K. Siratori

95 **Electronic Materials** A New Era in Materials Science
Editors: J. R. Chelikowsky and A. Franciosi

96 **Electron Liquids** By A. Isihara

97 **Localization and Confinement of Electrons in Semiconductors**
Editors: F. Kuchar, H. Heinrich, and G. Bauer

98 **Magnetism and the Electronic Structure of Crystals**
By V. A. Gubanov, A. I. Liechtenstein, and A.V. Postnikov

99 **Electronic Properties of High-T_c Superconductors and Related Compounds**
Editors: H. Kuzmany, M. Mehring, and J. Fink

100 **Electron Correlations in Molecules and Solids** By P. Fulde

101 **High Magnetic Fields in Semiconductor Physics III**
Quantum Hall Effect, Transport and Optics By G. Landwehr

102 **Conjugated Conducting Polymers** Editor: H. Kiess

103 **Molecular Dynamics Simulations** Editor: F. Yonezawa

104 **Products of Random Matrices in Statistical Physics**
By A. Crisanti, G. Paladin, and A. Vulpiani

105 **Self-Trapped Excitons** By K. S. Song and R. T. Williams

106 **Physics of High-Temperature Superconductors** Editors: S. Maekawa and M. Sato

107 **Electronic Properties of Polymers** Orientation and Dimensionality of Conjugated Systems
Editors: H. Kuzmany, M. Mehring, and S. Roth

108 **Site Symmetry in Crystals** Theory and Applications
By R. A. Evarestov and V. V. Smirnov

S. Maekawa M. Sato (Eds.)

Physics of High-Temperature Superconductors

Proceedings of the Toshiba International School of Superconductivity (ITS^2), Kyoto, Japan, July 15–20, 1991

With 211 Figures

Springer-Verlag
Berlin Heidelberg New York
London Paris Tokyo
Hong Kong Barcelona
Budapest

Professor Sadamichi Maekawa
Department of Applied Physics, Nagoya University,
Nagoya 464-01, Japan

Professor Masatoshi Sato
Department of Physics, Nagoya University,
Nagoya 464-01, Japan

Series Editors:
Professor Dr., Dres. h. c. Manuel Cardona
Professor Dr., Dr. h. c. Peter Fulde
Professor Dr., Dr. h. c. Klaus von Klitzing
Professor Dr., Dr. h. c. Hans-Joachim Queisser
Max-Planck-Institut für Festkörperforschung, Heisenbergstrasse 1,
W-7000 Stuttgart 80, Fed. Rep. of Germany

Managing Editor:
Dr. Helmut K. V. Lotsch
Springer-Verlag, Tiergartenstrasse 17, W-6900 Heidelberg, Fed. Rep. of Germany

ISBN-13: 978-3-642-84720-2 e-ISBN-13: 978-3-642-84718-9
DOI: 10.1007/978-3-642-84718-9

Library of Congress Cataloging-in-Publication Data. Toshiba International School of Superconductivity (1991 : Kyoto, Japan) Physics of high-temperature superconductors : proceedings of the Toshiba International School of Superconductivity (ITS^2) Kyoto, Japan, July 15–20, 1991 / S. Maekawa, M. Sato, eds. p. cm.–(Springer series in solid-state sciences ; 106) Includes bibliographical references and index. ISBN 0-387-55107-7 1. High temperature superconductors–Congresses. 2. High temperature superconductivity–Congresses. I. Maekawa, S. (Sadamichi), 1946– . II. Sato, M. (Masatoshi), 1946– . III. Title. IV. Series. QC611.98.H54T67 1991 537.6'23–dc20 91-47852

Softcover reprint of the hardcover 1st edition 1992

Typesetting: Camera-ready by authors
54/3140 – 5 4 3 2 1 0 – Printed on acid-free paper

Preface

This volume contains the proceedings of the *Toshiba International School of Superconductivity* (TIS^2), which was held at the Holiday Inn Kyoto in Kyoto, Japan, July 15–20, 1991. This was the week preceding the Third International Conference on Materials and Mechanisms of Superconductivity, High-Temperature Superconductors (M^2S-HTSC III), which was held in Kanazawa, Japan. About 200 young scientists and roughly 50 lecturers and chairmen had been invited from 25 countries.

The purpose of the school was to give young scientists, including graduate students, an opportunity to attend tutorial lectures on high-temperature superconductivity and to provide them with an introduction to the M^2S-HTSC Conference. One of the key lectures, given by Professor J.R. Schrieffer, was entitled "The Pairing Theory – Its Physical Basis and Its Consequences". A further aim of the school was to discuss a variety of aspects of high-temperature superconductors. Therefore, this volume comprises lecture notes and summaries of discussion meetings as well as research reports.

We are much indebted to the promoters of the international school, Dr. J.G. Bednorz and Professor K.A. Müller of IBM Zürich Research Laboratory, for their active contribution. At all stages, from identifying the scope to setting up the final program, we received valuable advice and suggestions. On behalf of all the participants, we would like to express our sincere gratitude to the Toshiba Corporation for its financial support, which made the school possible. Finally, we acknowledge the authors of the manuscripts in this volume for promptly preparing their contributions.

Nagoya
September 1991

S. Maekawa
M. Sato

School Organization

International School Promoters

J.G. Bednorz	(IBM Zürich)
K.A. Müller	(IBM Zürich)

Organizing Committee

S. Maekawa	(Nagoya University): Chairman
K. Kitazawa	(University of Tokyo)
H. Fukuyama	(University of Tokyo)
M. Tachiki	(Tohoku University)
Y. Endoh	(Tohoku University)
T. Fujita	(Hiroshima University)
M. Sato	(Nagoya University)
S. Uchida	(University of Tokyo)

Steering Committee

I. Hirabayashi	(ISTEC)
J. Inoue	(Nagoya University)
S. Maekawa	(Nagoya University)
T. Matsuura	(Nagoya University)
M. Sato	(Nagoya University)
T. Shinjo	(Kyoto University)

Contents

II.3 Superconducting Mechanism and Properties

Part III Experiment

III.1 Materials

III.2 Physical Properties and Electronic Structure

Part I

Plenary Lecture

The Pairing Theory – Its Physical Basis and Its Consequences

J.R. Schrieffer

Department of Physics, University of California, Santa Barbara, CA 93106, USA

ABSTRACT: The key developments which set the scene for the microscopic theory of superconductivity are discussed and the physical reasoning which lead to the pairing theory is presented. Consequences of the BCS theory are reviewed.

I. INTRODUCTION

Following the discovery of superconductivity in 1911, a 50-year search for the proper microscopic theory of this phenomenon was pursued by many of the foremost theoretical physicists at that time. To put in proper perspective the final resolution of this problem, it is helpful to recall a few key steps along the way.[1]

For two decades after its discovery, superconductivity was considered, in essence, to be a state of zero electrical resistance. That is, if an electric current is set up in a loop, the current will circulate indefinitely without dissipation exhibiting electrical perpetual motion. Not until 1933 with the discovery of the Meissner effect did it became evident that a totally different phenomena was actually involved, namely that of perfect diamagnetism. The essential point is that the currents which flow in a superconductor are very much like the diamagentic current which flows in the ground state of a hydrogen atom when it is placed in a magnetic field, namely, the 1s wave function is essentially, unaltered by the magnetic field and currents flow in the 1s orbital is due to the A term not the P term in the current density.

These diamagentic currents shield the external field only weakly in the hydrogen atom, but the shielding is nearly perfect in a bulk superconductor. When the magnetic field is removed the currents are again reduced to zero. This Meissner state or state of flux expulsion is the lowest free energy state of a bulk superconductor so long as the magnetic field is less than a critical field H_C. We now know that a state intermediate between this Meissner state and the normal state can be achieved in so-called type II superconductors with the magnetic field entering as quantized flux tubes which occupy only a portion of the volume, thereby leading to coexistence of magnetic fields and superconductivity in a bulk superconductor.

Soon after the discovery of perfect diamagnetism, Hans and Fritz London advanced a phenomenological relationship between the current density j and the magnetic vector potential A, now known as London's equation,

$$\vec{j}(\vec{r}) = -c\Lambda\vec{A}(\vec{r}). \tag{1}$$

The constant Λ is a phenomenological parameter which is related to the magnetic penetration depth λ in the superconductor by

$$\lambda^2 = 4\pi\Lambda, \tag{2}$$

which follows by combining London's equation with the last Maxwell equation

$$-\nabla^2 A = \frac{4\pi j}{c}. \tag{3}$$

Fritz London quickly suggested a possible quantum mechanical interpretation of equation 1. He proposed that if the ground state wave function of the superconductor is essentially unchanged by the application of the magnetic field (so-called London rigidity), then the London equation follows automatically. This can be readily seen by noting that if $\Psi_B \simeq \Psi_0$ then the current density satisfies

$$\langle \vec{j_B} \rangle_B = -\frac{\rho_s e}{m}\left\langle \vec{p} + \frac{e\vec{A}}{c}\right\rangle_0 = -\frac{\rho_s e^2}{mc}\vec{A}, \tag{4}$$

where ρ_s is the electron density in the superfluid. One can view the electrons in the superconductor as divided into two copenetrating fluids, one of superfluid density ρ_s and the other a normal fluid of density ρ_n having the traditional properties of electrons in a normal metal. Thus the total electron density is divided as

$$\rho = \rho_s + \rho_n, \tag{5}$$

where ρ_s vanishes at the transition temperature T_c and ρ_n vanishes at zero temperature. Combining equations (2) and (4) with $\rho_s = \rho$ at low temperature one obtains fairly good agreement with the observed magnetic penetration depth at low temperature in both conventional superconductors and particularly in high T_c materials.

The London equation was generalized to superconductors having non-uniform superfluid density by Ginzburg and Landau who postulated that the free energy difference between the normal and superconducting phase can be expressed near the transition temperature in terms of a complex c-number order parameter $\Psi(r)$ as

$$F_S - F_N = \int \left\{ \frac{a \mid \Psi(r) \mid^2}{2} + \frac{b \mid \Psi(r) \mid^4}{4} + \frac{c}{2} \mid (\nabla + i\frac{e^* \hbar}{c} \overrightarrow{A}) \Psi \mid^2 \right\} d\overrightarrow{r}. \quad (6)$$

The coefficient a is taken to vary linearly with temperature, vanishing at T_C and being negative below the transition temperature.

$$a = \left(\frac{T - T_C}{T_C} \right) a'. \quad (7)$$

Thus, the first two terms exhibit a stable minimum at $\Psi = 0$ for T greater than T_C while a double well potential appears for less than T_C with minima occuring at $\pm\Psi_0$ given by

$$\Psi_0(T) = \sqrt{\frac{\mid a(T) \mid}{b}}. \quad (8)$$

This phenomenological approach gives an excellent account of the macroscopic features of conventional superconductors and also works reasonably well when extended to the anisotropic quasi two-dimensional high temperature superconducting phases.

Other key developments which laid the groundwork for the pairing theory include 1) the discovery of the isotope effect, namely that the transition temperature T_C varys with the isotopic mass M as

$$T_C \sim \frac{1}{M^\alpha}, \quad (9)$$

with α being of order $\frac{1}{2}$. Early theories by Fröhlich and by Bardeen based on the electron-phonon interaction were consistent with such a behavior, however, the theories were too crude to account for superconductivity itself.

2) Drawing on a variety of experimental data, Pippard concluded that the local relationship between j and A in the London equation must

be generalized to a nonlocal form where the range of the kernal is given by the so-called coherence length ξ, whose magnitude in conventional superconductors is between $10^3 \mathring{A}$ and $10^4 \mathring{A}$ and is much shorter in high temperature superconductors being of order $15\mathring{A}$ in the ab plane and as small as $3\mathring{A}$ in the c direction of the layered high T_C materials,

$$\vec{j}(\vec{r}) = \int K(\vec{r} - \vec{r}')\vec{A}(\vec{r}')dr', \tag{10}$$

When the microscopic pairing theory was completed, the physical interpretation of the Ginzburg/Landau order parameter $\Psi(r)$ and the coherence length ξ became clear. Ψ is the center of mass wave function of a superconducting pair, with amplitude proportional to the square root of the number of condensed superconducting electrons, while ξ represents the mean spacing between mates of a given condensed pair.

3) Several authors including Fröhlich, Nakajima, Bardeen and Pines worked out the effective interaction between a pair of electrons due to their coupling to the phonon field. They found an effective attraction which suggested a possible mechanism to bring about the superconducting condensation. The form of the attractive potential was found to be consistent with the isotope effect given by equation (9).

At this point, in 1955, John Bardeen and Bernard Serin wrote a pair of splendid review articles summarizing the theoretical and experimental situation at that date. The stage was then set for the final assault on the problem.

II. The Pairing Theory

In the fall of 1955 John Bardeen, Leon Cooper and I joined in an attempt to construct a microscopic theory of superconductivity. Bardeen patiently explained to us the critical features which a successful theory must contain. He stressed the concepts a) the existence of an energy gap above the ground state of the superfluid which would lead to the London rigidity and account for the Meissner effect, b) the significance of Pippard's coherence length which is very large compared to the lattice spacing implying that only electrons within a very small energy of the Fermi surface are involved in the superfluid condensation, c) that the condensation must be thought of as occuring in momentum space and, d) that a very strong overlap of the relevant electronic states in real space are involved in the condensation rather than a dilute Bose gas type of condensation which was under consideration by Blatt and coworkers. Arguing that the proper condensation energy occurs when a fraction T_c/T_F of the electrons lower their energy by $k_B T_c$ where T_F is the Fermi temperature one obtains from the uncertainty principle good agreement with the coherence length as given by

$$\xi \sim \frac{\hbar v_F}{k_B T_c}. \tag{11}$$

Aside from a numerical factor of order unity, this relationship is consistent with the pairing theory. In addition M.R. Schafroth had proven that the Meissner effect can occur only if one goes beyond infinite order perturbation theory. That is, one must seek solutions which are not analytic in the coupling constant.

With this in mind Bardeen, Cooper and I began exploring various non-analytic approaches to the coupled electron-phonon system. In an attempt to understand how an energy gap had come about, Cooper studied the problem of two electrons interacting above a frozen Fermi sea via an attractive potential. The electrons in the Fermi sea were assumed not to be dynamically involved in the problem but simply act as blocking particles which act through the Pauli principle to forbid the two chosen electrons from enetering states below the Fermi momentum. In this case Cooper showed that regardless how weak the attractive potential, a bound state of the two particles is formed with binding energy given by

$$E_B = \omega_c exp[-2/N(0)V], \tag{12}$$

where ω_c is a cutoff energy for the electrons interacting via the attractive potential whose matrix element is V. $N(0)$ is the density of states in this small region above the Fermi surface. Cooper's result was important in two respects. Firstly, it was nonanalytic in the coupling constant $N(0)V$. Also, the cutoff frequency ω_c is of order the Debye frequency according to the Fröhlich interaction and therefore if E_B is of order the transition temperature T_c, then the isotope effect is obtained. While there is an energy gap for tearing apart members of Cooper's bound pair, there is no gap for the excitation of the center of mass coordinate of the pair. Rather, the energy of the pair varies as the kinetic energy of a composite system having a total mass twice that of an electron. Since the London derivation and many of the other consequences of superconductivity including an exponential temperature dependence of the thermal conductivity and heat capacity at low temperature all depend upon an energy gap for all electronic excitations, it is clear that this single pair approximation has no direct bearing on superconductivity. Rather, it is to be viewed as indicating an instability of the standard Landau Fermi liquid theory toward a new phase, the details of which may or may not have to do with some type of bound pairs.

The successful construction of a microscopic theory which resolves this instability was fraught with many apparently insurmountable difficulties. Namely, the very strong overlap of pairs would at first sight totally destroy the concept of individual bound pairs and replace them by a strongly correlated fermionic system. As mentioned above, Blatt and coworkers investigated the limit in which the size of the bound state ξ was small compared to the mean spacing d between pairs. As Bardeen stressed, this is exactly the opposite limit to that occuring in nature since roughly 10^6 pairs live within the volume ξ^3 occupied by a given pair.

We attempted many schemes to account for this very strong pair overlap and the influence of the dynamics of the Fermi sea on the paired dynamics considered by Cooper. Firstly, it was clear that particle-hole symmetry must be maintained; that is, the wave function of the many body system should look similar with regard to electronic coordinates above the Fermi surface and with regard to hole coordinates below the Fermi surface. In addition, the operators b_k^+ which create a pair in the momentum states $k - k$ shown

$$b_k^+ = c_{k\uparrow}^+ c_{-k\downarrow}^+, \tag{13}$$

do not satisfy Bose's statistics but rather satisfy mixed statistics. That b_k^+ and b's commute for different values of k, but they satisfy the Pauli principle for the same value of k as is evidenced by the amplitude for creating two pairs in the same state being identically zero.

$$b_k^{+2} = 0. \tag{14}$$

Equation (14) follows from Equation (13) and the standard computation relations for fermions. While at first sight the phase space for two k's being equal rather than k's being different is of zero measure and, hence, might be viewed as zero importance. In fact exactly the reverse is true and the constraint of Equation (14) is crucial in the microscopic theory of superconductivity. Hence, one cannot view the BCS superconductor as a boson condensation since the Pauli principle plays an absolutely crucial role in the problem. On the other hand, in the extremely strong coupling unit, when the size of the pairs are small compared to their mean spacing, a boson condensation picture is more appropriate, however the BCS theory itself does not adequately treat this extreme limit.

In searching for an appropriate solution to the many pair system, we constructed an effective model Hamiltonian for pairs of total momentum zero given by

$$H_{pair} = \sum_{ks} \epsilon_k n_{ks} - \sum_{kk'} V_{k'k} b^{+}_{k'} b_k, \tag{15}$$

where ϵ_k represents the Landau quasi particle energy and n_{ks} is the quasi particle occupation number, while $-V_{kk'}$ is the effective interaction between electrons due to the combination of the Fröhlich phonon attraction and the effective Coulomb repulsion between the charged Landau quasi particles. Since this Hamiltonian cannot be diagonalized by analytic means, we attempted to find approximate methods to determine its ground state.

Based on Bardeen's arguments that the pair overlap is very large, I thought that a statistical approximation might be appropriate as in Heisenberg magnets where the exchange interactions are very long range or the number of nearest neighbors is very large. In such problems one may replace the exchange coupling $J_{ij} \vec{S}_i \cdot \vec{S}_j$ by its mean field equivalent $J_{ij} S_i \langle S_j \rangle$, an approximation which works well when many dynamical degrees of freedom are coupled to each individual degree of freedom. Looking at Equation (15) ones sees that this is precisely the situation for coupling between the b operators, namely each k' is coupled to many pair operators k in the interaction term. However, since b destroys a pair of electrons, its expectation value in a state which has a sharp number of electrons is identically zero.

In attempting to handle this problem, I recalled a paper I had read by Professor Tomanaga who had proposed a mean field type of state for the pion dressing of a nucleon within the static approximation. A similar approach was used by Lee, Low and Pines in treating the polaron problem. Based on these works, I guessed the ground state of the superconductor to be of the form

$$\mid \Psi_0 \rangle = \prod_k (u_k + v_k b^{+}_k) \mid 0 \rangle, \tag{16}$$

where the product extends over all k states with u and v being expansion coefficients satisfying the normalization condition

$$\mid u_k \mid^2 + \mid v_k \mid^2 = 1, \tag{17}$$

so that the overall state is normalized to unity. The state $\mid 0 \rangle$ denotes the absence of all electrons.

Using this trial wave function I minimized the energy with respect to the u's and v's and found the minimizing condition to be expressed by

$$|u_k|^2 = \frac{1}{2}\left(1 + \frac{\epsilon_k}{E_k}\right), \tag{18a}$$

$$|v_k|^2 = \frac{1}{2}\left(1 - \frac{\epsilon_k}{E_k}\right), \tag{18b}$$

where E_k is given by

$$E_k = \sqrt{\epsilon_k^2 + \Delta_k^2}, \tag{19}$$

and Δ_k satisfies the energy gap equation

$$\Delta_k = \sum_{k'} \frac{V_{kk'}\Delta_{k'}}{2E_{k'}}. \tag{20}$$

Inserting these expressions into the variational energy, one finds the condensation energy is given by

$$E_{0S} - E_{0N} = \frac{-N(0)\Delta^2}{2} = -H_c^2/8\pi. \tag{21}$$

The relation of the pairing theory to the mean field theory of antiferromagnetism is quite clear, the difference being that instead of pairing particles in time reversed momentum states as in the superconductivity problem, one pairs a particle and a hole in momentum states k and $k+Q$, respectively, both with the same spin. The mathematics of these two problems is virtually identical with the condensation energy being of the same form as in Equation (21).

Bardeen quickly recognized that the excitation spectrum above this ground state had an energy gap and that the gap is precisely given by the Δ_k. Furthermore, one could check Equation (21) from the energy gap which Tinkham and Glover had observed in superconducting lead, and the observed thermodynamic critical field H_c. The accurate agreement between these quantities for lead was the first indication that the theory had predictive powers.

The theory can be readily extended to finite temperatures considering the same mean field approximation in the partition function rather than in the groundstate. In this way Equation (20) is generalized by including a factor $tanh\frac{\beta E_{k'}}{2}$ in the right hand side of this equation. Δ becomes temperature dependent and vanishes as T approaches T_c.

A convenient way of describing the fermionic excitations above the superconducting goundstate is to introduce the Bogoliubov-Valatan transformation by

$$\gamma_{k\uparrow} = u_k c_{k\uparrow} + v_k c^+_{-k\downarrow} R, \tag{22a}$$

$$\gamma_{-k\downarrow} = u_k c^+_{-k\downarrow} R - v_k c_{k\uparrow}. \tag{22b}$$

These γ operators when acting on the groundstate give identically zero, however their adjoints create quasi-particles. In contrast, the Landau operators $c_{k\uparrow}$ and $c_{-k\downarrow}$ and their adjoints both create excitations. It is for this reason that the γ operators are to be favored in calculating physical property of superconductors.

Remarkably with this minimal input of formalism and physics, one can account for essentially all the known properties of conventional superconductors, as well as the esoteric superconductors ^3He, atomic nuclei and neutron stars. The reason for the simplicity is, in essence, twofold; one, the largest correlation energies are handled already within the Landau Fermi liquid approach and, two, while the remaining interactions produce superconductivity causing very large changes in the many body wave function, these changes are limited to states near the Fermi surface and fortunately can be handled by a generalized mean field approximation due to the largeness to the number of pairs who center of mass falls within the volume occupied by a given pair wave function. As this number becomes of order unity, as presumably it does in the high temperature superconductors, the mean field approximation become less appropirate. However, it apparently continues to work sufficiently well that the fluctuation effects even in these strongly coupled systems are relatively modest.

A final point concerning the BCS theory is that it can be viewed in the context of the BCS state given by Equation (16) in which the number of electrons fluctuates about the mean number n_0 by a fractional amount of order 10^{-11}; alternatively, one can project this state onto a state of sharp electron number as first suggested by Dyson, then use this projected state in all calculations. As one can show, the n-projected state gives results which are fully consistent with those calculated with the full wave function (16) with corrections which are again of order 10^{-11} and hence negligible for extended systems.

In this connection we note that the symbol R in Equation (22) denotes the operator which removes one pair from the superconducting condensate. The reader will note that in this way the operators γ conserve charge, namely that γ removes one electron as indicated by the operator c on the right hand side of (22a) and by the operator $c+$ which creates an electron, but is compensated for by the removal of a pair by the operator R. Frequently the operator R is neglected since within the BCS-type state, this operator acts as a c number which is normalized to unity with fluctuation effects being on the scale of 10^{-11}.

If one forms a wave packet of quasiparticle states k with spin up, this corresponds to making a hole in this wave packet with amplitude u and an electron with amplitude v but in the latter case a condensed pair has been removed uniformly from the entire condensate extending out to the boundaries of the system. Since u is large above the Fermi surface and v is large below the Fermi surface, we see that the γ's approach particle-like and hole-like operators in these two domains. A remarkable fact, however, is that if one forms a wave packet made out of quasi particle states near the Fermi surface, then u and v are nearly equal so that in this region the accumulated charge in the wave packet is zero, however the conserved total charge given by Equation (22), namely $-e$ is removed uniformly from all space by removing one pair of electrons with probability $\frac{1}{2}$. Unfortunately, the literature has somewhat confused this issue and it is frequently stated that a quasi particle at the Fermi surface is half the time an electron and half the time hole so that the average charge is zero. While this is true for the core of the quasi particle corresponding to the charge in the wave packet mentioned above, there is a compensating charge given by the R operator which, when one includes the amplitude v^2 leads to depletion of one-half a pair and hence total charge is conserved in this system. This conservation of charge by the quasi particle is crucial to the identity of observed physical properties when calculated in the original BCS state given by Equation (16) and when calculated using the Dyson projected state mentioned above.

Another confusion which often if found in the literature is that the energy gap can be represented exactly as in a semiconductor such as silicon in which electrons can be created in the conduction band and holes created in the valence band. In superconductors this picture works in certain limited cases but gives incorrect results in general. The correct picture is rather that in which the excitation energies for quasi particles are positive, both above and below the Fermi momentum (see reference 1a, page 45). There is a unique quasi particle state for each value of k in a superconductor while there are, of course, two states–the valence band state and the conduction band state–in silicon for each value of k in the reduced zone. Thus, viewing the groundstate as being a filled sea of negative energy states, separated by a gap 2Δ from an empty set of states, is not correct.

III. Relation to Experiment

The above theory is basically a mean field type of approach, of course very different from the Hartree Fock scheme which does not give superconductivity. Calculation of observable effects are very simple and generally can be done in closed form. As we mentioned, quantum fluctuation effects are very small in low temperature superconductors and it is no surprise that excellent agreement between theory and experiment can be obtained for properties of conventional superconductors.

A major difference between traditional phenomenological theories and the BCS approach is that the coupling of the quasi particles to external electric, magnetic and sound fields are accompanied by so-called vertex functions which modify the coupling strength in a surprising fashion. This was first dramatically discovered when Hebel and Slichter discovered that their measurements of the NMR relaxation rate $1/T_1$ in aluminum showed a dramatic increase below T_c before falling to a small value in an exponential fashion at low temperature. Their experiments were being carried out in the same building in which we were developing the consequences of the theory and we delighted to find that the theory, in fact, was quite consistent with such a dramatic rise in that $1/T_1$ as T is lowered below T_c. This effect, and related effects in infrared and acoustic absorption are a direct consequence of correlations in the groundstate wave functions as evidenced by the quasi particle transformation in Equation (22). For example, in the NMR relaxation process, the electronic spin density operator at the nucleus

$$S(0) = \frac{1}{2}\sum_{kk'} c^{+}_{k's'} \vec{\sigma}_{s's} c_{ks} \;, \tag{23}$$

couples to the nuclear spin I via the hyperfine interaction. To calculate the spin-flip scattering rate of a quasi particle $\gamma_{\vec{k}}$ by a nucleus initially of spin-down, one must transform the c operators in Equation (23) by the inverse of the transformation Equation (22) Clearly this transformation involves the factor u and v bilinearly and these bilinear combinations are the so-called coherence factors entering the interaction vertices with the external fields. The nature of the coherence factor depends upon the time reversal and parity property of the coupling to the external field. In particular, sound waves couple with coherence factors of type I

$$m(k,k') = u_k v_{k'} + v_k u_{k'}, \tag{24a}$$
$$n(k,k') = u_k u_{k'} - v_k v_{k'} \tag{24b}$$

while coupling to spins and to the electromagnetic field involve coherence factors of type II.

$$l(k,k') = u_k u_{k'} + v_k v_{k'} \tag{25a}$$
$$p(k,k') = u_k v_{k'} - v_k u_{k'} \tag{25b}$$

These coherence factors enter in the Golden Rule for the scattering rate and enhance the spin-flip and electromagnetic absorption as T falls below T_C because of the large density of the states of the quasi particles near the gap edge while this density of states enhancement is precisely cancelled by a destructive coherence factor entering in the acoustic attenuation rate as T falls below T_C. This dramatic difference between type I and type II experiments was the first very strong experimental evidence for the detailed nature of the pairing correlation as stated in Equation (16). Subsequently many other detailed checks of the correctness of the theory have been carried out as well as predictions of new phenomena which follow as consequences of the theory. For details the reader is referred to the literature.

IV. Condensate Phase and It's Consequences

As mentioned in the introduction, the Ginzburg–Landau order parameter $\Psi(\overrightarrow{r})$ is a complex scalar whose absolute square is proportional to the density of the superfluid pairs. In addition, we noted that $\Psi(\overrightarrow{r})$ can be viewed as the center of mass wave function of pairs. Since the BCS state Equation (16) represents the situation in which the condensate is at rest, the center of mass momentum of the pairs is zero and hence Ψ is a constant whose phase can be chosen to be zero. However, if the condensate carries net momentum, then instead of pairing states $k\uparrow$ and $-k\downarrow$ one pairs $k\uparrow$ and $-k+q\downarrow$ so that each pair carries a net momentum q. In this case the Ginzburg–Landau order parameter, aside from a scale factor, is given by

$$\Psi_q(r) = e^{iqr}, \tag{26}$$

where r represents the center of mass coordinate of the pair. If the condensate is not simply undergoing a Galilean transformation to a moving frame, but rather carries a current which varies from point to point in space, the phase $\Theta(\overrightarrow{r})$ of the Ginzburg–Landau order parameter is a general function of r.

Two interesting and important applications of such states in which the pairing is not in time reversed states, are given by 1) the quantization of magnetic flux and 2) the Josephson effect.

In the case of flux quantization, it is most convenient to choose basis states not as plane waves but as solutions of the wave equation in cylindrical coordinates, namely

$$\chi_{k,m,\kappa} = j(kr)e^{im\phi}e^{i\kappa z}. \tag{27}$$

The ground state in the absence of magnetic field and/or current flow is given by pairing for each l state m with the time reversed state $-m$ and with reversed z-momentum, $-\kappa$. However, when the magnetic field enters the specimum as in a type II regime, the field is concentrated around a point which we choose to be the origin and fall off exponentially outside the penetration depth λ. In this case the lowest energy of the system is given by choosing a modified pairing m and $-m+\delta$ where δ is the number of flux quanta measured in units of the pairing flux quantum Φ_0 as given by

$$\Phi_0 = \frac{hc}{2e}. \tag{28}$$

which is one-half the direct flux quantum. The origin of the factor of 2 in the Equation (8) is, of course, the charge of the pair and corresponds to the effective charge e^* entering in the Ginzburg–Landau Equation (6) which is also equal to $2e$ the charge of a pair. As one can easily show, if one attempts to combine two flux quanta into one region of space, they experience a large repulsion, while if one attempts to divide a flux quanta into two portions, the kinetic energy of the surrounding fluid become large. Thus, there is a natural bunching of magnetic flux quanta into vortices carrying total flux given by Equation (28).

When many flux quanta enter the specimen they are arranged in a minimum energy configuration originally worked out by Abrikosov and form the so-called Abrikosov lattice, a triangular distribution of quanta. Such pair flux quantization has also been verified for the high T_c materials which supports the pairing nature of that state, although other theories have been proposed which also lead to such a flux quantum under suitable conditions.

A second important application of the phase of the condensate was that discovered by Josephson in the context of electron tunneling between two superconducting specimens. When zero voltage is applied across the junction, electron pairs can tunnel between S_1 and S_2 leading to a low energy coherent state of the combined pair of superconductors if the phase of the pair functions in S_1 and S_2 are identical. However, if the phase of these two pair of condensates differs by π, the energy is raised by the so-called Josephson coupling energy proportional to the square of the tunneling matrix element. As Josephson showed, the current can flow through the junction so long as no applied voltage occurs, and no resistance or dissipation is found. That is, one has a single coherent coupled superconducting material which carries a current without resistance. Mathematically this situation is described by the Josephson equations and

$$E_{12} = -E_0 \sin(\theta_1 - \theta_2), \tag{29a}$$

$$j = j_0 \sin(\theta_1 - \theta_2), \tag{29b}$$

where the phase difference varys with time according to

$$\dot{\theta}_1 - \dot{\theta}_2 = \frac{2eV(t)}{\hbar}. \tag{30}$$

For example, if a DC voltage is applied, it is clear that the phase difference varies linearly in time and an AC current flows in the junction. The reader is referred to the literature for many beautiful experiments concerning the Josephson effect, including the role magnetic flux plays when applied in the junction region.

V. Conclusion

While the microscopic theory of superconductivity remained a perplexing challenge for more than four decades, its ultimate solution took the form of a remarkably simple mean field type problem with the consequence that one could calculate in detail most of the observable physical properties in closed form. While in general the predicted results agree quite well with experiment, certain deviations were evident, particularly in the so-called "bad actor" materials which were known to be characterized by very strong electron–phonon couplings, such as lead and mercury. In such cases T_c becomes a reasonable fraction of the Debye temperature and strong coupling effects must be taken into account to obtain quantitative agreement between theory and experiment. Fortunately the Green function approach to the many body problem had advanced at that time to the point where one could treat such strong coupling effects in a straight forward way. This approach was first developed in superconductors by Gor'kov and by Nambu. Detailed application of this approach to superconductivity was carried out by G. Eliashberg. In fact, the first non trivial application of modern many body theory to an interacting system was the prediction of the current voltage characteristic for a normal-super and a super-superjunction, with account taken of the electron-phonon interaction in a proper field theoretic manner. The remarkable agreement between theory and experiment which was obtained was the ultimate proof that the electron-phonon interaction was at the heart of the attractive interaction which the BCS theory used to explain superconductivity in conventional metals.

It is important to note that the electron-phonon interaction which underlies the use of the BCS theory for conventional metals in no way restricts the use of the identical formalism in other systems. For example, in the case of ^3He the condensate is of the BCS form, but not with angular momentum zero (so-called "s-wave") but with $L = 1$ ("p-wave") with a spin of 1 for the pair, rather than spin zero for pairs in conventional metallic superconductors.

Soon after the BCS theory was constructed Bohr, Mottelson and Pines showed how correlations in nuclei could be quantitatively treated using the BCS approach, where the pairing is between time reversed angular momentum states in the shell model potential. Similar effects occur in neutron stars where the pairing is due to the strong interaction. Finally, it is widely believed that the first superconductor to occur in nature was the quark condensation which occurred approximately one second after the Big Bang. Thus, superconductivity is a phenomena which exists over more than 13 orders of magnitude in temperature and is accounted for by a very simple mean field approach.

This work was supported in part by NSF grant DMR89-16582.

References

1. For a more detailed review of the pairing theory and its applications see:

   a. J.R. Schrieffer, *Theory of Superconductivity*, Addison-Wesley, Reading, Mass. (1988).

   b. P.G. de Gennes, *Superconductivity of Metals and Alloys*, Addison-Wesley, Reading, Mass. (1966).

   c. *Superconductivity*, R.D. Pards ed., Marcel Dekker, N.Y. (1969).

   d. M. Tinkham, *Superconductivity*, Gordon and Breach, N.Y. (1965).

the first non-trivial application of modern many body theory to an interacting system was the prediction of the current-voltage characteristic for a normal-super and a super-super junction, to a correct inclusion of the electron-phonon interaction in a proper field theoretic manner. The remarkable agreement between theory and experiment which was obtained was the [illegible]

[illegible]

[illegible]

Part II

Theory

Electronic Excitations Within the Cu-O Planes

P. Fulde

Max-Planck-Institut für Festkörperforschung, Postfach 800665
W-7000 Stuttgart 80, Fed. Rep. of Germany

Abstract. Excitations of the strongly correlated electrons within the $Cu-O$ planes can be studied by numerical methods. The characteristic properties of the one hole (electron) spectral density can be rederived by analytic methods based on projection techniques, and as far as the coherent part of the motion of holes is concerned by a tight-binding-like approach using string states.

1 Introduction

Electrons in the $Cu-O$ planes are strongly correlated [1]. Direct evidence is provided by the semiconducting properties of La_2CuO_4. Were it not for the strong correlations the material would be metallic because there is one d hole per unit cell. Instead it is a semiconducting antiferromagnet with almost no changes in the conductivity when the Néel temperature is crossed.

In order to study the electronic excitations within the doped $Cu - O$ planes one can proceed along different lines, three of which we will discuss here. One is by numerical studies of small clusters of CuO_2 units, using a three-band Hubbard Hamiltonian (Emery model) for their description and applying periodic boundary conditions. This way the hole spectral function can be calculated for different doping concentrations (see e.g. Ref. [2]). Another description of the excitations is by the use of projection techniques. This method can be applied to strongly as well as weakly correlated electronic systems. When applied to the computation of the hole spectral function one finds the same structure as obtained from the numerical studies [3]. Finally one can tackle in particular the problem of coherent hole motion in a two-dimensional antiferromagnet by phenomenological methods. By introducing localized string states which are the solutions of a set of equations derived by Bulaevskii et al. [4] and by Shraiman and Siggia [5] one can study their coupling via paths on closed loops (Trugman paths [6]) and spin flip terms and obtain this way a simple physical description of the coherent motion of holes [7,8]. In distinction to the former case a simplified $t - J$ model Hamiltonian is used for the determination of the coherent hole motion.

2 Numerical Studies

Starting point is a three-band Hubbard Hamiltonian for the hybridizing $Cu\ 3d_{x^2-y^2}$ and $O\ 2p_{x(y)}$ orbitals of the $Cu - O$ planes. We will use a hole representation, because there is one hole per CuO_2 unit when the system is undoped. We set

$$H = \tilde{H}_0 + U_d \sum_i n_{d\uparrow}(i) n_{d\downarrow}(i) + U_p \sum_j n_{p\uparrow}(j) n_{p\downarrow}(j) \qquad (1)$$

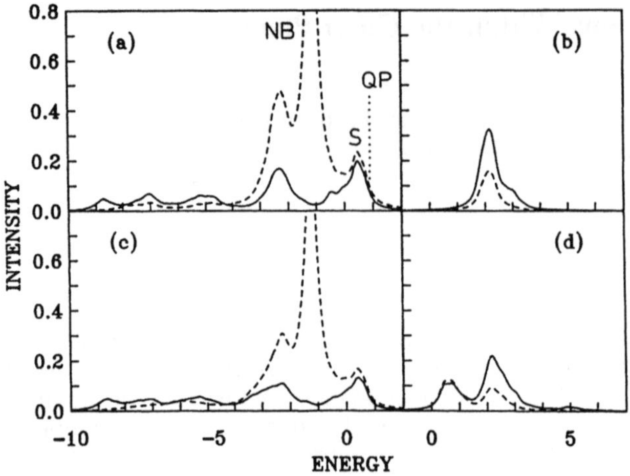

Fig. 1 Spectral density of a hole (a, c) and an electron (b, d) for half-filling (a, b) and for a doping concentration of 25% (c,d). The following parameter values are used (in units of t_{pd}) $\epsilon_p - \epsilon_d = 2, U_d = 6, U_p = 3$. Dashed (solid) lines show O (Cu) spectra. The peak labeled NB is from oxygen nonbonding bands while S labels a singlet contribution. The dotted peak QP shows the expected contribution from the coherent hole motion (from Ref. [2]).

where

$$\begin{aligned}\tilde{H}_0 &= \epsilon_d \sum_{i\sigma} n_{d\sigma}(i) + \epsilon_p \sum_{j\sigma} n_{p\sigma}(j) \\ &+ \sum_{<ij>\sigma} V_{ij}(d^+_{i\sigma} p_{j\sigma} + p^+_{j\sigma} d_{i\sigma}).\end{aligned} \quad (2)$$

Here ϵ_d and ϵ_p are the orbital energies of the d and p holes and $(\epsilon_p - \epsilon_d) > 0$. The V_{ij} describe the hybridization between the d and p orbitals and $V_{ij} = (-1)^{\alpha}_{ij} t_{pd}$ where $\alpha_{ij} = 0$ or 1, depending on the position of the O atom relative to the Cu atom. U_d and U_p denote the Coulomb interactions of holes in the $d_{x^2-y^2}$ and $p_{x(y)}$ orbitals. A cluster of four units of CuO_2 is considered with periodic boundary conditions. The hole spectral function is defined by

$$P_n(\omega) = -\frac{1}{\pi} Im < \psi_0(N) \mid c^+_{n\sigma} \frac{1}{\omega - E_0(N) + H + i\eta} c_{n\sigma} \mid \psi_0(N) > . \quad (3)$$

The operator $c^+_{n\sigma}$ is either a p *electron* (not hole) creation operator when n is a O site or a d electron creation operator when n is a Cu site.

Furthermore, $\mid \psi_0(N) >$ is the exact ground state of the N electron system with energy $E_0(N)$. The ground state of the cluster is determined by the Lanczos method. Results for $P(\omega)$ are shown in Fig. 1(a,b) for the case of half filling and when in addition 25 % of holes are added (this corresponds to a total of 5 holes in the cluster). The dominant peak labeled NB is from a nonbonding p band, i.e. a combination of p orbitals which does not hybridize with the Cu $3d_{x^2-y^2}$ orbital. Between it and the Fermi energy another structure is found which is labeled S. An analysis reveals that it is resulting from a state which is to a good approximation a singlet. Thereby the spin of a hole on a Cu and an O site couple to a spin singlet [9]. The situation resembles the one in a H_2 molecule at large interatomic separations where the Heitler-London theory applies. There is a corresponding triplet like structure on the low energy site of the nonbonding peak. The cluster size is not large enough in order to decide to what extent the peak resulting from the coherent motion of the hole is identical with the singlet peak. The quasiparticle peak is of width $2J$, where J denotes the antiferromagnetic coupling constant of the Cu spins (see below). Additional high excitation energy structure can be identified with $d^{10}p^4$, d^8p^6 etc. final states.

The spectrum for inverse photoelectron spectroscopy is calculated from an expression of the form of Eq. (3) but with $c_{n\sigma}$ and $c^+_{n\sigma}$ interchanged. One notices from Fig. 1 a that it is separated by a gap from the photoelectron spectrum (charge transfer gap [10]). It represents the upper Hubbard band with a $d^{10}p^6$ final state. For doped systems a new peak appears in the spectrum, because electrons are now injected not only into Cu $3d$ states as in the case of half-filling but also into empty O $2p$ states [2].

3 Application of Projection Techniques

The projection technique is suitable for studying excitations not only of weakly correlated but also of strongly correlated electronic systems [11]. We demonstrate this by considering the case of half filling, but the calculations can be extended to doped systems as well [12]. The excitation spectrum is calculated from the poles of the correlation function

$$C_{\underline{k}\alpha\sigma}(\omega) = <\psi_0 \mid p_{\underline{k}\alpha\sigma}\frac{1}{\omega - L + i\eta}p^+_{\underline{k}\alpha\sigma} \mid \psi_0 > . \qquad (4)$$

The $p^+_{\underline{k}\alpha\sigma}$ ($p_{\underline{k}\alpha\sigma}$) create (annihilate) holes on O sites in eigenstates to the momentum k, band index α and spin σ. The superoperator L (Liouvillean) acts on other operators A according to $LA = [H, A]_-$. One can show that Eq. (4) can be rewritten in the form

$$C_{\underline{k}\alpha\sigma}(\omega) = (\Omega \mid p_{\underline{k}\alpha\sigma}\frac{1}{\omega - L_0 - H_1 + i\eta}p^+_{\underline{k}\alpha\sigma}\Omega) \qquad (5)$$

where Ω plays the role of a wave operator and is defined by

$$\mid \Omega) = \lim_{z\to 0} \mid 1 + \frac{1}{z - L_0 - H_1}H_1). \qquad (6)$$

We have divided the Hamiltonian H given by Eqs. (1,2) into $H = H_0 + H_1$ where H_1 consists of the hybridization part of Eq. (2) and H_0 contains the remaining parts, i.e. the atomic energies. The superoperator L_0 refers to H_0.

The product $(A \mid B)$ in operator or Liouville space is defined by

$$(A \mid B) = <\Phi_0 \mid A^+B \mid \Phi_0>^c \qquad (7)$$

where $\mid \Phi_0 >$ is the ground state of H_0 and the upper script c implies taking the cumulant of the expectation value. The cumulant of the expectation value of a product of operators is defined by

$$< \Phi_0 \mid A_1^{n_1}A_2^{n_2}...A_m^{n_m} \mid \Phi_0 >= \frac{\partial^{n_1}}{\partial\lambda_1^{n_1}}...\frac{\partial^{n_m}}{\partial\lambda_m^{n_m}} \cdot \qquad (8)$$

$$\cdot \ell n < \Phi_0 \mid \Pi_{i=1}^m e^{\lambda_i A_i} \mid \Phi_0 >|_{\lambda_1=...\lambda_m=0}$$

and we refer to Kubo's original work for further discussions of their properties [12].

We solve Eq. (5) by limiting the Liouville space $\mathfrak{R}$ to a *relevant* part $\mathfrak{R}_0$ which we assume to be spanned by the set of operators $A_\nu = \tilde{A}_\nu \Omega$. This definition of the A_ν ensures that the interplay between the created hole and the changes in the ground state due to the presence of H_1 and represented by Ω is properly included in the calculations. We introduce a projector P onto $\mathfrak{R}_0$ by

$$P = \sum_{\nu\mu} \mid A_\nu)\chi_{\nu\mu}^{-1}(A_\mu \mid \tag{9}$$

$$\chi_{mn} = (A_m \mid A_n)$$

and approximate Eq. (5) by

$$C_{\underline{k}\alpha\sigma}(\omega) = (\Omega \mid p_{\underline{k}\alpha\sigma} P \frac{1}{\omega - P(L_0 + H_1)P + i\eta} P \, p_{\underline{k}\alpha\sigma}^{+} \, \Omega). \tag{10}$$

In order to determine this function the following matrix elements have to be computed:

$$R_{\nu\mu}(\omega) = (A_\nu \mid \frac{1}{\omega - P(L_0 + H_1) + i\eta} A_\mu). \tag{11}$$

This matrix is evaluated with the help of the projection technique [3]. According to it the following equation holds:

$$\sum_\nu \left[(\omega + i\eta)\delta_{\eta\nu} - \tilde{\omega}_{\eta\nu}\right] R_{\nu\mu}(\omega) = \chi_{\eta\mu} \tag{12}$$

with

$$\tilde{\omega}_{\eta\nu} = \sum_\lambda (A_\eta \mid (L_0 + H_1)A_\lambda)\chi_{\lambda\nu}^{-1}. \tag{13}$$

The dimension of the matrices depends on the number of operators contained in the set $\{A_\mu\}$. The latter are obtained by analyzing the different microscopic processes which occur when a hole moves in the $Cu - O$ plane. By using them as a guide we select seven variables $\tilde{A}_\nu$. They consist of the four $p_{\underline{k}\alpha\sigma}^{+}$ ($\alpha = 1, ...4$), a variable $\tilde{A}_5$ which describes the spin exchange between a hole on a O site and another one on a Cu site, a variable $\tilde{A}_6$ which denotes a dynamical $Cu - O$ charge fluctuation and finally $\tilde{A}_7$ which flips two neighboring Cu spins. Results for the function

$$P(\omega) = -\frac{1}{\pi N_0} \sum_{\underline{k}} Im C_{\underline{k}\alpha\sigma}(\omega + i\eta) \tag{14}$$

are shown in Fig. 2 (N_0 is the number of unit cells). One notices the similarity with the numerical studies shown in Fig. 1 a. In an improved theory one takes into account also variables which describe the string of frustrated spins which a hole generates when it moves in an antiferromagnet. This leads on to the next section.

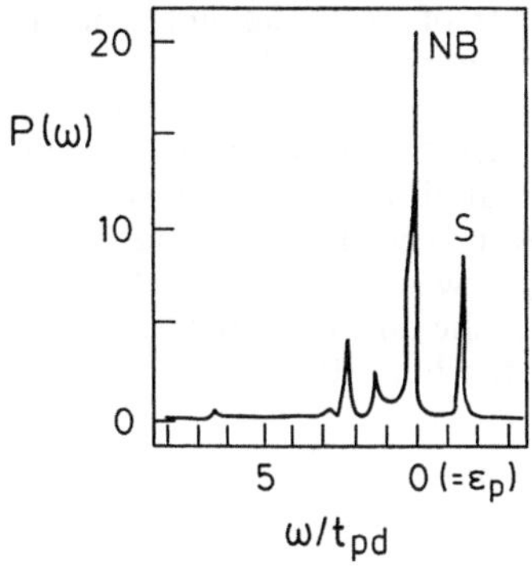

Fig. 2 Spectral density of a hole on a O site. Parameters vary slightly from those of Fig. 1. The meaning of NB and S is the same as in Fig. 1. The hump at $\omega/t_{pd} = 6.5$ is due to dynamical charge excitations. Its relative weight is of order $(t_{pd}/(\epsilon_p - \epsilon_d))^6$ (from Ref. [3])

4 Coherent Motion of a Hole

In order to discuss the coherent motion of a hole in a 2D Heisenberg antiferromagnet, we go over to the simplified $t - J$ model Hamiltonian

$$H_{t-J} = -t \sum_{<ij>\sigma} (\hat{a}^+_{i\sigma}\hat{a}_{j\sigma} + h.c.) + J \sum_{<ij>} (\underline{S}_i\underline{S}_j - \frac{n_i n_j}{4}). \quad (15)$$

The operators $\hat{a}^+_{i\sigma}$ are defined through $\hat{a}^+_{i\sigma} = a^+_{i\sigma}(1 - n_{i-\sigma})$ and the $a^+_{i\sigma}$ create electrons at sites i with spin σ. The $\underline{S}_i$ and n_i describe the spins and the occupation numbers of the singly occupied sites. The assumption is made that the low-energy excitation spectrum of H_{t-J} is equivalent to that of the Hamiltonian (1).

We consider N_0 sites and $(N_0 - 1)$ electrons and divide H_{t-J} into H_0 and H_1 with the latter given by

$$H_1 = \frac{J}{2} \sum_{<ij>} (S^+_i S^-_j + S^-_i S^+_j). \quad (16)$$

Without the hole the ground state of H_0 is the Néel state. When a hole which is created at site j moves through the system it generates spin disorder which increases the Ising energy. Thus an effective potential acts on the hole which localizes it to the vicinity of site j. Therefore for the ground state of H_0 in the presence of a hole centered at site j the ansatz can be made

$$| \phi_j >= \sum_\nu \alpha_\nu (\sum_p | j, \nu, p >). \quad (17)$$

The sum extends over all paths of lengths ν starting from site j. The coefficients α_ν must minimize the energy, which leads to the following set of equations:

$$\begin{aligned} -4t\alpha_1 &= E_B\alpha_0 \quad (18) \\ -t(3\alpha_{\nu+1} + \alpha_{\nu-1}) &= (E_B - J(\nu + \frac{1}{2}))\alpha_\nu . \end{aligned}$$

E_B is the binding energy. There exist two types of processes not considered until now which lead to an unbinding and hence delocalization of the hole. The first one is operative even when only H_0 is used and consists of a spiraling motion of the hole (Trugman path). By this kind of motion the hole annihilates again the spin disorder it has created in the first place. The main process leading to delocalization is however through H_1. The disordered spins generated by the hole motion are pairwise annihilated through H_1. One expects therefore an energy dispersion of the hole band with a width of order J [8]. The following ansatz is made for the hole state [7]:

$$| \psi(\underline{k}) >= \frac{1}{\sqrt{N_0}} \sum_j e^{i\underline{k}\underline{R}_j} | \phi_j > \tag{19}$$

where $\underline{R}_j$ is the position of site j. The energy dispersion for the hole motion follows from

$$E(\underline{k}) = \frac{< \psi(\underline{k}) | H_{t-J} | \psi(\underline{k}) >}{< \psi(\underline{k}) | \psi(\underline{k}) >}. \tag{20}$$

The matrix elements $< \phi_i | H_{t-J} | \phi_j >$ can be determined analytically and one finds

$$E(\underline{k}) = 4\tau_{1,1}(cosk_x + cosk_y)^2 + const. \tag{21}$$

if the lattice constant is set equal to unity. The factor $\tau_{1,1}$ is the next nearest neighbor hopping element of an effective tight binding Hamiltonian which is easily evaluated. The resulting $E(\underline{k})$ agrees well with that of numerical studies for the same Hamiltonian [13,14]. One should mention that these calculations can be generalized to the case where the hole is generated in the true ground state of the Heisenberg Hamiltonian, i.e. in the ground state of H_{t-J} at half filling, instead of in the Néel state. These calculations make use of the projection technique based on cumulants which was discussed in the previous section [15].

5 Conclusions

We have shown that the spectral density of the one particle excitations in the $Cu - O$ planes can be quantitatively detemined by numerical methods. The important features of the spectrum can be reproduced by analytical calculations when the projection method is used. The characteristic properties of the coherent hole motion follow from a tight binding type of approach using string states.

6 Acknowledgement

I would like to thank Drs. K.W. Becker, R. Eder, P. Horsch and P. Unger for numerous discussions.

References

[1] P. W. Anderson, Science 235, 1196 (1987)
[2] W. Stephan and P. Horsch in "Dynamics of Magnetic Fluctuations in High Temperature Superconductors" ed. by G. Reiter, P. Horsch and G. Psaltakis (Plenum, New York 1990)
[3] K. W. Becker, W. Brenig and P. Fulde, Z. Phys. B 81, 165 (1990)
[4] L. N. Bulaevskii, E. L. Nagaev and D. L. Khomskii, Zh. Eksp. Teor. Fiz. 54, 1562 (1968) [Engl. transl.: Sov. Phys.- JETP 27, 836 (1968)].
[5] B. I. Shraiman and E. D. Siggia, Phys. Rev. Lett. 60, 740 (1988)
[6] S. A. Trugman, Phys. Rev. B 37, 1597 (1988)
[7] R. Eder and K. W. Becker, Z. Phys. B 78, 219 (1990)
[8] C. L. Kane, P. A. Lee and N. Read, Phys. Rev. B 39, 6880 (1989)
[9] F. C. Zhang and T.M. Rice, Phys. Rev. B 37, 3759 (1988)
[10] J. Zaanen, G. A. Sawatzky and J. W. Allen, Phys. Rev. Lett. 55, 418 (1985)
[11] For a detailed discussion of the method see P. Fulde "Electron Correlations in Molecules and Solids " Springer Series in Solid-State Sciences, Vol. 100 (Springer, Heidelberg, New York 1991)
[12] R. Kubo, J. Phys. Soc. Jpn. 17, 1100 (1952)
[13] K. J. von Szczepanski, P. Horsch, W. Stephan and M. Ziegler, Phys. Rev. B 41, 2017 (1990)
[14] E. Dagotto, A. Moreo, R. Joynt, S. Bacci und E. Gagliano, Phys. Rev. B 41, 2585 (1990)
[15] K. W. Becker, R. Eder and H. Won (to be published)

Physical Parameters in High-Temperature Superconductors

S. Maekawa[1], *Y. Ohta*[1], *and T. Tohyama*[2]

[1]Department of Applied Physics, Nagoya University, Nagoya 464-01, Japan
[2]Department of Physics, Faculty of Education, Mie University, Tsu 514, Japan

Abstract: Transition metal oxides are classified into two categories: the charge transfer (CT) materials and the Mott-Hubbard (MH) ones. High temperature oxides are known to be in the CT materials. This is because of the subtle balance of the energy levels in copper and oxygen ions and of the unique crystal structure. In this lecture, we examine relations between crystal structure and electronic structure in a variety of the oxides and discuss what the key physical parameters are in the high temperature superconductors. Special attention is paid to the crucial difference in the electronic properties between CT and MH materials. Systematic variation in various electronic properties with materials provides us with valuable information on the mechanism of high temperature superconductivity.

1. Introduction

Search for the mechanism of high T_c superconductivity in the copper oxides has attracted much attention not only because T_c is high but because the copper oxides are strongly correlated electron systems so that they contain many essential problems in solid-state physics.[1)] There exists a variety of the high T_c superconductors. Therefore, it seems possible to ask the condition of the superconductivity through the examination of the characteristics of materials.

High T_c superconductors are composed of the CuO_2 plane as a structural unit. The superconductivity occurs when carriers are introduced into the plane. T_c exhibits a maximum uniquely at the carrier concentration $p \approx 0.15-0.2$ per CuO_2 unit in all hole-carrier superconductors. For p smaller than the value, the compounds are antiferromagnetic insulators, whereas for p larger than the value, they are metallic but not superconducting. We may say that the high T_c superconducting state exists between antiferromagnetic insulator and metal.

In spite of these common features among the high T_c superconductors, the maximum T_c's are dispersed from material to material. Therefore, one of the natural questions is what the common features mean and what the characteristic material dependences do. The answer to the question will give us a key for the mechanism of high T_c superconductivity. In addition, it will provide an opportunity of examining the controversial topics in solid-state physics which have been discussed for many years.

The study of the electronic structure in transition metal oxides has a long history. In the oxides, strong electron correlation forces electrons to localize in the atomic orbitals. On the other hand, the kinetic energy is reduced when electrons are itinerant. Therefore, these two antagonistic effects brought about much controversy. High T_c superconductivity may be considered to have renewed the study of the electronic structure in the oxides.

In this lecture, the electronic structure of high T_c copper oxides and the implications in the superconductivity are discussed.

2. Charge transfer compounds

2.1. Electronic structure in ionic model

As shown in Fig. 1, there exist three kinds of CuO_2 planes in high T_c copper oxides. It is empirically known that the presence of apex oxygens makes the CuO_2 plane easier to dope with holes; the planes with the square coordination (without apex oxygen) have not yet been doped with holes but those with the octahedral coordination have been made overdoped.[2] The compounds without carriers such as La_2CuO_4 and $YBa_2Cu_3O_6$ are called parent compounds.

Let us first examine the electronic structure in the parent compounds. By neglecting the covalency for a moment, the compounds are assumed to be ionic with O^{2-} and Cu^{2+}. Even though the covalency is strong in the CuO_2 plane, the ionicity between the layers is retained predominantly.

In the ionic model, the central role is played by the electrostatic potential, which is defined for a hole at the i-site as

$$V_i = \sum_{j \neq i} \frac{Z_j e^2}{|\vec{r}_j - \vec{r}_i|}, \tag{1}$$

where $\vec{r}_j$ is the position of the j-th ion with electric charge $Z_j e$. V_i is the electrostatic energy to bring a hole from infinity to the i-site and is called the Madelung energy. Assuming the lattice periodicity and total charge neutrality, we calculate V_i by the standard Ewald method for the lattice summation.

In Fig. 2, the schematic energy-level structure for a hole deduced from the ionic model is shown, where $2p_\sigma$ and $2p_z$ are the $2p$ orbitals on an oxygen in the CuO_2 plane (O(P)) and on an apex oxygen (O(A)), respectively, extended toward Cu ions in the plane (Cu(P)). When an extra hole is added in Cu(P), the energy increases by U_d because of the Coulomb repulsive interaction. Therefore, each hole is localized in a Cu ion. As a result, the compound is an ionic crystal.

The difference in the levels between O(A) and O(P) is written as [3]

$$\left.\begin{aligned} \Delta\varepsilon_A &= \Delta V_A/\epsilon(\infty), \\ \Delta V_A &= V_{O(A)} - V_{O(P)}, \end{aligned}\right\} \tag{2}$$

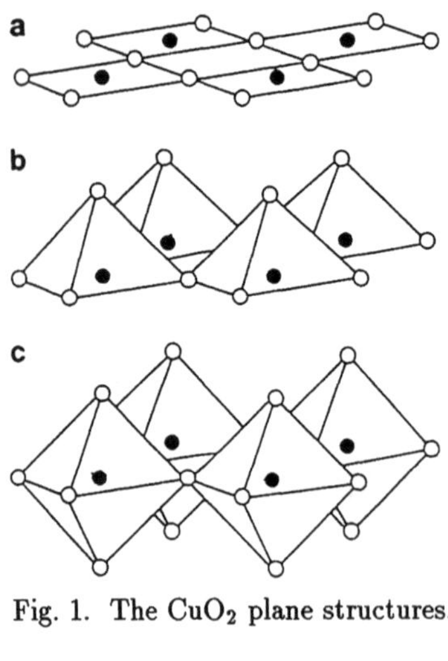

Fig. 1. The CuO_2 plane structures. ○: oxygen ion, ●: copper ion.

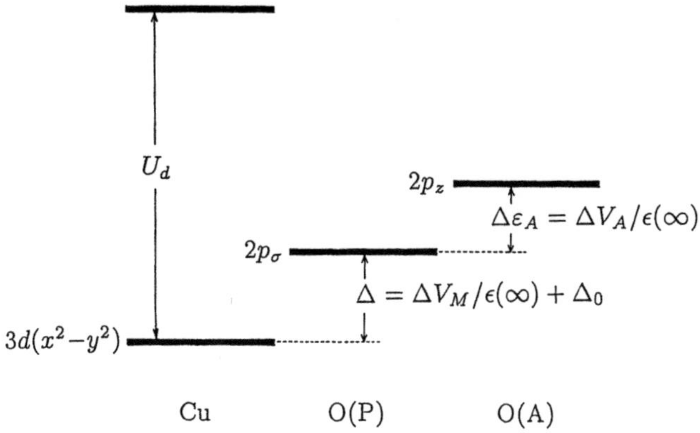

Fig. 2. Schematic energy-level structure for a hole in the CuO_2 plane plus the apex oxygen deduced from the ionic model.

where $\epsilon(\infty)$ is the dielectric constant at infinite frequencies and shows the reduction of the potential by the polarization of surrounding ions. When $\epsilon(\infty)$ is assumed to be constant and independent of materials, we find that ΔV_A is a measure of the stability of a hole induced into O(P). Another important parameter to specify the electronic structure is the difference Δ in the energy levels between $Cu3d(x^2-y^2)$ and $O2p_\sigma$ orbitals and is given by [4)]

$$\left.\begin{aligned} \Delta &= \Delta V_M/\epsilon(\infty) + \Delta_0\,, \\ \Delta V_M &= V_{\mathrm{O(P)}} - V_{\mathrm{Cu(P)}}\,, \end{aligned}\right\} \tag{3}$$

where ΔV_M is the difference in the Madelung site potentials for a hole between O(P) and Cu(P).[5)] Here, Δ_0 involves the second electron affinity of oxygen and the second ionization energy of copper, and may be considered to be independent of materials.

Copper oxides have the relation $\Delta < U_d$ as seen in Fig. 2, so that the energy gap is given by the charge transfer between copper and oxygen. Such materials are called the charge transfer (CT) materials.[6)] On the other hand, the materials with $U_d < \Delta$ such as V_2O_3 and Ti_2O_3 which are famous for the metal-insulator transition are called the Mott-Hubbard (MH) materials.[7)] Contrary to U_d, Δ is a function of the Madelung energy and thus is dependent on the crystal structure. Therefore, the electronic structure in the CT materials sensitively reflects the crystal structure. In Fig. 3, the CT gap (E_g) observed by the optical conductivity is shown as a function of ΔV_M. We find that E_g in the copper oxides is governed by the Madelung potential.[4)]

The Madelung potential ΔV_M in high T_c oxides is in the narrow energy range 46 eV $\lesssim \Delta V_M \lesssim$ 49 eV, irrespective of its large value,[5)] whereas ΔV_A is strongly dependent on materials. In Fig. 4, T_c is plotted as a function of ΔV_A in the parent compounds.[3)] We observe in the Figure that T_c is governed by ΔV_A and thus the ionic character. The detailed discussion will be given in Sec. 4. In Fig. 5, T_c is also plotted as a function of ΔV_M. ΔV_M is less dependent on materials than ΔV_A. Therefore, the correlation between T_c and ΔV_M is not always obvious. However, we note that the large value of ΔV_M is canceled out by the atomic quantity Δ_0.[4)] Such cancellation seems to be important for the copper oxides to become superconducting.

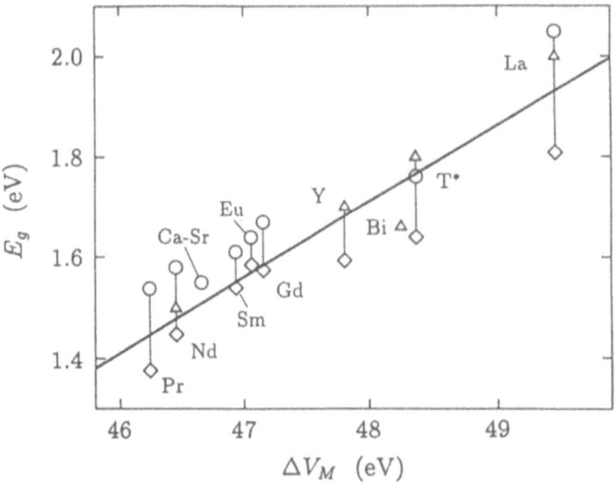

Fig. 3. The E_g vs. ΔV_M correlation.[4)] Included are L_2CuO_4 with T′ structure (L=Pr, Nd, Sm, Eu and Gd), La_2CuO_4 with T structure, $(La\text{-}Gd\text{-}Sr)_2CuO_4$ and $(La\text{-}Tb\text{-}Sr)_2CuO_4$ with T* structure, $(Ca\text{-}Sr)CuO_2$, $Bi_2Sr_2YCuO_8$ and $YBa_2Cu_3O_6$. For the experimental data, see refs. 48 and 49. The solid line is a guide to the eye.

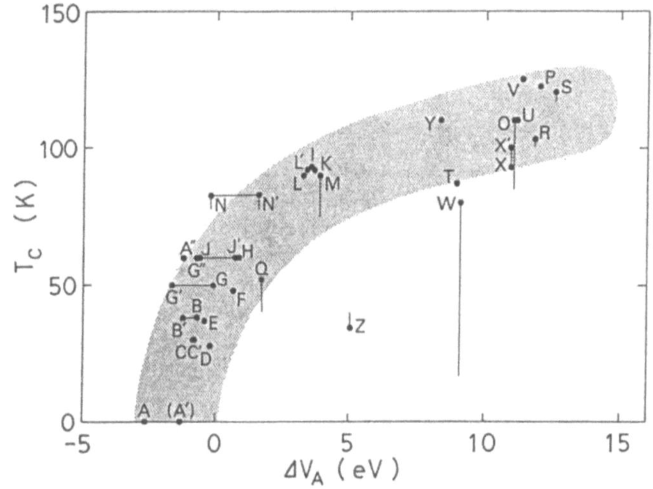

Fig. 4. The T_c vs. ΔV_A correlation in hole-carrier superconductors. The Madelung potential ΔV_A was evaluated for the corresponding parent compounds. A: $La_2SrCu_2O_{6.2}$, A′: $La_{1.9}Ca_{1.1}Cu_2O_6$, A″: $La_{1.6}Sr_{0.4}CaCu_2O_6$, B and B′: $La_{1.85}Sr_{0.15}CuO_4$, C and C′: $La_{1.85}Ba_{0.15}CuO_4$, D: $Nd_{0.66}Sr_{0.205}Ce_{0.135}CuO_4$, E: $Sm_1La_{0.75}Sr_{0.25}CuO_{3.95}$, F: $(Ba_{0.67}Eu_{0.33})_2(Eu_{0.67}Ce_{0.33})_2Cu_3O_{8+z}$, G, G′ and G″: $Y_{0.8}Ca_{0.2}Ba_2Cu_3O_{6.11}$, H: $YBa_2Cu_3O_{6.5}$, I: $YBa_2Cu_3O_7$, J and J′: $ErBa_2Cu_3O_{6.53}$, K: $ErBa_2Cu_3O_7$, L and L′: $Y_{0.9}Ca_{0.1}Ba_2Cu_4O_8$, M: $(Ca_{0.5}La_{0.5})(Ba_{1.25}La_{0.75})Cu_3O_{6+\delta}$, N and N′: $Pb_2Sr_2Y_{0.5}Ca_{0.5}Cu_3O_8$, O: $Pb_{0.5}Tl_{0.5}Sr_2CaCu_2O_7$, P: $Pb_{0.5}Tl_{0.5}Sr_2Ca_2Cu_3O_9$, Q: $Tl(Ba_{0.6}La_{0.4})_2CuO_{5-\delta}$, R: $TlBa_2CaCu_2O_{7-\delta}$, S: $TlBa_2Ca_2Cu_3O_{9-\delta}$, T: $Tl_2Ba_2CuO_{6+\delta}$, U: $Tl_2Ba_2CaCu_2O_{8+\delta}$, V: $Tl_2Ba_2Ca_2Cu_3O_{10+\delta}$, W: $Bi_2Sr_2CuO_{6+\delta}$, X: $Bi_2Sr_2Ca_{0.9}Y_{0.1}Cu_2O_{8.24}$, X′: $Bi_{2+x}Sr_{2+y}Ca_{1+z}Cu_2O_{8-\delta}$ ($x+y+z \simeq 0.1$), Y: $Bi_2Sr_2Ca_2Cu_3O_{10+\delta}$, and Z: $Bi_2Sr_2(Gd_{0.82}Ce_{0.18})_2Cu_2O_{10.24}$.

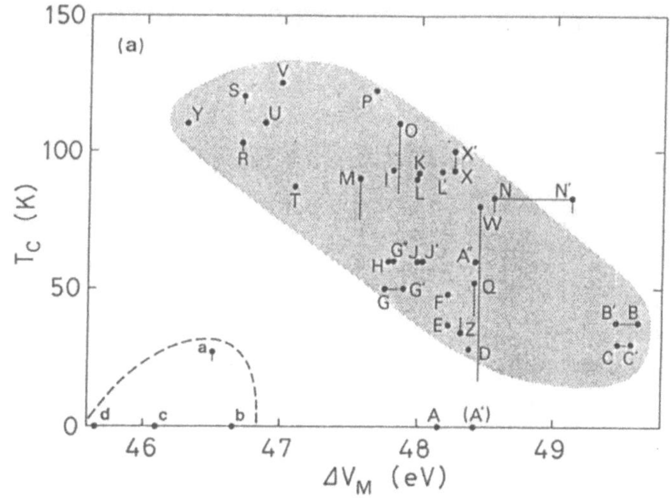

Fig. 5. The T_c vs. ΔV_M correlation.[3] The Madelung potential ΔV_M is calculated in the parent compounds. a:$Nd_{1.85}Ce_{0.15}CuO_4$, b:$Ca_{0.86}Sr_{0.14}CuO_2$, c:$Sr_2CuO_2Cl_2$, d:Sr_2CuO_3, and for the other compounds, see Fig. 4.

2.2. Electronic structure in cluster model approach.

Schematic figures of the electronic structure in transition metal oxides are shown in Fig. 6, where the covalency (hopping of electrons) together with the electron correlation causes the width of the ionic energy levels both in CT and MH insulators. One of the useful methods to treat the strong electron correlation is to calculate the electronic state exactly in a cluster with finite size. This method is called the cluster model approach. Since the electronic properties in the real crystals is extrapolated from those in a cluster with finite size, the method is known to be more useful in examining the high energy properties such as photoemission spectra [8,9] than the low energy properties.

Figure 7 shows Cu_2O_7 and Cu_2O_{11} clusters where $2p_\sigma$ orbitals in O(P), $2p_z$ in O(A) and $3d(x^2-y^2)$ and $3d(3z^2-r^2)$ orbitals in Cu(P) are explicitly shown. When two and three holes are introduced into the clusters, we obtain insulating and 50% hole-doped systems, respectively. Let us consider the Hamiltonian. Neglecting Cu$3d(3z^2-r^2)$ and O(A)$2p_z$ orbitals, for simplicity, we have

$$\begin{aligned} H = &- T_{pd} \sum_{mls} (d^\dagger_{ms} p_{ls} + h.c.) - T_{pp} \sum_{ll's} p^\dagger_{ls} p_{l's} \\ &+ U_d \sum_{m} n^d_{m\uparrow} n^d_{m\downarrow} + U_p \sum_{l} n^p_{l\uparrow} n^p_{l\downarrow} + \Delta \sum_{ls} n^p_{ls} , \end{aligned} \tag{4}$$

where the origin of energy is taken to be the level of Cu$3d(x^2-y^2)$ orbital. Here, the operators $d^\dagger_{is}$ and $p^\dagger_{is}$ create holes in Cu$3d(x^2-y^2)$ and O(P)$2p_\sigma$ orbitals with spin s at the i site, respectively, and $n^d_{ms}=d^\dagger_{ms}d_{ms}$ and $n^p_{ls}=p^\dagger_{ls}p_{ls}$. The m, l and l' summations are over the orbitals at the neighboring sites. T_{pd} and T_{pp} are the hopping parameters between the neighboring p and d orbitals and between the neighboring p orbitals, respectively. U_d and U_p denote the intra-orbital Coulomb repulsive interactions on copper and oxygen ions, respectively. It is not difficult to include Cu$3d(3z^2-r^2)$ and O(A)$2p_z$

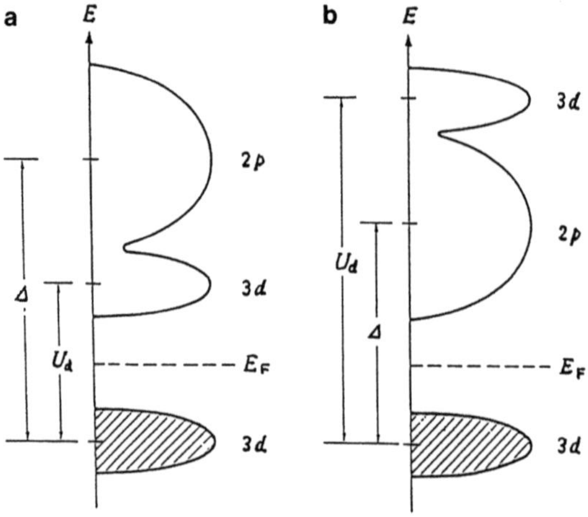

Fig. 6. One-particle view of the electronic structure in transition metal oxides. The figures show the hole bands. (a) Mott-Hubbard insulator ($\Delta > U_d$), (b) charge-transfer insulator ($U_d > \Delta$). The energy levels given by the ionic model have widths due to covalency and many-body effect.

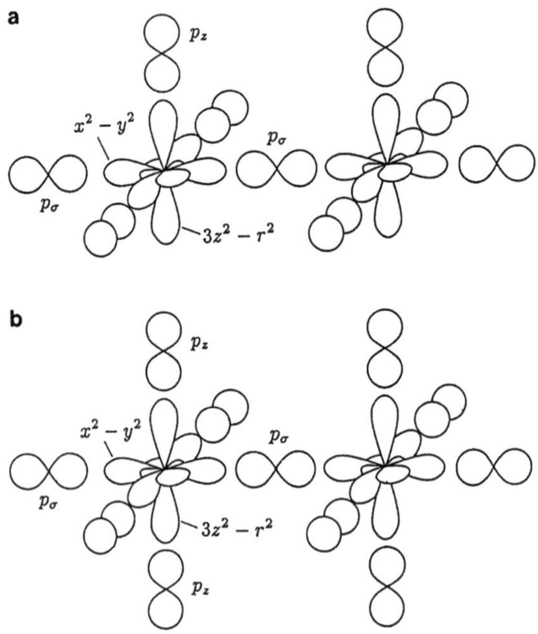

Fig. 7. Orbitals in (a) Cu_2O_9 and (b) Cu_2O_{11} clusters.

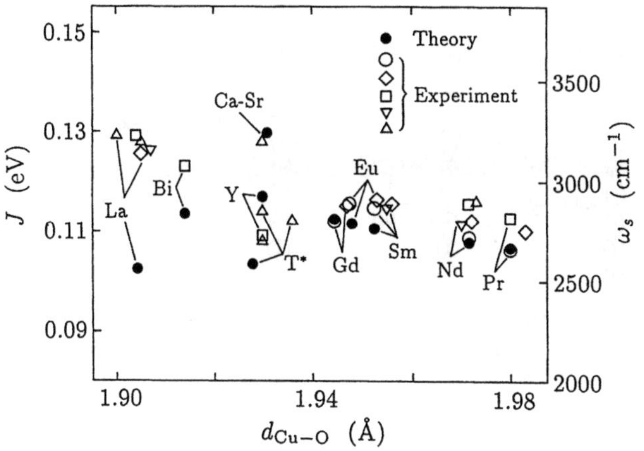

Fig. 8. Comparison between the theoretical superexchange interaction between Cu ions (J) in the CuO_2 plane and the experimentally observed peak position of magnetic Raman spectra (ω_s).[4] The relation ω_s=3.1 J is used. In the figure, d_{Cu-O} denotes the distance between nearest neighbor Cu and O ions in the CuO_2 plane. For most of the experimental data, see refs. 48, 49 and 50.

orbitals in the calculation.[10] Introducing holes into a cluster and diagonalizing the Hamiltonian exactly, we obtain the electronic sturucture. The material dependence is given by the shape of the cluster and the physical parameters in the Hamiltonian. Among the parameters, Δ is determined in the ionic model as discussed above. The material dependence of T_{pd} and T_{pp} is known from the band calculation.

We first examine the insulating state. For this purpose, we introduce two holes into the clusters in Fig. 7 and diagonalize the Hamiltonian. Then, we obtain the ground and first excited states, which are spin-singlet and spin-triplet, respectively. Therefore, the energy difference between the states corresponds to the antiferromagnetic exchange interaction (J) between neighboring Cu^{2+} spins given by the equation,

$$H_{\rm ex} = J \sum_{<i,j>} \vec{S}_i \cdot \vec{S}_j \, , \tag{5}$$

where $\vec{S}_i$ is the spin operator with S=1/2 at the i-site. In the case of large ΔV_A, O(A)$2p_z$ and Cu$3d(3z^2-r^2)$ orbitals give little effect to the electronic state. Let us determine the parameter values. The parameters U_d, U_p, $\epsilon(\infty)$ and Δ_0 are taken to be independent of materials and are given by U_d=8.5 eV, U_p=4.1 eV, $\epsilon(\infty)$=3.5, and Δ_0=−10.88 eV. The value of ΔV_A in eq. (2) is calculated by using the data of the crystal structure. The parameters T_{pd} and T_{pp} are given by T_{pd}=0.873 eV and T_{pp}=0.366 eV in Sm_2CuO_4 and the relations $T_{pd} \propto d^{-4}$ and $T_{pp} \propto d^{-3}$ for the other materials with d being the distance between atoms.

Figure 8 shows the calculated J together with the peak value ω_s of the magnetic Raman scattering spectra which are assumed to have the relation ω_s=3.1 J, where d_{Cu-O} denotes the distance between neighboring Cu(P) and O(P).[4] We note that J is almost constant irrespective of the material dependence of d_{Cu-O}. This can be characteristic of the CT materials, since the materials dependence of Δ cancels that of T_{pd} in J. This is in contrast with the typical MH material V_2O_3, in which J increases strongly with decreasing the lattice constant.[7]

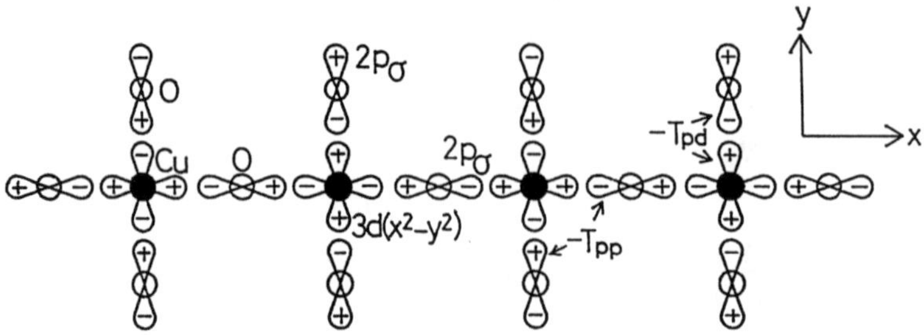

Fig. 9. A Cu_4O_{13} cluster used in the calculation of the optical conductivity, PES and BIS spectra, when 3, 4 and 5 holes are introduced into the cluster, 25% electron-doped system, insulator and 25% hole-doped system are obtained, respectively.

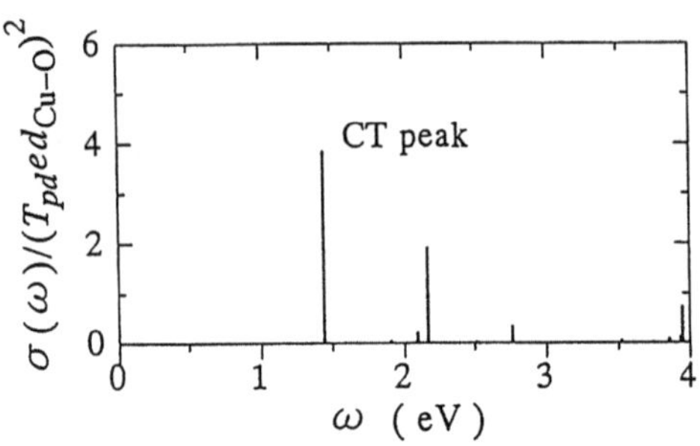

Fig. 10. Optical conductivity in the insulator calculated by using the cluster in Fig. 9. A peak at ω=1.4 eV gives the CT gap.

Let us next examine the optical conductivity in the cluster model approach.[11] The real part of the optical conductivity is given by the Kubo formula ($\hbar$=1) as

$$\sigma(\omega) = -\frac{1}{\omega} \operatorname{Im} \langle 0 \,|\, j_x \frac{1}{\omega + i0 + E_0 - H} j_x | \, 0 \rangle , \tag{6}$$

where $|\,0\rangle$ is the ground state with energy E_0 and j_x is the x component of the current operator which is defined as

$$\vec{j} = i\, e\, d_{\text{Cu-O}} \, [T_{pd} \sum_{mls} \vec{u}_{ml} \, (p_{ls}^\dagger d_{ms} - d_{ms}^\dagger p_{ls}) + \sqrt{2}\, T_{pp} \sum_{l'ls} \vec{v}_{l'l} \, p_{ls}^\dagger p_{l's}] . \tag{7}$$

Here, $\vec{u}_{ml}$ and $\vec{v}_{l'l}$ are the unit vectors directing from the m- and l'-sites to the l-site, respectively. The electric field of light is in the x direction. A Cu_4O_{13} cluster is given by Fig. 9. When, 3, 4 and 5 holes are introduced into the cluster, we have 25% electron-doped system, insulator, and 25% hole-doped system, respectively.

In the numerical calculation, the ground state $|\,0\rangle$ is obtained by using the modified Lanczos method. Then, $\sigma(\omega)$ is evaluated by using the Lanczos algorithm, where the continued fraction is truncated at 200 iterations. The optical conductivity of the insulator calculated in the cluster model approach is shown in Fig. 10. Although the conductivity consists of the line spectra in the cluster with finite size, we find that

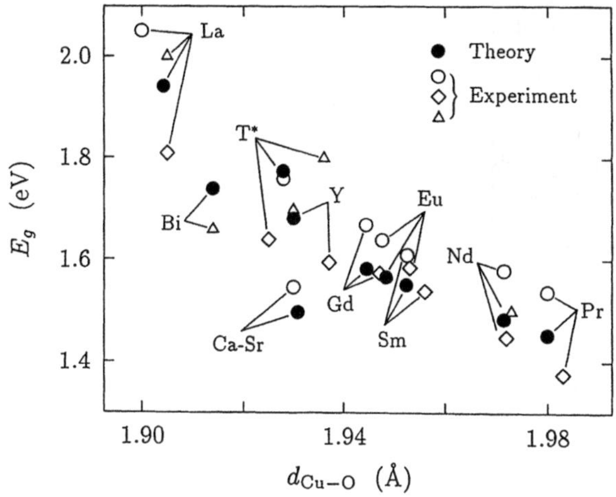

Fig. 11. Theoretical and experimental CT gap (E_g) obtained from the optical conductivity.[4] For the experimental data, see refs. 48 and 49.

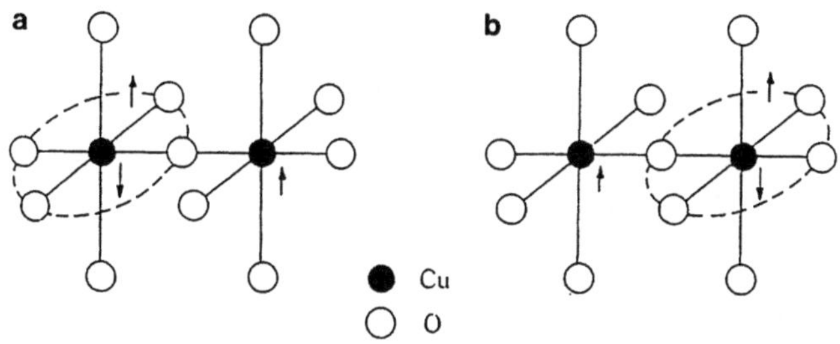

Fig. 12. Schematic figure of the motion of a spin-singlet state in the CuO_2 plane.

the peak at ω=1.4 eV corresponds to the CT excitation of holes from Cu3$d(x^2-y^2)$ to O(P)2p_σ orbitals. We note that the CT gap (E_g=1.4 eV) is different from Δ(=2.39 eV) in the ionic model. The difference between E_g and Δ comes from the many-body effects due to the coupling between the hole excited by light and spins on copper ions.

In Fig. 11, the theoretical E_g's are compared with the experimentally observed ones.[4] E_g decreases with increasing d_{Cu-O}. This is due to the Madelung potential ΔV_M as seen in Fig. 3. Because the material dependence of T_{pd} does not affect E_g, E_g reflects the three-dimensional crystal structure instead. It is interesting to see that although the covalency is strong in the CuO_2 plane, the essential feature of the electronic structure is given by the ionic model. In the next section, we will examine the electronic structure in the electron- and hole-doped systems.

3. Carrier-doping

Let us introduce holes into the CuO_2 plane. Since $\Delta < U_d$ as seen in Fig. 2, holes go to O(P)2p_σ orbitals and hybridize Cu3$d(x^2-y^2)$ orbitals when $\Delta V_A \gg 0$. As a result, there appear spin-singlet states between the holes and spins of Cu^{2+} ions.[11-13] Therefore, the motion of a doped-hole is expressed as that of the spin-singlet state as shown in Fig. 12 schematically.

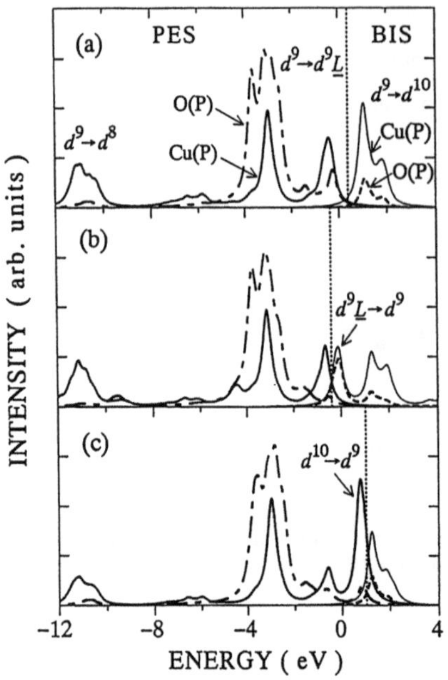

Fig. 13. PES and BIS spectra calculated in the Cu_4O_{13} cluster. (a) insulator, (b) 25% hole-doped system, and (c) 25% electron-doped system. The solid and dot-dashed lines denote $Cu3d(x^2-y^2)$ and $O(P)2p_\sigma$ PES, and the thin-solid and dashed lines denote $Cu3d(x^2-y^2)$ and $O(P)2p_\sigma$ BIS spectra. The spectra of copper and oxygen in the middle of the cluster were calculated in order to avoid the boundary effect. The spectra with the Lorentzian convolution with the half-width γ=0.2 eV are shown. The vertical dotted lines indicate the Fermi level defined as the center of the first ionization and first affinity states.

The dramatic change of the electronic structure caused by carrier-doping is found in the one-particle spectral density.[14)] The spectral density is directly observed by the valence-band photoemission (PES) and inverse photoemission (BIS) experiments: the spectral densities to take off an electron from and add one to the system correspond to PES and BIS spectra, respectively.

Figure 13 shows the calculated results of $Cu3d(x^2-y^2)$ and $O(P)2p_\sigma$ PES and BIS spectra in the Cu_4O_{13} cluster (Fig. 9).[15)] Here, the spectra of copper and oxygen ions in the middle of the cluster was calculated in order to avoid the boundary effect. The Figures (a), (b) and (c) correspond to the cases of insulator, 25% hole-doped system and 25% electron-doped systems, respectively. In these figures, the solid and dot-dashed lines denote $Cu3d(x^2-y^2)$ and $O(P)2p_\sigma$ PES and the thin solid and dashed lines denote $Cu3d(x^2-y^2)$ and $O(P)2p_\sigma$ BIS, respectively. The similar calculations have been done by several authors.[16,17)]

The difference between PES and BIS near the energy zero in the insulating state gives the CT gap. In the hole-doped system, on the other hand, a new peak appears in the gap and the peaks near the gap in the insulator are depressed by the doping. The new peak is called the gap state,[14–21)] which is brought about by reconstructing the states near the gap. Considering that the carrier-doping causes the high T_c superconductivity, we see that the gap state is responsible for the superconductivity.

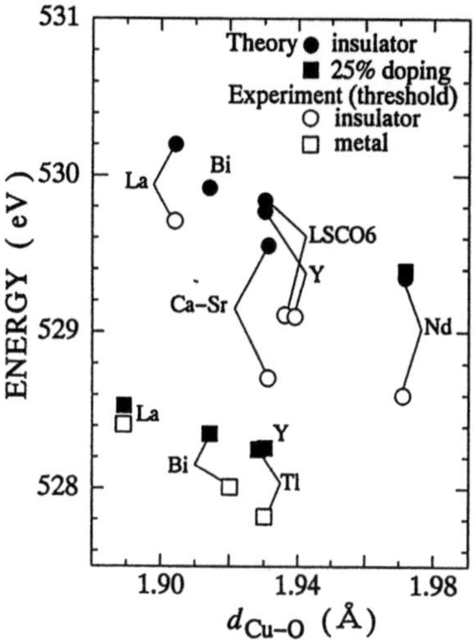

Fig. 14. O1*s* threshold energy for various undoped and carrier-doped compounds. The white circles and white squares denote the experimental results in undoped and carrier-doped samples, respectively.[24] The solid circles and solid squares denote the theoretical results calculated in the Cu_4O_{13} cluster (Fig. 9) in undoped and 25% carrier-doped systems, respectively.[25]

The gap state is the spin-singlet state and thus is in contact with Kondo state in the rare-earth compounds. The Kondo state occurs based on the singularity of the Fermi surface, whereas the gap state emerges in the CT gap on the carrier-doping in the self-consistent way with the Fermi surface.

The gap state occurs also in the electron-doped system. If we assume that an electron is introduced into the Cu3$d(x^2-y^2)$ orbital neglecting the effect of the Cu4*s* orbital on the doping,[22,23] for simplicity, the Cu(P) ion becomes the $3d^{10}$ state which is the spin-singlet state. Figure 13(c) shows the spectra in the electron-doped system. We find that the gap state emerges in PES spectra.

In Fig. 14, white circles and white squares show the O1*s* threshold energies of core-level electron-energy-loss spectroscopy (EELS) for insulators and hole-doped samples, respectively.[24] Also shown by solid circles and solid squares are the theoretical values obtained in the cluster in Fig. 9 for insulators and 25% hole-doped system, respectively.[25] Note that the theoretical values give the peak positions, whereas the threshold values in experiment give the energies of the raising edges at half height. The threshold energy gives the transition of electrons from O1*s* level to unoccupied O2*p* orbitals. In the insulators, the transition from O1*s* level to Cu3$d(x^2-y^2)$ orbital is seen since the latter orbital is hybridized by O2p_σ orbitals. We note that the threshold energy depends on materials as E_g does. This is because of the difference of the Madelung potentials between O(P) and Cu(P) sites. When holes are introduced into the insulators, the gap states appear so that the transition of electron from O1*s* level to the states becomes possible, resulting in the shift of the threshold energy. The similar shift of the energy upon hole-doping has been observed in the soft X-ray absorption experiments (XAS).[26,27] In the electron-doped Nd compounds, there is little shift of the energy upon electron-doping.[24] This is consistent with theory.

Optical conductivity also changes dramatically upon carrier-doping. Figure 15(a) and (b) show the conductivity calculated in the cluster in Fig. 9 in the cases of 25%

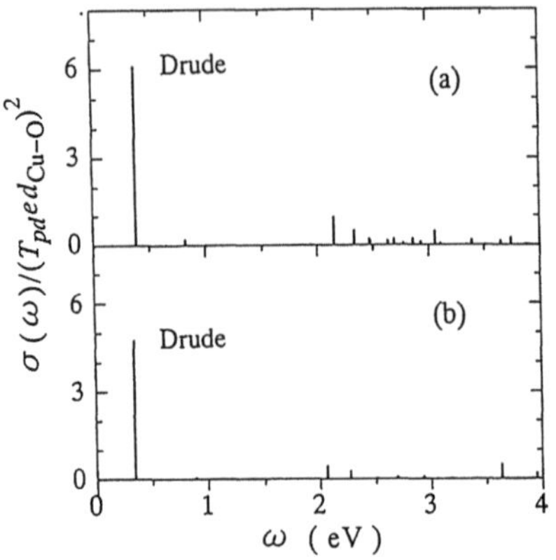

Fig. 15. Optical conductivity calculated in the Cu_4O_{13} cluster (Fig. 9). (a) 25% hole-doped system, and (b) 25% electron-doped system. In both systems, peaks at ω=0.4 eV correspond to the Drude ones.

hole-doped and 25% electron-doped systems, respectively.[11] When holes are introduced into the insulator, much of the spectral weight of the CT peak is transferred to the peak at ω=0.4 eV, which corresponds to the Drude one. Consulting the one-particle spectral density in Fig. 13, we may understand that such transfer of the spectral weight of the optical conductivity implies that holes localized on copper sites in the undoped state contribute to the formation to the Drude peak together with doped-holes in the CuO_2 plane. The effects of the electron-doping on the conductivity are very similar to those of the hole-doping. The similarity of the conductivity between electron- and hole-doped systems has been observed experimentally.[28]

We have found that the electron- and hole-dopings bring about the formation of the gap states in the low and high energy sides in the CT gap, respectively. This theoretical result is in accord with EELS[24] and XAS[26,27] experiments. On the other hand, it has been shown in PES experiments that the gap state emerges near the center of the CT gap in both electron- and hole-doped systems.[29,30] At the present stage, it is not clear why the position of the state in the latter experiments is different from the theoretical results.

The dispersion relation and the Fermi surface in the hole-doped high T_c superconductors have been observed by the angular-resolved photoemission experiments.[31–33] The de Haas-van Alphen experiments[34,35] are now being applied to study the Fermi surface. Although there is still some inconsistency among the experimental data, the photoemission spectra show the existence of the large Fermi surface in the materials, indicating that not only the doped-holes but also holes localized in the insulating state are mobile and contribute to the Fermi surface. This fact agrees with the optical conductivity and the one-particle spectral density, where the mobile gap state emerges in the CT gap by reconstructing the electronic state. Therefore, we expect that the large Fermi surface is also seen in the electron-doped system.

4. Correlation between T_c and electronic structure

A great variety of high T_c materials has been discovered.[3] Therefore, one of the interesting directions to study the superconductivity is to examine the properties in the materials, to extract key parameters for the superconductivity, and to reveal the mech-

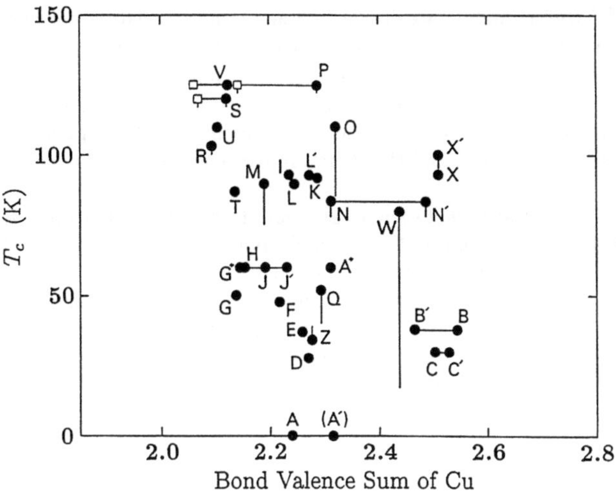

Fig. 16. Correlation between T_c and bond valence sum (BVS) of Cu(P).[40] The materials are shown in Fig. 4.

anism. Much study has been done in this direction. It should be noted, however, that in order to make this kind of study valuable, the parameters must be connected with the electronic structure. As mentioned in the introduction, high T_c superconductors must have CuO_2 plane and the carrier concentration of $p \approx 0.15-0.2$ per CuO_2 unit in the plane. We examine what the other key parameters are for high T_c superconductivity.

Let us first consider the correlation between T_c and the crystal structure. The correlation between T_c and ΔV_A introduced in Sec. 2 is one of them. Here, ΔV_A determines the energy level of O(A)$2p_z$ orbital. Compounds with a larger ΔV_A have a higher T_c. T_c appears at $\Delta V_A \simeq -2$ eV and increases rapidly with increasing ΔV_A. This means that holes should be on O(P) but not on O(A) for the compounds to be superconducting. When ΔV_A is small even if positive, holes sit not only on O(P) but on O(A) so that the interaction between the holes and Cu spins weakens. On the contrary, if ΔV_A is large, holes stay on O(P) and makes spin-singlet states with the spins. When the spin-singlet state is stable, the gap state is so and thus T_c is high. Therefore, ΔV_A is a measure of the stability of the spin-singlet state and the gap state.[3]

As mentioned above, the presence of apex oxygen makes CuO_2 plane easier to dope with holes. However, the holes should not go to the apex sites for the compounds to be high T_c superconductors. We may say that the superconductivity exists on such subtle balance of the effects of apex oxygens.

T_c has also a correlation with the Madelung potential ΔV_M (Fig. 5). Because ΔV_M governs the CT gap and the gap state carries the superconductivity, ΔV_M is considered to be an important parameter for the CuO_2 plane which is the stage of the superconductivity.

The correlation between T_c and the bond valence sum (BVS) has been discussed by several groups.[36–40] BVS is a function of interatomic distance and is considered to be a measure of the oxidation state. We write BVS as follows:

$$v_i = \sum_j \exp[(R_0 - R_{ij})/0.37]\,, \tag{8}$$

where R_{ij}(Å) is the distance between i and j atoms. Here, R_0 is a phenomenological parameter and is taken to be 1.679 Å for $Cu^{2+}-O^{2-}$ pair. In Fig. 16, we have plotted

T_c as a function of BVS.[40] Unfortunately, we do not find any clear correlation between T_c and BVS. This is because T_c is dependent on the whole crystal structure but not the local arrangement of atoms, whereas BVS is a quantity to reflect local crystal structure.

Several other correlations between T_c and crystal structure have been proposed. T_c is higher in the compounds with a shorter lattice constant in the CuO_2 plane.[41] However, no clear correlation between T_c and the structure perpendicular to the planes was not seen.[42]

Also proposed were the correlation between T_c and various physical quantities. It was shown that the muon spin relaxation rate,[43] the quadropole resonance frequency of Cu(P),[44] the edge of the reflectivity of light,[45] the Hall coefficient[46] and the pressure coefficient of T_c[47] have unique correlations with T_c. However, these physical quantities depend on the carrier concentration. Therefore, we should first ask what the other electronic parameters are in these quantities. Such question is expected to be one of the directions to study the mechanism of high T_c superconductivity.

5. Conclusion

Since the discovery of high T_c superconductivity, five years have passed. Although high T_c superconductors involve many difficult problems because of the strongly correlated electron systems, a variety of the materials stimulates to examine the properties of the materials, to extract key parameters for the superconductivity, and to reveal the mechanism. This type of the approach requires various collaborations such as those between theory and experiment and between physicists and chemists. The more collaborations are now growing in the field of high T_c superconductivity.

Acknowledgement

The work has been supported by a Grant-in-Aid for Scientific Research on Priority-Areas from the Ministry of Education, Science and Culture of Japan.

References

1) See, for example, *Strong Correlation and Superconductivity*, ed. H. Fukuyama, S. Maekawa and A. P. Malozemoff (Springer Series in Solid-State Sciences **89**, 1989).
2) J. Kondo: in Ref. 1, page 57.
3) Y. Ohta, T. Tohyama and S. Maekawa: Phys. Rev. B **43** (1991) 2968.
4) Y. Ohta, T. Tohyama and S. Maekawa: Phys. Rev. Lett. **66** (1991) 1228.
5) J. B. Torrance and R. M. Metzger: Phys. Rev. Lett. **63** (1991) 1515.
6) J. Zaanen, G. A. Sawatzky and J. W. Allen: Phys. Rev. Lett. **55** (1985) 418.
7) D. B. McWhan, A. Menth, J. P. Remeika, W. F. Brinkman and T. M. Rice: Phys. Rev. B **7** (1973) 1920.
8) A. Fujimori: in Ref. 1, page 300.
9) L. H. Tjeng, H. Eskes and G. A. Sawatzky: in Ref. 1, page 33.
10) T. Tohyama and S. Maekawa: J. Phys. Soc. Jpn. **59** (1990) 1760.
11) T. Tohyama and S. Maekawa: J. Phys. Soc. Jpn. **60** (1991) 53.
12) F. C. Zhang and T. M. Rice: Phys. Rev. B **37** (1988) 3759.
13) H. Matsukawa and H. Fukuyama: J. Phys. Soc. Jpn. **58** (1989) 2845 and 3687.

14) S. Maekawa, Y. Ohta and T. Tohyama: to be published in the Proceedings of the 2nd ISSP International Symposium on Physics and Chemistry of Oxide Superconductors (1991).
15) T. Tohyama and S. Maekawa: Physica C (1991).
16) C. A. Balseiro, M. Avignon and E. R. Gagliao: Solid State Commun. **72** (1989) 763.
17) W. Stephan and P. Horsch: *Dynamics of Magnetic Fluctuations in High Temperature Superconductors*, ed. G. Reiter, P. Horsch and G. Psaltakis (Plenum, New Yor, 1990)
18) M. Matsumoto, M. Sasaki and M. Tachiki: Phys. Rev. B **43** (1991) 10247 and 10264.
19) D. S. Hirashima, H. Jichu, Y. Kuroda, T. Matsuura, K. Miura and Y. Ono: *Research Report on Mechanism of Superconductivity II* (1991) p.446.
20) P. C. Pattnaik and D. M. Newns: Phys. Rev. B **41** (1990) 880.
21) C. A. R. Sa de Melo and S. Doniach: Phys. Rev. B **41** (1990) 6633.
22) T. Tohyama, Y. Ohta and S. Maekawa: Physica C **158** (1989) 525.
23) K. Okada, Y. Seino and A. Kotani: J. Phys. Soc. Jpn. **60** (1991) 1040.
24) M. Alexander *et al.*: Phys. Rev. B **43** (1991) 333.
25) T. Tohyama and S. Maekawa: to be published.
26) C. T. Chen *et al.*: Phys. Rev. Lett. **66** (1991) 104.
27) S. Tan *et al.*: Phys. Rev. B **42** (1990) 1037.
28) S. Uchida *et al.*: Phys. Rev. B **43** (1991) 7942.
29) H. Namatame *et al.*: Phys. Rev. B **41** (1990) 7205.
30) J. Allen *et al.*: Phys. Rev. Lett. **64** (1990) 595.
31) T. Takahashi *et al.*: Nature **334** (1988) 691.
32) J. C. Campuzano *et al.*: Phys. Rev. Lett. **64** (1990) 2308.
33) R. Manzke *et al.*: Europhys. Lett. **9** (1988) 477.
34) F. M. Mueller *et al.*: preprint (Los Alamos).
35) G. Kido *et al.*; preprint (Tohoku University).
36) D. M. de Leeuw *et al.*: Physica C **166** (1990) 133.
37) R. J. Cava *et al.*: Physica C **165** (1990) 419.
38) J. L. Tallon: Physica C **168** (1990) 85.
39) S. Tanaka, N. Fukushima, H. Nin and K. Ando: Jpn. J. Appl. Phys. **29** (1990) L1987.
40) Y. Ohta, T. Tohyama and S. Maekawa: to be published in Mod. Phys. Lett. B **5** (1991).
41) H. Nobumasa, K. Shimizu and T. Kawai: Physica C **167** (1990) 515.
42) T. Ito, H. Takagi, S. Ishibashi, T. Ido and S. Uchida: Nature **350** (1991) 596.
43) Y. J. Uemura *et al.*: Phys. Rev. Lett. **66** (1991) 2665.
44) G. Zheng *et al.*: to be published in Solid State Commun.
45) S. Tajima, T. Nakahashi, S. Uchida and S. Tanaka: Physica C **156** (1988) 80.
46) N. P. Ong *et al.*: in Ref. 1, page 204.
47) C. Murayama *et al.*: Nature **339** (1989) 293.
48) S. L. Cooper *et al.*: Phys. Rev. B **42** (1990) 10785.
49) T. Arima *et al.*: Phys. Rev. B **44** (1991) 917.
50) S. Sugai *et al.*: Phys. Rev. **42** (1990) 1045.

Binding and Structural Aspects of High-T_c Superconductivity

H. Eschrig

Zentralinstitut für Festkörperphysik und Werkstofforschung,
Helmholtzstr. 20, O-8027 Dresden, Fed. Rep. of Germany

Abstract. All layered cuprate superconductors are formed of several extended anionic complexes embedded in a cationic lattice frame. Layered anionic $[CuO_2]^{2-}$-complexes are undispensibly present. There is a second anionic complex, varying from system to system, and doping the $[CuO_2]^{2-}$-complex. As another generic feature, this second complex tends to quasi-onedimensional bonding. A further generic feature is the presence of 'bridging' oxygen mediating the doping charge transfer and causing a very strong deformation potential for the doping electron states. These features can lead to an effective interlayer coupling which is capable of producing s-channel superconductivity.

1. Introduction

Cuprate superconductors are complex materials in almost every conceivable respect. They appear in a crystal structure not only with many atoms per unit cell but also with all types of bonding — metallic, covalent, and ionic — mixed up. They show strong electron-electron and electron-lattice correlations as well. They have internal structural degrees of freedom causing phase coexistence (and phase transitions), twinning, microdomains, inhomogeneous 'single crystals', etc. etc. Complex high-temperature phase diagrams make them condense into complex microstructures.

All in all, the cuprates do not make the impression of a preferable model system for physical investigations. Yet, as long as it is not clear to the end which of their features are responsible for the superconducting properties, all of them, step by step, should be modelled separately and in combination by theory.

In the following, details of structure and bonding and of structural distortions are discussed in section 2. Consequences on the electronic structure and on electron-lattice interaction are considered in section 3. Models of electron-lattice interactions and of charge transfer excitations are treated in section 4, and a possible mechanism of superconductivity is discussed in section 5.

2. Structure and Bonding

Some 'parent structures' of cuprate superconductors are shown in fig.1. Conclusions on ionicity and bonding can be drawn immediately from the experimental

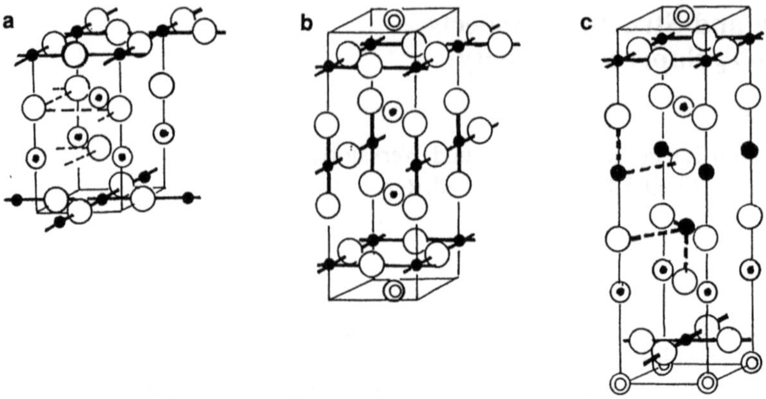

Fig.1. 'Parent structures' of cuprate superconductors:
a) La_2CuO_4, b) $YBa_2Cu_3O_7$, c) $Bi_2Sr_2CaCu_2O_8$.
Thick lines and broken lines denote covalent bonds within anionic complexes.
○ oxygen, ⊙ lanthanum, barium, strontium, • copper,
● bithmuth, tallium, ◎ yttrium, calcium.

Table 1. Ion radii and covalent atom radii.

ion	La^{3+}	Y^{3+}	Ba^{2+}	Sr^{2+}	Ca^{2+}	Cu^{2+}	Bi^{3+}	Tl^{3+}	O^{2-}
radius	1.04	0.88	1.29	1.10	0.94	0.92	1.00	1.06	1.46

atom	Bi	Tl	O
covalent radius	1.36	1.44	0.66

bond lengths. Standard ion radii of the appearing ions are given in table 1. A distance between an Y^{3+} ion and an O^{2-} ion, e.g. should be nearly equal to 2.34 Å, the sum of the corresponding ion radii. With an actual distance of 2.40 Å in $YBa_2Cu_3O_7$ this is well fulfilled, so that Y is to be expected in the ionized Y^{3+} state. The same conclusion can be drawn with respect to Ba^{2+}, Sr^{2+}, and Ca^{2+}. These cations form a lattice frame of purely ionic bonds to the anionic oxidic complexes.

Copper-oxygen distances are found equal to roughly 2.4 Å, corresponding to an ionic bond, and 1.8 Å, corresponding to a covalent bond. The important covalent Cu-O bonds are indicated by thick lines in fig.1.

A bit more complicated is the situation with the La-O, Bi-O, and Tl-O bonds, the corresponding lengths in the reference structures of fig.1. would be 2.7 Å, too long for ionic bonds and covalent bonds as well. (Note that the Bi-O covalent bondlength in the anionic $[BiO_3]^{2-}$ complex of $BaBiO_3$ is 2.15 Å[1].)

This is the reason for distortions of the corresponding layers: the corresponding oxygens move out of their symmetry positions to form shorter bonds.

The results of these distortions are zig-zag chains of contracted peroxidic bonds in the LaO planes and chains of contracted BiO (TlO) bonds in the BiO (TlO) planes, so that each Bi (Tl) atom has three nearest oxygen neighbours and three farer off [2,3]. Note that attempts to explain the distortion of the LaO planes by Fermi surface nesting failed. The explanation as a bondlength mismatch seems natural.

Summarizing the discussion of the structure we have cationic lattice frames with extended anionic complexes embedded. These anionic complexes contain covalent bonds and hence the current carriers. The cuprate complex $[CuO_2]^{-2+\delta}$ is quasi-twodimensional. Sandwiched between sheets of those complexes there is another polarizable anionic complex with a tendency to quasi-onedimensionality. Charge transfer (δ) between these complexes leads to metallization of the CuO_2 sheets. There are indications that the quasi-onedimensionality of the intermediary complexes plays a role in superconductivity: a different type of distortion of Ba doped La_2CuO_4 with oxygen dumpbells instead of chains is not superconducting [4], and $Tl_2Ba_2CuO_6$ with Tl vacancies ordered in chains is superconducting while the same material with the same stoichiometry but disordered Tl vacancies is not [3].

3. Electronic Structure

The undoped insulating state is $[CuO_2]^{2-}$. Bandstructure calculations yield the Fermi level in the middle of an antibonding $dp\sigma$-band [5]. Experimentally there appears, however, a conductivity gap of 1.5–2 eV which is generally attributed to strong copper $3d$ onsite correlations [6] and is qualified as of charge transfer nature [7,8].

If these planes are doped with holes, they become metallic with seemingly a Fermi liquid behaviour [9,10] and a Fermi surface not far from ordinary band structure predictions. This can be understood as reinforcement of occupation number fluctuations and hence band-like behaviour due to a special interplay of Madelung and intraatomic potentials [11]. The still present correlations seem to lead to a mass renormalization by a factor of 3–4.

There is general agreement that these CuO_2-planes carry normal conductivity and are essential in superconductivity.

There is an out-of-plane orbital coupling of the conduction states both with the apical oxygen and between pairs of planes separated by Y and Ca layers in the Y-Ba-Cu system and the Bi or Tl systems, respectively. Both coupling strengths are of the same order of magnitude, $t_\perp \approx 0.1$ eV. The matrix element $t_\perp$ depends sensitively upon the plane-apical oxygen distance. Hence, the coupling of the $[CuO_2]^{2-}$ layers to the apical oxygen in the lanthanum system is considerably smaller.

All cuprate structures have a mirror or glide mirror plane between CuO_2 layers. Hence, the coupled in c-direction conduction states obey a $(k_z, -k_z)$

degeneracy and may be classified as symmetric or antisymmetric with respect to this mirror plane. In the La_2CuO_4 based system the $k_z = 0$ state is symmetric. All other cases contain the CuO_2 planes pairwise within one unit cell so that there appear several conduction bands with symmetric as well as antisymmetric $k_z = 0$ states.

In the Y-Ba-Cu system, the doping anionic complex is formed by the well known CuO_3 chains (or Cu_2O_5 double chains). Their role as a charge reservoir, i.e. their doping effect on the CuO_2 planes is by now generally accepted, although the precise amount of the charge transfer between the CuO_2-planes and the chains is not established, yet. For the related Bi and Tl systems an analogous second band complex near the Fermi level resulting from the Bi(Tl)-O planes has been obtained by many bandstructure calculations, e.g. [12] and observed experimentally [13]. The $La_{2-x}M_xCuO_{4-\delta}$ (M=Sr,Ba,Ca) and the recently discovered double layered $La_{2-x}Sr_xCaCu_2O_{6-\delta}$ ($T_c = 60$K) [14] are a bit unclear in this respect. LDA- and LSDA band structure calculations [15] for the undoped system yield apical oxygen states well below the Fermi energy while the highly precise FLAPW results of Freeman et al. (see [5]) yield states with an apical oxygen p_z admixture very close to the Fermi level. At least for high doping level $x \approx 0.5$ such states appear near the Fermi level. Taking into account correlation effects at the apical oxygen positions, the corresponding band complex might be upward shifted resulting in a partially filled band. Similar arguments might be given for the electron doped HTSC with respect to the out-of-plane O(2)-bands. In the Sr doped La-system, a chain-like behaviour with a quasi-onedimensionality of the electronic subsystem results from the tilting deformation of the oxygen octahedra (towards a La-position) present locally even in the nominally tetragonal phase near $x \approx 0.15$ [16], where T_c reaches its maximal value (see fig. 1a).

This universality of quasi-onedimensionality in the doping complex has been overlooked hitherto. We regard it as essential as the universality of a bridging (apec) oxygen between the two anionic complexes [17]. The quasi-onedimensional band is antibonding in nature with a $2p$-component from the bridging oxygen and has a strong deformation potential for that oxygen displaced in c-direction. For the Y-Ba-Cu system, its presence seems to be seen in the dynamical conductivity as a polarization dependent mid-infrared contribution [18].

4. Electron-Lattice Interaction in Chains

In the following we will restrict ourselves mainly to the well studied Y-Ba-Cu structure, generalizations to the other systems are straightforward although more complicated in detail. The electronic structure of a CuO_3-chain in the Y-Ba-Cu structure can be modelled considering the chain as a dimer chain of alternating O-ions and CuO_2-molecules. The band under consideration is formed of $p\sigma$-states on the O-sites and antibonding molecular $dp\sigma$-states on the CuO_2-sites. The Hamiltonian matrix consists of two on-site energies ε_0 and

ε_m (of the O-site and the molecular site, respectively) and one transfer integral t, consisting of $dp\sigma$-overlap between both sites and $p_z p_y$-overlap between the oxygens. Deformation corresponding to a symmetric z-displacement (u) of the two molecular O-ions is considered. At a first glance one would attribute the strong deformation potential found in selfconsistent calculations [19] to a strong sensitivity of ε_m on the displacement, which was also assumed in [17]. A detailed analysis of the corresponding matrix elements resulting from selfconsistent LCAO calculations [20] shows, that due to a balance between the changes in Madelung potential and dp hybridization matrix element, ε_m is only weakly dependent on the displacement. The dependence of t, on the other hand, comes from the direct $p_z p_y$-overlap and from the change in hybridization of the molecular $dp\sigma$-state changing the mixing rate between d- and p-states. In addition we consider y-displacements of O(1) along the chain as suggested by thermal diffuse scattering data [21].

For the chain we adopt the effective AB-Hamiltonian (A = CuO_2, B = O) with alternating parity

$$H = H_{el} + \frac{K_u}{2} \sum_{m \epsilon A} u_m^2 + K_y \sum_{m' \epsilon B} y_{m'}^2 , \tag{1}$$

$$H_{el} = \sum_{n,s} \varepsilon_n c_{n,s}^+ c_{n,s} - t_{n,n+1} \left(c_{n+1,s}^+ c_{n,s} + h.c. \right) , \tag{2}$$

where

$$t_{m,m+1} = \begin{cases} - & [t_0 - \alpha_1 u_m - \alpha_2 y_{m+1}] \quad m \epsilon A \\ & t_0 - \alpha_1 u_{m+1} + \alpha_2 y_m \quad m \epsilon B \end{cases} , \tag{3}$$

and analogously

$$\varepsilon_m = \begin{cases} \epsilon_A & - \quad \alpha_A u_m - \beta (y_{m+1} - y_{m-1}) \\ \epsilon_B & - \quad \alpha_B (u_{m+1} + u_{m-1}) \end{cases} . \tag{4}$$

Taking into account the constraint $\sum_{m'} y_{m'} = 0$ resulting from the fixed chain length requirement, two self-consistency equations are obtained from the variation of (1) with respect to u_m and y_m [22].

Using eq.(2) for the parametrisation of the LCAO-deformation study we found: t_0 = 1.67 eV, α_1 = 1.32 eV/Å, ε_A = -3.59 eV, ε_B = -5.3 eV, $\alpha_A = 2$ α_B = -0.816 eV/Å and from the corresponding phonon modes at 505 cm^{-1} and 400 cm^{-1} K_u = 7.78 eV/$Å^2$ and K_y = 4.87 eV/$Å^2$.

Generally, at least in mean-field approximation for commensurate bandfilling factors $\nu = 1/2, 1/3, 1/4, \ldots$ the chain groundstate is given by a commensurate charge density wave (CDW) connected with a corresponding superstructure and a Peierls gap E_g. The energy gain has maxima for commensurate CDWs. Therefore we suppose that the charge transfer between planes and chains exhibits a 'lock-in' at one of these special bandfilling ratios ν. Experimentally, the magnitude of ν is not precisely known, because the interpretation of the Hall or optical data of strongly correlated systems like the cuprates is nontrivial. The Peierls gap and the lattice distortion decrease with increasing commensurability.

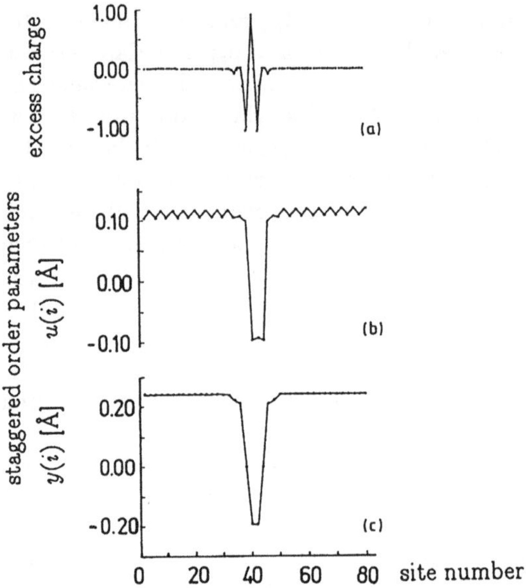

Fig.2. Excess charge and staggered order parameters of a bipolaron vs. site number i ($\alpha_2 = \alpha_1, \beta = 1$ eV/Å, $N = 80$ sites).

The homogenously Peierls distorted chains of the last paragraph with apical oxygen attached have a rich structure of charged nonlinear excitations which can be obtained by a rigorous treatment of the Hamiltonian for a large but finite chain length. Considering the half-filled (antibonding) band case for fixed $y_m \equiv 0$, self-consistent bipolaronic solutions appear only for $\alpha_A > 0$. ($\alpha_A \geq 0.75$ eV/Å^2 for the present parameter set). However, allowing y_m displacements too, for reasonable values of α_2 and $\beta \sim 1$ eV/Å 'two-component' bipolarons have been found for the present case $\alpha_A < 0$. Here the stabilization of the bipolaron is caused by the enhancement of the CDW-gap due to the second order parameter $\sim y_{m'}$. The latter effect was firstly noticed for the case of an AB-chain with even parity and half-filled bonding band in [23]. The excess charge distribution as well as the staggered order parameters $(-1)^m u_m, (-1)^{m'} y_{m'}$ of the bipolaron are shown in Fig. 2.a-c, respectively. Note that the bipolaron is confined to several lattice sites ($2b \approx 3.88$ Å), i.e. its width is comparable with the coherence length $\xi_{ab} \approx 15\div 20$ Å. From the intermediate size of the bipolaron, an enhanced but not too large mass $M_{BP} < 10 m_e$ can be expected.

5. Implications for Superconductivity

Now, taking mobile conduction electrons from neighbouring CuO_2-planes with field operators $a_{\mathbf{k}\sigma}$ and bipolarons in chains with field operators $b_{\mathbf{p}}$ we may write down the interaction Hamiltonian of [24]

$$H_{int} = \frac{g}{\sqrt{\Omega}} \sum_{\mathbf{p},\mathbf{k}} (u_{\mathbf{k}} b^{\dagger}_{\mathbf{p}} a_{\mathbf{p}/2+\mathbf{k},\uparrow} a_{\mathbf{p}/2-\mathbf{k},\downarrow} + h.c.), \qquad (5)$$

which, in the limit of localized bosons with infinite mass M_{BP} was considered by Enz and Galasiewicz [25].

According to our self-consistent numerical calculations the formation energy of such a bipolaron E_{BP} will be in the order of several tenths of an eV. The coupling constant g of eq.(5) is of the order of the square of the hopping integral $t_{\perp}$ between the apical oxygens and the CuO_2-planes. Hence we are in the weak-coupling limit of the Friedberg-Lee theory at least at low temperatures.

In the case of the Y-Ba-Cu structure, the electronic states of the chain band are necessarily symmetric with respect to the mirror plane containing the Cu-O chain. Antisymmetric combinations of apical oxygen $2p_z$ states decouple and appear well below the Fermi level.. Hence, the coupling in eq.(5) is with symmetric planar conduction states only (see the discussion in section 3). This symmetry constraint leads naturally to *interlayer pairing* via the interaction eq.(5).

The spatial extent of the bipolaron is several interatomic distances, (see fig. 2), i.e. comparable with or slightly larger than the supeconducting in-plane (a,b) coherence length ξ_{ab}. In the perpendicular c-direction its size is restricted by the apex-apex distance $\simeq$ 3.66 Å also in good agreement with the experimental value $\xi_c \geq$ 3-4 Å. Thus, according to the proposed scenario a finite boson density occurs below T_c within the chains, i.e. in other words a redistribution of charge between planes and chains starts just at T_c affecting the positions of the apical oxygens.

Strong experimental support for the charge redistribution effect stems from recent positron annihilation data [26], where a decrease (increase) of positron lifetime for pure (Zn-doped) $YBa_2Cu_3O_7$ beginning at T_c has been interpreted in terms of such a charge redistribution. According to the calculation of these authors the positron wave function is localized in the chain (plane) region for the pure (doped) case. Thus, the positron probes directly the charge redistribution of the corresponding subsystems.

In the Peierls distorted state, a site separation for subsequent apical oxygen positions is to be expected. The occurrence of bipolarons below T_c destroys their regular positions within the former ordinary superstructure. Such an effect, the disappearance (or decrease of the site separation) has been reported by recent EXAFS-data [27]. However, the reported reentrance of the regular superstructure at low temperatures $\simeq$ 10 K remains an unresolved problem.

Of course, one has to consider the concurring mechanism of ordinary s-channel superconductivity (exchange of neutral bosons).

Acknowledgement. Most of this work was done together with S.-L. Drechsler. Discussions with K. A. Müller, C. P. Enz, R. Monnier, W. Weber, J. Harris, G. Eilenberger, J. Fink, and many others were a great help for the author. We would like to thank J.Malek for his help with the numerical calculations. Financial support from the KFA Jülich is greatful acknowledged.

References

[1] R. J. Cava, B. Batlogg, J. J. Krajewski, R. Farrow, L. W. Rupp Jr., A. E. White, K. Short, W. F. Peck Jr., and T. Kometani, Nature **332** (1988)814.

[2] W. Dmowski, B. H. Toby, T. Egani, J. Subramanian, J. Gopalakrishnan, and A. Sleight, Phys. Rev. Lett. **61** (1988) 2608.

[3] A. W. Hewat, E. A. Hewat, P. Bordet, J. Capponi, C. Chaillout, J. Chenevas, J.-L. Hodeau, and M. Marezio, Proc. Int. Sem. HTSC Dubna 89, World Scientific, Singapore 1990, p. 24, and quotations therein.

[4] A. R. Moodenbaugh, Y. Xu, M. Suenaga, T. J. Folkerts, and R. N. Shelton, Phys. Rev. B**38** (1989)4596.

[5] L. Mattheiss, Phys. Rev. Lett. **58** (1987)1028; A. J. Freeman and J. Yu, Helvetica phys. acta **61** (1988)401.

[6] P. W. Anderson, Science **235** (1987)1196.

[7] A. M. Oles and J. Zaanen, Physica C**153–155** (1988)1229.

[8] J. Zaanen and A. M. Oles, Phys. Rev. B**37** (1988)9423.

[9] R. Claessen, R. Manzke, H. Carstensen, B. Burandt, T. Buslaps, and M. Skibowski, Phys. Rev. B**39** (1989)7316.

[10] C. C. Olsen, R. Liu, D. W. Lynch, B. V. Veal, Y. C. Chang, P. Z. Jiang, J. Z. Liu, A. P. Paulikas, A. J. Arko, and R. S. List, Physica C**162 – 164** (1989)1697.

[11] H. Eschrig, Physica C**159** (1989)545.

[12] S. Massida, J. Yu, and A. J. Freeman, Physica C**152** (1988)251

[13] G. Mante, R. Claessen, T. Buslaps, S. Harm, R. Manzke, M. Skibowski, and J. Fink, Z. Phys. B**80** (1990)181.

[14] R. J. Cava, B. Batlogg, R. B. van Dover, J. J. Krajewski, J. V. Waszczak, R. M. Fleming, W. F. Peck Jr.,L. W. Rupp Jr., P. Marsh, A. C. James, and L. F. Schneemeyer, Nature **345** (1990)602.

[15] R. Kasowski,M.-H. Tai, J. D. Dow, Phys. Rev. B**41** (1990)8949; A. Fujimori, Phys. Rev. B**39** (1989)793.

[16] T. Egami and W. Dmowski, Proc. of the Drexel Int. Conf on HTSC, Philadelphia, USA, July 1987, Progr. in HTSC **3** 247 World Scientific Singapore (1988).

[17] H. Eschrig and S.-L. Drechsler, Physica C**173** (1991)80.

[18] K. F. Renk, H. Eschrig, U. Hofmann, J. Keller J. Schützmann, and W. Ose, Physica C**165** (1990)1; B. Koch, H. P. Geserich, and T. Wolf, to be published.

[19] E. T. Heyen, S. N. Rashkeev, I. I. Mazin, O. K. Andersen, R. Lin, M. Cardona, O. Jepsen, Phys. Rev. Lett. **65**, (1990)3048.

[20] H. Eschrig, R. Monnier, G. Schadler, and D. Schmicker, to be published.

[21] Y. Koyama and Y. Hasebe, Phys. Rev. B**37** (1988)5831.

[22] S.-L. Drechsler, J. Malek, and H. Eschrig, to be published.

[23] F. Fischer and H. Köppel, to be published.

[24] J. Ranninger and S. Robaszkiewicz, Physica B**135** (1985)468; J. Ranninger, R. Micnas, and S. Robaszkiewicz, Phys. Rev. B**36** (1987)180; R. Friedberg and T. D. Lee, Phys. Rev. B**40** (1989)6745.

[25] C. P. Enz and Z.M.Galasiewicz, Solid State Commun. **66** (1988)49; C. P. Enz, Helvetica phys. acta **62** (1989)122.

[26] Y. C. Jean, C. S. Sundar, A. Bharathi, J. Kyle, H. Nakanishi, P. K. Tseng, P. H. Hor, R. L. Meng, Z. J. Huang, C. W. Chu, Z. Z. Wang, P. E. A. Turchi, R. H. Howell, A. L. Wachs, and M. J. Fluss, Phys. Rev. Lett. **64** (1990)1593.

[27] S. D. Conradson, I. D. Raistrick, and A. R. Bishop,to appear in Science (1990), J. M. de Leon, S. Conradson, I. Batistic, and A. Bishop, Phys. Rev. Lett. to be published.

Electronic State in Oxide Superconductors

M. Tachiki, S. Ishihara, M. Sasaki, and H. Matsumoto

Institute for Materials Research, Tohoku University, Sendai 980, Japan

Abstract. The electronic state of oxide superconductors is investigated by use of the p-d mixing model. The carrier concentration dependence of the electron density of states is discussed by considering composite electronic states associated with $Cu-O$ bonds.

1.INTRODUCTION

The cuprate oxide superconductor is characteristic in the fact that its superconductivity appears only in certain metallic region close to the metal-insulator transition [1]. The highly correlated electronic state realized in this region is one of the central subjects of interest in order to understand the mechanism of high T_c superconductivity. Experimentally and theoretically, there established consensus that a p-d mixing model on the $Cu-O$ plane is a good starting model to discuss the electronic state of these materials [2],[3]. There are a lot of controversy for description of the electronic state near the Fermi level(FL), even starting from the same p-d mixing model.

Experimentally, there seems to occur two drastic changes of electron density of states, depending on doping of charge carriers. In insulator phase, one observes charge transfer gap of $1.5 \sim 2eV$ [4]-[6]. Carriers are doped mainly in oxygen p bands [7]. Differently from rigid shift of the FL expected from a simple band model, there develop certain electronic states at the FL inside charge transfer gap, by transferring some portion of density of states from occupied states and so-called midgap states are formed. Experimental results of photoemission [8]-[10], EELS [7],[11], XAS [9],[12],[13] and optical conductivity [6],[14],[15] (specially, development of midinfrared structure) are consistently explained when one assumes formation of midgap states with doping. As for the origin of the midgap states, there are controversy among interpretations as impurity states, intrinsic correlated states and conventional band-like states. The other drastic change is collapsing of upper Hubbard levels for further doping. The soft X-ray absorption spectrum of the oxygen K edge [12],[13] and the optical conductivity [6] show rapid decrease of spectral intensity around energy of the charge transfer gap from certain carrier concentration ($x \sim 0.05$ for $La_{2-x}Sr_xCuO_4$). Together with the collapsing of charge transfer excitations, the midinfrared intensity shifts to low energy region and two components Drude structure is formed, at which the superconductivity appears. In this paper, we show that the above mentioned phenomena is a result of strong correlation among electrons and that both of strong covalency between p and d electrons and short distant Coulomb interaction (the so-called V-term) play important roles for those phenomena.

In the next section, we summarize the result obtained from calculation of the local limit, which is supposed to be good for low carrier concentration [16]. In Sect. 3, we reinterprete the midgap state as a composite electronic state associated with $Cu-O$ bond and show that the midgap state is controlled mainly by such a state. We show some drastic change to occur due to the presence of short distant Coulomb interaction from certain carrier concentration. Sect. 4 is devoted to concluding remarks.

2. APPROACH FROM THE LOCAL LIMIT–LOW CARRIER CONCENTRATION

The model is given by

$$H = \int dx[\frac{1}{2}\epsilon_\eta \eta^\dagger(x)\eta(x) + \epsilon_p p^\dagger(x)p(x) + \eta^\dagger(x)p_t(x) + p^\dagger(x)\eta(x)] \quad , \tag{2.1}$$

where d electron freedom is restricted to the transition $n_d = 1 \leftrightarrow n_d = 2$ and for a d electron annihilation operator $d_\sigma(x)$, $\eta_\sigma(x) = d_\sigma(x)n_{-\sigma}(x)$ with $n_\sigma(x) = d_\sigma^\dagger(x)d_\sigma(x)$, $p_\sigma(x)$ is a p electron annihilation operator, and $p_t(x)$ is the abbreviation of $p_t(x) = t(-i\vec{\nabla})p(x)$ with

$$t^2(\vec{k}) = 4t_0^2(\sin^2\frac{1}{2}k_x a + \sin^2\frac{1}{2}k_y a) \quad . \tag{2.2}$$

The equations of motion for $p(x)$ and $\eta(x)$ are given by

$$i\frac{\partial}{\partial t}p(x) = \epsilon_p p(x) + \eta_t(x) \tag{2.3a}$$

$$i\frac{\partial}{\partial t}\eta(x) = \epsilon_\eta(x) + (-\frac{1}{2}\sigma^\mu)n_\mu(x)p_t(x) \tag{2.3b}$$

with $\sigma^\mu n_\mu(x) = -n(x) + \vec{\sigma}\cdot\vec{n}(x)$, where $n(x) = \sum_\sigma d_\sigma^\dagger(x)d_\sigma(x)$, $\vec{n}(x) = \sum_{\sigma,s} d_\sigma^\dagger(x)(\vec{\sigma})_{\sigma,s}d_s(x)$ and $\vec{\sigma}$ is the Pauli matrices. The equation (2.3) indicates that there are two discrete energy levels mixed with t, and the state η changes by emitting and absorbing a p electron together with spin and charge fluctuations. We calculate the electron density of states by including corrections arising from multiple scattering of electrons due to fluctuations. The interaction vertices are determined by use of the local algebra of restricted d electron operators. The result is shown in Fig. 1. The bold lines show the d electron density of states and the fine solid lines show the p electron density of states. There are certain tail structure in the occupied p electron states. With hole doping, there develop certain electronic states at the FL. This state can

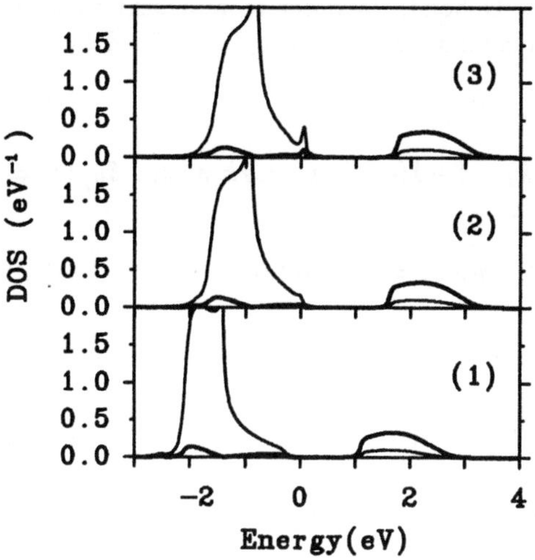

Fig. 1. The p and d electron density of states. The fine and bold solid line shows p and d electron components respectively. (1) is an insulating case. The calculated hole numbers(n_h) are (2)0.022 and (3)0.050 respectively. For all cases, $p-d$ mixing interaction t_0 and temperature are 0.6 and 0.02 eV respectively.

be regarded as a spin fluctuation dressed electron state. These results show that, due to strong correlation, rearrangement of electron states is induced at the FL, and accumulation of electron states at the FL leads to a stabler phase. This situation is quite different from a simple band theory. Unfortunately, the approximation used to obtain Fig.1 breaks down, when holes are doped further. In the next section, we develop a new approach to overcome this difficulty and reinterprete the so-formed midgap state in a different way.

3. COVALENT ELECTRONIC STATE AND COMPOSITE OPERATORS

The new state at the FL arises because electronic motion is always accompanied by fluctuations. This indicates that freedoms of the p and d electron are more closely intertwined. Specially in the region close to insulator, an electronic state of a d electron much affect motion of a p electron, since whether or not surrounding d electron states are occupied gives quite big difference to p electron motion. When we consider a p electronic excitation, we need to consider which states neighboring Cu-ions are. This kind of consideration becomes more and more important when covalency between Cu and O is stronger. We can imagine p electrons floating on Cu-ion background; even though no condition at an oxygen site changes, energy of the p electron fluctuates due to modification of environment caused by Cu-ions. This kind of situation can be treated in the case of cuprate oxide superconductors by considering electronic states associated with covalent $Cu-O$ bond. In order to represent such an electronic excitation, we introduce a composite operator $p_\mu(x)$ defined by

$$p_\mu(x) = p_\gamma(x)\delta n_\mu(x) \quad , \tag{3.1}$$

where $p_\gamma(x) = \gamma(-i\vec{\nabla})p(x)$ with $\gamma^2(\vec{k}) = 1 - \frac{1}{2}(\cos k_x a + \cos k_y a)$ and $\delta n_\mu(x) = n_\mu(x) - < n_\mu(x) >$. We use notations

$$p_s(x) = \vec{\sigma} p_\gamma(x)\vec{n}(x) \quad , p_0(x) = p_\gamma(x)\delta n(x) \quad . \tag{3.2}$$

The equations of motion for $p(x)$, $\psi_\eta(x)$ $(= \sqrt{\frac{2}{n}}\eta(x))$, $p_s(x)$ and $p_0(x)$ are summarized as

$$i\frac{\partial}{\partial t}p(x) = \epsilon_p p(x) + t_n \gamma(-i\vec{\nabla})\psi_\eta(x) \tag{3.3a}$$

$$i\frac{\partial}{\partial t}\psi_\eta(x) = t_n p_\gamma p(x) + \epsilon_\eta \psi_\eta(x) + \frac{t_n}{n}(p_0(x) - p_s(x)) \tag{3.3b}$$

$$i\frac{\partial}{\partial t}p_s(x) = \epsilon_p p_s(x) + t_n(-h_s(x) - \psi_s(x) + \varphi_s) \tag{3.3c}$$

$$i\frac{\partial}{\partial t}p_0(x) = \epsilon_p p_0(x) + t_n < 2-n > \psi_\eta(x) + t_n(-h_0(x) - \psi_0(x) + \varphi_0) \quad , \tag{3.3d}$$

where

$$t_n = 2t_0\sqrt{\frac{n}{2}} \tag{3.4}$$

$$h_s(x) = \vec{n}\gamma_1^2(-i\vec{\nabla})\psi_\eta(x)\vec{n}(x) \quad , \quad h_0(x) = \gamma_1^2(-i\vec{\nabla})\psi_\eta(x)\delta n(x) \tag{3.5a}$$

$$\psi_s(x) = \vec{\sigma} p_\gamma(x)\delta\vec{n}_+(x) \quad , \quad \psi_0(x) = p_\gamma(x)\delta n_+(x) \tag{3.5b}$$

$$\varphi_s(x) = \vec{\sigma} p_\gamma(x)\delta\vec{n}_-(x) \quad , \quad \varphi_0(x) = p_\gamma(x)\delta n_-(x) \tag{3.5c}$$

with

$$\gamma_1^2(\vec{k}) = \frac{1}{2}(\cos k_x a + \cos k_y a) \tag{3.6}$$

$$\delta n_{+\mu}(x) = p_\gamma^\dagger(x)\sigma_\mu\psi_\eta(x) - < p_\gamma^\dagger(x)\sigma_\mu\psi_\eta(x) > \tag{3.7a}$$

$$\delta n_{-\mu}(x) = \psi_\eta^\dagger(x)\sigma_\mu p_\gamma(x) - < \psi_\eta^\dagger(x)\sigma_\mu p_\gamma(x) > \quad . \tag{3.7b}$$

For electron fields $\psi_n(x)$ $(n = 1, 2, \cdots)$ including composite operators, whose equations of motion are given by

$$i\frac{\partial}{\partial t}\psi_n(x) = j_n(x) \quad , \tag{3.8}$$

a propagator $< T\psi_n(x)\psi_m^\dagger(y) >$ is expressed in general as

$$< T\psi_n(x)\psi_m^\dagger(y) > = \frac{i}{(2\pi)^3}\int d\omega d^2k e^{-i\omega(t_x - t_y) + i\vec{k}\cdot(\vec{x}-\vec{y})}\Big(\frac{1}{\omega - \Sigma(\omega,\vec{k})}I\Big)_{nm} \tag{3.9}$$

with I_{nm} being determined from

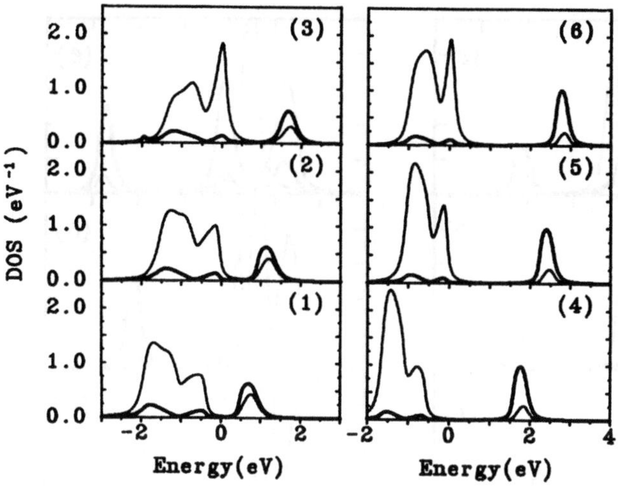

Fig. 2. The p and d electron density of states. The fine and bold solid lines show p and d electron components respectively. For all cases ((1) through (6)), $p-d$ mixing interaction t_0 and temperature are 0.5 and 0.05 eV respectively. In (1),(2) and (3), bare charge transfer gap (input value for numerical calculation) is 1.2 eV. (1) is an insulating case. The calculated hole numbers(n_h) are (2)0.047 and (3)0.256 respectively. In (4),(5) and (6), we take 2.5 eV as bare charge transfer gap. (4) is an insulating case. The calculated hole numbers(n_h) are (2)0.078 and (3)0.353 respectively.

$$I_{nm}(\vec{x}-\vec{y}) = < \{\psi_n(\vec{x}), \psi^\dagger_m(\vec{y})\} > \tag{3.10}$$

and $(\Sigma I(x-y))_{nm}$ being determined from

$$(\Sigma I)_{nm}(x-y) = < \{j_n(x), \psi^\dagger_m\} > + < Tj_n(x)\delta j^\dagger_m(y) > \quad , \tag{3.11}$$

where

$$\delta j^\dagger_m(y) = j^\dagger_m(y) - \psi_{m'}(y)\Sigma^\dagger(i\vec{\nabla})_{m'm} \quad . \tag{3.12}$$

The detailed formulation will be given elsewhere. By use of Eq. (3.3) and Eq. (3.9)–(3.11), we can calculate the density of states. It should be noted that there are three different states overlap around ϵ_p. Fig. 2 shows the result of numerical calculation including dynamical effects induced by one loops from the sources in Eq. (3.5). The bold solid lines are for d electron density of states and the fine solid lines are for p electron density of states. The difference between (1)~(3) and (4)~(6) is the values of bare charge transfer gap. The peak structure at the FL is mainly composed from composite state p_s. Those state are constructed from the composite electronic state by mixing with η level and p levels. In the present approximation, the local nature of the composite state is emphasized, so that one observes the level splitting even in the insulator phase. With hole doping, increase of intensity at the FL is observed. If the peak structures in (4)~(6) are smeared away more, they produce similar behaviors as

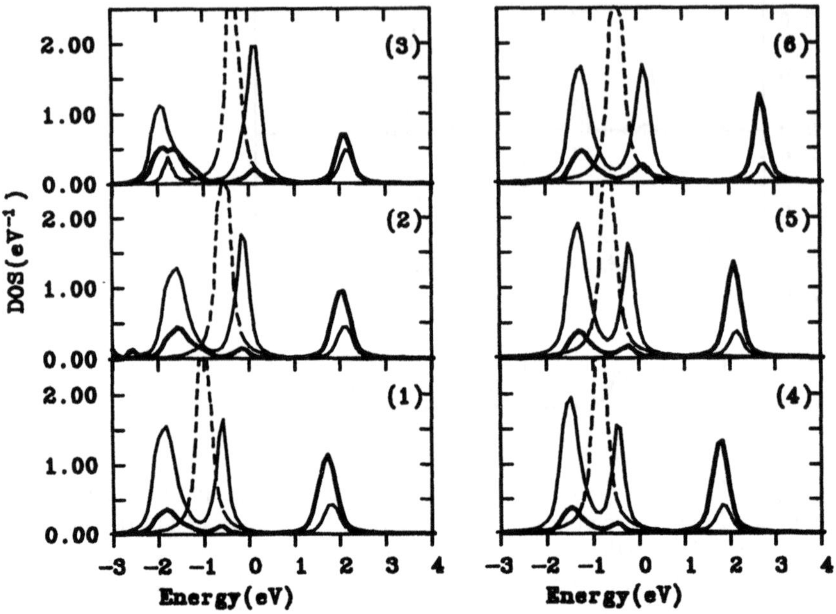

Fig. 3. The p and d electron density of states. The bold solid line shows d electron component. The fine solid and dotted lines show $x^2 - y^2$ and xy symmetry p components respectively. For all cases ((1) through (6)), $p-d$ mixing interaction t_0 and temperature are 0.6 and 0.1 eV respectively. (1),(2) and (3) are $V = 0.5$ case and (4),(5) and (6) are $V = 0.0$ case. (1) is an insulating case. The calculated hole numbers(n_h) are (2)0.29 and (3)0.82 respectively. (4) is an insulating case. The calculated hole numbers(n_h) are (5)0.13 and (6)0.52 respectively.

in Fig. 1. Although small reduction of density of states in the upper Hubbard level is observed for larger hole doping, change is not so drastic even for reduced charge transfer gap $\big((1) \sim (3)\big)$.

A calculation including the short distant Coulomb interaction,

$$H_I = \int dx n(x)[V_x(-i\vec{\nabla})n_{p_x}(x) + V_y(-i\vec{\nabla})n_{p_y}(x)] \tag{3.13}$$

with $V_x(\vec{k}) = 2V_0 \cos k_x a$ and $V_y(\vec{k}) = 2V_0 \cos k_y a$, is done in the same method. We neglected, for simplicity, dynamical corrections from $< Tj_n(x)\delta j^\dagger_m(y) >$. The result is shown in Fig. 3, where (1)~(3) is with V-effects and (4)~(6) without V-effects. Up to certain hole density, the behavior is similar to the previous ones, although the splitting and peak positions are slightly different due to the absence of dynamical effects. From certain hole concentration, there occurs rapid collapsing of the upper Hubbard intensity. Energetically, the level energy is shifted up to higher energy. The main reason of this phenomenon is level crossing of the renormalized η level and the composite state p_s. The mixing between them results in the interchange of spectral intensity for higher and lower energy levels.

4. CONCLUDING REMARKS

We have seen that doping dependence of change of the electronic state in oxide superconductors is understood from the correlated p-d model. There the notion of the composite electronic state which is associated with the $Cu-O$ bond and which is described by a certain composite operator is shown to be useful for understanding change of the density of states. Specially, the electronic states near the FL are composed of these composite electronic states, and both charge transfer and Madelung interactions play important roles to determine electronic states. For medium concentration, there appear rapid modification of density of states due to the screening effect of the Madelung interaction. The fact that the superconductivity appears in this region may suggest importance of the charge fluctuation accompanying to such drastic intensity transfer of electronic states from high energy region to low energy region.

ACKNOWLEDGMENTS

This work was supported by a Grant-in-Aid from the Ministry of Education, Science and Culture, Japan. The authors would like to express their thanks for valuable discussions to Dr. T. Koyama and Dr. S. Takahashi. One of the authors (M. S.) would like to thank NKK Corporation for his financial support to make him continuing this work.

REFERENCES.

[1]. J. B. Torrance, Y. Tokura, A. I. Nazzal, A. Bazinge, T. C. Huang and S. S. P.Parkin, Phy. Rev. Lett. **61**, 1127(1988); J. B. Torrance, A. Bezinge, A. I. Nazzal, T. C. Huang, S. S. P. Parkin, D. T. Keane, S. J. LaPlaca, P. M. Horn and G. A. Held, Phy. Rev. **B40**, 8872(1989);

[2]. V. J. Emery, Phys. Rev. Lett. **58**, 2794(1987).

[3]. F. Milla, Phys. Rev. **B38**, 11358(1988).

[4]. J. Orenstein, G. A. Thomas, D. H. Rapkine, A. J. Millis, L. F. Schneemeyer and J. V. Waszczak, Physica **153-155**, 1740(1988); G. A. Thomas, J. Orenstein, D. H. Rapkine, M. Capizzi, A. J. Millis, R. N. Bhatt, L. F. Schneemeyer and J. V. Waszczak, Phys. Rev. Lett. **61**, 1313(1988).

[5]. M. Suzuki, Phys. Rev. **B39**, 2312(1989).

[6]. S. Tajima, S. Tanaka, T. Ido and S. Uchida, *in Advances in Superconductivity II : Proceedings of the 2nd International Symposium on Superconducutivity (ISS '89)*, edited by T. Ishiguro and K. Kajimura (Springer-Verlag, Berlin, 1990); S. Uchida, T. Ido, H. Takagi, T. Arima, Y. Tokura and S. Tajima, Phys. Rev. **B43**, 7942(1991).

[7]. N. Nücker, J. Fink, J. C. Fuggle, P. J. Durham and W. M. Temmerman, Phys. Rev. **B37**, 5158(1988).
[8]. A. J. Arko, R. S. List, R. J. Bartlett, S.-W. Cheong, Z. Fisk, J. D. Thompson, C. G. Olson, A.-B. Yang, R. Liu, C. Gu, B. W. Veal, J. Z. Liu, A. P. Paulikas, K. Vandervoort, H. Claus, J. C. Campuzano, J. E. Schirber and N. D. Shinn, Phys. Rev. **B40**, 2268(1989).
[9]. H. Matsuyama, T. Takahashi, H. Katayama-Yoshida, T. Kashiwamura, Y. Okabe, S. Sato, N. Kosugi, A. Yagishita, K. Tanaka, H. Fujimoto and H. Inokuchi, Physica **C160**, 567(1989).
[10]. Y. Fukuda, K. Terashima, Y. Nakanishi, T. Suzuki, M. Nagoshi, Y. Syono and M. Tachiki, Physica **C162-164**, 1315(1989).
[11]. D. Shindo, K. Hiraga, S. Nakajima, M. Kikuchi, Y. Syono, N. Kobayashi, K. Hojou, T. Soga, S. Furuno and H. Otsu, Physica **C159**, 794(1989).
[12]. C. T. Chen, F. Sette, Y. Ma, M. S. Hybertsen, E. B. Stechel, W. M. C. Foulkes, M. Schluter, S-W. Cheong, A. S. Cooper, L. W. Rupp, Jr., B. Batlogg, Y. L. Soo, Z. H. Ming, A. Krol, and Y. H. Kao. Phys. Rev. Lett. **66**, 104(1991).
[13]. H.Romberg, M. Alexander, N. Nücker, P. Adelmann, and J. Fink, Phys. Rev. **B42**, 8768(1990).
[14]. T. Timusk, S. L. Herr, K. Kamarás, C. D. Porter, D. B. Tanner, D. A. Bonn, J. D. Garrett, C. V. Stager, J. E. Greedan and M. Reedyk, Phys. Rev. **38**, 6683(1988).
[15]. G. A. Thomas, S. L. Cooper, J. Orensteine, D. H. Rapkine, J. V. Waszczak and L. F. Schneemeyer, Physica **C165&166**, 1257(1990).
[16]. H. Matsumoto, M. Sasaki and M. Tachiki, Solid State Commun. **71**, 829(1989) S. Ishihara, H. Matsumoto and M. Tachiki, Phys. Rev. **B42**, 10041(1990); H. Matsumoto, M. Sasaki and M. Tachiki, Phys. Rev. **B43**, 10247(1991); H. Matsumoto, M. Sasaki and M. Tachiki, Phys. Rev. **B43**, 10264(1991).

Key Factors for the Superconducting and Normal State Properties of Cuprates

H. Kamimura

Department of Applied Physics, Faculty of Science, Science University of Tokyo, 1-3 Kagurazaka, Shinjuku-ku, Tokyo 162, Japan

Abstract. This paper explains the key factors which play essential roles in determining (i) the multiplet state of CuO_6 in hole-doped La_2CuO_4 or that of CuO_4 in electron-doped Nd_2CuO_4, (ii) the mechanism of superconductivity and (iii) the normal state properties of the above cuprate superconductors.

1. Introduction

Recently Kamimura and Eto [1] have showed from the first principles cluster calculation that the many-electron multiplet state of CuO_6 or CuO_4 cluster embedded in the above cuprates is the $^3B_{1g}$ state in the superconducting and normal phases. Based on this result, Kamimura and his coworkers [2 to 7] have recently calculated most of the normal state properties and explained the observed results successfully. In this paper I discuss the key factors which govern the superconducting and normal state properties.

2. Key Factors for the Electronic Structures of Cuprates

2.1 Hole-doped Cuprates

There are three key factors which play essential roles in determining the electronic states of hole-doped cuprates which consist of the layers of CuO_6 octahedra or CuO_5 pyramids in their crystal structures.

These are 1) the electron correlation, 2) the existence of three kinds of orbitals; Cu dx^2-y^2, Cu dz^2, and O p_σ orbitals, and 3) apical oxygens in CuO_6 octahedra or CuO_5 pyramids.

2.2 Electron-doped Cuprates

In this case there are no apical oxygens. Then the following two key factors play the essential roles in determining the electronic states: 1) electron correlation, and 2) the three kinds of orbitals; Cu dx^2-y^2, Cu 4s, and O p_σ orbitals.

Let me explain how the above key factors contribute to the determination of the electronic structures. On doing so I choose the simplest hole- and electron-doped cuprates with complete stoichiometry as the objects of study, because it is better to work on the simplest ideal system in order to explore the essential points of superconductivity. These are hole-doped $La_{2-x}Sr_xCuO_4$ and electron-doped $Nd_{2-x}Ce_xCuO_4$, which I denote hereafter La-Sr-Cu-O and Ne-Ce-Cu-O, respectively. Since the electron correlation is very strong in cuprates (~10 eV), we have first to calculate the many-electron multiplet states of CuO_6 in La-Sr-Cu-O or those of CuO_4 in Nd-Ce-Cu-O, rather than the one-electron orbital states in a periodic system from the first principles.

Springer Series in Solid-State Sciences, Vol. 106 Physics of High-Temperature Superconductors
Eds.: S. Maekawa and M. Sato

For this purpose the multi-configuration selfconsistent field variational method with configuration interactions which I denote MCSCF-CI is the most suitable method [1,8,9]. In this method the effect of electron correlation is fully taken into account. Only problem is that this method is at present applicable to a cluster embedded in a crystal so that we need a certain kind of approximation [5] or assumptions when we extend the results of MCSCF-CI to a periodic system.

3. The Ground State of CuO_6 Octahedron in $La_{2-x}Sr_xCuO_4$

When Sr or Ba ions are substituted for La ions in La_2CuO_4, there are two cases concerning the dopant hole states;

i) holes occupy oxygen p_σ orbitals, or

ii) holes occupy Cu dz^2 orbitals.

For the former the multiplet state of CuO_6 is $^1A_{1g}$ while it is $^3B_{1g}$ for the latter.

In order to determine which of the above two states has lower energy, we have to perform variational calculations for each of $^1A_{1g}$ and $^3B_{1g}$ states and to compare their lowest states with each other. In adopting the MCSCF-CI method, and applying it to La-Sr-Cu-O we have noted that the distance between apical oxygens and Cu in CuO_6 decreases with increasing the hole concentration. This has been pointed out theoretically by performing the total-energy calculation within the local density approximation (LDA) with use of the ab initio pseudopotentials and the virtual crystal approximation [10]. Experimentally this trend has been reported first by EXAFS experiments by Boyce et al.[11], and recently by the neutron diffraction experiments by Kamiyama et al.[12]. Similar trend in the decrease of the apical O-Cu distance in CuO_5 pyramids in YBCO has been reported by Cava et al.[13].

By using this result Kamimura and Eto[1] have performed the MCSCF calculation for $^1A_{1g}$ and $^3B_{1g}$, in which the one-electron orbitals of Cu 1s, 2s, 2p, 3s, 2p, 3d and 4s and O 1s, 2s and 2p are taken as linear combinations of Gaussian orbitals and different variational functions have been adopted for these two states. Further, to take account of the Madelung potential from the exterior ions outside the CuO_6 cluster, the point charges are placed at exterior ion sites.

The difference between the optimized energies of the $^1A_{1g}$ and $^3B_{1g}$ is shown as a function of the Cu-apical O distance in Fig.1 [1]. It is clearly seen from this figure that the ground state of the CuO_6 octahedron in $La_{2-x}Sr_xCuO_4$ changes from the $^1A_{1g}$(low spin)

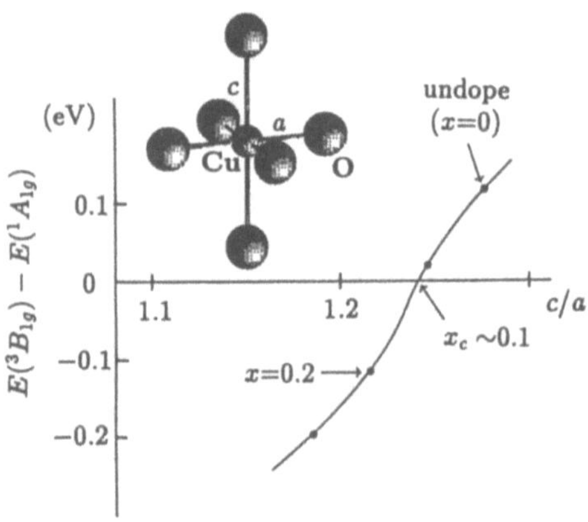

Figure 1: The energy difference between $^3B_{1g}$ and $^1A_{1g}$ states vs. the Cu-apical O distance c in CuO_6 octahedron. $^3B_{1g}$ state becomes the lowest at and above x_c which corresponds to about 0.1 in the $La_{2-x}Sr_xCuO_4$ formula. The inset indicates the CuO_6 elongated octahedron. The distance a is fixed as 1.89 A.

state to the $^3B_{1g}$(high spin) state at x_c, which is about x=0.1 in the present calculation. When we consider the effect of doped Sr ions on the Madelung potential, this change of low spin to high spin occurs at a lower concentration which is close to the observed onset concentration of superconductivity, i.e. x∼0.05.

The reason for the change of a low spin to a high spin state is as follows: When the Cu-apical O distance decreases by doping and an elongated octahedron approaches a regular octahedron, the energy difference between the Cu dx^2-y^2 and dz^2 orbitals becomes smaller so that the Hund's coupling effect becomes stronger. Thus the $^3B_{1g}$ state becomes lower at and above x_c.

A number of cluster calculations have been reported[14-18] before Kamimura and Eto's work, but these variational calculations have been performed only for the $^1A_{1g}$ state but not or the $^3B_{1g}$ state. Because of this and also of the neglect of the apical oxygen effect those calculations have not been able to find the low-spin to high-spin change at a reasonable value of x except ref.18.

4. The Change of a Low Spin State to a High Spin State at the Onset of Superconductivity in La-Sr-Cu-O

Based on the result shown in Fig.1, Kamimura and Eto [1] have assumed that the doping concentration at which the transition of the ground state of CuO_6 octahedron from a low spin state to a high spin state takes place corresponds to the onset of superconductivity. Then they have been able to clarify the microscopic origin of the T versus x phase diagram observed for $La_{2-x}Sr_xCuO_4$, as I will describe in section 7.

5. High Spin Ground State for CuO_4 in Nd-Ce-Cu-O

Kamimura and Eto [1] have also applied the MCSCF-CI method to a CuO_4 cluster embedded in Nd-Ce-Cu-O, compound, and have showed that, upon doping with Ce ions, a dopant electron is accommodated in the Cu 4s orbital which has the same symmetry as the Cu dz^2 orbital so that the effect of Hund's coupling acts to make the $^3B_{1g}$ state lower in energy by 1.1 eV, as compared with the $^1A_{1g}$ state.

In this context both the ground states of CuO_6 in La-Sr-Cu-O and of CuO_4 in Nd-Ce-Cu-O are the same in the superconducting phase , that is the $^3B_{1g}$ state . Therefore, we may imagine that the mechanism of superconductivity can be the same both for the hole-doped and electron-doped superconductors.

6. The Spin-Polaron Mechanism of Superconductivity

According to Kamimura and Eto's result, the Cu dx^2-y^2 holes are localized around the Cu sites and their spins are coupled antiferromagnetically by the superexchange interactions via the intervening O ions. When the dopant holes are accommodated in the Cu dz^2 orbitals in La-Sr-Cu-O or Cu 4s orbitals in Nd-Ce-Cu-O, the RVB like state proposed by Anderson[19] are partially destroyed by the Hund's coupling between the Cu dx^2-y^2 and dz^2 spins and as a result the spin-polarons are created , as proposed by Kamimura Matsuno and Saito[20] and later proved by Ishida and Kamimura[21] by the computer simulation work. It is interesting to see in Ishida and Kamimura's work[21] that a single spin- polaron hardly moves due to the spin polarization around it but when the same numbers of spin-polarons are created in the two different sublattices, they can itinerate in a crystal by the transfer interaction in a dz^2 band.

Then Kamimura et al.[2, 20, 22] showed that the spin-polarons interact via a weak but long-ranged attractive antiferromagnetic interaction. Assuming that the spin-polarons form the Cooper pairs, they calculated T_c as a function of x for $La_{2-x}Sr_xCuO_4$ and obtained a good agreement with the experimental result by Torrance et al[23]. On the other hand, Ushio and Kamimura[24] recently pointed out another possibility that the spin-polarons may form the Cooper pairs via the BCS type phonon-mediated attractive interaction. In this context I can conclude that the $^3B_{1g}$ state contributes to superconductivity, although several mechanisms are considered as a possible origin of superconductivity.

7. The Appearance of Two Kinds of Phase Diagrams

Based on the results mentioned above, we can predict two kinds of phase diagrams in the temperature (T) versus concentration (x) plane. In the very low concentration a three-dimensional antiferromagnetic interaction exists among the Cu dx^2-y^2 spins so that an insulating phase appears. In a hole-doped La-Sr-Cu-O system the holes are accommodated in O p_σ orbitals below the onset of superconductivity. Thus the superexchange interactions are partially destroyed and as a result the antiferromagnetic ordered state disappears at x_o. Further, in this case Eto, Saito and Kamimura[25] have showed that the Cu spins adjacent to a hole-doped oxygen site tend to become parallel so that a spin- glass phase appears. When the Sr concentration reaches a critical concentration at which the superconductivity occurs, the ground state of the CuO_6 octahedra change from a low spin to a high spin state, by transferring the holes from O p_σ orbitals to Cu dz^2 orbitals. Then the superconductivity occurs by a mechanism mentioned in a previous section. In this way a phase diagram shown in Fig.2(a) should appear for La-Sr-Cu-O system. This is consistent with the experimental result[26].

On the other hand, in an electron doped Nd-Ce-Cu-O system the electrons are accommodated in the Cu 4s orbitals immediately after the doping of Ce ions. Since there are no holes in the O orbitals in CuO_2 layers, the antiferromagnetic ordering state are kept up to higher Ce concentrations, compared with the La-case, Because of the antiferromagnetic spin ordering, the spin-polarons cannot move up to the concentration x_o at which the antiferromagnetic ordering disappears. Thus at this critical concentration a metal-insulator transition takes place and the superconductivity occurs simultaneously. Thus, in this case a phase diagram shown in Fig.2(b) should appear. This is also consistent with the experimental result[26].

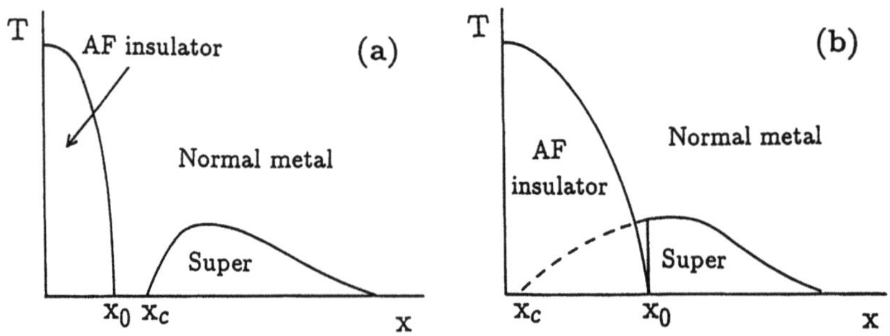

Figure 2: Two kinds of phase diagrams in the temperature vs. dopant concentration plane. (a) corresponds to the case of $x_c > x_o$ while (b) corresponds to the case of $x_c < x_o$, where x_c and x_o are the concentration at which the $^3B_{1g}$ ground state appears and the antiferromagnetic order(AF) disappears, respectively.

8. Key Factors for Understanding the Normal State Properties

The most important key factor in the normal state of the present model is the coexistence of the conduction and spin states, just like in the superconducting phase. As I have showed in Figs. 2(a) and (b), both the multiplet states of CuO_6 and CuO_4 cluster are the $^3B_{1g}$ state in the normal states of hole- and electron-doped cuprates. Extending this result to a periodic cuprate system, the electronic structure of $La_{2-x}Sr_xCuO_4$ is developed in the following way: According to Kamimura and Eto [1], the Hubbard U interaction for the dx^2-y^2 and dz^2 states is about 10 eV. On the other hand, the band width of the Cu dx^2-y^2 band is much reduced from that of the one-electron band ($\sim$ 4 eV) to about 1 eV by the spin frustration effect, and thus it becomes of the order of the band width of the Cu dz^2 band. Since the band widths of both the Cu dz^2 and the modified dx^2-y^2 bands are much smaller than the on-site Coulomb interactions, the Hubbard I approximation [27] may be applied to both the Cu dx^2-y^2 and dz^2 bands in the band structure of $La_{2-x}Sr_xCuO_4$ calculated by Shiraishi et al.[28] with the local density approximation (LDA). The electronic structures thus obtained for the superconducting and normal states are shown, respectively, in Figs. 3(b) and (c). For comparison the electronic structure in the hole concentration regime below the onset of superconductivity is also shown in Fig.3(a).

According to this result the conduction state before the onset of superconductivity corresponds to the oxygen $p\sigma$ bands (Fig.3(a)). At the onset of superconductivity the holes start to occupy the Cu dz^2 upper Hubbard band in addition to the O $p\sigma$ bands. Then, with increasing the Sr concentration the holes occupy mainly the Cu dz^2 upper Hubbard band (Fig.3(b)). In the normal phase the conduction state corresponds to the Cu dz^2 upper Hubbard band.

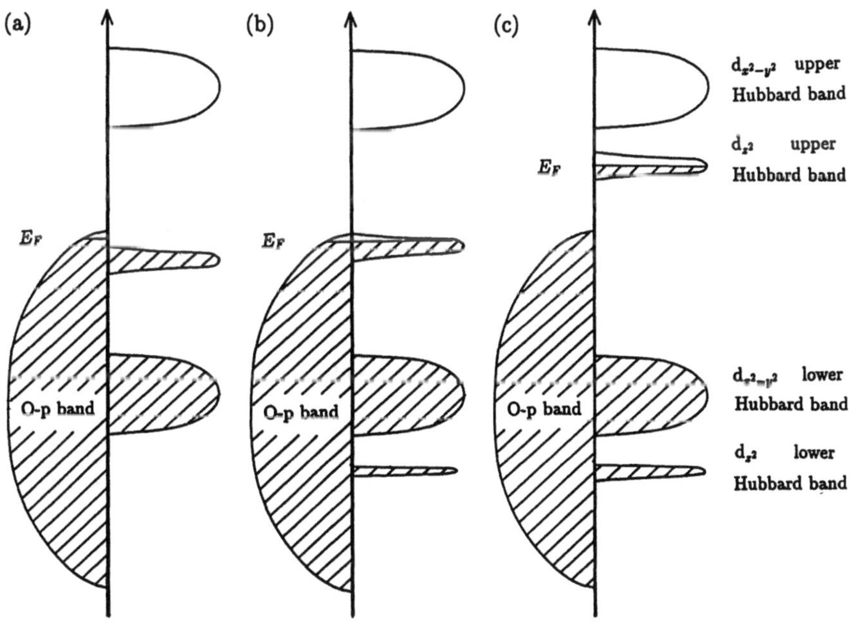

Figure 3: The electronic structures of $La_{2-x}Sr_xCuO_4$ in the three hole concentration regimes: (a)in the spin-glass phase below the onset of superconductivity, (b)in the superconducting phase, and (c) in the normal metallic phase.

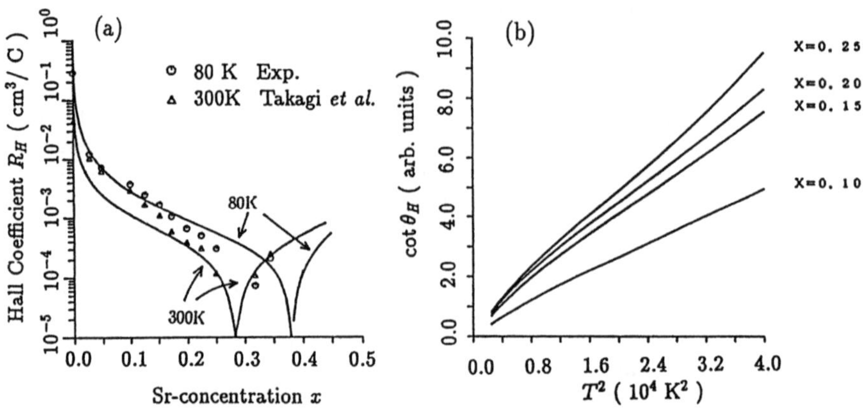

Figure 4: Theoretical results of the Hall effect in $La_{2-x}Sr_xCuO_4$[4]: (a) The calculated hole-concentration dependence of the Hall coefficient R_H for T=80 and 300K, together with the experimental results by Takagi et al.[29]. (b) The calculated temperature dependence of the Hall angle θ_H. Here cotθ_H is plotted against T^2.

As regards the spin state, on the other hand, a spin state coexists with a conduction state in the whole hole-concentration regime. The spin state is constructed from the Cu dx^2-y^2 localized spins which interact with each other by the antiferromagnetic superexchange interaction and form a two- dimensional quantum spin state, although the spin correlation length in the spin state varies, critically depending on the hole concentration.

As far as the spin-independent normal state properties are concerned, these properties are mainly determined by the behavior of the itinerant holes in the conduction state of the Cu dz^2 upper Hubbard band. In this context Kamimura and his coworkers [2 to 7] have recently calculated the concentration and temperature dependences of the resistivity, the Hall effect, the thermoelectric power, the electronic specific heat, and the de Haas van Alphen effect in the normal state of $La_{2-x}Sr_xCuO_4$, by using the energy dispersion of the Cu dz^2 upper Hubbard band which are obtained from the dz_2 one-electron band in the band structure by Shiraishi et al.[28] with use of the Hubbard I approximation, and have succeeded in explaining the most of the experimental results quantitatively. For example, the calculated x- and T-dependences of the Hall effect[4] are shown in Figs. 4(a) and (b), respectively, where the Hall coefficient R_H for T = 80 and 300K are plotted as a function of x in (a) and the cotangent of the Hall angle θ_H is plotted at a function of T^2 in (b).

As seen in Fig.4(a), the x- and T- dependences of R_H observed by Takagi et al.[29] including the sign change are reproduced by the theory satisfactorily. Further the T^2 dependence of cotθ_H has been recently confirmed by Ong[30] for powdered $La_{2-x}Sr_xCuO_4$ samples. It is concluded from the present theory that the anomalous behaviors of the Hall effect in $La_{2-x}Sr_xCuO_4$ are due to the competition between the electron- and hole-like curvatures in the warped cylindrical Fermi surfaces. An example of such Fermi surfaces is shown for x=0.15 in Fig.5[4]. According to the present theoretical treatment, the behaviors of the Hall effect may change from material to material, depending on the shape of its Fermi surfaces.

As regards the spin-dependent properties such as the uniform spin susceptibility, NMR, etc., those are determined by a coupled system of the conduction and spin states. In this respect Matsuno, Ushio and Kamimura [7] have recently showed that the interplay among the Zeeman interaction in the Cu dz^2 upper Hubbard band, the Hund's coupling between

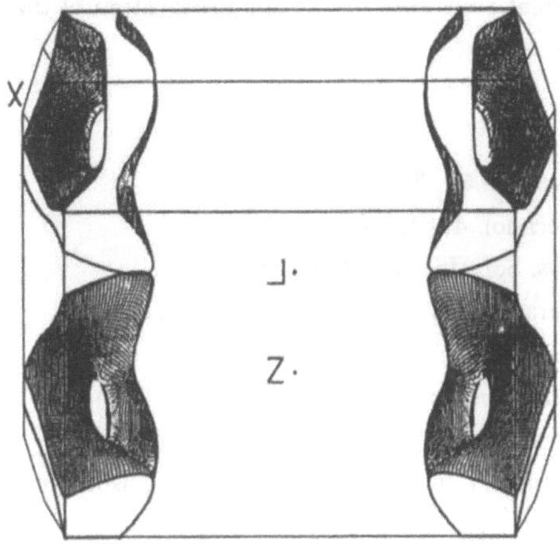

Figure 5: The Fermi surface of the Cu dz^2 upper Hubbard band for x=0.15 in the Brillouin zone for an ordinary unit cell of La_2CuO_4 system[4]. This is a view from the (0,1,1) direction between the y and z axes.

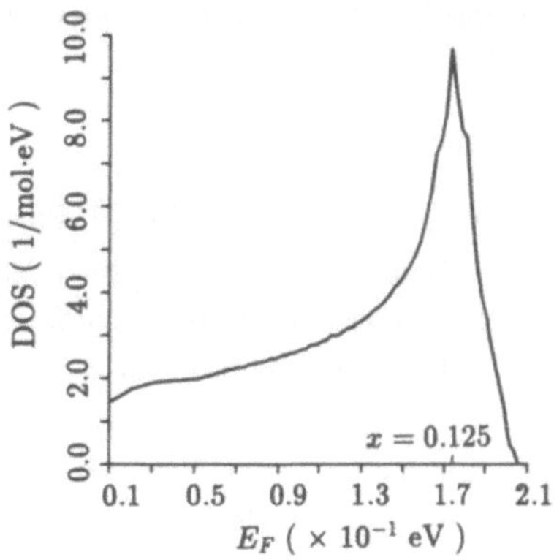

Figure 6: Density of states of the Cu dz^2 band calculated by fitting to the LDA band by Shiraishi et al.[28]. This band is completely filled by electrons below the onset of superconductivity. A van Hove singularity appears at the energy corresponding to the Fermi energy for x=0.125.

dz^2 holes and dx^2-y^2 localized spins and the antiferromagnetic superexchange interaction between the localized spins plays an essential role in determining the x- and T-dependences of the uniform spin susceptibility.

Here I should like to remark that, in the above-mentioned normal state properties the characteristic shape of the density of states of the Cu dz^2 upper Hubbard band which is shown in Fig.6 plays an important role in determining the T- and x-dependences of normal state properties. Its characteristic feature is the appearance of a sharp peak at the energy corresponding to the Fermi energy for x=0.125 in $La_{2-x}Sr_xCuO_4$. This sharp peak is related to a logarithmic van Hove singularity in 2D band, but modified by the transfer interaction between the adjacent CuO_2 layers. Furthermore the Fermi surfaces for x=0.125 has a nesting feature. Based on this feature Kamimura et al[6] have recently suggested a possibility that a kind of CDW instability may occur around x=0.125 in $La_{2-x}Sr(\text{or Ba})_xCu0_4$. This may be an origin of the peculiar behaviors observed in $La_{1.875}Ba_{0.125}CuO_4$.

9. Summary

In this paper I have showed how the coexistence of the Cu dz^2 itinerant holes and the Cu dx^2-y^2 localized spins play the important roles in determining the superconducting and normal states.

Acknowledgments

It is a pleasure to acknowledge the collaboration of Mikio Eto, Kunio Ishida, Shunichi Matsuno, Takashi Nakayama, Atsushi Oshiyama,Riichiro Saito, Tatsuo Schimizu, Nobuyuki Shima, Kenji Shiraishi and Hideki Ushio on the work reviewed in this article. I am also greatly indebted to Yuji Suwa for his great assistance during the preparation of the manuscript.

References

1. H. Kamimura and M. Eto: J. Phys. soc. Jpn.**59**, 3053 (1990).
2. H. Kamimura et al: Supercond. Sci. Technol. **4**, S121 (1991).
3. T. Schimizu and H. Kamimura: J. Phys. Soc. Jpn.**59**, 3691 (1990).
4. H. Ushio, T. Schimizu and H. Kamimura: J. Phys. Soc. Jpn. **60**, 1445 (1991).
5. H. Ushio and H. Kamimura: Solid. St. Commun. **76**, 1115 (1990).
6. H. Kamimura et al: Proc. M^2HTSC-III, to be published.
7. S. Matsuno, H. Ushio and H. Kamimura: Proc. M^2HTSC-III, to be published.
8. H. Kamimura and H. Aoki: The Physics of Interacting Electrons in Disordered Systems (Oxford Univ. Press, Oxford, 1989).
9. M. Eto and H. Kamimura: Phys. Rev. Lett. **61**, 2790 (1988).
10. N. Shima, K. Shiraishi, T. Nakayama, A. Oshiyama and H. Kamimura: Proc. 1st. Int. Conf. on Elec. Mats, p.51 (MRS, Pittsburgh, 1988).
11. J.B. Boyce et al: Phys. Rev. B**35**, 7203 (1987).
12. T. Kamiyama et al: Physica C**172**, 120 (1990).
13. R. J. Cava et al: Physica C **153-155**, 560 (1988).
14. Y. Gun, J. Langlois and W.A. Goddard III: Science, **239**, 896 (1988).
15. Y. Asai: J. Phys. Soc. Jpn. **58**, 3264 (1989).
16. A. Fujimori: in Mechanisms of High Temperature Superconductivity, eds. H. Kamimura and A. Oshiyama, Springer Series in Materials Science Vol.11, p.176 (Springer, Heidelberg, 1989), and related references therein.
17. H. Eskes, H. Tjeng, and G.A. Sawatzky: in Mechanisms of High Temperature Superconductivity, eds. H. Kamimura and A. Oshiyama, Springer Series in Materials Science Vol.11, p.20 (Springer, Heidelberg, 1989), and related references therein.
18. J.F. Annett et al: Phys. Rev.**40**, 2620 (1989).
19. P.W. Anderson: Science **235**, 1196 (1987).
20. H. Kamimura, S. Matsuno and R. Saito: in Mechanisms of High Temperature Superconductivity, eds. H. Kamimura and A. Oshiyama, Springer Series in Materials Science Vol.11, p.8 (Springer, Heidelberg, 1989).
21. K. Ishida and H. Kamimura: Phys. Rev. B**42**, 8312 (1990).
22. H. Kamimura, S. Matsuno and K. Ishida: Physica C**162-164**, 781 (1989).
23. J.B. Torrance et al: Physica C**162-164**, 291 (1989).
24. H. Ushio and H. Kamimura: Proc. M^2HTSC-III, to be published.
25. M. Eto, R. Saito and H. Kamimura: Solid St. Commun.**71**, 425 (1989).
26. H. Takagi, S. Uchida and Y. Tokura: Phys. Rev. Lett. **62**, 1197 (1989).
27. J. Hubbard: Proc. Roy. Soc. A**276**, 238 (1963).
28. K. Shiraishi, A. Oshiyama, N. Shima, T. Nakayama, and H. Kamimura: Solid St. Commun. **66**, 629 (1988).
29. H. Takagi et al.: Phys. Rev. B 40, 2254 (1989).
30. N.P. Ong: Lecture in this school. see also Proc. 3rd M^2S HTSC, Physica C, to be published.

Evidence in Favor of a Nematic Ordering in $La_2CuO_{4+\delta}$

L.P. Gor'kov

L.D. Landau Institute for Theoretical Physics, Academy of Sciences of the USSR, SU-142432 Chernogolovka, Moscow district, USSR

Abstract. An attempt to interpret the peculiarities of the phase diagram of $La_2CuO_{4+\delta}$ in the plane (T,δ) in terms of the nematic ordering is reviewed.

The complexity of the High-T_c superconducting oxides properties made obvious that without the exhaustive understanding of their normal phase any progress in the identification of the new superconductivity mechanisms is hardly possible. From the experimental side, the main difficulties seems to be in having reproducible samples of good quality. On the other hand, the physical intuition prompts that namely these difficulties may be relevant being somehow related to phase transitions peculiar to the materials. From such a standpoint the parent compound La_2CuO_4 is of the special interest owing to its simpler lattice structure compared with the others. In addition, it has the single CuO_2 plane per cell.

That the $La_2CuO_{4+\delta}$ system is spatially nonuniform and at low temperatures consists of a mixture of metallic and insulating phases was apparently first pointed out in 1 and 2. The schematic (T,δ) phase diagram [3,4] is shown in Fig.1. The hatched region covers the concentration interval in which phase separation occurs. The symmetry of the left part of the diagram, the OI phase, is known reliably - Cmca(D_{2h}^{18}) while for the metallic side (the phase OII) the symmetry remains a matter of controversy [3,4]. However, the very existence of a first order transition below approximately 300K is also supported by the observation of hysteresis[5].

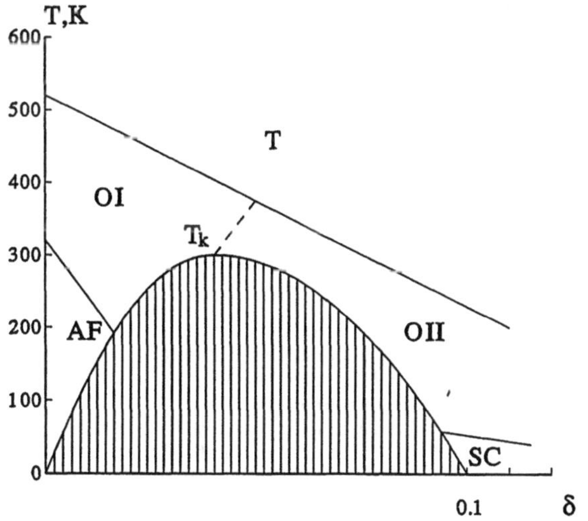

Fig.1 Phase diagram of $La_2CuO_{4+\delta}$ (Ref.3). The hatched region is the range of a phase separation.

The principal possibility of the phase segregation upon the dissolution of oxygen is a consequence of the thermodynamics. Its realization, however, means a spatial redistribution of the oxygen which would be possible at such low temperatures only by virtue of a high enough mobility of oxygen [6]. The growth of nucleating centers of the insulating phase was attributed in 2 to an attraction of clusters of localized moments as a result of the elastic strains (from the side of the metallic phase), Another mechanism was mentioned in 5 and 7, this time for holes added into the antiferromagnetic state; the growth of "metallic" clusters results from the decrease in the cost of the magnetic energy as holes come closer together. The actual mechanism for the phase separation has to take into account the energy change when an extra oxygen ion becomes embodied into the La_2CuO_4 cell to prevent the loss of the electrical neutrality.

Nevertheless, it is tempting to relate the entire phase diagram in Fig.1 to magnetic phenomena. In fact, all the characteristic temperatures here are comparable with the magnetic scale. Thus, in a pure La_2CuO_4 we have $T_N \simeq 250$-$300K$ and $T_{T-O} \simeq 500K$; the exchange integral J is estimated to be $J \simeq 0.1$ eV; and the critical point in Fig.1 is $T_K \simeq 300K$ [3,5]. The observation of an antiferromagnetic phase in the stoichiometric La_2CuO_4 proves that there is a local moment associated with the configuration (d^9) of the copper ion Cu^{2+}. The phase separation occurs at comparatively low values $\delta \lesssim 0.08$. It is hard to imagine that the local moment would disappear completely outside the antiferromagnetic region. In order to relate the transition in Fig.1 to magnetic phenomena it was assumed [8] that outside the region of the antiferromagnetic state there can be an ordering which does not violate the time reversal invariance ($t \to -t$): a spin nematic state manifested in appearance of nonzero anisotropic correlations $\langle S_\alpha(1) S_\beta(2) \rangle \neq \delta_{\alpha\beta} F(1-2)$.

The T-O structural transition in La_2CuO_4 is commonly described in terms of phonon modes connected with rotations of elongated oxygen octahedra. Unusual are the exceedingly small changes in the lattice structure induced by the transition: $2(a-b)/(a+b) \neq 10^{-2} \div 10^{-3}$ in the ortho-phase, with the octahedron rotation angle $\phi_0 \sim 5 \times 10^{-2}$ rad. In [8] the implications for the phase diagram in Fig.1 were examined which follow from the suggestion that a nematic spin order parameter is predominant in the $La_2CuO_{4+\delta}$ system and that the structural changes result from slight relativistic magnetoelastic interactions ($\sim v^2/c^2$). It was assumed that the physical source for a nematic ordering lies in a long enough range of antiferromagnetic interaction ("frustrations") giving rise to an instability of the Neel-type ordering [9].

Remind first the concept of the nematic ordering in the exchange approximation [10]. Two types of binary spin correlations exist [11]:

$$\langle S_\alpha(1) S_\beta(2) \rangle = T_{\alpha\beta} \phi(\vec{r}_1, \vec{r}_2) \qquad (1)$$

with $T_{\alpha\beta} = T_{\beta\alpha}$ and $T_{\alpha\alpha} = 0$ ("ordinary nematics"),and

$$\langle S_\alpha(1) S_\beta(2) \rangle = e_{\alpha\beta\gamma} P_\gamma(\vec{r}_1, \vec{r}_2) \qquad (2)$$

called "p-nematic" in 11. The complete classification of the different symmetries was recently given in 12 for the second order namatic transitions in the tetragonal systems including the La_2CuO_4 structure. (All phase transitions are considered as the 3D phenomena).

In accordance with the above ideas it is now necessary to restore the form of the two spins average (1) under the assumption that all structural distortions below T_{T-O} result from magnetoelastic interactions. In the tetragonal notations (the x and y axes run along the sides of a CuO_2 square), the body-centered lattice of La_2CuO_4 doubles below T_{T-O}. The octahedra deformation is conveniently characterizied by the angle of rotation $\delta\vec{\Omega}(\vec{r}_n)$. In the ortorhombic axis (x', y') rotated 45°, one has

$$\delta\Omega_{y'}(\vec{r}_n) = \tilde{\omega}_{y'}\exp(i\vec{Q}_0\,\vec{r}_n) \tag{3}$$

where the vector $\vec{Q}_0$ is $(\pi/a, \pi/a, 0)$ in tetragonal axes, or $(\sqrt{2}\pi/a)(1,0,0)$ in the axes (x',y'). The vector $\tilde{\omega}_{y'}$ correspondingly transforms as an even representation y' (i.e. as a product zx') of the subgroup D_{4h} of the vector $\vec{Q}_0$. Since $P_J(r_1,r_2)$changes sign under the permutation of $\vec{r}_1$ and $\vec{r}_2$ we are left with the nematic order parameter of the form (1):

$$\langle S_\alpha(1)S_\beta(2)\rangle = T_{\alpha\beta}\cos[\pi\vec{Q}_0\vec{R}/2]\phi(\vec{r}_n) \tag{4}$$

where $\vec{R}=\vec{r}_1+\vec{r}_2$, $\vec{r}_n=\vec{r}_1-\vec{r}_2$, and $\phi(\vec{r}_n)$ realizes an identical representation of the space group D_{4h} because there is only one Cu ion per unit cell of La_2CuO_4. In accordance with 12, the general form of $T_{\alpha\beta}$ in eq. (4) is biaxial. In addition to that, the spins coordinate system which freely rotates in the exchange approximation, becomes fixed by two anisotropy terms, say, K_1T_{zz} and $K_2T^2_{xy}$, where both constants depend on T and δ. On all these reasons the correspondence between the structural order parameter (3) and the $T_{\alpha\beta}$-components is not simple and unique. The analyses in 8 has shown that the choice of $T_{\alpha\beta}$ in the form

$$T_{\alpha\beta}=a_\alpha b_\beta+a_\beta b_\alpha$$

where $\vec{a}$ and $\vec{b}$ are mutually orthogonal vectors, $(\vec{a},\vec{b})=0$, is not compatible with the phase diagram in Fig.1 and at any values of the anisotropy constants, K_1 and K_2, does not provide the smooth transformation of the line the second order transition into the first order transition. On the other hand, it would be possible to interpret in such terms an abrupt change in the octahedra incline observed for $(La_{1-x}Ba_x)_2CuO_{4-\delta}$ in the interval $0.035<x<0.1$ [13].

Meanwhile, the uniaxial form of $T_{\alpha\beta}$

$$T_{\alpha\beta}=n_2n_\beta-1/3\delta_{\alpha\beta} \tag{5}$$

after making use of a special guess concerning the signs of the anisotorpy terms, is compatible with the fact of the existence of a critical point, T_K.

The order of magnitude for the elastic deformations (the orthorombicity of the new lattice) is explained as the secondary efect. Namely, the order parameter (5) in the second order generates terms of the form $T_{\alpha\nu}T_{\nu\beta}\sim n_\alpha n_\beta+1/3\delta_{\alpha\beta}$ even in the exchange approximation. The relativistic terms provide the magnetoelastic interactions with a strain tensor $\varepsilon_{\alpha\beta}$. For instance, there is the contribution of the form $\varepsilon_{x'x'}T^2_{x'z'}$. The value (a-b)/(a+b) is therefore proportional to the square of the order parameter and would be small everywhere below T_{T-O} as $\sim v^2/c^2<<1$. If, instead, the driving parameter is of the form (3) there would be elastic strains proportinal to $(\delta\Omega)^2$, i.e. strains are small as a measure of small $\delta\Omega<<1$ squared.

From the above arguments one sees that the nematic ordering could be consistent with the experimental data in Fig.1. It is also worth of mentioning that the nematic phase would show the low-lying modes related with the free rotations of the spin degrees of freedom in the exchange approximation[10]. To our level of knowledge, not so much can be concluded from the existing neutron data for the $La_2CuO_{4+\delta}$ crystals.

The actual hallmark of the nematic ordering would be the reaction to the applied magnetic field. First, the temperature dependence of the static magnetic susceptibility and its anisotropy is expected [10]. It seems that modern experiments tend to give a negative answer about that point. However, carefull magnetic susceptibility measurements are not available for the $La_2CuO_{4+\delta}$ samples. Secondly, the interpretation suggested above includes as the essential ingredient the magnetic anisotropy. Therefore, effects similar to the spin-flop transitions in antiferromagnetics are to be observable in strong enough magnetic fields and would result in the abrupt changes in the orientation of $T_{\alpha\beta}$. Such reorientation would be seen, in its turn, as a change in the crystalline deformations. Unifortunately, so far the only such a phenomenon studied in the antiferromagnetic La_2CuO_4 concerns the "canted" spins component of the neighbouring planes (The induced ferromagnetism in high fields along the c-axes). Anyway the search of the magnetic field dependence of the structural transformations (T_{T-O} (H), for instance) deserves an attention.

References

1 V.A.Borodin et al., Zh. Eksp. Teor. Fiz. **46**, (Appendix), 211 (1987) [JETP Lett. **46**, Supplement, S181 (1987)]
2 L.P.Gor'kov and A.V.Sokol, Pis'ma Zh. Eksp. Teor. Fiz. **46**, 333 (1987) [JETP Lett. **46**, 420(1987)]
3 B.Dabrowski et al., Physica C **162-164**, 99 (1989).
4 C.Chaillout et al., Physica C **162-164**, 57 (1989).
5 M.F.Hundley et al., Phys. Rev. B **41**, 4062(I) (1990).
6 L.P.Gor'kov, Physica C **162-164**, 12 (1989).
7 V.J.Emery, S.A.Kivelson, and H.Q.Lin, Phys. Rev. Lett. **64**, 475 (1990).
8 L.P.Gor'kov, JETP Lett. **52**, 500 (1990).
9 L.P.Gor'kov and A.V.Sokol, JETP Lett. **52**, 504 (1990).
10 A.F.Andreev and V.I.Marchenko, Usp. Fiz. Nauk **130**, 39 (1980) [Sov. Phys. Usp. **23**, 21 (1980)].
11 A.F.Andreev and I.A.Grishchuk, Zh. Eksp. Teor. Fiz. **87**, 467 (1984) [Sov. Phys. JETP **60**, 267 (1984)].
12 V.V.Barzykin, L.P.Gor'kov and A.V.Sokol, Europhys. Lett., to be published (1991).
13 T.Suzuki and T.Fujita. Physica C **159**, 111(1989).

Strong Correlation Models for High-T_c Superconductors

T.M. Rice

Theoretische Physik, ETH-Hönggerberg, CH-8093 Zürich, Switzerland

Abstract. A single band model with strong repulsive onsite correlations on the one hand, can be derived from a priori electronic structure calculations and on the other hand, explains many global features in the normal state. The simplest such model is the t-J model and in this talk I will review the properties of this model. There are many open questions both about the predictions of this model and its adequacy to explain the key experimental facts such as the anomalous properties of the normal state, their evolution with doping and last but not least the superconducting state.

Introduction

The high-T_c superconductors continue to pose the most pressing problem in solid state theory. A lot has been learned but in many cases this knowledge has only deepened the mystery. The goal is clear – to find and to solve the simplest model [1,2] that can describe the key features of the generic phase diagram of the cuprates. These are the rapid suppression of the antiferromagnetic order with doping leading to an anomalous normal state with unusual transport, optical and magnetic properties that have been reviewed earlier in the lectures of Drs. Uchida, Schlesinger, Slichter and Ong, and last but by no means least to superconductivity with a substantially higher energy scale than any other material.

Further the experimental evidence points to a continuous and smooth evolution with doping away from a Mott insulating behavior which occurs for the stoichiometric compositions with formal valence Cu^{2+}. As the simplest example, consider the crystal structure itself. This is after all determined by the chemical bonding pattern and since it changes only slightly with doping, a radical change in the electronic structure is ruled out. Other experiments which offer direct evidence in favor of this hypothesis are for example NMR investigations, reviewed elsewhere in this volume, which show that the hyperfine coupling constants do not change with doping and that there is a single quantum fluid which can be observed by monitoring the Knight shift on either ^{63}Cu or ^{17}O nuclei. X-ray absorption spectroscopy also offers direct evidence that distinct electronic configurations of formal valence Cu^{2+} and Cu^{3+} appear upon doping [3]. This shows that the Coulomb repulsion continues to be relevant on the low energy scale in the doped materials just as it is in the Mott insulators. The BCS theory of superconductivity, which is so successful for the standard superconductors, is based on the fact that the Coulomb repulsion scales to zero as the energy scale reduces in standard metals so that the total interaction electron-phonon plus rescaled electron-electron terms becomes attractive at low energies. Therefore the cuprate superconductors which are obtained by hole doping of a Mott insulator would seem at first glance unlikely materials to be superconductors at any value of T_c, never mind record large values of T_c!

The use of strongly correlated models has a firm basis also from 'a priori' methods. For example Hybertsen and coworkers [4] demonstrated that the low energy effective Hamiltonian for a CuO_2-plane could be parameterized either as a 1-band Hubbard model or a t-J model. They started from a multiband model with $3d_{x^2-y^2}$-orbitals on Cu-sites and $2p_{x(y)}$-orbitals on O sites using a priori methods to estimate the various parameters that entered (e.g. hopping matrix elements, on-site and inter-site Coulomb interactions). The reduction to a low energy effective Hamiltonian is performed by comparing to the low energy spectrum of a cluster. In the effective models, there is a single state per Cu-atom on a square lattice and one obtains either a Hubbard model

$$H_H = \sum_{\langle ij \rangle,\sigma} t(i-j)\left(d^{+}_{i\sigma} d_{j\sigma} + h.c.\right) + U \sum_i n_{i\uparrow} n_{i\downarrow} \tag{1}$$

or a t-J model

$$H_{t-J} = \sum_{\langle ij \rangle \sigma} t(i-j)\, P_0 \left(d^{+}_{i\sigma} d_{j\sigma} + h.c.\right) P_0 + J \sum_{\langle ij \rangle} \left(\vec{S}_i \cdot \vec{S}_j - n_i n_j/4\right). \tag{2}$$

The hopping matrix elements $t(n.n.) = 0.44eV$ and $t(n.n.n.) = -0.06eV$. In the Hubbard model parameterization the onsite Coulomb term $U = 5.4eV$ which makes it a strong coupling problem. In the t-J model parameterization the hopping matrix elements are the same and the exchange constant $J = 0.13eV$. The t-J model which is contrained to the singly occupied Hilbert subspace by the projector $P_0 = \Pi_i(1 - n_{i\uparrow}\, n_{n\downarrow})$, is inherently a strong coupling model.

The two models are closely related in strong coupling limits where $J/t \ll 1$ and $U/zt \gg 1$ and the parameters quoted above place these models in the strong coupling regime. These models are the simplest which are consistent with the strongly correlated Hilbert subspace and the symmetries of the interactions. So a lot of effort has gone into attempts to understand the predictions of these models and to compare them to experiment. It is however far from complete at the present time. In this talk I will concentrate on our understanding of the zero temperature phase diagram of the strongly correlated t-J model. The properties at finite temperatures and their relation to the anomalous properties of the normal state of the cuprate superconductors will be topic of other lectures at this school.

An alternative strategy discussed in the previous talk by Prof. Schrieffer is to investigate intermediate coupling strength, i.e. $U/zt \approx 1$ as more appropriate for the cuprate superconductors and indeed a very recent study of the electronic structure by Bacci et al. [5] proposes just such an intermediate value of the ratio U/zt as more correct. These two approaches should be consistent and the basic understanding should not be so sensitive to the actual value of this ratio.

The past year has seen a lot of progress on the model in 1-dim. only and it is appropriate to begin with a review of this progress in Sect. 2. The real case we are interested in is the course that of a 2-dim. square lattice and here there is still considerable uncertainty. Sect. 3 will review recent understanding in this area and Sect. 4 is devoted to some concluding remarks.

2. Phase Diagram in 1-Dimension

The 1-dim. Hubbard model with only $t(n.n.) \neq 0$ was solved by Lieb and Wu [6] using Bethe-Ansatz methods. They obtained the total energy and the excitation energies for all values of the parameters $U/t, n$ (n: electron density/site). The main conclusions are as follows:

i) For all n, the only singularities as a function of U/t occur at the value $U = 0$. For a finite value of U/t the system scales to strong coupling in the sense that the properties at small and large values of U/t are qualitatively similar.

ii) The chemical potential $\mu(n)$ is a continuous function of n for all U/t except at the half-filled density, $n = 1$ where it has a discontinuity whose magnitude varies with U and is $\propto U$ for $U/t \gg 1$. At $n = 1$ the system is insulating and incompressible with a gap for charge excitations. The low energy properties are described by a Heisenberg $s = 1/2$ model which as $U/t \gg 1$ reduces to a $n.n.$ Heisenberg chain with $J = 4t^2/U$.

Turning to the t-J model again the case $n = 1$ is a Heisenberg chain and an insulator. The limit $J/t \to 0$ is identical to the Hubbard model in the limit $U/t \to \infty$. The t-J model in 1-dim. is not soluble for general values of J/t but at the special value $J/t = 2$, a Bethe-Ansatz solution was found by Bares and Blatter [7]. Actually this model was solved earlier in different contexts by Sutherland [8] and by Schlottmann [9]. The main conclusions are as follows:

i) The ground state energy is an analytic function of the density over the whole range $0 \leq n \leq 1$ and interpolates between free fermions as $n \to 0$ and a Heisenberg model as $n \to 1$.

ii) Neither a phase transition nor phase separation occurs in the whole density range $0 \leq n \leq 1$.

The Bethe-Ansatz solutions do not by themselves allow the direct determination of the correlation functions. Recently however a powerful method has been used to make the connection between the Bethe-Ansatz solutions and the correlation functions. This is, based on a relation between the key exponent, K_ρ, and the charge velocity (v_c) and compressibility (κ). This relation is valid for a wide class of 1-dim. models which were called Luttinger liquids by Haldane [10]. The key point is that the long wavelength density fluctuations can be described by a harmonic theory whose coefficients can be determined from total energy calculations. These fluctuations control the asymptotic behavior of the correlation functions whereas the spin fluctuations are responsible only for logarithmic corrections (cf. the discussion by Giamarchi and Schulz [11]). The exponent, K_ρ, satisfies the relation

$$K_\rho = \pi n^2 \kappa v_c/2 \quad , \tag{3}$$

where the quantities on the r.h.s. of (3) are all accessible from the Bethe-Ansatz solutions. The asymptotic form of the density-density and spin-spin equal time correlation functions are

$$\begin{aligned}
\langle n(r)\, n(0)\rangle &\sim A_0 r^{-2} + A_1 \cos(2k_F r) r^{-1-K_\rho} \ell n^{-3/2}(r) + A_2 \cos(4k_F r)^{-4K_\rho} \\
\langle S^z(r)\, S^z(0)\rangle &\sim B_0 r^{-2} + B_1 \cos(2k_F r) r^{-1-K_\rho} \ell n^{1/2}(r)\,.
\end{aligned} \tag{4}$$

These correlations are longer-range than free fermions if $K_\rho < 1$. On the other hand if $K_\rho > 1$, it is the pairing correlation functions which are longer range since these are determined by similar formula but with the interchange $K_\rho \leftrightarrow K_\rho^{-1}$.

Recently complete plots of K_ρ as a function of n and U/t were presented for the Hubbard model [12, 13] and for the t-J model with $J/t = 2$ [14] using the Bethe-Ansatz solutions to evaluate the r.h.s. of Eq. (3). These results show $K_\rho \leq 1$ always, indicating enhanced magnetic correlations rather than pairing correlations. As $n \to 1$, $K_\rho \to 1/2$, both in the strong coupling Hubbard and t-J models. This indicates that doping with holes leads to shorter range magnetic correlations than in the Heisenberg model ($K_\rho = 0$).

The full phase diagram for the t-J model was recently presented by Ogata et al. [15] who used results from numerical diagonalization of finite clusters to evaluate the r.h.s. of Eq. (3). Fig. 1 shows contours of constant K_ρ in the

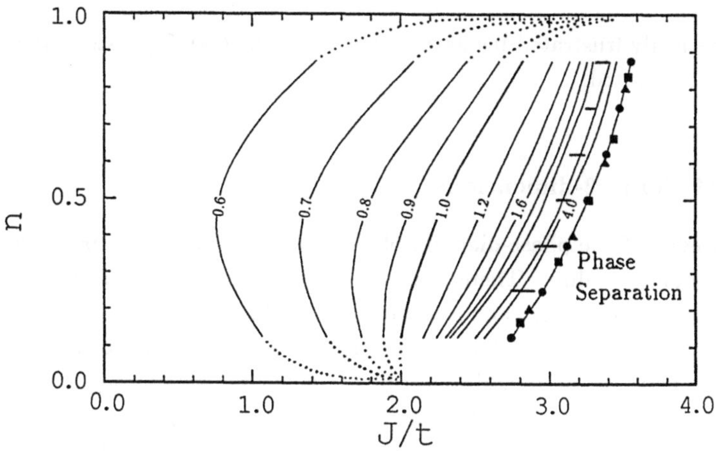

Fig. 1 Phase Diagram of the t-J model in 1-dim. showing the phase separation line and contours of constant correlation exponent, K_ρ.

$(n, J/t)$ plane. For values of $J/t < 1$ the results are quite similar to those of the Hubbard model. In particular in the region of relevance for high-T_c ($n \approx 0.8$; $J/t \approx 0.3$), $K_\rho < 1$ and magnetic correlations dominate. There is a region with predominantly pairing correlations (i.e, $K_\rho > 1$) but this appears as a precursor of phase separation which occurs for $J/t \gtrsim 3$. Phase separation was discussed by Emery et al. [16] and for large J/t and $n < 1$ involves a complete separation into a chain of spins ($n = 1$) and a vacuum.

The 1-dim. t-J model is amenable to Quantum Monte-Carlo simulation. Recent work by Imada [17] and particularly Assaad and Würtz [18] show nicely the evolution of the correlation function from predominantly magnetic correlations for $J/t \lesssim 1$ to pairing correlations and phase separation for $J/t > 2$. At least, in 1-dim. the phase diagram of the Hubbard and t-J models are clearly determined.

In a 1-dim. system singlet correlations are accompanied by a spin-gap, i.e. an energy gap to triplet excitations. One question then is can we find a lightly doped 1-dim. model which shows a spin gap and therefore s-wave singlet superconducting fluctuations. In renormalized perturbation theory (or 'g-ology') [19] this requires an attractive backward scattering matrix element (i.e., $g_1 < 0$). Recently Luchini et al. [20] examined the effect of introducing frustration into the t-J model by adding a *n.n.n.* J'-term. Such a spin model was shown by Majumdar and Ghosh [21], to have a ground state composed of singlet-dimers when $J' = 1/2J$. Luchini et al. introduced doping in this t-J-J' model and found that, while the doping relieved the frustration and caused a transition back to a Luttinger liquid, a finite region with a spin gap remained even for $J/t \ll 1$. Note, however, that the spin-gap region was obtained by doping a magnetically frustrated insulator, whereas in the high-T_c system, the insulator is not frustrated.

3. The t-J Model in 2-dimensions

Our knowledge of the phase-diagram of the t-J model in 2-dim. is much less advanced than in 1-dim. There are new features that can appear in a 2-dim. system which do not or cannot occur in the 1-dim. model.

One such example is the possibility of ferromagnetism in the Nagaoka limit, $J/t \to 0$. The Nagaoka theorem [22] concerning ferromagnetism of a single hole applies to the square lattice. Recently, there has been considerable interest in determining the extent of the ferromagnetic region. Yokoyama and Shiba [23] used Gutzwiller-projected wavefunctions and obtained a small region bounded by $n > n_c (\approx 0.62)$ and $J/t \leq (J/t)_c \approx 0.1$, where ferromagnetism was favored. A high temperature expansion in powers of (βt) by Yedidia [24] for the case $J/t = 0$ gave a larger critical value of $1 - n_c = 3/11$ or $n_c \approx 0.73$. Very recently Putikka and coworkers [25], extended this expansion to finite values of J/t and found also that $(J/t)_c \approx 0.1$ for the ferromagnetic order.

Emery and coworkers [16] estimated the phase separation line in 2-dim. by examining cluster calculations to obtain the energy/electron. They obtained a very different form for the phase boundary line than that which was found in 1-dim. (see Fig. 1). Emery et al. [16] proposed that phase separation occurs at all values of J/t. The phase separation at large values of J/t and all values of n caused by the large energy cost of breaking bounds in this parameter range, joins smoothly on to phase separation at small J/t and large n caused by the Nagaoka effect. Putikka et al. examined the phase boundary line using the high temperature series expansion. At values of $J/t > 1$ their curve agrees well with that of Emery at al., but at values of $J/t \lesssim 1$ they do not find phase separation. This leaves a window between the phase separation line ending at $J/t \gtrsim 1$ and the Nagaoka ferromagnetic region at $J/t < 0.1$ where some other form of phase transition could occur. This is also the parameter region relevant to the high-T_c superconductors.

The search for possible broken symmetry states in the 2-dim. t-J model has been actively pursued in recent years using a variety of methods. A direct numerical solution using Quantum Monte Carlo methods is not possible because of the notorious fermion sign problem. Variational methods have naturally been employed. One important class of wavefunctions consists of a Hartree-Fock-Bogoliubov wavefunction projected down to Hilbert subspace with double occupancy forbidden, i.e a wavefunction of the form

$$| \Psi \rangle = P_0 \, | \Psi^0_{HFB} \rangle . \tag{5}$$

It is interesting to examine first the limit $n \rightarrow 1$ where the t-J model reduces simply to a Heisenberg model. A Heisenberg model is purely a spin problem and can be solved by wavefunctions which span the spin subspace rather than fermion wavefunctions which span the much larger unprojected fermion wavefunctions. Affleck et al. [26] found that in fact a class of local $SU(2)$ transformations $| \Psi^0 \rangle$ leaves $| \Psi \rangle$ invariant. In other words very many seemingly different fermion wavefunctions $| \Psi^0 \rangle$ are related through this large class of local $SU(2)$ transformations.

Two methods are available to evaluate expectation values with wavefunctions of this type (5). An analytic approximation can be made to the projector using the Gutzwiller form leading to a renormalized mean field theory, (Zhang et al. [27]). The Gutzwiller approximation scheme has been recently shown by Metzner and Vollhardt [28] to be exact in a limit of an infinite dimension. Alternatively direct numerical evaluation is possible for adequatly large clusters (e.g. see Gros [29]). The two methods agree quite well so that one can regard the evaluation problem as solved.

The key results can be summarized as follows':

a) At $n = 1$ a projected Fermi sea does not have a very good energy ($-1.04J/$ site) unlike the case in 1-dim.

b) A class of paramagnetic states (i.e. no AF long range order) has a considerably better energy ($-1.14J$) at $n = 1$. Out of this class which are all related by local $SU(2)$-transformations, two members are especially interesting namely the Affleck-Marston state [30] and the d-wave RVB state [27, 29] since these are favored at small doping, i.e. $1 - n \ll 1$.

c) The Affleck-Marston state uses for $| \Psi^0 \rangle$ a Slater determinant made from 1-electron solutions of the flux problem with (1/2)-flux quanta per plaquette. An interesting generalization to the general n-case proposed by Anderson et al. [31] uses Hofstadter states with a flux commensurate with the density. These commensurate flux phases were shown by Lederer et al. [32] to be favorable for the magnetic or J-term but unfavorable for the kinetic or t-term. They concluded that they were not stable in the interesting region $0.1 \leq J/t \leq 1$, a result confirmed by direct evaluation [33, 34]. Recently Ogata et al. [35] examined a modified form with a prefactor which introduced a flux for the empty sites or holons but this did not change the energy much. Such commensurate flux phases are of special interest since they are closely related to the anyon models proposed by Laughlin [36], Wiegmann [37] and others for high-T_c superconductors. At least for the t-J model such states do not seem to be energetically favored.

d) The d-wave RVB state looks more promising. Recently two groups Lee and Feng [38] and Chen et al. [39] examined variational states which were a mixture of AF long range order and d-wave pairing. Such states give a further improvement in the energy at $n = 1$ to a value ($-1.16J$/site) which is very close to the best energy obtained for a spin-state ($-1.17J$/site) [40]. With doping there is a rapid suppression of the AF-order so that for example for a value of $J/t = 0.3$ a transition to a pure d-wave paired state (i.e. d-wave RVB state) occurs at a value of $1 - n \simeq 0.04$ which in turn is stable against a Fermi liquid state until a value of $1 - n \simeq 0.4$. Within the class of wavefunctions Eq. (5), these states are favored for the relevant values of $J/t \sim 0.3$.

The key question of how reliable the variational methods are is difficult to answer confidentally. Naturally one looks to unbiased methods for support. The Lanczos diagonalization calculation for finite clusters is such a method but it is restricted to small clusters ($\leq$ 20 sites) which makes it difficult to extrapolate to large systems and learn about possible symmetry-breaking. One thing however is clear namely that the long range AF order disappears rapidly upon doping for values of $J/t \sim 0.3$. Bonča et al. [41] and Poilblanc et al. [42] looked at the spin-spin correlations for 2 holes in a 4×4 cluster and found a range much shorter than the cluster diagonal. The question of possible broken symmetry is tricky. There are signs of d-wave pairing [41, 43]. However, a recent study of the single-particle Green's function for the case of 2-holes in an 18-site cluster by Stephan and Horsch [44] did not seem to show any signs of an energy gap and rather favored a Fermi liquid state with a coherent band width of $0(J)$. Incoherent excitations were spread over the total bandwidth $0(t)$. This points to an important role for short range correlations around the holes and these may not be properly included in the variational wavefunctions.

4. Conclusion

In this lecture, I have concentrated on a limited set of questions concerning the simplest strongly correlated models and the nature of their $T = 0^o K$ phase diagram as a function of their parameter J/t (or U/t) and n. There is considerable experimental and theoretical support now for these models which was only briefly discussed here. In addition the progress being made on understanding the anomalous normal state properties based on gauge theories of strongly correlated electrons is discussed elsewhere at this school. The answer to the questions posed here remains open. Only when we have a better understanding of the phase diagram in 2-dim. of these simplest strongly correlated models will we know whether they will suffice to explain high-T_c superconductivity or whether other interactions actually dominate the low energy properties and determine the nature of the ground state.

Acknowledgements

I am very grateful to many colleagues for collaboration and discussions on these topics.

References

1. P.W. Anderson, Science, 235, 1196 (1987); In Frontiers and Borderlines in Many Particle Physics (ed. R.A. Broglia and J.R. Schrieffer), Proc. E. Fermi Summer School, 1, (1988).
2. F.C. Zhang and T.M. Rice, Phys. Rev. B 37, 3754 (1988).
3. C.T. Chen et al., Phys. Rev. Lett. 66, 104 (1991).
4. M.S. Hybertsen, E.B. Stechel, M. Schlüter and D.R. Jennison, Phys. Rev. B 41, 11068 (1990).
5. S.B. Bacci, E.R. Gagliano, R.M. Martin and J.F. Annett, Univ. of Ill. preprint (1991).
6. E.H. Lieb and F.Y. Wu, Phys. Rev. Lett. 20, 1445 (1968).
7. P.-A.Bares and G. Blatter, Phys. Rev. Lett. 64, 2569 (1990).
8. B. Sutherland, Phys. Rev. B 12, 3795 (1975).
9. P. Schlottman, Phys. Rev. B 36, 5177 (1987).
10. F.D.M. Haldane, J. Phys. C 14, 2585 (1981); Phys. Rev. Lett. 45, 1538 (1980).
11. T. Giamarchi and H.J. Schulz, Phys. Rev. B 39, 4620 (1989).
12. H.J. Schulz, Phys. Rev. Lett. 64, 2831 (1990).
13. F.D.M. Haldane and Y. Tu, UCSD preprint (1990).
14. N. Kawakami and S.-N. Yang, Phys. Rev. Lett. 65, 2309 (1990).
15. M. Ogata, M.U. Luchini, S. Sorella and F.F. Assaad, Phys. Rev. Lett. 66, 2388 (1991).
16. V.J. Emery, S.A. Kivelson and H.Q. Lin, Phys. Rev. Lett. 64, 475 (1990).
17. M. Imada, J. Phys. Soc. Japan 59, 4121 (1990).
18. F.F. Assaad and D. Würtz, Phys. Rev. B (in press).
19. J. Solyom, Adv. Phys. 28, 201 (1979).
20. M.U. Luchini, M. Ogata and T.M. Rice, ETH-preprint (1991).
21. C.K. Majumdar and D.K. Ghosh, J. Math. Phys. 10, 1388 (1969).
22. Y. Nagaoka, Solid State Comm. 3, 409 (1965).
23. H. Yokoyama and H. Shiba, J. Phys. Soc. Japan 56, 3570 (1987).
24. J.S. Yedidia, Phys. Rev. B 41, 9397 (1990).
25. W. Putikka and M.U. Luchini (private communication).
26. I. Affleck, Z. Zou, T. Hsu and P.W. Anderson, Phys. Rev. 38, 745 (1988).
27. F.C. Zhang, C. Gros, T.M. Rice and H. Shiba, Super. Sci. Tech. 1, 36 (1988).

28. W. Metzner and D. Vollhardt, Phys. Rev. Lett. 62, 329 (1989).
29. C. Gros, Phys. Rev. B 38, 931 (1988); Ann. Phys. 189, 53 (1989).
30. I. Affleck and J.B. Marston, Phys. Rev. B 37, 3664 (1988).
31. P.W. Anderson, B.S. Shastry and D. Hristopoulos, Phys. Rev. B 40, 8939 (1989).
32. P. Lederer, D. Poilblanc and T.M. Rice, Phys. Rev. Lett. 63, 1519 (1989).
33. S. Liang and N. Trivedi, Phys. Rev. Lett. 64, 232 (1990).
34. D. Poilblanc, Y. Hasegawa and T.M. Rice, Phys. Rev. B 41, 1949 (1990).
35. M. Ogata, B. Douçot and T.M. Rice, Phys. Rev. B 43, 5582 (1991).
36. R.B. Laughlin, Science 242, 525 (1988); Phys. Rev. Lett. 60, 2677 (1988).
37. P.B. Wiegmann, Phys. Scripta T 27, 160 (1988).
38. T.K. Lee and S. Feng, Phys. Rev. B 38, 11809 (1988).
39. G.J. Chen, R. Joynt, F.C. Zhang and C. Gros, Phys. Rev. B 42, 2662 (1990).
40. E. Manousakis, Rev. Mod. Phys. 63, 1 (1991).
41. J. Bonča, P. Prelovśek and I. Sega, Phys. Rev. B 39, 7074 (1989).
42. D. Poilblanc, E. Dagotto and J. Riera, Phys. Rev. B 43, 7899 (1989).
43. E. Dagotto, J. Riera and H.P. Young, Phys. Rev. B 42, 2347 (1990).
44. W. Stephan and P. Horsch, Phys. Rev. Lett. 66, 2258 (1991).

On the Spin Excitations of High-T_c Oxides

H. Fukuyama[1], *T. Tanamoto*[2], *and K. Kuboki*[3]

[1]Institute for Solid State Physics, University of Tokyo,
Roppongi, Minato-ku, Tokyo 106, Japan
[2]Research & Development Center, Toshiba Co., Ltd.,
Kawasaki 210, Japan
[3]Department of Physics, Kobe University, Nada-ku, Kobe 657, Japan

Abstract. The proximity of the superconducting state to the antiferromagnetic state makes the problem of spin fluctuations one of the central issues in exploring the mechanism of high T_c superconductivity. The experimental results of the magnetic spin excitations probed through nuclear magnetic relaxation (NMR) and neutron scattering are discussed with special emphasis on the existence of the apparent gap in the excitation spectrum (spin gap), and its possible explanation is proposed based on the mean field theory for the $t-J$ model.

1. Introduction

The experimental results of the magnetic spin excitations probed by NMR and neutron scattering have been playing vital roles in the understanding of the electronic state in high T_c Cu-oxides both below and above the critical temperature, since these experiments can disclose the distribution of the spectral weight of spin excitations in the Brillouin zone [1]. Several authors [2,3,4] have analyzed these experimental results by use of the dynamical spin susceptibility $\chi(q,\omega)$ derived by the random phase approximation (RPA), the self-consistent renormalization (SCR) or the phenomenological consideration. Bulut et al. [2] examined the rate of the nuclear magnetic relaxation based on RPA treatment of the Hubbard model and concluded that the antiferromagnetically correlated one-component spin system can explain the overall features. In their study the effect of the band structure is fully taken into account in the evaluations of the dynamical spin susceptibility of non-interacting electrons, $\chi_0(q,\omega)$, and the existence of rather complicated features associated with the incommensurate peak in $\chi_0(q,\omega)$ has been noted. At the same time they indicated a difficulty in the choice of values of parameters in fitting to the experimental results.

A similar conclusion of the antiferromagnetically correlated one-component spin system was reached by Millis et al. [3], who employed a phenomenological model. The nuclear spin relaxation rate together with the T-linear resistivity due to spin fluctuations has also been investigated within the self-consistent renormalization (SCR) theory by Moriya et al. [4] In this treatment as well as in the study of Millis et al, the roles of spin fluctuations with $\vec{q} \sim \vec{Q} \equiv (\pi, \pi)$ (lattice spacing is taken as unity in this paper) are stressed, and Im $\chi(q, \omega)$ in $\omega \to 0$ is assumed to have a maximum as a function of $\vec{q}$ at $\vec{q} = \vec{Q}$.

While these theories have been successful in explaining some of the characteristic features of NMR rate in normal state [5], there exists an important unresolved question of the apparent indication of the spin gap. Yasuoka et al. [6] have stressed the fact that T_1T as a function of T has a Curie-Weiss behavior at high temperatures ($\gtrsim$ 100K) but a clear minimum exists just above the critical temperatures of superconductivity. The existence of a similar spin gap has also been noted in neutron scattering. Shirane et al. [7,8] has observed a dramatic reduction of the integrated intensity at 6meV for $La_{1.85}Sr_{0.15}CuO_4$ as a function of temperature. Such a reduction of the intensity was not observed for 12meV and it was argued that these results could be due to the opening of the gap in the spin excitations. Gehring et al. [9] found that the value of this gap is strongly dependent on doping but that it is not directly correlated with the critical temperature of superconductivity. The experiment of Rossat-Mignod et al.[10] on $YBa_2Cu_3O_{6+x}$ also observed a similar phenomenon. They studied Im$\chi(\vec{q}, \omega)$ at $\vec{q} = \vec{Q}$ as a function of ω and an apparent gap structure has been identified at low temperatures, though they noted that it is not a sharp gap since a finite slope with a value larger than twice the experimental resolution is seen.

In this paper we address ourselves to this problem of the spin gap. The theoretical model we employ for this purpose is the $t-J$ model, whose relevance to high T_c oxides has been discussed in various context [11,12,13,14,15]. The reasons of our preference of the $t-J$ model in spite of the fact that the energy spectrum of the original $d-p$ model can equally well reproduced by the $t-J$ model (t ~0.44eV, J ~0.13eV) and the Hubbard model (t ~0.43eV, U ~5.4eV) as clarified by Hybertsen et al. [14] are as follows; 1) the RPA for the Hubbard model is very suitable to take the lattice effect into account [2] but with the choice of such a bare value of U it results in magnetic instability for unphysically wide range of doping and the adoption of the renormalized value of U can not be avoided. The effect of such renormalization is, however, hard to assess. (Moreover it is natural to expect that the range of interaction is strongly modified. Actually the J-term in the $t-J$ model can be considered to be the result of this renormalization.) 2) It will be very appropriate and interesting if the SCR type of the treatment of the Hubbard model is carried out with the explicit lattice structure taken into consideration but it appears to be very hard to perform in practice. 3) On the other hand the RPA to the $t-J$ model can take full account of lattice structures on one hand but at the same time results in magnetic instability only very limited range of doping in accordance with experiments as shown in the following.

In Sec.2 the model and the approximation are introduced and in Sec.3 the typical results of numerical calculations of NMR rate and the spectral weight of magnetic excitations are given. We discuss associated problems in Sec.4.

More detailed description of the present study will be reported elsewhere [16].

We take unit of $\hbar = k_B = 1$.

2. Model and Approximation

The $t-J$ model is given by

$$H = -t \sum_{<ij>\sigma} [\tilde{c}^\dagger_{i\sigma} \tilde{c}_{j\sigma} + \text{h.c.}] + J \sum_{<ij>} \vec{S}_i \cdot \vec{S}_j - \mu_e \sum_{i\sigma} c^\dagger_{i\sigma} c_{i\sigma}, \tag{1}$$

where $< ij >$ denotes the nearest-neighbor bond, and μ_e is a chemical potential of electrons, and t and J are the transfer integrals and the antiferromagnetic superexchange interactions between nearest-neighbor sites, respectively, and $\tilde{c}^\dagger_{i\sigma} \equiv c^\dagger_{i\sigma}(1-n_{i,-\sigma})$ creates the electron with spin σ at site i only when the site i is unoccupied, namely, the double occupancy is excluded.

We use the slave-boson method where the electron operators, $c_{i\sigma}$, are expressed by those of the spinon, f_σ, and the holon, b_i; $c_{i\sigma} = b^\dagger_i f_{i\sigma}$, under the local constraint, $b^\dagger_i b_i + \sum_\sigma f^\dagger_{i\sigma} f_{i\sigma} = 1$ at each site i. Then the Hamiltonian is written as

$$H = -t \sum_{<ij>} [b^\dagger_j b_i \chi_{ij} + \text{h.c.}] - \frac{J}{2} \sum_{<ij>} \chi^\dagger_{ij} \chi_{ij} - \mu \sum_{i\sigma} f^\dagger_{i\sigma} f_{i\sigma}$$

$$- \sum_i \lambda_i (b^\dagger_i b_i + \sum_\sigma f^\dagger_{i\sigma} f_{i\sigma} - 1), \tag{2}$$

where $\chi_{ij} \equiv \sum_\sigma f^\dagger_{i\sigma} f_{j\sigma}$ is the bond operator for the spinon and the last term represents the local constraint.

We employ a mean-field approximation with respect to the bond order parameters for spinons, $< \chi_{ij} >$, and holons, $< b^\dagger_i b_j >$, expecting the absence of the magnetic long range order in the case of finite doping which we are interested in. (This corresponds to introducing the fictitious large-$N_{\rm F}$ limit of the electron spin indices, i.e., $f_{i,\sigma}(\sigma = \pm) \to f_{i\alpha}(\alpha = 1 \cdots N_{\rm F})$. In the limit $N_{\rm F} \to \infty$ the decoupling in terms of bond order parameters will be exact. We hope that the real system with $N_{\rm F} = 2$ is in the same universality class as that with $N_{\rm F} = \infty$ when moderately doped.)

In this approximation the Hamiltonian is given by

$$H^{(0)}_{\rm MF} = - \sum_{<ij>} [F^*_{ij}\chi_{ij} + {\rm h.c.}] - \lambda_{\rm F} \sum_{i\sigma} f^\dagger_{i\sigma} f_{i\sigma}$$

$$- \sum_{<ij>} [B^*_{ij} b^\dagger_i b_j + {\rm h.c.}] - \lambda_{\rm B} \sum_i b^\dagger_i b_i, \qquad (3)$$

where $F_{ij} \equiv t < b^\dagger_i b_j > +(J/2) < \chi_{ij} >$, $B_{ij} \equiv t < \chi_{ij} >$ and $\lambda_{\rm F} \equiv \mu + \lambda + J(1-\delta)$ and $\lambda_{\rm B} \equiv \lambda$. Here we have replaced the local constraint by a global one, i.e., λ_i is taken to be independent of i ($\lambda_i = \lambda$) and thus considered as a chemical potential. The chemical potential for the spinon, $\lambda_{\rm F}$ (holon, $\lambda_{\rm B}$), is determined in accordance with the doping rate, δ, as $(1/N)\sum_{i\sigma} < f^\dagger_{i\sigma} f_{i\sigma} >= 1-\delta$, and similarly $(1/N)\sum_i < b^\dagger_i b_i >= \delta$, where N is the total number of lattice sites.

The Hamiltonian eq.(3) (and a similar but slightly different models) has been studied extensively and various kinds of solutions are found depending on the parameters, t/J and δ. [17,18,19] In the following we focus on the fluxless state, which Nagaosa and Lee [20] proposed as a model to high T_c oxides. In this state $< \chi_{ij} >$ and $< b^\dagger_i b_j >$ are real and constant for all neighboring bonds. Hereafter we take $t = 4J$ (which will be a reasonable choice for high T_c oxides) and in this case the fluxless state is stabilized for moderate doping, i.e., $\delta \gtrsim 0.1$.

In the fluxless state the energy dispersions of the spinon and the holon are written as $\varepsilon^{\rm F}(\vec{k}) = -2F(\cos k_x + \cos k_y)$ and $\varepsilon^{\rm B}(\vec{k}) = -2B(\cos k_x + \cos k_y)$, respectively, where F and B to be evaluated self-consistently are the values of F_{ij} and B_{ij} for neighboring bonds in the fluxless state, respectively. Then the Fermi surface determined by $\varepsilon^{\rm F}(\vec{k})$ is large for small δ as shown in Fig.1 and has a significant deviation from

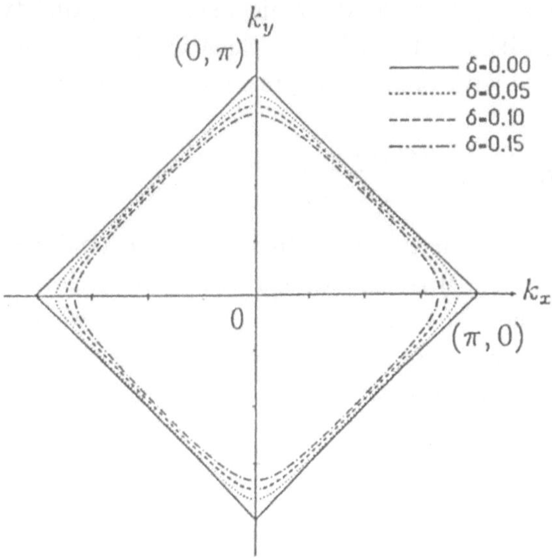

Fig.1 The Fermi surfaces of free spinons for several choices of δ.

a circle. While such a precise shape of the Fermi surface of spinons is not directly reflected in the transport properties investigated by Nagaosa and Lee [20], which are governed by holons and gauge fields, it plays an essential role in the study of magnetic properties as will be seen in the following.

3. NMR Rate and the Spectral Weight

The dynamical spin susceptibility $\chi(q,\omega)$ of the $t-J$ model treated within the mean field approximation defined in the preceding section is given as follows

$$\chi(\vec{q},\omega) = C(\vec{q},\omega)\chi_0(\vec{q},\omega), \tag{4}$$

where

$$C(\vec{q},\omega)^{-1} = 1 + J(\cos q_x + \cos q_y)\chi_0(\vec{q},\omega) \equiv 1 + J_q\chi_0(\vec{q},\omega), \tag{5}$$

with $J_q = J(\cos q_x + \cos q_y)$ and

$$\chi_0(\vec{q},\omega) = \sum_k \frac{f(\xi_{\vec{k}+\vec{q}}) - f(\xi_{\vec{k}})}{\xi_{\vec{k}} - \xi_{\vec{k}+\vec{q}} + \omega + i0}, \tag{6}$$

with $\xi_{\vec{k}} = \varepsilon^{\mathrm{F}}(\vec{k}) - \lambda_{\mathrm{F}}$ and $f(\varepsilon) = 1/(\exp(\varepsilon/T) + 1)$ being the Fermi distribution function. The $\chi_0(\vec{q},\omega)$ is the dynamical spin susceptibility of non-interacting fermions with band width, $8F$. Equation (4) together with eq.(5) is to be contrasted with the RPA expression for the Hubbard model

$$\chi_{RPA} = \frac{\chi_0(q,\omega)}{1 - U\chi_0(q,\omega)} \tag{7}$$

The difference is clear; $\chi(q,0)$ is enhanced (reduced) at around $q = Q = (\pi,\pi)(q=0)$ while $\chi_{RPA}(q,0)$ is enhanced in the whole region of q. Because of this the uniform spin susceptibility $\chi(0,0)$ is only weakly dependent on temperature even if $\chi(Q,0)$ is appreciably enhanced.

By use of $\chi(q,\omega)$ the nuclear magnetic relaxation rate, T_1^{-1}, is given by

$$\frac{1}{T_1T} = \sum_q F(\vec{q})\frac{1}{\omega}\mathrm{Im}\chi(\vec{q},\omega)|_{\omega\to 0} \equiv R(T), \tag{8}$$

where $F(\vec{q})$ is the form factor at a particular nuclear site.

We have calculated NMR rate at Cu and O nuclei based on the theory of Mila and Rice [21] that the NMR rate at oxygen sites is determined by the transferred hyperfine coupling. The results of numerical calculations for the doping rate $\delta \sim 0.1$ show: 1) $R(T)$ at Cu-sites obeys the Curie-Weiss law for $T \gtrsim 0.1J$, i.e. $R(T)^{-1} = K(T+\theta)$ with θ linearly dependent on δ, 2) $R(T)^{-1}$ at Cu-sites has a minimum at around $T \sim (0.1-0.2)J$, 3) $R(T)$ at O-sites is only weakly dependent on temperature.

Examples to show these results of numerical calculations are shown in Figs.2(a)(b) and (c). In the present choice of $t/J = 4$ the magnetic instability is limited to $\delta \lesssim 0.09$. The experimental data by Yasuoka et al. [6] and by Asayama et al. [22,23] are in accordance with 1) while the result 2) can be consistent with the experimental indication by Yasuoka et al. [6] of the apparent spin gap. (In the paper by Yasuoka et al. [6] the experimental data were analyzed to obey the form $T_1^{-1} = a + bT$ but those data can also be fitted as well as in the present form in the temperature range of experiment.)

The physical origin of this apparent spin gap behavior is very simple; at relatively higher temperatures, $\mathrm{Im}\chi(q,\omega)/\omega(\omega \to 0)$ has a peak at $q = Q$, but as temperature is reduced, this peak at $q = Q$ is sharply suppressed, and below some temperature it is no longer a peak but becomes a local minimum. This feature at around $q = Q$ is similar to the RPA result of the Hubbard model (with a proper choice of t, U, δ)

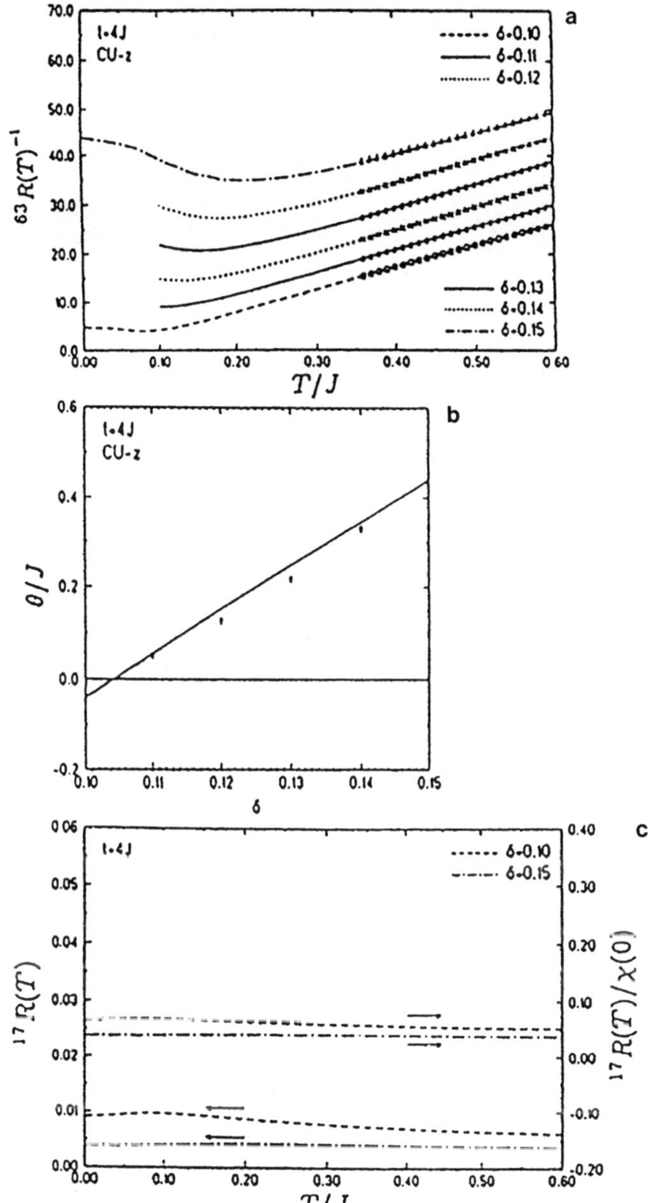

Fig.2 a) The temperature dependences of the inverse of the NMR rate at Cu-site for several choices of δ with the magnetic field parallel to the c-axis. The data obey the Curie-Weiss law, $^{63}R(T)^{-1} = K(T + \theta)$, at high temperatures. b) The δ-dependences of θ. c) The temperature dependences of NMR rate at O-site, $^{17}R(T)$, and those scaled by the uniform susceptibility, $\chi(0)$.

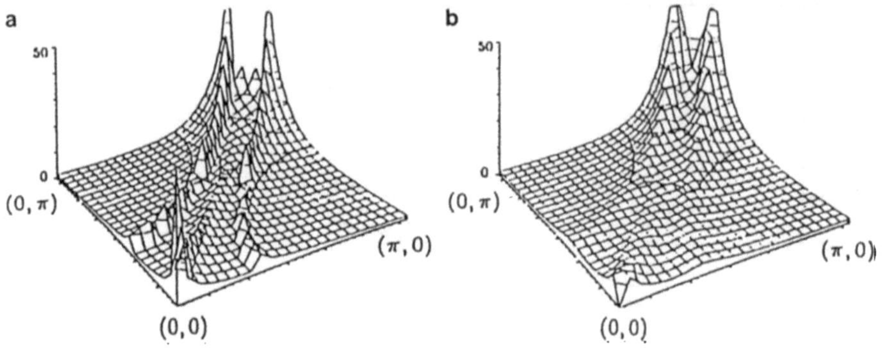

Fig.3 The q-dependences of $\mathrm{Im}\chi(q,\omega)/\omega(\omega \to 0)$ in the case of $\delta = 0.15$ for (a) $T = 0.01J$ (b) $T = 0.1J$.

by Bulut et al. [2,24] since $C(Q,\omega)^{-1} = 1 - 2J\chi_0(Q,\omega)$. This is in contrast to the assumption made in the phenomenology of Millis et al. [3] and the SCR theory of Moriya et al. [4] This strong variation at $q = Q$ results in the reduction of the enhancement of spectral weight of the spin excitations leading to the minimum of T_1T. An example of numerical calculations to demonstrate this tendency is seen in Figs.3 (a) and (b), where the q-dependences of $\mathrm{Im}\chi(q,\omega)/\omega(\omega \to 0)$ are plotted for $T = 0.01J$ and $T = 0.1J$ in the case of $\delta = 0.15$.

The qualitative feature of $\mathrm{Im}\chi(q,\omega)$ at $\omega \to 0$ is also maintained in the ω-dependence; i.e., at low temperatures $\mathrm{Im}\chi(\vec{q},\omega)$ has a local minimum at $(\vec{q} = \vec{Q})$ in low ω but this is transformed into a maximum as ω is raised. The same feature can typically be seen in its ω-dependences at $q = Q$ with varying temperatures, as seen in Fig.4 where the doping rate is fixed at $\delta = 0.15$.

This result is also in qualitative agreement with the neutron scattering data [7,8,9,10].

Hence it will be concluded that the apparent spin gap observed both in NMR rate and neutron scattering can be understood as due to the band structure effects, which are appreciable in $\chi(q,\omega)$ at around $q = Q$ if the doping rate is low.

The foregoing studies are on the state above the critical temperature, T_c, of superconductivity. We have also examined [25] the NMR rate in the superconducting state. An example for the case of the d-wave state is shown in Fig.5 for a choice of $\delta = 0.10$, where $R_0(T)$ and $R(T)$ are NMR rate at the Cu-site evaluated without the exchange correction (i.e. in terms of $\mathrm{Im}\chi_0(q,\omega)$) and that with the exchange correction, respectively. (In the calculations we have taken temperature independent

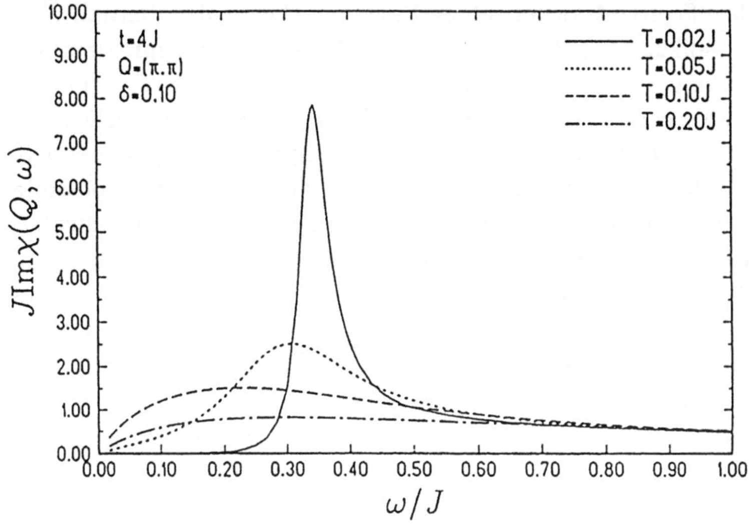

Fig.4 The ω-dependences of $\mathrm{Im}\chi(Q,\omega)$ for several choices of temperature in the case of $\delta = 0.15$, where $Q = (\pi, \pi)$.

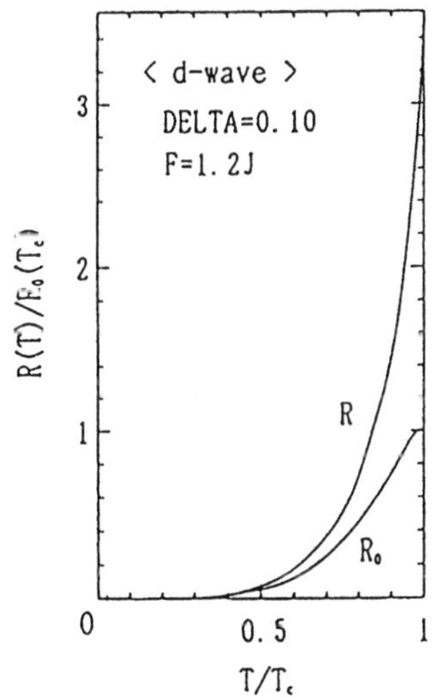

Fig.5 The NMR rate at Cu-site in the d-wave superconducting state for $\delta = 0.10$, where $R_0(T)$ and $R(T)$ are those without and with the exchange correction, respectively.

values of the effective transfer integral and the chemical potential, $F = 1.2J$, $\lambda_F = -0.22J$, based on our results of mean field approximation [16]). Note the very sharp reduction of the rate just below T_c, which is due to the enhancement of the rate above T_c by the exchange corrections characterized by $R(T_c)$ on one hand and its suppression by the superconducting order parameter below T_c on the other hand. These possible effects associated with the exchange enhancement have been indicated by Monien and Pines [26]. This feature of strong reduction of the rate is also in qualitative agreement with experiments [23,27]. Through the theoretical analysis we noted that the comparison of the NMR rate in the superconducting state of Cu and O sites, as has been actually performed by Hammel et al. [28] on $YBa_2Cu_3O_7$, can be a powerful tool to determine the distribution of the spin fluctuation spectral weight in the Brillouin zone [29]. This is the subject to be discussed in detail elsewhere [25].

Acknowledgement

The author thanks H. Yasuoka and H. Kohno for very useful and clarifying discussions.

References

1) For review, e.g. a) *Towards the Theoretical Understanding of High T_c Superconductors* eds. S. Lundqvist, E. Tosatti, M.P. Tosi and Yu Lu (World Scientific, 1988); b) *Strong Correlations and Superconductivity* eds. H. Fukuyama, S. Maekawa and A.P. Malozemoff (Springer, 1989); c) *The Los Alamos Symposium-1989 High Temperature Superconductivity* eds. K.S. Bedell, D.E. Meltzer, D. Pines and J.R. Schrieffer (Addison-Wesley, 1990).
2) N. Bulut, D.W. Hone, D.J. Scalapino and N.E. Bickers: Phys. Rev. **B41** (1990) 1797; Phys. Rev. Lett. **64** (1990) 2723.
3) A.J. Millis, H. Monien and D. Pines: Phys. Rev. **B42** (1990) 167.
4) T. Moriya, Y. Takahashi and K. Ueda: J. Phys. Soc. Jpn. **59** (1990) 2905.
5) For a review on experiments of NMR, C.H. Pennington and C.P. Slichter; in *Physical Properties of High Temperature Superconductors II* ed. D. M. Ginsberg (World Scientific, 1990) p.269.
6) H. Yasuoka, T. Imai and T. Shimizu: p.254 in ref. [1.b].

7) G. Shirane, R.J. Birgeneau, Y. Endoh, P. Gehring, M.A. Kastner, K. Kitazawa, H. Kojima, I. Tanaka, T.R. Thurston and K. Yamada: Phys. Rev. Lett. **63** (1989) 330.
8) G. Shirane: p.236 in ref. [1.b].
9) J. Rossat-Mignod, L.P. Regnault, C. Vettier, P. Burlet, J.Y. Henry and G. Lapertot: Physica **B 169** (1991) 58.
10) P.M. Gehring, J.M. Tranquada, G. Shirane, J.R.D. Copley, R.W. Erwin, M. Sato and S. Shamoto: to appear in Phys. Rev. B.
11) P.W. Anderson, Science **235** (1987) 1196.
12) F.C. Zhang and T.M. Rice: Phys. Rev. **B37** (1988) 3759.
13) H. Eskes and G.A. Sawatzky: Phys. Rev. Lett.**61** (1988) 1415; L.H. Tjeng, H. Eskes and G.A. Sawatzky: p.33 in ref. [1.b].
14) M.S. Hybertsen, E.B. Stechel, M. Schluter and D.R. Jennison: Phys. Rev. **B41** (1990) 11068.
15) H. Matsukawa and H. Fukuyama, J. Phys. Soc. Jpn. **58** (1989) 2845; ibid **58** (1989) 3687, **59** (1990) 1723; p.45 in ref.[1,b].
16) T. Tanamoto, K. Kuboki and H. Fukuyama, ISSP Tech. Rep. no.2421 [to appear in J. Phys. Soc. Jpn. **60** (1991) no.9].
17) I. Affleck and J.B. Marston: Phys. Rev. **B37** (1988) 3774; J.B. Marston and I. Affleck: Phys. Rev. **B39** (1989) 11538.
18) P. Lederer, D. Poilblanc and T.M. Rice: Phys. Rev. Lett. **63** (1989) 1519.
19) O. Narikiyo, K. Kuboki and H. Fukuyama; J. Phys. Soc. Jpn. **59** (1990) 2443.
20) N. Nagaosa and P.A. Lee, Phys. Rev. Lett. **64** (1990) 2450.
21) F. Mila and T.M. Rice: Phys. Rev. **B40** (1989) 11382; Physica **C157** (1989) 561.
22) K. Asayama, Y. Kitaoka, K. Fujiwara, K. Ishida and S. Ohsugi: Report in The 2nd ISSP International Symposium on Physics and Chemistry of Oxide Superconductors (Tokyo, 1991).
23) S. Ohsugi, Y. Kitaoka, K. Ishida and K. Asayama: J. Phys. Soc. Jpn. **60** (1991) 2351.
24) N. Bulut and D.J. Scalapino, preprint.
25) H. Kohno and H. Fukuyama, in preparation.
26) H. Monien and D. Pines, Phys. Rev.**B41** (1990) 6297.
27) T. Imai, T. Shimizu, H. Yasuoka, Y. Ueda and K. Kosuge, J. Phys. Soc. Jpn. **57** (1988) 2280.
28) P.C. Hammel, M. Takigawa, R.H. Heffner, Z. Fisk and K.C. Ott, Phys. Rev. Lett. **63** (1989) 1992.
29) This view has been also emphasized by H. Yasuoka [private communications].

Gauge Field Theory of the Normal State Properties of High-T_c Superconductors

N. Nagaosa

Department of Applied Physics, The University of Tokyo,
Bunkyo-ku, Tokyo 113, Japan

Abstract. The anomalous properties of the high-T_c superconductors in the normal phase are analyzed in terms of the uniform RVB picture where the spin and charge degrees of freedom are separated. The gauge field is introduced to retain the local gauge invariance which is related to the local constraint. The gauge flux is related to the spin chirality, and its fluctuation gives rise to the T-linear resistivity as well as the temperature dependent Hall constant. The backflow effect induced by the gauge field leads to the composition laws for various physical quantities. The crossover from the RVB state to the usual fermi liquid state is also discussed. The spinon pairing and the holon condensation occur simultaneously at the superconducting transition temperature.

1. Introduction

The unusual normal state properties of high-T_c superconductors have been extensively studied both experimentally and theoretically. The origin of these anomalous properties is believed to be the strong correlations between the electrons on the Cu d-orbital and the quasi-two dimensional structure of the system. At half filling, the system is a Mott insulator with spin $S = 1/2$ at each orbital, which is described by the antiferromagnetic (AF) Heisenberg Hamiltonian. When the holes are doped, the AF long range order diminishes rapidly and disappears. The Hamiltonian which describes the motion of the doped holes as well as the exchange interaction between the spins is called t-J model and is given by[1]

$$H = -\sum_{ij,\sigma} t_{ij}\tilde{C}^{\dagger}_{i,\sigma}\tilde{C}_{j,\sigma} + \sum_{<ij>} J\vec{S}_i \cdot \vec{S}_j, \tag{1}$$

with the constraint that no orbital is doubly occupied. This constraint is taken into account by expressing the electron operator $\tilde{C}_{i,\sigma}$ as $f_{i,\sigma}b^{\dagger}_i$ in the slave boson formalism. The fermion operator $f_{i,\sigma}$ annihilates the spinon with spin σ and the boson operator $b^{\dagger}_i$ creates the holon. Therefore the total number of the fermions and the bosons are conserved at each site i to be unity, which is expressed by $f^{\dagger}_{i,\uparrow}f_{i,\uparrow} + f^{\dagger}_{i,\downarrow}f_{i,\downarrow} + b^{\dagger}_i b_i = 1$ where the three terms on the lefthand side correspond to the three possible states for site i.

2. Uniform RVB State and its Physical Properties

There are many possible mean field theories to the Hamiltonian eq.(1) expressed in terms of $f, f^\dagger$ and $b, b^\dagger$. Compared with the experimental results, we proposed that the uniform RVB state[2],[3] is the most promising candidate. In this state, we have the nonzero uniform order parameters $\chi_{ij} =< f^\dagger_{i,\sigma} f_{j,\sigma} >$ and $B_{ij} =< b^\dagger_i b_j >$. The spinon pairing $\Delta_{ij} =< f_{i,\uparrow} f_{j,\downarrow} - f_{i,\downarrow} f_{j,\uparrow} >$ and the holon condensation $b_i =< b_i >$ is assumed to be zero in the normal state. The effective Lagrangian describing the low energy dynamics of this uniform RVB state is given in the continuum approximation by[2],[3]

$$L = \sum_\sigma \int \mathrm{d}^2\vec{r} \Big[f^*_\sigma(\vec{r})(\partial_\tau - a_0 - A_0 - \mu_F) f_\sigma(\vec{r}) + f^*_\sigma(\vec{r}) \frac{(i\nabla - \vec{a} - \vec{A})^2}{2m_F} f_\sigma(\vec{r}) \Big]$$
$$+ \int \mathrm{d}^2\vec{r} \Big[b^*(\vec{r})(\partial_\tau - a_0 - \mu_B) b(\vec{r}) + b^*(\vec{r}) \frac{(i\nabla - \vec{a})^2}{2m_B} b(\vec{r}) \Big], \qquad (2)$$

where the lattice constant is put to be unity. The energy is measured in units of J because $1/m_B$ and $1/m_F$ are considered to be of order J. The spinon fermi surface is large, i.e, it encloses the area $1 - x$ where x is the concentration of the doped holes, while the holon is distributed around the zero momentum with the characteristic momentum $\sim T^{1/2}$. The phase fluctuation of the order parameters constitutes the spatial components of the gauge field while the Lagrange multiplier field becomes its time component[4]. The spin chirality density is expressed as $\nabla \times \vec{a}$, whose average is zero in the mean field solution. Its fluctuation is described by the propagator of the transverse part of the gauge field $D^T = (\Pi_F + \Pi_B)^{-1} = (|\omega_n|/q + \chi_d q^2)^{-1}$, which is obtained by integrating out the fermion and boson degrees of freedom[2],[3]. Here Π_F and Π_B are the current-current correlation function of the fermions and bosons, respectively, and $\chi_d = \chi_F + \chi_B$ is the sum of the Landau diamagnetic susceptibilities of the fermions and bosons. This fluctuation of the spin chirality acts on the motion of the holons and spinons as the random magnetic field. The mean free time of the boson τ_B and fermion τ_F due to the scattering by the gauge field are given by $\tau_B^{-1} \cong T/m_B\chi_d$ and $\tau_F \cong (T/\chi_d)^{4/3} E_F$, respectively, in the lowest order with respect to the coupling constant[3]. Because the kinetic energy of the fermions are of order $E_F \sim J$ due to the fermi degeneracy, the coupling with the gauge field is always in the weak coupling regime, while that of boson is of the order of the temperature T which leads to the intermediate or strong coupling. The dimensionless coupling constant g for the boson is given by $g = 1/m_B\chi_d$, which is larger than or of the order of unity. This strong coupling effect is analyzed by the trajectory integral method[3] and the results are (a) the inverse of the mean free time τ_B^{-1} saturates as $\tau_B^{-1} \cong T$ for $g > 1$, (b) the linear response of the boson to the external magnetic field H is reduced by the factor $1/(1+g)$ because the characteristic area enclosed by the trajectory is reduced by the random interference due to the gauge field. We now discuss the physical quantities of the total system[2],[3]. The dynamics of the fermions and bosons are correlated by the constraint or equivalently through the coupling with the gauge field. This is expressed as $J_F + J_B = 0$. The fermion (boson) current $J_F(J_B)$ is $\Pi_F(\Pi_B)$ times the average gauge field $< a > +A$ $(< a >)$ which the fermions (bosons) feel. Putting this into $J_F + J_B = 0$, we obtain $< a >$ and the physical current J as $J = J_F = -J_B = \Pi A$ with $\Pi = \Pi_F\Pi_B/(\Pi_F + \Pi_B)$. From this, the conductivity σ of the total system is $\sigma_F\sigma_B/(\sigma_F + \sigma_B)$ which means that the resistivity ρ is the sum $\rho_B + \rho_F$. From the above discussion, $\rho_B >> \rho_F$ and $\rho \cong \rho_B \cong m_B T/x$, which remains T-linear irrespective of the possible temperature dependence of the diamagnetic susceptibility χ_d and the coupling constant g as long as $g > 1$. When the magnetic field is applied, the fermions (bosons) feel the magnetic field $\chi_B H/(\chi_F + \chi_B)$ $(-\chi_F H/(\chi_F + \chi_B))$ and the Hall constant R_H is given by $R_H = (\chi_F R_H^B + \chi_B R_H^F)/(\chi_F + \chi_B)$ where $R_H^B = 1/[(1+g)x]$[3]. R_H^F is the Hall constant for the fermion system which depends on the details of the band structure but its magnitude is expected to be smaller than x^{-1} and we approximateR_H by $R_H \cong x^{-1}\chi_F/(m_B^{-1} + \chi_F + \chi_B)$. Well above T_c, $\chi_F >> \chi_B$ and the T-dependence of χ_F, which is expected from the pairing fluctuation of the fermions, can explain the decreasing R_H as a function of T. Experimentally, the high temperature limit of R_H is reduced from x^{-1} by a factor of 2 to 4 depending on the material which can be regarded as $1 + g$.

The physical electron Green function $G_\sigma(\vec{k}, \Omega)$ is given by the convolution of the Green functions of the fermion and boson in the mean field approximation. Taking into account the interaction with the gauge field, we obtain the following results. The spectral function $\mathrm{Im}G_\sigma(\vec{k}, \Omega)$ consists of a continuum for $\Omega < -|\mu_B| - (|\vec{k}| - k_F)^2/2m_F$ and a peak at $\Omega = \vec{k}^2/2m_F - \mu_F$. The width of the peak is $(Tm_B)^{1/2}m_F^{-1}$, which comes from the momentum distribution of the bosons, or $\Omega^{2/3}J^{1/3}$ from the coupling with the gauge field, whichever is larger. This width violates the Landau criterion for the quasi-particle. On the other hand, the location of the peak in $\vec{k}$ space for small $|\Omega|$ is determined by the large "spinon" fermi surface which satisfies Luttinger's theorem.

3. Spinon Pairing and the Holon Condensation

We now turn to the effects of the fermion pairing and the bose condensation which have been assumed to be absent in the above considerations. The phenomenological Ginzburg-Landau(GL) functional for the free energy F is given by[5]

$$F = \frac{H_{\mathrm{cF}}^2}{8\pi}\int \mathrm{d}^2 r\left[2\xi_{\mathrm{F}}^2\left|\left(\nabla + i2\vec{a} + i2\vec{A}\right)\psi\right|^2 + 2\mathrm{sign}(T - T_D^{(0)})|\psi|^2 + |\psi|^4\right]$$

$$+\frac{H_{\mathrm{cB}}^2}{8\pi}\int \mathrm{d}^2 r\left[2\xi_{\mathrm{B}}^2\left|\left(\nabla + i\vec{a}\right)\phi\right|^2 + 2\mathrm{sign}(T - T_{BE}^{(0)})|\phi|^2 + |\phi|^4\right]$$

$$+\int\left[\frac{1}{8\pi e^2}\left(\nabla\times\vec{A}\right)^2 + \chi_F\left(\nabla\times(\vec{a}+\vec{A})\right)^2 + \chi_B\left(\nabla\times\vec{a}\right)^2\right]\mathrm{d}^2 r, \quad (3)$$

where ψ and ϕ are the normalized order parameters of the spinon pairing and the holon condensation, respectively. We consider the case of $A = 0$. When one of the order parameter, e.g. ψ, is fluctuating around 0, the integration of that order parameter does not give the mass to the gauge field, and the resulting effective free energy for the other order parameter is the usual GL free energy with the role of A being replaced by a. Then, the vortex excitation around which the phase of the order parameter changes 2π has the finite energy E_V because $\int|(\nabla\phi + ia)|^2 \mathrm{d}r^2$ is finite due to the screening by the gauge field ($a \sim 1/r$) while $\int|\nabla\phi|^2\mathrm{d}r^2$ is logarithmically divergent. Therefore, the average of the order parameter $<\phi>$ is always zero at finite temperature, i.e., there is no Kosterlitz-Thouless transition. The correlation length ξ is approximately the average distance between the vortices which is proportional to $\exp(E_V/2T)$. Therefore, a crossover occurs from the RVB state to the fermi liquid as ξ exceeds the thermal length of the fermion $\xi_F \sim hv_F/T$. When $\xi >> \xi_F$, the holon can be effectively regarded as condensed and the system is indistinguishable from the fermi liquid. Strictly speaking, however, $<\phi>= 0$ at any finite temperature as long as $<\psi>= 0$. On the other hand, if one of the order parameter remains nonzero, there occurs the Meissner efffect and the gauge field aquires the mass m. Then the above scenario breaks down because $\int m^2 a^2 \mathrm{d}r^2$ is logarithmically divergent when $a \sim 1/r$ for large r. Then the isolated vortex energy is logarithmically divergent, and the Kosterlitz-Thouless transition is possible. Summarizing the above discussion, the only possible phase transition is that both $<\psi>$ and $<\phi>$ becomes nonzero simultaneously at the superconducting transition temperature T_c.

In conclusion, we have discussed the uniform RVB state as the most promising candidate to describe the normal state properties of high-T_c superconductors because it can explain the unusual transport properties in consistent with the large fermi surface observed in angle-resolved photoemission experiments. The effects of the spinon pairing and the holon condensation has been also discussed using the phenomenological GL free energy functional. This analysis can be extended also to the superconducting properties of this spin-charge separated system and the results will be reported elsewhere.

The author thanks Profs. P.A.Lee, S.Sachdev, H.Fukuyama, M.Imada, S.Uchida, Y.Tokura and Drs. T.K. Ng, C.L.Kane and B.Shraiman for discussions and sending him the preprints.

References

[1] P.W.Anderson, Science 235, 1196(1987); P.W.Anderson and Z.Zou, Phys. Rev. Lett. 60, 132(1988).

[2] L.B.Ioffe and A.I.Larkin, Phys. Rev. B39, 8988 (1989); L.B.Ioffe and G.Kotliar, Phys. Rev. B42, 10360 (1990).

[3] N.Nagaosa and P.A.Lee, Phys. Rev. Lett. 64, 2450(1990) and Phys. Rev. B43, 1233(1991).

[4] G.Baskaran and P.W.Anderson Phys. Rev. B37, 580(1988); G.Baskaran, Phys. Scr. T27, 53(1989).

[5] N.Nagaosa and P.A.Lee, unpublished; S.Sachdev, unpublished.

Microscopic Theory of Luttinger Liquids in One Dimension

N. Kawakami[1] *and S.-K. Yang*[2]

[1]Yukawa Institute for Theoretical Physics, Kyoto University, Kyoto 606, Japan

[2]National Laboratory for High Energy Physics, Ibaraki 305, Japan

Abstract. We present the microscopic theory of Luttinger liquid for correlated electron systems in one dimension. Our description is based on conformal field theory and the Bethe-ansatz solution. Exact critical exponents for various correlation functions are obtained for two typical examples of the Luttinger liquid: the Hubbard model and the t-J model. We discuss the universal properties of the Luttinger liquid.

1. Introduction

Recent studies of the high-T_c superconductivity have attracted much interest in the non-Fermi liquid behavior in low-dimensional electron systems. Particularly, in one dimension (1D), the large quantum fluctuation gives rise to the characteristic behavior which is quite distinct from the Fermi-liquid nature. A decade ago, Haldane proposed the idea of Luttinger liquid to unify a large class of 1D quantum liquids [1]. It has recently been established that the microscopic foundation of Luttinger liquids is provided by $c = 1$ conformal field theory with $U(1)$ symmetry.

In this article, the microscopic theory of Luttinger liquids is presented for 1D correlated electron systems based on conformal field theory [2-4]. We discuss the low-energy behavior of various correlation functions and determine their critical exponents exactly for two typical examples of the Luttinger liquid: the Hubbard model [2,3,5] and the t-J model for $t = J$ [4]. We demonstrate that the universal behavior of the Luttinger liquid is characterized by the hyper scaling relations among the critical exponents. Each critical exponent, however, depends strongly on the model parameter. We shall see that the correlation exponents of the Hubbard model and the t-J model behave quite differently away from half filling although both models belong to the same universality class.

After introducing the Luttinger liquid in sect.2 we illustrate conformal field theory in sect.3. We obtain general expressions for the correlation exponents in sect.4, and discuss the characteristic properties of the Hubbard model and the t-J model in sect.5. Brief summary is given in sect.6.

2. Luttinger Liquid

The concept of Luttinger liquids applies to bose and spin systems as well [1]. In this article, we restrict our attention to electron systems. Let us first remind that bulk quantities in 1D correlated electron systems behave analogously to the Fermi liquid: the T-linear specific heat, the Pauli susceptibility, and so on. Therefore Luttinger-liquid properties may not be clearly observed in the bulk quantities.

In comparison with the Fermi liquid the most striking feature inherent in the Luttinger liquid appears as the power-law singularity in the momentum distribution near the Fermi point $k = k_F$,

$$< n_k > = < n_{k_F} > -\text{const.}|k - k_F|^\theta \text{sgn}(k - k_F) \tag{1}$$

which is characterized by the critical exponent θ. This reflects the fact that the low-energy excitation in the Luttinger liquid is not of the quasi-particle type, but of the collective type.

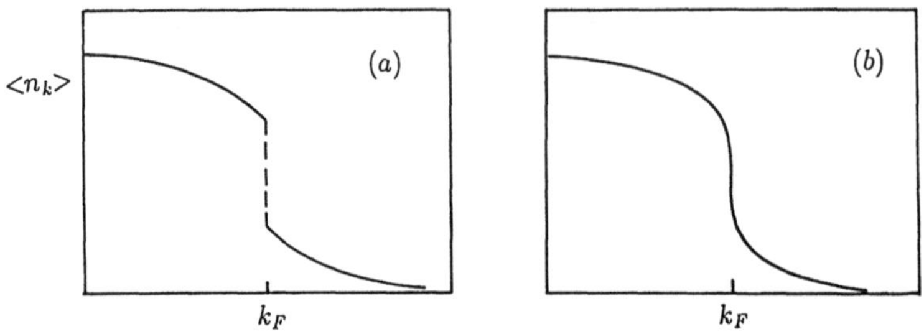

Fig. 1. Momentum distribution function for (a) Fermi liquid and (b) Luttinger liquid

By the same reason, correlation functions such as the density correlator also exhibit the power-law behavior in the low-energy region with anomalous critical exponents (see eq.(13)). These power-law behaviors characteristic of Luttinger liquid have been observed in the Tomonaga-Luttinger model which is classified as the weak-coupling model [6]. As for strongly correlated systems, the problem has not been settled for a long time. This is due to the notorious difficulty in calculating the correlation functions by the Bethe-ansatz method, although the Hubbard model, for instance, was solved by this method many years ago [7].

Recently many attempts to calculate the correlation functions for 1D electron systems have been made by means of the numerical calculation [8-10]. Among others the numerical-diagonalization approach with the Bethe-wave function discovered a strong evidence that the $U \to \infty$ Hubbard model is the Luttinger liquid [10]. The precise evaluation of correlation functions, however, has remained still difficult for arbitrary U. Moreover the microscopic description of the Luttinger liquid has been highly desirable.

Stimulated by these numerical works, several groups have successfully obtained the critical exponents in the Hubbard model [2,3,5] and the t-J model [4] last year, and found the universal properties of Luttinger liquid. At the same time the microscopic foundation of the Luttinger liquid has been established by conformal field theory.

3. Conformal Field Theory

Recent rapid progress in conformal field theory has enabled us to calculate the correlation functions exactly in two-dimensional critical phenomena [11]. Although there is no true phase transition in the 1D quantum systems, the critical behavior shows up at absolute zero since the correlation length becomes infinite at this point. As a result, correlation functions exhibit the power-law behavior at $T = 0$ instead of the exponential decay observed at finite temperatures.

It has been recognized that at criticality there appears conformal invariance (local-scale invariance) in addition to global-scale invariance. In particular the symmetry of conformal invariance plays a remarkable role in (1+1) space-time dimension since it contains the infinite number of transformation generators [11]. The generators L_n of conformal transformations form the infinite-dimensional Lee algebra (Virasoro algebra)

$$[L_m, L_n] = (m-n)L_{m+n} + \frac{c}{12}(m^3 - m)\delta_{m+n,0} , \tag{2}$$

where n and m are integers. On the right-hand side there exists the so-called central extension term with the central charge c. This number c turns out to specify the universality class of a given critical system. Roughly speaking, the value of c is proportional to the effective number of excitation modes. There is another important parameter in the theory, scaling dimension, which directly controls the power-law decay of the correlation function of the scaling field $\phi(r)$,

$$< \phi(r)\phi(r') > \sim |r - r'|^{-2x} . \tag{3}$$

Therefore in order to determine the correlation functions we have to calculate the scaling dimension x of scaling fields as well as the central charge c.

The most practical way to compute the above conformal parameters is the finite-size scaling analysis [12,13]. Let us consider the 1D quantum system of linear size L under periodic boundary conditions. We assume the one-component excitation whose velocity is denoted as v. It is known that the central charge c enters in the universal finite-size correction term in the ground-state energy

$$E_0 \sim \epsilon_0 L - \pi v c/(6L) , \tag{4}$$

while the scaling dimension appears in the excited-state energy

$$E - E_0 \sim 2\pi v(x+n)/L , \tag{5}$$

where the non-negative integers n coming from the particle-hole excitations feature the infinite conformal tower which forms the irreducible representation of the Virasoro algebra (2). The parameters c and x can be obtained if one knows the gap due to the finite-size effect in the spectrum of the hamiltonian at criticality.

We now ask whether conformal theory can immediately apply to 1D electron systems. It is well known that 1D electron systems carry low-energy charge and spin excitations with linear dispersion relations, but with different velocities. In such a case, generically, the system is not Lorentz invariant, and the applicability of conformal field theory is not obvious. If these excitations are decoupled in the low-energy regime, however, we can still apply the conformal technique. As we will see in a moment, this is indeed the case for the Hubbard model and the t-J model, where the charge and spin degrees of freedom are separated in the continuum limit. By analyzing the exact spectrum obtained from the Bethe-ansatz solution, therefore, we can describe the universal behavior of the 1D electron systems microscopically on the basis of conformal field theory [2-4].

4. Critical Exponents of Correlation Functions

Following the idea mentioned above, we perform the finite-size scaling analysis of the energy spectrum employing the Bethe-ansatz solution of the Hubbard model and the t-J model at $t = J$. Before dwelling upon each model, we would like to summarize the results of the finite-size corrections in general formulas so as to apply them straightforwardly to any 1D correlated metallic system.

The finite-size corrections to the ground-state energy read

$$E_0 \sim \varepsilon_0 N_a - \frac{\pi v_c}{6N_a} - \frac{\pi v_s}{6N_a} , \tag{6}$$

where N_a is the number of the lattice sites, and v_c and v_s are charge and spin velocities of the elementary excitations, respectively [2-4,14]. Referring to (4) we see that both of the charge and spin sectors are described by $c = 1$ conformal field theory. On the other hand the corrections to excitations are classified as

$$E - E_0 \sim \frac{2\pi v_c}{N_a}(x_c + n_c^+ + n_c^-) + \frac{2\pi v_s}{N_a}(x_s + n_s^+ + n_s^-) , \tag{7}$$

with the scaling dimensions

$$x_c = \frac{N_c^2}{2\alpha_c} + \frac{\alpha_c}{2}(D_c + \frac{D_s}{2})^2, \qquad x_s = \frac{1}{4}\Big(N_s - \frac{N_c}{2}\Big)^2 + D_s^2 , \tag{8}$$

where non-negative integers $n_c^\pm$ and $n_s^\pm$ counting the number of particle-hole excitations represent the structure of the infinite conformal tower for charge and spin sectors. Here we have denoted the change of the number of electrons (down spin electrons) as N_c (N_s), and the number of particles moving from left "fermi point" to right one as D_c (D_s) for charge (spin) excitations, respectively. This excitation carries the momentum

$$P = 4k_F D_c + 2k_F D_s + \frac{2\pi}{N_a}\sum_{\alpha=c,s}(N_\alpha D_\alpha + n_\alpha^+ - n_\alpha^-) . \tag{9}$$

We remark that the quantum numbers introduced are subject to the constraint $D_c = (N_c + N_s)/2 \mod 1$ and $D_s = N_c/2 \mod 1$.

We emphasize that the scaling dimensions in (8) are taking the typical form of the $c = 1$ conformal field theory with $U(1)$ symmetry in the charge sector and $SU(2)$ symmetry in the spin sector [15]. There is only one parameter α_c in the scaling dimensions, which will be identified below as the $4k_F$ exponent for the charge density correlator. This fact leads to the fundamental universal relations in the Luttinger liquid.

We now consider the following equal-time correlation functions:

(a) charge and spin density correlators

$$N(r) =< n(r)n(0) >, \quad \chi(r) =< S_z(r)S_z(0) >, \tag{10}$$

(b) singlet and triplet pair correlators

$$\begin{aligned} P_s(r) &=< c_\uparrow^\dagger(r+1)c_\downarrow^\dagger(r)c_\uparrow(1)c_\downarrow(0) > , \\ P_t(r) &=< c_\uparrow^\dagger(r+1)c_\uparrow^\dagger(r)c_\uparrow(1)c_\uparrow(0) > , \end{aligned} \tag{11}$$

(c) electron correlator

$$G_\sigma(r) =< c_\sigma^\dagger(r)c_\sigma(0) > . \tag{12}$$

Let us start with the charge density correlation function. We write down the long-distance asymptotic form of the correlator,

$$N(r) \sim \text{const.} + A_0 r^{-2} + A_2 r^{-\alpha_s}\cos 2k_F r + A_4 r^{-\alpha_c}\cos 4k_F r . \tag{13}$$

The critical exponents α_s and α_c are read off from (8) by identifying the quantum numbers N_α and D_α for the operator in the correlation function. First we verify that the $4k_F$ oscillation exponent is given by α_c in (8). The quantum numbers N_c and N_s for $N(r)$ should be taken zero since the operator $n(r)$ does not change the electron number. The quantum numbers are thus assigned as $(N_c, N_s, D_c, D_s) = (0,0,1,0)$, from which we verify that the momentum $P = 4k_F$ and the corresponding exponent is given by α_c. Similarly the $2k_F$ oscillation piece is shown to come from $(N_c, N_s, D_c, D_s) = (0,0,0,\pm 1)$. We thus find the relation between these two exponents [2-5],

$$\alpha_s = 1 + \alpha_c/4 \ . \tag{14}$$

The spin correlation function $\chi(r)$ has the same form as (13) except that the $4k_F$ part is absent. The critical exponent for the $2k_F$ part is equal to α_s of the charge density correlation. The leading uniform term in the singlet and triplet pair correlation functions are given by $(N_c, N_s, D_c, D_s) = (2,1,1/2,-1)$ and $(2,2,0,0)$,

$$P_s(r) = P_t(r) \sim r^{-\beta}, \qquad \beta = 1 + 4/\alpha_c \ . \tag{15}$$

The long-distance behavior of the electron correlator has the leading k_F oscillation term $G_\sigma(r,0) \sim r^{-\eta} \cos k_F r$, which is controlled by $(N_c, N_s, D_c, D_s) = (1,1,0,1/2)$. The Fourier transform of this function yields the momentum distribution function (1). One obtains the typical power-law singularity of Luttinger liquids with the critical exponent

$$\theta = (\alpha_c - 4)^2/(16\alpha_c) \ . \tag{16}$$

This completes our derivation of the correlation exponents based on the finite-size scaling analysis.

Finally we remark that the expressions (14)-(16) for the critical exponents as well as the finite-size scaling formula (6)-(9) apply generally to any 1D metallic system, although we have derived through the analysis of the Bethe Ansatz solution of the Hubbard model and t-J models. Below we calculate α_c in (8) and get the correlation exponents for the Hubbard model and t-J model explicitly as functions of the electron concentration and the interaction strength.

5. Luttinger Liquid Properties

We discuss two examples of correlated electron systems which are solved by the Bethe-ansatz. The hamiltonian for the Hubbard model is well known,

$$\mathcal{H} = -t \sum_{<ij>,\sigma} C_{i\sigma}^\dagger C_{j\sigma} + U \sum_i n_{i\uparrow} n_{i\downarrow} \ . \tag{17}$$

Another important model is the t-J model,

$$\mathcal{H} = -t \sum_{<ij>,\sigma} C_{i\sigma}^\dagger C_{j\sigma} + J \sum_i (\vec{S}_i \cdot \vec{S}_{i+1} - \frac{1}{4} n_i n_{i+1}) \ , \tag{18}$$

with $J > 0$, where the double occupancy of every site is forbidden. Formally, this model is derived from the strong correlation limit ($U \gg t$) of the Hubbard model but with the limitation $J \ll t$. In the t-J model, however, we regard t and J just as two free parameters (see T. M. Rice in this proceedings). We deal with the 1D t-J model for $t = J$ since it can be diagonalized right at this point by the Bethe ansatz method [16,17]. We are thus examining the t-J model with an extremely large coupling J.

5.1 Hubbard Model

Let us begin with the correlation functions of the Hubbard model. All the critical exponents are calculated exactly by the finite-size scaling analysis based on the Bethe ansatz solution [2,3,5]. In Fig. 2(a) the $4k_F$ CDW exponent α_c is shown as functions of the electron concentration ν ($\nu = 1/2$ at half filling), from which other correlation exponents can be readily obtained through the scaling relations (14)-(16). Strong dependence of α_c on the Coulomb interaction as well as the electron filling is clearly observed. As $U \to \infty$, α_c approaches 2 in agreement with the result for the spinless fermion. In the opposite limit $U \to 0$, α_c converges to 4 for the electron concentration $0 < \nu < 1/2$, which is consistent with the result of the Tomonaga-Luttinger model [6]. It should be noticed that α_c takes the value close to 2 near half filling as long as the Coulomb interaction exists. This is due to the fact that at half filling the system is Mott insulator for any strength of the Coulomb interaction and a doped hole behaves as if it is a spinless hole in the $U \to \infty$ limit.

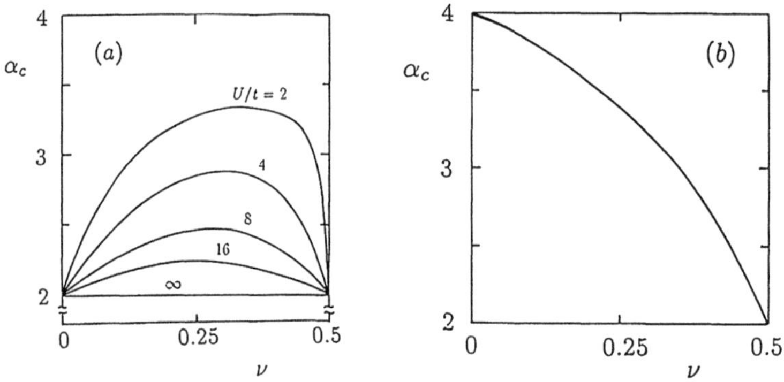

Fig. 2. Critical exponents for the $4k_F$ oscillation exponent α_c for (a) the Hubbard model and (b) the t-J model

From α_c one can evaluate the superconducting correlation exponents β and the Luttinger anomaly exponent θ for the momentum distribution. We see that correlation exponents for the singlet and triplet pair correlations range $2 \le \beta \le 3$, while θ takes the values $0 \le \theta \le 1/8$. Note that $\theta = 1/8$ is realized as $U \to \infty$ [10]. In this case all the exponents listed above do not depend on the electron concentration, which is sharply contrasted to the t-J model.

5.2 t-J Model

Let us compare these results with those of the t-J model. When J is small, the t-J model is reduced to the $U \to \infty$ Hubbard model. Hence the comparison of the t-J model ($t = J$) with the $U \to \infty$ Hubbard model may clarify the effect of the large antiferromagnetic coupling J. In Fig. 2(b) we show the concentration dependence of the $4k_F$ correlation exponent for the t-J model at $t = J$. In the vicinity of half filling all the exponents of the t-J model take the values expected in the strong correlation limit of the Hubbard model: $\alpha_c = 2, \alpha_s = 3/2, \beta = 3, \theta = 1/8$. This is because the exclusion of the double occupation gives the most dominant effect near half filling in the t-J model, which makes the motion of doped holes behave like spinless fermions as in the highly correlated Hubbard model. In the $U \rightarrow \infty$ Hubbard model, $\alpha_c(= 2)$ stays constant as ν decreases from half filling, and hence $\alpha_s = 3/2$, $\beta = 3$, $\theta = 1/8$ for any filling [2,3,5,10]. On the other hand, in the t-J model the critical exponents take the values for the non-interacting system such as $\alpha_c(= 4)$ in the low-density limit $\nu \rightarrow 0$.

This non-interacting behavior of the t-J model in the low-density limit seems peculiar since the model is originally supposed to have the strong correlation. Our result implies that the hole motion in the t-J model is not like spinless fermions for large hole-doping, but is considerably influenced by the spin fluctuation through the strong antiferromagnetic coupling J. We think that the large antiferromagnetic coupling favors the antiparallel-spin electron pairs to sit on the nearest-neighbor sites, which renders the hole motion quite different from spinless particles. In the superconducting correlations we see that the large spin coupling as well as the hole doping in the t-J model play a conspicuous role to enhance the superconducting correlation in comparison with the highly correlated Hubbard model. In spite of this fact, however, the spin correlation always dominates the superconducting correlations for arbitrary electron filling both for the Hubbard model and the t-J model. For arbitrary J the numerical calculations have been performed. See ref. [18] for exact diagonalization of the finite system and ref. [19] for simulations.

5.3 Universal Relations

Although the critical exponents depend strongly on the microscopic model parameter, the hyper-scaling relations (14)-(16) are valid generically for 1D correlated electron systems and then characterize the Luttinger-liquid properties. In order to obtain these universal relations, it has been crucial that the charge and spin degrees of freedom are separated and described by two independent $c = 1$ conformal field theories. Another universal aspect of the Luttinger liquid is that the critical exponents are related exactly to the bulk quantities. In terms of the compressibility χ_c and the specific-heat coefficient γ_c coming from the charge sector, we find [2-5]

$$\alpha_c = 2\pi^2 \chi_c / (3\gamma_c) \, . \tag{19}$$

Using the scaling relations (14)-(16) all the critical exponents can be expressed in terms of these these bulk quantities. The universal formula (19) also characterizes the Luttinger liquid.

There are many interesting correlated electron systems which cannot be diagonalized by the Bethe ansatz method. Even in such non-integrable systems, by computing the bulk quantities numerically it may be possible to evaluate the critical exponents very accurately using the universal relation (14)-(16) and the formula (19).

6. Transport Properties

Another interesting quantity characterizing the Luttinger liquid is the conductivity. Kohn pointed out many years ago that the low-frequency conductivity is directly related to the shift of the energy levels due to twisted boundary conditions [20]. The important point is that using the pseudo-gauge transformation the ring system threaded by the magnetic field is mapped onto the system without the external field. The effect of the vector potential is then absorbed into twisting the boundary conditions. Therefore from the energy change under twisted boundary conditions, it is possible to evaluate the conductivity in principle. This has been demonstrated recently using the Bethe-solvable spin systems [21], and subsequently the conductivity for the Hubbard model and the t-J model has been evaluated exactly [22].

Under twisted boundary conditions with the twisting phase ϕ, the shift of the ground-state energy from the periodic case (ϕ=0) is

$$E_0(\phi) - E_0(0) = \mathcal{D}_c \phi^2/N + O(\phi^2) \; . \tag{20}$$

The charge stiffness $\mathcal{D}_c$ is directly related to the dc part of the conductivity $\sigma(\omega)$

$$\mathrm{Re}\; \sigma(\omega) = (2\pi e^2/\hbar)\mathcal{D}_c \; \delta(\hbar\omega) \; . \tag{21}$$

The finite-size corrections under twisted boundary conditions can be computed in the same way as in section 4. Consequently we get the exact expressions for the conductivity

$$\mathcal{D}_c = \alpha_c v_c/(4\pi) = \pi^2 \chi_c/(18\gamma_c^2) \; , \tag{22}$$

where we have used (19). In Fig.3 we show the results for the conductivity both for the Hubbard model and the t-J model. We readily see that the conductivity is suppressed as the system approaches half filling. This indicates that the system becomes closer to the insulating state. See refs. [5,22] for detailed discussion. To conclude this section we stress that formula obtained for the conductivity is again valid for any 1D correlated electron system, and hence characterizes the universal properties of Luttinger liquids.

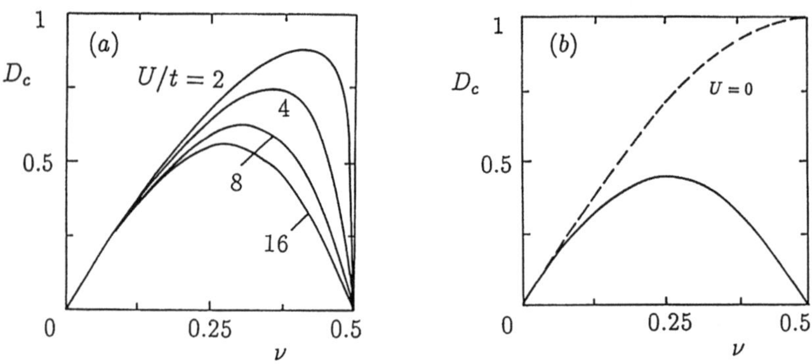

Fig.3. Conductivity for (a) the Hubbard model and (b) the t-J model

7. Summary

We have shown that the low-energy critical behavior of the Hubbard model and the t-J model is described as the Luttinger liquid. In order that the electron systems be identified as the Luttinger liquid, it is quite crucial that the the charge and spin degrees of freedom are decoupled in the low-energy region, and both sectors are described independently by the $c = 1$ conformal field theory. In conclusion we have presented the microscopic foundation of the Luttinger liquid for 1D correlated electron systems on the basis of conformal field theory and Bethe-ansatz solution.

References

[1] F.D.M. Haldane, Phys. Rev. Lett. **45** (1980) 1358; J. Phys. **C14** (1981) 2585; Phys. Lett. **81A** (1981) 153

[2] N. Kawakami and S.-K. Yang, Phys. Lett. **148A** (1990) 359

[3] H. Frahm and V.E. Korepin, Phys Rev. **B42** (1990) 10553

[4] N. Kawakami and S.-K. Yang, Phys. Rev. Lett. **65** (1990) 2309; J. Phys. Condensed Matter, *in press*

[5] H.J. Schulz, Phys. Rev. Lett. **64** (1990) 2831

[6] See, for review, J. Solyom, Adv. Phys. **28** (1979) 201

[7] E.H. Lieb and F.Y. Wu, Phys. Rev. Lett **20**(1968) 1445

[8] S. Sorella, E. Tosatti, S. Baroni, R. Car and M. Parrinello, Int. J. Mod. Phys. **B2** (1988) 993

[9] M. Imada and Y. Hatsugai, J. Phys. Soc. Jpn. **58** (1989) 3752

[10] M. Ogata and H. Shiba, Phys. Rev. **B41** (1990) 2326;
M. Ogata, T. Sugiyama and H. Shiba, Phys. Rev. **B43**(1991) 8401

[11] A.A. Belavin, A.M. Polyakov and A.B. Zamolodchikov, Nucl. Phys. **B241** (1984) 333
See for a collection of reprints, *Conformal Invariance and Applications in Statistical Mechanics*, edited by C. Itzykson, H. Saleur and J.-B. Zuber (World Scientific, Singapore, 1988)

[12] H.W.J. Blöte, J.L. Cardy and M.P. Nightingale, Phys. Rev. Lett. **56** (1986) 742;
I. Affleck, Phys. Rev. Lett. **56** (1986) 746

[13] J.L. Cardy, Nucl. Phys. **B270** [FS16] (1986) 186

[14] F. Woynarovich, J. Phys. **A22** (1989) 4243

[15] L.P. Kadanoff and A.C. Brown, Ann. Phys. **121** (1979) 318;
S.-K. Yang, Nucl. Phys. **B285** [FS19] (1987) 183

[16] B. Sutherland, Phys. Rev. **B12** (1975) 3795

[17] P. Schlottmann, Phys. Rev. **B36** (1987) 5177

[18] M. Ogata, M. Luchini, S. Sorella and F. F. Assad, Phys. Rev. Lett. **66** (1991) 2388; T. Pruschke and H. Shiba, Phys. Rev. **B** to appear

[19] M. Imada, J. Phys. Soc. Jpn. **59** (1990) 4121

[20] W. Kohn, Phys. Rev. **133** (1964) A171

[21] B.S. Shastry and B. Sutherland, Phys. Rev. Lett. **65** (1990) 243

[22] N. Kawakami and S.-K. Yang, Phys. Rev. Lett. **65** (1990) 3063; Kyoto preprint YITP/K-918(1991)

Generalization of BCS Superconductivity to Non-Phonon Mediated Interactions: The Excitonic Interaction

W.A. Little

Physics Department, Stanford University, Stanford, CA 94305, USA

Abstract. It has long been recognised that, in principle, superconductivity can be induced by mechanisms other than the phonon mechanism. This follows from a simple generalization of the BCS theory, or better the Eliashberg version of that theory, that makes it clear that certain electronic excitations can provide such an attractive interaction. Because of the much higher energies of such excitations, superconductivity at much higher temperatures than those of phonon-mediated superconductors, should be possible. This raises the question whether this mechanism might be responsible for the high-transition temperatures of the recently discovered cuprate superconductors. In considering this question, we discuss the similarities and differences between the phonon mechanism, and such a non-phonon, exciton mechanism, the effects of these upon the superconducting transition temperature, and their effects upon the nature and symmetry of the superconducting state. We discuss the characteristic signature of the non-phonon interaction and review the experimental evidence for and against a possible role such a non-phonon mechanism might play in the superconductivity of the cuprates.

1. Introduction

In the few years immediately after the introduction of the BCS theory [1], the scientific community had come to understand why superconductivity occurred, at that time, exclusively at low temperatures. The theory showed that the transition temperature should be an exponentially small fraction of the Debye energy. This, indeed, was found to be the case. Moreover, the coupling constant λ, which occurred as $-1/\lambda$ in the exponential, and which determines T_c, was believed also to be limited in magnitude to a value $<1/2$. These two factors conspired to delude many into believing that high temperature superconductivity would be for ever impossible, or worse, that the search for it would become the opium of the masses [2]. However, in 1964, in a paper on the possibility of synthesizing an *organic* superconductor [3], the present author showed that high temperature superconductivity should be possible within the BCS model by utilizing a mechanism in which the virtual excitation of electronic states would replace the phonon mechanism. This work attracted a great deal of attention, particularly in the Soviet Union, and led Ginzburg [4] to propose a two-dimensional analogue to the one-dimensional, polymeric compound we had suggested, which again would use such an electronic excitation to provide the attractive mechanism for the superconductivity. Over the past two decades, both Ginzburg's group in Moscow and our's at Stanford have studied model systems in which this mechanism might be operative, and have

Springer Series in Solid-State Sciences, Vol. 106 Physics of High-Temperature Superconductors
Eds.: S. Maekawa and M. Sato

studied the many factors that might influence, or interfere with the effectiveness of this interaction. While no superconductors are presently known in which the excitonic mechanism, as this mechanism has been named, is known to be operable for sure, it is one model of a high temperature superconductor that has been scrutinized to a depth and detail that is far beyond that of any other. In light of this it is useful to examine the possible role the exciton mechanism might have in explaining the recent remarkable discoveries of superconductivity at high temperatures in the cuprates [5,6].

Several reviews are available that provide a general overview and reference to the large number of papers on the subject, many of which are in the Soviet literature. A brief review is included in Novel Superconductivity, edited by Wolf & Kresin [7] but perhaps the most comprehensive is one published in 1979 by H. Gutfreund and myself [8]. Since then there have been a few key papers that have focussed on previously unresolved issues [9,10] and one, particularly instructive paper on the stability of such interacting systems [11]. Our colleagues in theoretical physics in the Soviet Union have played a major role in the resolution of many of the problems in this field and contributions of Ginzburg, Gorkov, Migdal, Dzyaloshinskii, Kresin, Kirzhnits, Khomski to name a few, permeate the field and have been crucial to the development of the subject.

The possibility of obtaining superconductivity at temperatures approaching ambient by utilizing an electronic excitation rather than a phonon mechanism was first raised in 1964 and while it is now generally accepted that the replacement of the Debye energy, $h\omega$ in the weak coupling BCS expression for T_c, $kT_c = 1.14\ h\omega \exp(-1/\lambda)$, by an electronic transition of much higher energy, could lead to high T_c, this was by no means accepted in the first several years after its proposal. The key point which is now agreed upon is that if one ignores the Coulomb interaction then the coupling constant λ which appears in this expression for T_c does not contain the mass of the particles that mediate the interaction. The only factor which does contain this mass, m is the the prefactor $h\omega$, which is proportional to $m^{-1/2}$ and it is this which gives rise to the Isotope Effect. If a sufficient attraction among the conduction electrons can be obtained by the virtual movement of some other sub-set of electrons, rather than ions then substantially higher values of T_c should be obtainable. Such an interaction has been called the "exciton" interaction in reference to the electronic excitations that are involved. It should be noted that it does not imply the use of the more limited concept of excitons described by Mott-Wannier or Frenkel excitons, but simply that the interaction is mediated by the excitation of some sub-set of electrons.

2. Role of Limited Dimensionality

Much of the early work was dominated by concern over the role fluctuations and limited dimensionality would have on the interaction. This arose because the original proposal had been based on the concept of designing a polymeric system with certain electronically polarizable side-chains to provide the electronically mediated attraction. This was the first time that superconductivity in a system of limited dimensionality had been considered. It was soon shown that the arguments which prohibited classical phase transitions in one-dimensional (1-D) systems should apply equally well to the quantum phase transition to the superconducting state [12,13]. As a result interest shifted rapidly to the possibility of realizing the same attractive mechanism in a 2-D system, where fluctuations would tend to play a less crucial role. This was the sandwich structure proposed by Ginzburg. It was followed several years later by the proposal of a more specific example of a 2-D, metal-semiconductor sandwich by Allender, Bray and Bardeen [14] in 1972.

While fluctuations can be of great importance in isolated systems of low dimensionality and do provide the correct explanation of the decay of persistent currents in small samples [15,16], in the real world, such 1-D or 2-D systems have weak interactions with other chains or planes adjacent to them and this greatly mitigates the effects of the limited dimensionality. We recognize now that such fluctuations were something of a red herring. This view was confirmed by the discovery by Gamble et al. [17] of 2-D superconductivity in TaS_2 and intercalated derivatives of this is 1970, and the even more remarkable discovery of superconductivity in the quasi-1-D, polymeric system $(SN)_x$ containing no metal atoms, by R. L. Greene et al. [18] in our laboratories in 1975.

The exciton interaction raised a number of other problems in addition to those associated with the role of dimensionality and these, too must be considered.

3. Factors of Unique Importance to the Excitonic Interaction

While the exciton mechanism is similar in many respects to the phonon mechanism, the two mechanisms are by no means identical. In the first place, the electron is much lighter than the ions of the lattice, and this has important implications in regard to the applicability of perturbation theory used to determine the interaction. Second, the electrons involved in the conduction process are, of course, identical to those involved in the excitations which are responsible for the interaction. In computing their interaction, not only must the direct interaction between them be taken in to account as for the electron-ion interaction, but also the exchange interaction. This adds complexity and has the effect usually of weakening the interaction. We consider these in turn. Thirdly, because the phonons and the excitons are different one must consider their mutual interaction and possible interference.

3.1 Vertex Corrections

Of particular importance is the question whether the interaction with any electronic degrees of freedom can be described adequately in lowest order of perturbation theory. This is possible for the phonon interaction because of the, so called, Migdal approximation [19], which shows that the vertex corrections such as those of Fig. 1 are proportional to $(m/M)^{1/2}$, where m is the mass of the electron, and M the mass of the ions. The question is, if one replaces the phonons by an electronic excitation so M is replaced by some mass of order m, are these corrections then of order unity i.e. $(m/m)^{1/2}$? We have examined this point [20,8] in detail and have shown that if the two electron subsystems are physically separated in space so that the exciton coupling term is strongly momentum dependent then the magnitude of any vertex corrections are strongly reduced. If, on the other hand, the exciton and conduction systems co-exist in the same region of space then this is not necessarily so and one should view with caution results obtained using the simple electron-phonon vertex.

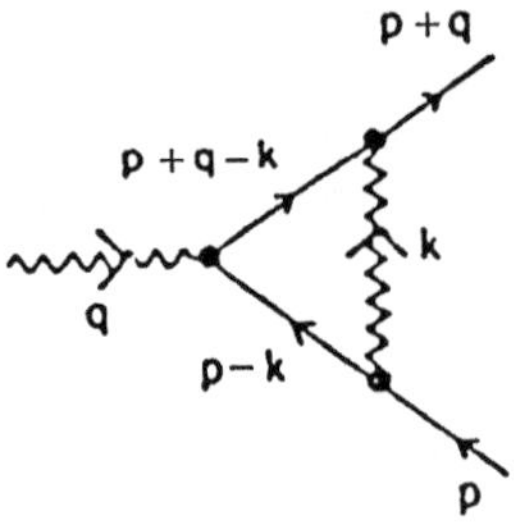

Figure 1. First vertex correction beyond the Migdal Approximation

More recently Cai et al. [21] have studied this point in great detail and, in particular, the effect of vertex corrections on the value of the transition temperature for electron or exciton coupled systems. They conclude that for moderately strongly coupled systems with $\lambda = 1$ to 2 the value of T_c is reduced by no more than 20% from the value obtained without taking into account such vertex corrections. For weak coupling systems ($\lambda = 0.4$) the effects are somewhat larger but by no means catastrophic.

3.2 Exchange

The exciton interaction differs from the phonon interaction in another, fundamental way [22]. The electrons and the ions which participate in the phonon modes are distinguishable. This is not so for the case of the electron-exciton interaction. The conduction electrons and the exciton electrons are identical and consequently in computing their interaction, exchange must be considered. Such matrix elements are given by an expression of the form $2J_{ik} - K_{ik}$, where J_{ik} is the Coulomb interaction between the conduction and exciton electrons and K_{ik} is the exchange term. In general, if both i and k are drawn from the same set of orbitals one finds that K_{ik} and J_{ik} are of the same order of magnitude. So the neglect of K_{ik} would overestimate the total element by as much as a factor of 2. This element appears quadratically in the expression for the effective interaction and consequently a gross error is committed if exchange is not treated with reasonable care. If, once more, the conduction and exciton systems are physically separated in space, as we have considered in several models [20, 8] K_{ik} vanishes and one obtains the simple Coulomb term. On the other hand in calculations of the plasmon-mediated interaction [23, 24] where the plasmon and the electrons coexist in the same region of space, the exchange term could be of major importance.

3.3 Stability

A number of authors have argued that high temperature superconductivity via the excitonic interation is impossible because the electron-lattice system would become unstable for an interaction sufficiently strong to yield such high transition temperatures. This argument is based on the examination of the electron-electron interaction, which can be described in terms of the dielectric function, $\varepsilon(q, \omega)$:
$V(q,\omega) = 4\pi e^2/(q^2 \varepsilon(q,\omega))$, where $\varepsilon(q, 0)$ is the dielectric function resulting from the phonon and all electronic excitations combined. This expression can be decomposed

into its phonon and electron components and the interaction then averaged over the Fermi surface to give: $N(0)V = \mu - \lambda$, where μ is the contribution to the net interaction due to the Coulomb repulsion. It was argued [25, 26] that $\varepsilon(q, 0)$ would have to be positive in order to prevent an instability in the conduction electron gas and this would require $\mu - \lambda$ to be positive and thus $\lambda < \mu$, prohibiting superconductivity. This general conclusion is incorrect for several reasons. This has been shown in a comprehensive article on the subject of the sign of the static dielectric function by Dolgov, Kirzhnits, and E. G. Maksimov [11].

First, if one takes the above argument at face value, the above would apply only if the electron gas were to be uniformly distributed in the unit cell. In general, this is not the case and therefor, the local fields can be very different from the average field. This is not the situation which was considered in [25, 26]. The effect of these local fields is to require the inclusion of Umklapp processes in the calculation, and these lift the above restriction.

Second, the energy asociated with the exciton excitations generally will be different from that of the Fermi energy. This allows one to transform away some of the Coulomb term μ to yield an effective Coulomb pseudopotential, μ^* over the same range of energies as encompassed by the exciton excitations, μ^* is then given by $\mu^* = \mu(1 + \mu \ln(E_f/\omega))^{-1}$, where, as for the phonon case, ω represents the excitation energy and E_F the Fermi energy. For the phonon case the logarithmic factor is typically of order 5 but for excitons it can be of order 1 or 2, in which case a significant reduction in μ will still occur, allowing negative values for $\lambda - \mu^*$.

Third, as was shown by Ginzburg [27] and Kirzhnits et al [28], and we will show shortly, it is not, in fact, $V(q,\omega)$ which enters the integral equation for the BCS gap function but a better behaved, relative of V, $U(q,\omega)$, which is subject to different constraints and, consequently, the simple arguments on the nature of the bare interaction, should not be applied directly to the interaction term in the gap equation.

Fourth, as Dolgov et al. have stressed [11], arguments in regard to causality and stability, such as used in the above argument, cannot be derived via Kramers-Kronig relations for the dielectric function except for $\varepsilon(q, 0)$ with $q = 0$. It is the reciprocal of the dielectric function $1/\varepsilon(q, 0)$, which is the proper response function to consider, and the restrictions imposed upon it do not prohibit negative values for $\varepsilon(q, 0)$, in general. Furthermore, they show from the above considerations that the condition for stability of an interacting system based on the total energy of a polarized system, $\varepsilon E^2/8\pi$ being positive, cannot be used to conclude that ε itself must be positive.

Finally, there is ample evidence in the case of phonons that $\lambda - \mu^*$ can be substantially greater than zero, ranging up to values of 2.7 or more in Pb and Bi and their alloys [29], and almost as high for $\lambda - \mu$.

3.4 Effects of Phonons

Phonons can, in principle, interfere with the exciton interaction in two different ways. First, by contributing through virtual phonons, a repulsive part in the region where the excitonic interaction is attractive and thus nullifying the effect of the excitonic attraction, and second, by the depairing effect of real phonons which might disrupt the ordered state created by the excitonic interaction. In fact, as we shall show, neither of these factors are significant.

Virtual Phonons. In any real physical system the exciton system would coexist with the phonon degrees of freedom. It had been argued that the energy region where the excitons would act, the virtual phonons would generate a strongly repulsive term which would nullify any contribution from the excitons. This was based on the use of the Bohm-Pines expression for the electron-electron interaction,

$$V(q,\omega) = \frac{g^2 E(q)}{\omega^2 - E^2(q)}$$

which becomes positive for values of ω larger than the phonon energies E(q). That such repulsion does not occur was shown by Ginzburg [27] and Kirzhnits, Maksimov and Khomskii [28]. They showed that if one transforms the Eliashberg equations to the form of the BCS gap equation, it is not V(q,ω) which appears in the integral equation but a smooth function U(ξ,ξ′) involving the energies ξ and ξ' of the electrons, where

$$U(q,\omega) = \frac{-g^2}{E(q) + |\xi| + |\xi'|}$$

This gives an attractive interaction for all excitations, regardless of their energy. This has been confirmed by a rigorous proof by Bergmann and Rainer [30] given in 1973, that all phonon (or exciton) modes contribute to an increase in T_c regardless of their energy. In 1984 we closed a subtle loophole in the argument [31] and the subject can now be considered settled.

Real Phonons. It had also been argued [32] that just as for the Mössbauer effect, the coherent superconducting state would be destroyed by real phonons and this would prevent the occurrence of high T_c superconductivity for temperatures approaching the Debye temperature. Early approximate treatments supported this conjecture [33, 34]. However, careful use of the full Eliashberg equations show again that the phonon contributions can only help to raise T_c. The underlying formal reason can be seen from the Eliashberg expression for the gap function Δ(ω).

$$\Delta(\omega) = \frac{1}{Z(\omega)} \int \frac{d\omega'}{\omega'} \mathrm{Re}\left[\Delta(\omega')\right] \int_0^{\omega'} d\omega_q \alpha^2(\omega_q) F(\omega_q)$$

$$\times \Big\{ \left[N(\omega_q) + f(-\omega')\right] \left[(\omega' + \omega_q + \omega)^{-1} + (\omega' + \omega_q - \omega)^{-1}\right]$$

$$- \left[N(\omega_q) + f(\omega')\right] \left[(-\omega' + \omega_q + \omega)^{-1} + (-\omega' + \omega_q - \omega)^{-1}\right] \Big\} .$$

The major contribution to this integral comes from those regions of small ω'. If one neglects ω' relative to ω_q in the two last brackets one sees that the phonon occupation numbers cancel. The only contribution from the phonon occupation which remain come from the terms neglected in this approximation and the terms in Z(ω), and these, at long wavelengths, cancel exactly [35]. The net effect is that the phonons do not interfere significantly with the exciton interaction.

4.0 Strength of the Excitonic Interaction

The above arguments show that there is no fundamental principle which prohibits the attainment of high temperature superconductivity using an exciton interaction. The problem is purely one of finding or devising a system with a sufficiently strong interaction. Unfortunately, little serious consideration was given to this important point in the early work on the exciton interaction, particularly in regard to two dimensional systems. An exception, was the work of Allender, Bray and Bardeen, cited earlier, who did make an effort to come up with reasonable numerical estimates of material parameters, but more needs to be done. In an effort to address this issue, in 1973 [36] we derived a set of criteria for obtaining a strong exciton interaction in 2-D systems based on the nature of the electron-exciton interaction. These are the following:

(a) The separation in space between the electron and exciton system should be less than the Fermi wavelength.
(b) The Fermi momentum should therfore be "small".
(c) The transition density of the exciton system must be large close to the conduction system.
(d) The exciton system must be densely packed. Exciton "impurities" in general, will give only a negligible increase in T_c.
(e) The exciton system must have a low energy of excitation to give an adequate coupling term.

We found, theoretically, that a very strong exciton interaction could be obtained by using transition metal complexes in which the exciton and conduction electron systems share the same atomic site [8, 9, 20]. As an example of this we considered a 1-D chain of a transition metal contained in a porphyrinic organic ligand, where the linked d_{z^2} orbitals of the metal form a 1-D conduction band and the $d_{xz,yz}$ orbitals of the same atom overlap the π-electron orbitals of the organic ligand to form the exciton system. In this case the conduction and exciton electrons come within less than an Angstrom of one another and their interaction is consequently large. In this case, however, the problems of exchange and vertex corrections must be considered before drawing any conclusions on the efficacy of the interaction leading to superconductivity. Later work showed that the coupling constant in such systems is strongly dependent on the relative energies of the metal and ligand states prior to hybridization [10]. These energies must be degenerate to give the strongest interaction. This is an important point and may be of critical importance in determining whether a particular material can have a significant excitonic contribution to the superconductivity. While this was originally directed at possible 1-D conductors, the remarks are relevant to the cuprates, where an interaction exists between the Cu-transition metal atom and the oxygen orbitals, but, with the difference that the conduction occurs now largely on the oxygen sites.

The above remarks are of a general nature. It is useful to ask how one would determine whether or not the exciton interaction is a contributor in a particular superconductor and whether it is responsible for the superconductivity in high transition temperature cuprates. Kresin, [37] has considered in detail how one might differentiate between phonon and non-phonon mechanisms and has shown that this can be done through the analysis of tunneling data and the temperature dependence of of the electronic heat capacity. While this was considered at a time when no high temperature superconductors were known, it is relevant to the study of present day superconductors as well. It may not be necessary to examine these finer details, however, for one can understand much by taking a broad look at these superconductors to see if they meet any of the criteria which would classify them as BCS-like superconductors. We address this in the next section.

5. Exciton Superconductivity in the Cuprates?

In a series of papers we have examined the vast experimental literature of the cuprate superconductors. From this we have found certain experimental constraints on theories of these superconductors [38,39,40]. Our conclusion is that the cuprates appear to have properties remarkably close to a classic BCS superconductor, but with much higher transition temperatures. This is not to say that everything about them is understood, but they do appear to have most of the characteristics expected of a weak coupled, BCS superconductor, having a well defined Fermi surface, a temperature dependent energy gap $\Delta(T)$, BCS-like coherence factors (u_k, v_k), singlet pairs in s-wave states, type II behavior in a magnetic field, and a specific heat anomaly of about the right size at T_c. On the otherhand, there are anomalies. At face value the ratio of the gap to T_c is much too high to be a weak-coupled superconductor, the NMR relaxation rates do not show characteristic BCS behavior, and some consider the normal state to be decidedly not normal. We shall examine some of these and comment on certain other factors.

The question whether the coupling is weak or strong is central to a determination of the mechanism responsible for the superconductivity. The two factors, weak coupling and a transition temperature of the order of 100K can only point to a mechanism involving excitation energies well above those of the phonon energies.

5.1 Coupling Strength

In principle, the coupling strength can be determined by measuring certain ratios, involving the gap, $2\Delta(0)/kT_c$, the specific heat anomaly, $\Delta C/\gamma T_c$ and the thermodynamic critical field, $\gamma T_c^2/H_c^2(0)$, knowing the constant γ in the electronic contribution to the specific heat. This was shown by Geilikman and Kresin [41], Kresin and Parkhomenko [42] and, later Marsiglio and Carbotte [43] using the strong coupled Eliashberg [44] version of the BCS theory. Of these ratios, the first has received the most attention. In a weak coupled, isotropic BCS superconductor, the ratio, $2\Delta(0)/kT_c$ has the value 3.5. In the cuprates it is found to lie between 6.0 and 7.0. This would appear to imply very strong coupling. On the otherhand, evidence from optical data [43] and the other ratios show no evidence of such strong coupling. Before offering a possible explanation of this it is worth noting that the figure of 3.5 for a weak coupled BCS superconductor assumes the system to be isotropic and free from depairing factors. The cuprates do not meet these criteria, they contain magnetic ions, they lie close to an antiferromagnetic transition and they are highly anisotropic. So let us consider the effects of anisotropy first.

Anisotropy. In an anisotropic material the gap function varies as a function of the angle with respect to the crystal axes. In the cuprates, it is expected that the gap should be large in the a-b, or a-planes, while it is expected to be smaller perpendicular to this plane. In determining the ratio, $2\Delta(0)/kT_c$ in such a situation, it is important to use the proper average for the gap function. For an anisotropic BCS superconductor with a separable, angular dependent interaction it can readily be shown that this average is always *smaller* than for the isotropic superconductor and thus lies *below* 3.5. This has also been proved for the general case [45]. However, if one only measures the gap in the a-b, or a-plane, rather than a series of measurements that determine the proper angular average, then this one value can, of course, lie above the BCS value. Experimentally, in many cases, it is not clear that the proper average has been taken for the determination of this ratio. On the other hand it still appears that the observed values in the cuprates probably lie above the weak coupling, isotropic BCS superconductor value. Anisotropy, though, is not the only factor that affects this ratio. Another is temperature dependent damping of the quasi-particle states.

Temperature Dependent Damping. Ruvalds [46] has shown that in a two-dimensional system where nesting of the Fermi surface can occur, the simple isotropic picture of the Fermi liquid breaks down. In the isotropic Fermi liquid, damping of the quasi-particle states vanishes at the Fermi surface, but grows like $(k - k_F)^2$ away from the surface. This gives rise to a weak temperature dependent damping of the quasi-particle states that is proportional to T^2. On the other hand, in the 2-D nesting system, Ruvalds has shown that damping grows like $(k - k_F)$ away from the ?Fermi surface, leading to a larger, and a linear temperature dependence in the damping. Varma [47] has argued along different lines but comes to the same conclusion that there is a linear, temperature dependent damping of these states. The net effect of this is that damping will be larger at the transition temperature than at $T = 0$, and because such damping has a depairing effect upon the pair condensate, the ratio, $2\Delta(0)/kT_c$ will be larger than for the undamped BCS case. This is probably the reason for the anomalously large value observed for this ratio in the cuprates.

5.2 Coupling Range

The range of energies over which the pairing interaction is attractive is an excellent clue to the mechanism. A crude measure of this is given by the Isotope Effect.

The Isotope Effect. In its simplest form, the prefactor to the exponential term in the BCS expression for T_c depends upon the ionic mass through a factor $m^{-1/2}$, while the interaction λ is independent of this mass. This prefactor is a measure of the range of energies over which the interaction is attractive. The measurement of the variation of T_c for different isotopes then can give one an indication of this energy range. The absence of an isotope effect, or a reduction in its value, is often quoted as an indication of the presence of another mechanism. But this is a weak argument. The Coulomb repulsion must also be taken in to account and even in weak coupled, conventional superconductors, where we know from tunneling data that the phonon mechanism and only the phonon mechanism is responsible for the superconductivity, the exponent for the mass dependence of the isotope effect ranges in value from zero to -0.5 [49].

On the other hand, some useful information on the mechanism responsible for the high transition temperature of the cuprates can be gathered from the isotope effect, but more in the sense of circumstantial evidence. If, as we find, there is some isotopic dependence upon the transition temperature and it is of the same sign as that predicted by the BCS theory, then this is an indication that the phonon mechanism is not detrimental to the other mechanism, but rather supportive of it. This provides some support for an excitonic interaction, which like the phonon mechanism involves charge fluctuations, not spin fluctuations, and evidence against a magnetic or, strictly speaking, a spin dependent interaction in the cuprates.

Energy Dependence of the Gap Function, $\Delta(\omega)$. Much more useful data could be got from a determination of the superconducting gap parameter $\Delta(\omega)$ from tunneling studies if these could be done over the range of energies spanned by the interaction. There are many expermental difficulties here but this could give the least ambiguous data of the nature of the interaction.

Some limited information can be obtained from relatively low energy studies of this gap function. It can be shown from the Eliashberg equation for a phonon superconductor, that the gap function, $\Delta(\omega)$ changes sign just above the energy of the highest phonon modes, and then rapidly approaches a value $-\mu^*/(\lambda - \mu^*)$ times the zero frequency gap, $\Delta(0)$ above these energies. On the other hand if another interaction is present such a sign change for $\Delta(\omega)$ need not occur. Such a measurement could give evidence of the presence of a non-phonon mechanism.

Tunneling Studies. It is difficult to obtain data on the gap functions at energies above about 100mV by conventioal tunneling, due to damage to the tunnel junction at these relatively high voltages. On the other hand it may be possible to obtain related data using proximity effect tunneling. If such studies could be done they would give the most valuable information on the mechanism because the spectral distribution of the excitations responsible for the pairing is imprinted in the gap function and this would thus allow their identification.

Optical Studies. We have suggested recently [50] that using what we refer to as, gap modulation spectroscopy, one should be able to get similar information to that obtainable from tunneling but over a much wider frequency range. A combination of this technique using samples for the optical studies in which superconductivity is induced by the proximity effect in a noble metal on the cuprate surface, holds some promise.

6. Conclusion

With improving experimental techniques and better sample preparation one can expect to obtain data in the next year or two which will be able to identify the mechanism responsible for the high transition temperture superconductivity in the cuprates. There appears to be a good chance that this will show that some type of electronic or excitonic interaction is indeed responsible. Innovative and imaginative experimental work will be needed to achieve this goal.

7. Acknowledgements

I am indebted to J. P. Collman, C. Caley, M. Holcomb, and G. Friedmann for valuable discussions, comments and criticism. I acknowledge financial support for this work from the Department of Energy (grant DEFG03-86ER45245).

8. References

1. J. Bardeen, L. N. Cooper and J. R. Schrieffer, Phys. Rev. **108**, 1175 (1957)
2. B. T. Matthias, Physics Today, April (1972), p18-19
3. W. A. Little, Phys. Rev. **134**, A1416 (1964)
4. V. L. Ginzburg, Zh. Eksp. Teor. Fiz. **47**, 2318 (1964) [Sov. Phys. JETP **20**, 1549 (1965)]
5. J. G. Bednorz and K. A. Müller, Z. Phys. **B64**, 189 (1986)
6. M. K. Wu, J. R. Ashburn, C. J. Torng, P. H. Hor, R. L. Meng, L. Gao, Z. J. Huang, Y. Q. Wang, and C. W. Chu, Phys. Rev. Lett. **58**, 908 (1987)
7. W. A. Little, Novel Superconductivity, S .A. Wolf and V. Z. Kresin Eds.(Plenum Press,New York, 1987) p341-353.
8. H. Gutfreund and W. A. Little, "Prospects of Excitonic Superconductivity" in Highly Conducting One Dimensional Solids, J. T.Devreese, R. P. Evrard, and V. E. Doren, Eds (Plenum , New York 1979) p305 - 372; See also V. L. Ginzburg, Contemp. Phys. **9**, 355 (1968)
9. W. A. Little, Int. Jour. Quantum Chem. **15**, 545 (1981)
10. W. A. Little, Jour. de Physique, Colloque C3, **44**, 819 (1983)
11. O. V. Dolgov, D. A. Kirzhnits, and E. G. Maksimov, Rev. Mod. Phys. **53**, 81-93 (1981)
12. R. A. Ferrell, Phys. Rev. Lett. **13**, 330-335 (1964)
13. T. M. Rice, Phys. Rev. **140A**, 1889-1891 (1965)
14. D. Allender, J. Bray, and J. Bardeen, Phys. Rev. **B7**, 1020-1029 (1973)
15. W. A. Little, Phys. Rev. **156**, 396 (1967)
16. J. S. Langer and V. Ambegaokar, Phys. Rev. **164**, 498 (1967)
17. F. Gamble, F. J. DiSalvo, R. A. Klemm and T. H. Geballe, Science **168**, 568 (1970)
18. R. L. Greene, G. B. Street, and L. J. Suter, Phys. Rev. Lett. **34**, 577 (1975)
19. A. B. Migdal, Sov. Phys. - JETP, **7**, 996 (1958)
20. D. Davis, H. Gutfreund, and W. A. Little, Phys. Rev. **13**, 4766 (1976)
21. J. Cai, X. L. Lei, and L. M. Xie, Phys. Rev. **B39**, 11618-11623 (1989)
22. W. A. Little, J. Polymer Sci. **C29**, 17 (1970)
23. V. Kresin, Phys Rev. **B35**, 8716 (1987)
24. J. Ruvalds, Phys. Rev. **B35**, 8868 (1987)

25. N. L. Cohen and P. W. Anderson, in Superconductivity in d- and f-Band Metals, D. H. Douglass (Ed.) AIP, New York (1972).
26. J. C. Phillips, Phys. Rev. Lett. **29,** 1551 (1972)
27. V. L. Ginzburg, JETP Lett. **14**, 396 (1971); Sov. Phys. Usp. **13**, 335-352 (1972)
28. D. A. Kirzhnits, E. G. Maksimov, and D. I. Khomskii, J. Low Temp. Phys. **10**, 79 (1973)
29. J. P. Carbotte, Rev. Mod. Phys. **62**, 1027-1157 (1990)
30. G. Bergmann and D. Rainer, Z. Phys. **263**, 59 (1973)
31. V. Z. Kresin, H. Gutfreund, and W. A. Little, Solid State Comm. **51**, 339 (1984)
32. P. W. Anderson, Editorial Comment, Physica **2**, 151 (1966)
33. J. Appel, Phys. Rev. Lett. **21**, 1164 (1968)
34. P. B. Allen, Solid State Comm. **12**, 379 (1973)
35. A. E. Karakozov, E. G. Maksimov, and S. A. Mashkov, Sov. Phys.-JETP **41**, 971-976 (1976)
36. W. A. Little, J. Low Temp. Phys. **13**, 365 (1973)
37. V. Z. Kresin, Phys. Rev. **B30**, 450 (1984)
38. W. A. Little, Science **242**, 1390-1395 (1988)
39. W. A. Little, Czech. J. Phys. **40**, 790 -799 (1990)
40. W. A. Little in Ceramic Superconductors (World Scientific Co. Pte, Ltd., Singapore) 1991, p. 79-91.
41. B. T. Geilikman and V. Z. Kresin, Phys. Lett. **40A**, 123 (1972)
42. V. Z. Kresin and V. P. Parkhomenko, Soviet Physics. Solid State **16**, 2180 (1975)
43. F. Marsiglio and J. P. Carbotte, Phys. Rev. **B33**, 6141 (1986)
44. G. M. Eliashberg, Soviet Phys. JETP **7**, 696 (1960); see also D. J. Scalapino, in Superconductivity, Volume 1, Ed. R. D. Parks, Marcell Dekker, Inc., New York (1969), p 473.
45. V. L. Pokrovskii and M. S. Ryvkin, Soviet Phys. JETP 16, 67-75 (1963)
46. Z. Schlesinger, R. T. Collins, F. Hotzberg, C. Field, N. E. Bickers & D. J. Scalapino, Nature **343**, 242-243 (1990)
47. A. Virosztek and J. Ruvalds, Phys Rev. B42, 4064 (1990)
48. C. M. Varma, P. B. Littlewood, S. Schmitt-Rink, E. Abrahams and A. E. Ruckenstein, Phys. Rev. Lett. **63**, 1996 (1989)
49. G. Gladstone, M. A. Jensen and J. R. Schrieffer in Superconductivity, Volume 2, Ed. R. D. Parks, Marcell Dekker, Inc., New York (1969), p 767-771.
50. W. A. Little and J. P. Collman, Proc. Natl. Acad. Sci. U. S.A. 85,4596 (1988)

Formation, Motion, and Superconductivity of Large Bipolarons

D. Emin

Solid State Physics Division, 1152 Sandia National Laboratories, Albuquerque, NM 87185, USA

Abstract. Bipolarons can form in materials with exceptionally strong electron-lattice interactions. Although small (singlet site) bipolarons readily localize, large (multisite) bipolarons, in analogy with large polarons in ionic solids, are presumed to move itinerantly. Perovskite-based materials are prime candidates for the formation of large bipolarons. As mobile charged bosons, large bipolarons condense into a superconducting state. Large bipolarons possess distinctive properties that are in accord with observations in novel superconductors.

1. Introduction

Superconductivity is usually considered to be a property of metals. However, the carrier densities of some novel superconductors (e.g., 10^{19} cm^{-3} in superconducting doped-$SrTiO_3$) can be well below that of normal metals [1]. Indeed, superconductivity in these materials even disappears when the carrier density is raised too high. Furthermore, carrier-related properties in these superconductors depart significantly from those of conventional metal-based superconductors. The cuprate superconductors as well as other superconductors (e.g., based on WO_3, $BaBiO_3$ and TiO_2) appear to be of this type.

The insulating parents of these novel superconductors are noted for their unstable lattices. Lattice instabilities affect charge carriers by providing especially effective electron-lattice interactions. For this reason, as well as their ionic (polar) character, transition-metal oxides and perovskites are among the classes of materials in which polarons and/or bipolarons have long been thought to form.

This article will address 1) the conditions under which charge carriers form mobile (large) bipolarons, 2) the nature and properties of bipolaronic superconductivity, and 3) the electronic, optical, vibrational and magnetic properties of large bipolarons. The novel superconductors are prime candidates for the formation of large bipolarons. Unusual observations of the novel superconductors appear consistent with distinctive properties expected of a bipolaronic superconductor.

2.1 Preliminaries

The name polaron denotes a quasiparticle composed of a self-trapped carrier and the atomic displacements associated with the self-trapping. Self-trapping occurs with the binding of carriers in a potential well produced by atoms displaced from their carrier-free equilibrium positions. Self-trapping is energetically stable when the binding energy of the carriers exceeds the strain energy of producing the atomic displacements.

Springer Series in Solid-State Sciences, Vol. 106 **Physics of High-Temperature Superconductors**
Eds.: S. Maekawa and M. Sato

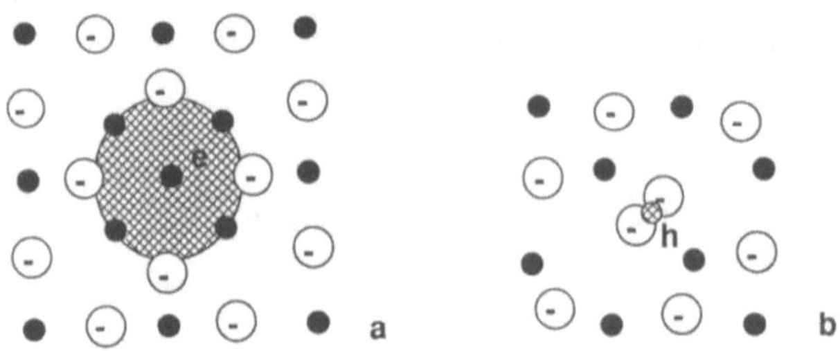

Fig. 1. Large (a) and small (b) polarons in an ionic solid.

The strain-related lowering of the energy of two noninteracting self-trapped carriers is enhanced if the carriers reside at the same location rather than being separated. This feature fosters the coalescence of two carriers into a pair. If this driving force for merger is sufficient to overcome the Coulomb repulsion of the carriers, two carriers find it energetically favorable to form a singlet bipolaron. The formation of a bipolaron is analogous to the formation of a covalent bond in that a stable bound state for a pair of electrons is produced by altering atomic positions. Furthermore, as in bond formation, the requirement that no more than two electrons occupy a nondegenerate state impedes forming multipolarons with more than a pair of self-trapped carriers centered about a site.

The nonlinearity of self-trapping phenomena results in there being two distinct types of self-trapped electronic states [2,3]. The self-trapped carrier may be confined to the smallest unit compatible with the atomicity of the solid (e.g., an atom or a bond). Then, the polaron or bipolaron is referred to as *small*. Alternatively, the self-trapped state extends over several units and the polaron or bipolaron is referred to as *large*. Small and large polarons are illustrated in Fig. 1.

Charge carriers typically form polarons or bipolarons in four classes of solids. First, strong electron-phonon interactions and polaron formation generally occur in ionic solids. Here the strong electron-phonon interactions arise from the long-range Coulombic interactions of charge carriers with the anions and cations [4,5]. In fact, polarons received their name from their formation in ionic (*polar*) solids. Second, carriers associated with oxygen anions generally possess very strong short-range electron-lattice interactions. This strong interaction arises because the binding of the outermost electrons of an O^{2-} ion requires the presence of the positive charge of surrounding cations. As a result, the highest occupied electronic states of an oxygen ion are exceptionally sensitive to the positions of the adjacent cations. That is, oxygen-related orbitals possess a very strong electron-lattice interaction. Third, the relative ease of producing atomic displacements in unstable structures (e.g., ferroelectric-like materials) promotes formation of polaronic carriers. Finally, polaron and bipolaron formation is fostered in low-dimensional electronic systems because of the relative ease of the electronic localization through which charge carriers self-trap to form polarons [2]. The insulating parents of the high-temperature superconductors are obvious candidates for the formation of polaronic charge carriers.

A self-trapped carrier only moves when the atoms associated with the self-trapping alter their positions. Because of the severe compactness of a small polaron, its intersite motion requires some atoms to move distances much greater than their zero-point amplitudes. As a result, small polarons readily localize (e.g., self-trapped holes in alkali halides). Therefore, small bipolarons do not move coherently and cannot give rise to superconductivity. By contrast, since large polarons extend over several sites they can adjust continuously to atomic motion. Therefore, they move in a coherent manner (e.g., self-trapped electrons in the alkali halides). Since large bipolarons are (in analogy with large polarons) mobile, they may form a superconducting groundstate.

2.2 Formation of a Large Bipolaron

The conditions under which two excess charges will merge to form a stable large bipolaron in a deformable continuum have been studied within the adiabatic (strong coupling) limit [3,6]. Beyond one-dimensional (polymeric) systems, large (multiunit) bipolaron formation is found in ionic systems with exceptionally strong long-range (Fröhlich) electron-lattice interactions. Then, as illustrated in Fig. 2, the two carriers will be bound together in a multisite (large) state, sharing a modified hydrogenic well. The short-range (deformation-potential-like) portion of the electron-lattice interaction fosters the stabilization of the large bipolaron with respect to separation into two large polarons by deepening the central portion of the shared potential well. Since electronic localization is fostered by reduced electronic dimensionality, large-bipolaron formation is easier to achieve in quasi-two-dimensional electronic systems than in similar systems of three dimensions [3,6]. Finally, as illustrated in Fig. 2, the promotion energy associated with having more than two carriers occupy a common potential well tends to destabilize a multipolaron of more than two carriers [7]. Thus, with a strong electron-lattice interaction, carriers tend to form bipolarons but not "grander" polarons.

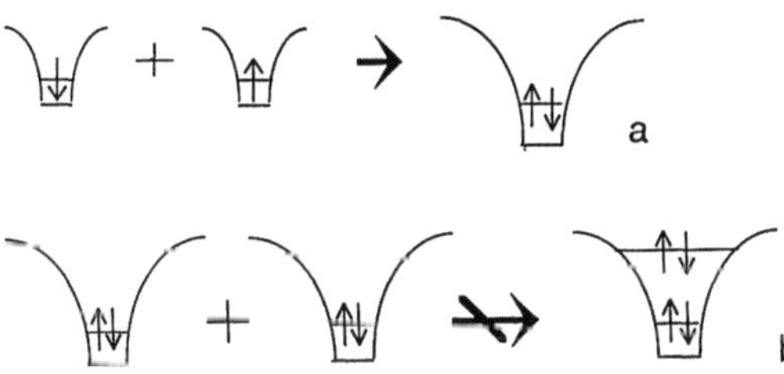

Fig. 2. Stability of a bipolaron (a) and instability of a quadpolaron (b).

The strength of the long-range portion of the electron-lattice interaction is measured by the ratio of the static to optical dielectric constants: $\varepsilon_0/\varepsilon_\infty$. The formation of a large bipolaron is found to require that $\varepsilon_0/\varepsilon_\infty > 2$ [3,6]. The insulating parents of the novel superconductors have $\varepsilon_0/\varepsilon_\infty > 5$. Such large values of $\varepsilon_0/\varepsilon_\infty$ correspond to long-range electron-lattice interactions, proportional to $1/\varepsilon_0 - 1/\varepsilon_\infty$, roughly twice as strong as typical polaron materials (e.g., alkali halides). These materials also possess the strong short-range electron-lattice interactions characteristic of oxygen-related orbitals. Thus, the insulating parents of the novel superconductors are prime candidates for large-bipolaron formation.

2.3 Energy Spectrum of a Bipolaron

At small momenta, $\hbar k$, a large polaron or bipolaron is a freely moving quasiparticle. Its energy is then $\varepsilon(k) \approx (\hbar k)^2/2m$, where m is the effective mass of the large bipolaron. At larger momenta the energy of the quasiparticle saturates since the self-trapped carrier cannot move faster than the atoms that provide the self-trapping. Thus, one may envision the large bipolaron's energy spectrum having the form illustrated by curve a of Fig. 3.

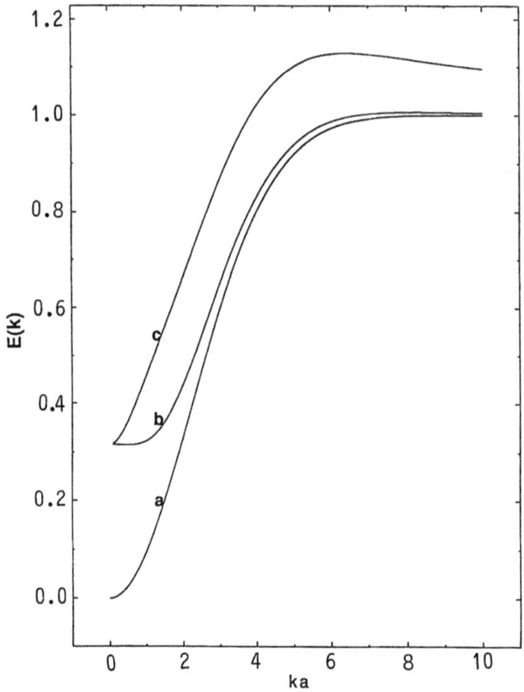

Fig. 3. Dispersion relations of (a) an isolated large bipolaron and of collective excitations in (b) a three-dimensional system and in (c) a layered system.

3.1 Bipolaronic Superconductivity

Bipolaronic superconductivity is a property of the groundstate of interacting large bipolarons. At long range, large bipolarons interact with one another via their Coulomb repulsion. Since the self-trapped carriers move very slowly as they follow the atomic displacements, the Coulomb interaction between two bipolarons is reduced by the low frequency dielectric constant, ε_0 [5]. As a result of their Coulomb interactions, the long-wavelength excitations of mobile bipolarons are plasmon-like. For example, neglecting the short-range repulsion between bipolarons the excitation spectrum in a three-dimensional system is [7,8]:

$$E(k) = [(\hbar\omega_p)^2 + \varepsilon^2(k)]^{1/2}, \quad (1)$$

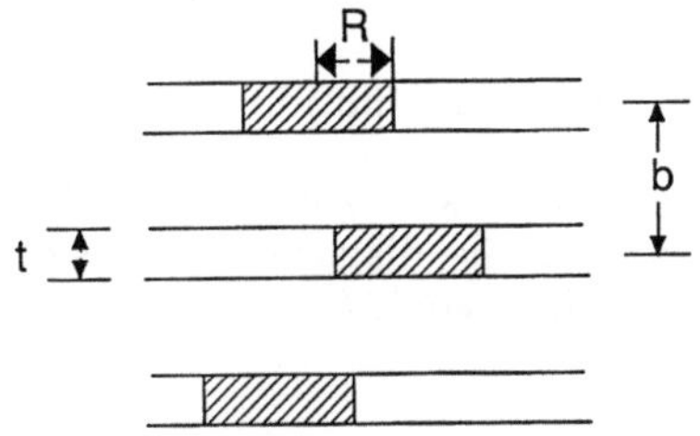

Fig. 4. Bipolarons of radius R confined to layers of thickness t separated by distance b.

where $\hbar\omega_p = \hbar[4\pi n q^2/\varepsilon_0 m]^{1/2})$ is the bipolarons' plasma energy, k is the wavevector of the bipolarons' collective excitation and q is the charge (2e) of a bipolaron. Curve b of Fig. 3 illustrates the excitation spectrum for interacting large bipolarons in a three-dimensional system. In a layered system, shown in Fig. 4, $(\hbar\omega_p)^2$ of Eq. (1) is multiplied by the factor (kb/2) coth (kb/2), where b is the separation between conducting layers [7]; in the cuprates b is the separation between groups of contiguous conducting (CuO_2) layers. As shown by curve c of Fig. 3, the dispersion relations of layered systems differ from those of 3-d systems. With $n \approx 10^{21}$ cm^{-3}, $\varepsilon_0 \approx 20$ and $m \approx 1000$ free-electron masses, $\hbar\omega_p$ is about three orders of magnitude smaller than the the plasma energy of electrons in metals.

For charge to flow without resistance in the large-bipolarons' groundstate, the excitation spectrum must be such that the flow will not be impeded by the creation of excitations. Landau's condition for resistanceless flow is that the minimum value of E(k)/k be nonzero [9]. Because of the presence of the gap in the energy spectrum, the excitation spectrum of interacting large bipolarons satisfies this condition: that is, E(k) remains finite as $k \to 0$. In addition, the presence of this energy gap, $\hbar\omega_p$, ensures the Meissner effect [7].

Since self-trapped carriers adiabatically follow the atomic motion, the bipolaronic groundstate is characterized by long-range coherence of the *atomic motions*. This correlated motion of atoms is lost when excited states are occupied. Concomitantly, the electronic charge distribution of the self-trapped carriers in the normal state is inhomogeneous on a time scale of the atomic motions. The length scale below which the excitations appear as independent bipolarons is the correlation length, $\xi \sim (\hbar/2m\omega_p)^{1/2}$. ξ is typically only several Ängstroms.

A distinctive feature of bipolaronic superconductivity is the limited range of carrier concentrations over which it can occur [7]. With too low a density the bipolaronic groundstate will be insulating because the carriers localize. In an ideal ordered solid this localization will occur as the Coulomb interactions between heavy massed carriers cause them to form a Wigner insulator [7,10]. The maximum density for Wigner localization falls as delocalization becomes easier: as ε_0 rises or the bipolaron's mass, m, falls. This feature may explain why doped $SrTiO_3$ (with $\varepsilon_0 = 20{,}000$ near 1 K) becomes superconducting with a carrier density of only 10^{19} cm^{-3} [1]. An upper limit to the carrier density for bipolaronic superconductivity occurs because bipolarons can only form if their density is low enough so that they do not overlap one another. Thus, bipolaronic superconductivity can only occur within the range of bipolaron densities wherein bipolarons form with a fluid groundstate. For this range of bipolaron densities to even exist, m must be small enough and ε_0 large enough [7].

3.2 A Bipolaron's mass

A self-trapped carrier only moves when atoms alter their positions. In the adiabatic (strong-coupling) approach to polaron motion, the effective mass of a bipolaron is a weighted sum of the masses of the atoms whose motion enables the self-trapped electronic carrier to move. For example, the diagonal elements of a bipolaron's effective mass tensor in an isotropic medium are [3,6]:

$$m = \Sigma_k M_i (\Delta d_i/a)^2, \tag{2}$$

where Δd_i is the shift of the equilibrium position of the i-th atom (with mass M_i) that accompanies the motion of the bipolaron by a lattice constant, a. Displacements of atoms in contact with a self-trapped carrier most strongly affect its motion. Thus, displacements associated with the short-range electron-lattice interaction tend to dominate the bipolaron's effective mass. The effective mass for the short-range (deformation-potential-type) electron-lattice interaction with acoustic phonons is of the order of the bipolaron's binding energy divided by the square of the sound velocity. Thus, a bipolaron's effective mass is of the order of 1000 free-electron masses. Measurements of the conductivity, microwave relaxation time and carrier density (Hall coefficient) in semiconducting La_2CuO_4 yield an effective mass of an individual carrier of this magnitude [11]. The effective mass varies inversely with the bipolaron's size, the number of displaced atoms in contact with the self-trapped carrier. The bipolaron's effective mass is also proportional to the mass of the atoms that dominate the sum of Eq. (2). Thus, changing the isotope of the atoms that provides the dominant contribution to the effective mass alters the effective mass.

3.3 The Transition Temperature

Bipolaronic superconductivity is described in terms of a two-fluid picture. The resistanceless superconducting fluid results from occupancy of the ground-state. Following Landau [9], the carrier density participating in normal-state flow is [7]

$$n_n = \int dk \, \{e^{\beta[E(k) - \mu]} - 1\}^{-1}$$

$$\approx e^{-\beta(\hbar\omega_p - \mu)} \int dk \, \exp\{-\beta[(\hbar k)^2/2m_c]\}, \tag{3}$$

where $\beta = 1/k_B T$ and μ is the bipolarons' chemical potential. The second line of Eq. (3) is obtained when $\beta[E(k) - \mu] >> 1$ and $E(k) = \hbar\omega_p + (\hbar k)^2/2m_c$. At T_c, $n_n = n$ and $\mu = 0$. Then, $k_B T_c$ is found from Eq. (3) to be of the order of $h\omega_p$. Thus, T_c is nearly proportional to $(n/\varepsilon_o m)^{1/2}$.

This result provides a qualitative explanation of several striking facts. First, T_c in the CuO_2 materials rises with the number of contiguous CuO_2 layers until a saturation value of T_c is reached. With the thickness of a disklike bipolaron being determined by the number of contiguous CuO_2 planes to which it is confined, a bipolaron's mass falls (and T_c rises) as the number of contiguous CuO_2 sheets is increased. This increase will saturate once the net thickness of the contiguous CuO_2 layers is great enough so that it no longer determines the thickness of the bipolaron (i.e., when the bipolaron becomes three-dimensional). Second, T_c falls in CuO_2-based superlattices as the number of insulating layers between contiguous CuO_2 layers is increased. This result is expected since the overall carrier density tends to fall as insulating layers are placed between contiguous CuO_2 layers. Finally, the very large low-temperature static dielectric constant of $SrTiO_3$ (ε_o= 20,000 at 1 K) is consistent with its T_c rising to only about 1 K as its carrier density is increased to 10^{21} cm^{-3} [1].

Isotopic substitutions will generally affect T_c through m, the mass of a large bipolaron. The bipolaron's mass is proportional to the mass of those atoms near to the self-trapped carrier that are primarily displaced. Thus, a small shift of T_c with oxygen isotope in the cuprate superconductors can be understood if the carrier is associated with occupation of oxygen orbitals and the dominant atomic displacements are of the adjacent metal cations. The situation in the $BaBiO_3$ system may be quite different. With bipolarons associated with Bi sites, the nearest-neighbor atomic displacements are of the oxygen atoms adjacent to the Bi sites. Thus, both small and large isotope effects can be rationalized [3,6].

4.1. Transport Properties of Large Bipolarons

The transport properties of a large bipolaron depend critically on how it is scattered. Here the scattering of a large bipolaron (or polaron) by acoustic phonons is reviewed [12]. A large bipolaron, like a multi-site impurity state, only interacts effectively with phonons of half-wavelengths much greater than the diameter of the self-trapped state, 2R. Since these phonons are of low momentum, $\hbar q << \hbar/R$, they are ineffective in scattering a large bipolaron. In these scattering events acoustic phonons may be regarded as bouncing off a large bipolaron with momentum transfer $2\hbar q$. The net scattering rate of a bipolaron of effective mass m_p by acoustic phonons is $1/\tau \approx k_BT/|m_p|sR$ when $k_BT > \hbar s/4R$, where the acoustic-phonon dispersion relation at low frequencies is $\omega_q = s|q|$, where s is the sound velocity. There are two important features of this result. First, since the difficulty of scattering a heavy particle increases with its mass, the scattering time, τ, is proportional to the bipolaron's mass. Thus large-polaronic carriers can have high mobilities despite possessing large masses (e.g, as in alkali halides). Second, the scattering rate of the heavy particle is proportional to the density of low-momentum particles that scatter it. For this reason $1/\tau$ is proportional to k_BT. The resistivity is thus $\rho = [q^2\Sigma_k f_k (\tau_k/m_k)]^{-1} = k_BT/nq^2sR$ with $n = \Sigma_k f_k \, \mathrm{sign}(m_k)$, where f_k is the density of carriers of wavevector k [3,6].

At finite frequencies the conductivity of large bipolarons has two distinctive features [13,14]. First, since large bipolarons move itinerantly, the conductivity manifests a Drude fall-off with increasing frequency. As noted above, $1/\tau$, the relaxation rate, is proportional to temperature. Second, at higher frequencies there is an absorption band arising from exciting self-trapped electronic carriers to higher levels of their self-trapping potential wells and to states that are resonant with the continuum. The density of states associated with the continuum rises with excitation energy while the matrix element for photon absorption to a resonant-continuum state falls for final-state wavelengths falling below the diameter of the self-trapped state, 2R. As with impurity states, these effects produce a nearly temperature-independent asymmetric absorption band (as phonon broadening is relatively small). The intensity of this polaronic absorption is proportional to the density of large bipolarons. As a large bipolaron is increased in size, its electronic levels become shallower, its absorption band then narrows and shifts toward lower photon energies. In particular, increasing a bipolaron's thickness by increasing the number of contiguous CuO_2-sheets of a cuprate should shift the absorption band toward lower energies. In the superconducting state the Drude tail disappears (since $\tau \rightarrow \infty$) while the absorption band remains.

The infrared absorption of large polarons (or bipolarons) differs qualitatively from that of small (severely localized) polarons (or bipolarons) [15]. The small polaron absorption band arises from the photon-induced transfer of a self-trapped carrier from the site of its localization to a neighboring site. Since the absorption energy, the difference of the electronic energies of these two states, is strongly dependent on the atomic configuration at the instant of excitation, the small-polaron absorption band broadens significantly with increasing temperature.

The large-bipolarons' Hall coefficient is

$$R = (1/nq)\, \Sigma_k f_k\, [1- m_{xx}m_{yy}/m_{xy}m_{xy}]\, \mathrm{sign}(m_{xx}m_{yy})\, /\Sigma_k f_k\, \mathrm{sign}(m_{xx}) \qquad (4)$$

for a magnetic field applied in the z-direction, where the m_{xy} are components of the effective mass tensor: $m_{xy} = \hbar^{-2}\, \partial^2E(k)/\partial k_x \partial k_y$ [15]. In obtaining this result, $\tau_k/|m_{xx}(k)|$ is again treated as k-independent. With E(k) depending only quadratically on the magnitude of k, the Hall coefficient reduces to its free-carrier value, $1/nq$, since then $1/m_{xy} = 0$. However, the bipolarons' Hall coefficient can fall increasingly below its free carrier value with rising temperature. This situation occurs with the bipolaron dispersions shown in Fig. 3, as the thermal population of states for which E(k)'s rise with increasing k is less than quadratic increases.

The Seebeck coefficient of large bipolarons is very distinctive. As such, it is especially useful in establishing the presence of large bipolarons. For example, the Seebeck coefficient measurements can provide a means of determining whether a carrier's charge is $|e|$ or $2|e|$. In particular, in a semiconductor with itinerant charge carriers the dc conductivity and Seebeck coefficient are approximated by $\sigma = \sigma_0 \exp(-E/k_BT)$ and $S = (k_B/q)(E/k_BT + A)$, respectively, where E is the energy for generating carriers and A is near temperature-independent constant of order unity. Eliminating E/k_BT between these two relations yields an expression for the carriers' charge:

$$q = [A + \ln(\sigma_0/\sigma)]k_B/S. \qquad (5)$$

With the limited data available on the insulating parents of the superconductors, I find values of q close to $2|e|$ in La_2CuO_{4+x} [16] and near 1.7e in $YBa_2Cu_3O_{6.25}$ [17]. Although preliminary, these results suggest paired carriers.

The Seebeck coefficient in a bipolaronic superconductor's normal state is [15]

$$S = (1/qT)\, [\Sigma_k f_k\, \mathrm{sign}(m_{xx})\, E(k)\, /\Sigma_k f_k\; - \mu]. \qquad (6)$$

Since the bipolarons act as bosons, μ can be no greater than the lowest carrier energy, $\mu < 0$, while $E(k) > \hbar\omega_p$. Therefore, just above T_c, $S \approx \hbar\omega_p/2|e|T_c$. Furthermore, S decreases as the temperature is increased further. In addition, as the temperature is raised, states of negative effective mass make an increasing contribution to S. As a result, the sign of the Seebeck coefficient can change with rising temperature. This effect is much more likely to occur in layered materials than in isotropic systems due to their distinctive dispersion relation: curve c of Fig. 3. Thus, the large-bipolarons' Seebeck effect is qualitatively different from that of electrons in metals where $S \propto T$. However, the bipolarons' Seebeck coefficient is in general agreement with that observed for in-plane motion in single-crystal cuprate superconductors [18-19]. Furthermore, estimating $\hbar\omega_p$ as $2|e|T_cS$ gives a value $\approx 10^{-3}$ eV, in accord with the estimate of Sec. 3.1.

In a normal metal, the Seebeck coefficient, the entropy per charge of a carrier, is sensitive to the aligning effect of an applied magnetic field. In particular, the magnetic-field dependent contribution to the Seebeck coefficient is $(\kappa/|e|)(g\mu_BH/k_BT)^2/2$. However, no magnetic-field dependence of the Seebeck coefficient is observed in the normal state of superconducting $YBa_2Cu_3O_7$ despite a predicted 300% change at 30 T [20]. This result is consistent with the charge carriers having no permanent magnetic moments, e.g., being singlet bipolarons.

4.2 Distinctive Spectroscopic Properties of Large Bipolarons

Large bipolarons produce distinctive signatures in vibrational, electronic, optical and magnetic spectroscopies. Here the physical origins of these unusual effects are succinctly reviewed.

Large bipolarons distinctively affect the vibrational motions of a solid's atoms [21]. In particular, within the adiabatic approach the electronic energy of the self-trapped carrier (a function of atomic positions) contributes to the potential energy experienced by the atoms. If the phonon half-wavelength is less than the length of the self-trapped state, the phonon-related atomic displacements are nonuniform across the length of a self-trapped carrier. Then the concomitant sloshing of the self-trapped charge shifts the local vibrational frequency by an amount proportional to the square of the electron-lattice coupling strength [19]. With a sufficiently strong electron-lattice coupling, this effect yields different vibrational frequencies in the vicinity of the self-trapped carriers than in carrier-free regions. Thus, large bipolarons can contribute distinct modes to the vibrational spectrum. However, such modes only appear for half-wavelengths that are shorter than the bipolarons' diameter. "Ghost" modes that only exist at short wavelengths appear as carriers are introduced in the cuprate matertials [22].

As noted in Sec. 3.1, bipolaronic superconductivity is associated with long-range coherence of the atomic motions. Observations of increased atomic channeling as T is lowered below T_c are consistent with this feature [23].

Bipolarons are local charge centers. As such, they may provide sites for positron annihilation and thereby reduce the positron lifetime [13]. Such an occurrence would explain why the addition of holes to CuO_2 sheets (e.g., by replacing La with Sr in La_2CuO_4) decreases the positron lifetime [24]. Since the diversity of bipolaronic centers is suppressed upon cooling below T_c, the net efficiency of positron annihilation may be concomitantly reduced. This speculation is consistent with the observed progressive increase of the positron lifetime as the temperature is lowered below T_c [24]. Neither the reduction of the positron lifetime with the addition of holes nor the increase of the positron lifetime as T is lowered below T_c is predicted nor observed for standard metallic superconductors.

Photoemission probes the distribution of self-trapped states that exist at the time of a photons' absorption [13]. As the temperature is lowered below T_c, bipolarons progressively condense into a single collective groundstate. As a result, the photoemission spectra for a bipolaronic superconductor becomes more peaked (about the electronic binding energy of the minimum-energy bipolaron) as T is lowered below T_c. Distinctively, in bipolaron formation the self-trapped carriers themselves induce the formation of the potential wells in which they are bound.

Hence, for bipolarons the density of carrier-related states is proportional to the carrier density [13]. This result is generally observed in the cuprates [25].

A singlet large bipolaron has no spin and presumably quenched orbital angular momentum. Nonetheless, because of the multisite extent of the bipolaron, a magnetic field can induce significant Van Vleck paramagnetism [26]. This occurrence depends on matrix elements of the component of orbital angular momentum parallel to the magnetic field between different electronic states. For the disklike bipolarons envisioned in the cuprates, symmetry only permits Van Vleck paramagnetism with magnetic fields aligned parallel to the CuO_2 planes. Then, the paramagnetic susceptibility [$\approx (2e^2/m_ec^2)(R^2 - t^2)^2/(R^2 + t^2)$] rises with the ratio of the bipolaron's radius to thickness, R/t, to dominate the diamagnetism.

In their normal state, bipolarons are distributed in size and shape but become uniform in their collective groundstate. Because of the nonlinear dependence of a bipolaron's susceptibility on R and t, the net susceptibility of bipolarons must fall as they condense into their groundstate. In addition, the energy gap precludes altering the bipolarons' groundstate susceptibility with low-energy excitations. The observation that the magnetic susceptibility of cuprate superconductors only changes at T_c when the magnetic field is aligned parallel to the CuO_2 planes is consistent with the carrier being large bipolarons [27].

Acknowledgement

This work was performed under auspices of the U.S. Department of Energy and funded in part by its Office of Basic Energy Sciences, Division of Materials Sciences under contract #DE-AC04-76DP00789.

References

1. J. F. Schooley, W. R. Hosler and M. L. Cohen, Phys. Rev. Lett. **12** , 474 (1964).
2. D. Emin and T. Holstein, Phys. Rev. Lett. **36** , 323 (1976).
3. D. Emin and M. S. Hillery, Phys. Rev. B **39** , 6575 (1989).
4. L. D. Landau, Phys. Z. Sovietunion **3** , 644 (1933).
5. H. Fröhlich, in Polarons and Excitons G. G. Kuper and G. D. Whitfield, Eds. (Plenun, New York, 1963) p. 1.
6. D. Emin, Phys. Rev. Lett. **62**, 1544 (1989).
7. D. Emin and M. S. Hillery, in preparation.
8. L. L. Foldy, Phys. Rev. **124** , 649 (1961).
9. L. D. Landau, Zh. Eskp. Teor. Fiz. Nauk, **5 9**, 592 (1941)
10. E. P. Wigner, Phys. Rev. **46**, 1002 (1934)
11. D. Reagor, E. Ahrens, S.-W. Cheong, A. Migliori and Z. Fisk, Phys. Rev. Lett. **64**, 2048 (1989)
12. H.-B. Schüttler and T. Holstein, Phys. Rev. Lett. **51** , 2337 (1983); Ann. Phys. (N.Y.), **166**, 93 (1986).
13. D. Emin, in Electronic Structure and Mechanisms for High Temperature Superconductors, J. Ashkenazi and G. Vezzoli, Eds. (Plenun, New York, 1991).
14. D. Emin, in preparation
15. D. Emin, Adv. Phys. **24**, 305 (1975).
16. M. F. Hundley, private communication.

17. B. Fisher, J. Genossar, L. Patlagan, I. O. Lelong and J. Ashkenazi, Physica C **162-164**, 1207 (1989) Fig. 2.
18. G. H. Chen, J. H. Wang, D. N. Zheng, Y. F. Fan, S. L. Jia, Q. S. Yang, Y. M. Ni and Z. X. Zhao, Mod. Phys. Lett. **3**, 295 (1989).
19. A. K. Bhatnagar, R. Pan, D. G. Naugle, P. J. Squattrito, A. Clearfield, Z. Z. Sheng, Q. A. Shams and A. M. Herman, Solid. State. Comm. **73**, 53 (1990)
20. R. C. Yu, M. J. Naughton, X. Yan, P. M. Chaikin, F. Holtzberg, R. L. Greene, J. Stuart and P. Davies, Phys. Rev. **37**, 7963 (1988).
21. D. Emin, Phys. Rev. B **43**, 8610 (1991).
22. L. E. Rehn, R. P. Sharma, P. M. Baldo, Bull. Am. Phys. Soc. **36**, (1991) p. 989.
23. H. Rietschel, L. Pintschovius, and W. Reichardt, Physica C **162-164**, 1705 (1989).
24. Y. C. Jean, J. Kyle, H. Nakanishi, P. E. A. Turchi, R. H. Howell, A. L. Wachs, M. J. Fluss, R. L. Meng, H. P. Hor, J. Z. Huang and C. W. Chu, Phys. Rev. Lett. **60**, 1069 (1988).
25. H. Matsuyama, T. Takahashi, H. Katayama-Yoshida, T. Kashiwakura, Y. Okabe and S. Sato, Physica C **160**, 567 (1989); T. Takahasi, H. Matsuyama, H. Katayama-Yoshida, K. Seki, K. Kamiya and H. Inokuchi, Physica C **170**, 416 (1990).
26. D. Emin, Phys. Rev. B **43**, 2633 (1991).
27. M. Takigawa, P. C. Hammel, R. Heffner and Z. Fisk, Phys. Rev. B **39**, 7371 (1989).

Application of Marginal Fermi-Liquid Phenomenology to the Superconductive State of Cu-O Materials

C.M. Varma

AT&T Bell Laboratories, 600 Mountain Ave., Murray Hill, NJ 07974, USA

Abstract. Calculations of the properties of the superconductive state using the marginal Fermi-liquid hypothesis are reviewed. While decent results for T_c and $2\Delta(0)/T_c$ can be obtained, the more spectacular finding is the absence of a coherence peak in nuclear relaxation rate and a peak in the optical conductivity below T_c in agreement with experiment. They come about because of a sharp decrease in quasi-particle relaxation rate below T_c as the continuum of normal state low energy excitations is self-consistently eliminated in the superconductive state. This is a very strong evidence that pairing is caused by the excitations that give rise to the marginal Fermi-liquid properties.

Some problems remain, even at the phenomenological level, in extending the hypothesis to the superconductive state. These are also discussed.

1. Introduction

Every transport property[1] studied in the normal state of Cu-O superconductive materials is unlike expected of a Landau Fermi-liquid. This has led to an examination of the conditions in which the quasi-particle concept — one of the most profound and useful concepts in condensed matter physics — breaks down.

In an attempt to focus on the central issue in the diverse anomalous normal state properties, a phenomenological hypothesis about the excitation spectrum was suggested.[2,3] The hypothesis of the "marginal Fermi-liquid"(MFL) phenomenology is that over a wide range of momenta, the charge(ρ) and spin (σ) polarizabilities have a low energy scale given by the temperature of measurement rather than any intrinsic scale in the Hamiltonian:

$$\mathrm{Im}P(q,\omega) \quad \begin{matrix} -N(0)\omega/T, & \omega \ll T \\ -N(0)\ , & \omega_c \gg \omega \gg T \end{matrix} \tag{1}$$

Here ω_c is a cut-off energy. As a results of Eq.(1), the frequency dependence of the electron self-energy is given by

$$\Sigma(\omega) = \lambda_+\omega\left(\log\frac{x}{\omega_c} + i\frac{\pi}{2}x\,\mathrm{sgn}\,\omega\right), \tag{2}$$

where $x = \max(|\omega|, T)$ and

$$\lambda_+ = \lambda_\rho + \lambda_\sigma\ , \tag{3}$$

where $\lambda_{\rho,\sigma}$ are the coupling constants in the charge and spin channels respectively. Eq.(2) leads to a quasi-particle weight z, which depends logarithmically on energy or temperature,

Springer Series in Solid-State Sciences, Vol. 106 Physics of High-Temperature Superconductors
Eds.: S. Maekawa and M. Sato

$$1/z = |1+\lambda+\ln \frac{y}{\omega_c}| \tag{4}$$

where $y=\max[|E_k-\mu|,T]$ with E_k the real part of the quasi-particle energy. It is the weakest imaginable way in which the quasi-particle concept of the conventional Fermi-liquid theory could be violated, suggesting the term "marginal" in MFL.

Recall that in Landau theory, for $\omega \ll qv_F$, $\mathrm{Im}P_{\rho,\sigma}(q,\omega) \sim N(0)(\omega/qv_F)$ multiplied by appropriate Landau parameters. Correspondingly $\mathrm{Im}\Sigma(\omega,k_F) \sim \omega^2 \mathrm{sgn}\omega$, $\omega^2 \ln\omega\,\mathrm{sgn}\omega$, in 2d.

In Eq.(1) only the frequency dependence has been emphasized. If there are any nesting features in the band-structure, there will doubtless be a q-dependence. The suggestion is that such dependence does not represent either the new or the essential physics.

The MFL hypothesis reconciles the experimental observation of a Fermi-surface similar to the band-structure calculations with the anomalous scattering observed in transport measurements. The linear contribution to the resistivity, and to the tunneling conductance, the ω^{-1} like optical conductivity, the constant Raman scattering intensity, and the non-Korringa contribution to the nuclear relaxation rate T_1^{-1} on Cu(but not the different temperature dependence of T_1^{-1} of Cu and O)are all naturally explained by the MFL hypothesis. Angle resolved photoemission results[4] have been found consistent with the predictions of Eq.(2). Recent neutron scattering results[5] have been observed to be directly consistent with Eq.(1) in the spin-channel.

A microscopic justification of the MFL phenomenology has been recently proposed.[6] In some models, and for sufficiently strong interactions, excitonic charge and spin modes between quasi-coherent one-particle states at the chemical potential and high energy incoherent one-particle states which are residual of the Hubbard bands of the insulating phase give rise to novel singularities in certain low energy scattering amplitudes. The resulting new phase has MFL properties. A new prediction following from this theory is that one-particle spectrum has besides the quasi-coherent MFL spectrum, an incoherent contribution at and around the chemical potential. Another is that the chemical potential shifts towards the middle of the insulating gap both for electron and hole doping compared to a naive rigid band filling of the empty states of the insulator. A brief discussion of this theory is presented in the proceedings of the Kanazawa meeting.[6]

At this school, I wish to describe only the application and extensions of the MFL phenomenology to the superconductive state. There is a considerable body of experimental evidence that although the transition temperature is extraordinarily high, the superconductive state is not extraordinarily different from conventional superconductors like Pb or Sn. From a variety of measurements of the London penetration depth,[1] we know that the superfluid density $\rho_s(T)$ is characteristic of an s-wave superconductor. The NMR measurements are consistent with this conclusion. The most remarkable properties of the superconductive state is an apparent disrespect of coherence factors of the BCS theory. The NMR relaxation rate T_1^{-1} does not have the Hebel -Slichter peak. In BCS theory, the optical conductivity $\sigma(\omega)$ at low frequencies has a behavior similar to T_1^{-1}. In Cu-O materials $\sigma(\omega)$ by contrast has a bump below T_c — lower the frequency the larger the bump. The thermal conductivity also has a bump below T_c.

The other remarkable properties are quantitative. The value of the zero temperature gap $\Delta(0)$ appears unusually large:$2\Delta(0)$

$/T_c \simeq 5$-7 as compared to the weak-coupling value 3.52. Also the gap appears to rise very abruptly below T_c. Finally, it is of interest to know whether the order of magnitude of T_c can be obtained using parameters which fit the normal state.

2. Pairing Interaction, T_c and $2\Delta(0)/T_c$

It is natural to inquire whether the excitations responsible for the anomalous normal state properties are also responsible for pairing. Recent microscopic theory which yields the polarizability, Eq.(1) in the particle-particle spectra, also yields attractive interaction in the particle-particle channel. The simple physics of this is that the particle-particle scattering amplitude which is the pairing kernel in a theory of superconductivity in the s-wave channel is given approximately by $V(q)\mathrm{Re}\,\varepsilon^{-1}(q,\omega)$ where $V(q)$ is the bare interaction. From $\mathrm{Im}P_\rho(q,\omega)$ of Eq.(1) one can obtain a contribution to the density-density correlation function $\tilde{\varepsilon}^{-1}(q,\omega)$. $\mathrm{Re}(\tilde{\varepsilon}^{-1}(q,\omega)-1)$ is then found to be negative for $\omega<\omega_c$, see Fig.(1). Contributions from other processes, not included in Eq.(1) will in general ensure that $\mathrm{Re}\,\varepsilon^{-1}(q,0)>0$ as required for stability. The absorptive part of such processes presumably extends on a scale of the bandwidth $W>>\omega_c$ so that the attractive contribution for $\omega<\omega_c$ can win out for pairing. This is similar to the conventional theory, where electron-electron repulsion essentially only reduces the effective electron-electron attraction via phonons. By similar arguments, the spin fluctuation part $P_\sigma(q,\omega)$ is deleterious for s-wave pairing. This should also be explicitly included in deducing the properties of the superconductive state.

Since $\omega_c<<W$, ε_F one may use the strong-coupling theory of superconductivity due to Eliashberg and Schrieffer, Scalapino and Wilkins. There is one important and crucial difference compared to the case when the pairing kernel comes form exchanging phonons, which to a good approximation, do not alter from above T_c to below T_c. If $P(q,\omega)$ is the important low frequency excitation spectrum of particles and holes which themselves undergo the superconductive instability, its low frequency part, $\omega\lesssim\Delta(T)$, will be strongly modified by pairing. Quite obviously $P(q,\omega)$, T_c, $\Delta(T)$ etc. should be determined self-consistently. The interesting part of $P(q,\omega)$ can be represented as in Fig.(2) i.e., $P=GG\Gamma GG$; where the obvious internal variables and summations are implied. G is the exact single particle Green's function of the normal or superconductive phase

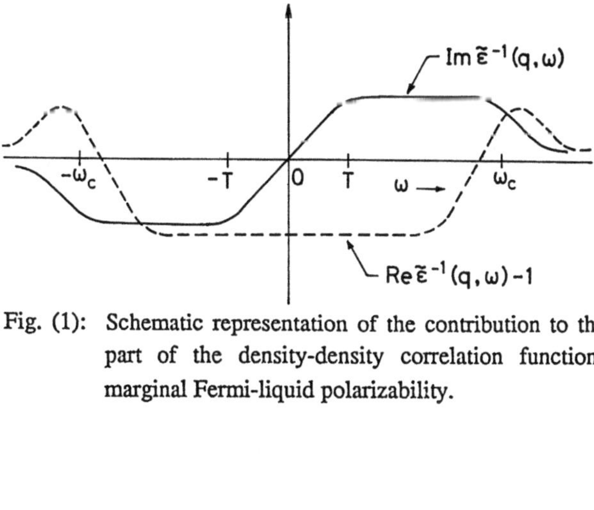

Fig. (1): Schematic representation of the contribution to the Imaginary and the Real part of the density-density correlation function $\varepsilon^{-1}(q,\omega)-1$ from the marginal Fermi-liquid polarizability.

Fig. (2): Exact representation of the polarizability in terms of the two-particle vertex Γ and the one-particle Green's functions.

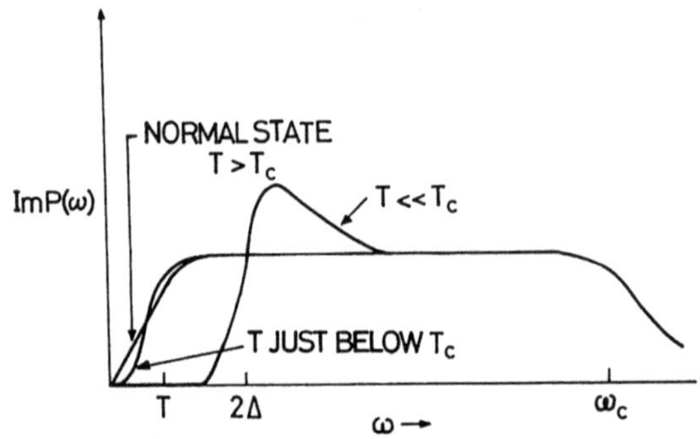

Fig. (3): Schematic change in ImP(q,ω) as a function of ω for ω >> $v_F q$ for temperatures below T_c.

and Γ is the exact two particle vertex. In the normal phase G contains MFL self-energy given by Eq.(2). In the superconductive state one must include both the "normal" and the "anomalous" i.e. the pairing self-energy. In going to the superconductive state both G and Γ change. Calculating the change in Γ requires at the least that we know what Γ is in the normal phase. This requires a microscopic theory of MFL. The recent theory does provide Γ but it has not been used yet to calculate changes in Γ in going to the superconductive state. The calculations done so far for P(q,ω) in the superconductive state have assumed that Γ does not change, only G changes. This is doubtless in correct. I believe it is much worse near T_c than for $T \to 0$. This is because the change in G alone causes a gap 2Δ(0) in ImP(q,ω) for $T \to 0$; (see Fig.(3)) changes in Γ cannot change this. Near T_c, the detailed shape of ImP(q,ω) for $\omega \to 0$, is important for several properties including Δ(T)(but, I think not T_c).

With the above state approximations, Kuroda[7] and I have solved the Eliashberg equations together with the equation for P(q,ω) analytically for T near T_c and T near 0, while Littlewood[8] and Carbotte et al.[9] have numerically calculated the behavior for all temperatures. The results of these calculations depend on three parameters λ_ρ, λ_σ and ω_c. ω_c is known from fitting the normal state properties to be $2\text{-}3\times10^3$K for the Cu-O superconductors. From the normal state resistivity , we know $u\lambda_+ = u(\lambda_\rho+\lambda_\sigma) \simeq 1$ where u<1 parameterizes the fraction of scattering that is Umklapped,(and leads to decay of total momentum). With $u \simeq 1/2$ and $(\lambda_\rho-\lambda_\sigma) \simeq 1.2$, we can calculate $T_c \simeq 10^2$K, with $2\Delta(0)/T_c \simeq 7$. These parameters lead however to a weak first-order transition at T_c, i.e. Δ changes discontinuously. The reason for this is that the energy lowering due to a finite Δ and development of a gap in ImP(q,ω) tend to reinforce each other in a non-linear fashion. It remains to be seen whether a proper theory of ImP(q,ω) near T_c replaces this behavior with just a very rapidly rising Δ(T) near T_c. One can also follow the locally stable solution which shows a second order transition. Then Δ(T) does rise much more rapidly below T_c than in BCS theory. Also, a calculation of T_1^{-1} shows a rapid diminution below T_c rather than a Hebel Slichter peak. We found $(d\ell nT_1^{-1}/d\ell nT)_{T_c} \simeq 3$ compared to the experimental value of about 5. (Of course, the first-order transition solution has a discontinuous drop in T_1^{-1} at T_c). The reason for the decrease in T_1^{-1} rather than the peak is the diminution in $ImP_\sigma(q,\omega)$ for $\omega \to 0$ of T just below T_c. This eliminates the large constant contribution to T_1^{-1} in the normal state which arises from $ImP_\sigma(q,\omega) \sim \omega/T$ for $\omega \ll T$.

An important consequence of the self-consistency requirement (which is incorporated in the calculations above) is the change in inelastic scattering from above to below T_c[8-10]. As always inelastic scattering (just as well as elastic time-reversal breaking scattering) is pair-breaking. This is an unusually strong effect here since the inelastic scattering rate is $\sim T$. Thus T_c is strongly depressed compared to what it would be in the absence of inelastic scattering. On the other hand $\Delta(0)$ is determined in the absence of inelastic scattering. Hence the large value of $2\Delta(0)/T_c$.

The change in inelastic scattering has dramatic consequences on the optical conductivity and the thermal conductivity. The point is that long wavelength properties depend on the inelastic scattering rate $\tau^{-1}(\omega)$. The sharp decrease in $\tau^{-1}(\omega)$ below T_c leads to a bump in both these quantities below T_c. On the other hand, the nuclear relaxation rate T_1^{-1} which is a local probe and so sums overall q does not involve τ^{-1}. The absence of a coherence peak in T_1^{-1} while at the same time peak in $\sigma(\omega)$ below T_c at low frequencies is a very strong confirmation of the idea that a continuum of particle-hole excitations exists in the normal state and that it promotes pairing. The self-consistent elimination of this continuum below T_c leads to a rapid decrease of T^{-1}.

3. One particle excitation spectra in the superconducting state

Consider the one particle self-energy in the superconductive state for $T \ll T_c$. This is represented exactly in Fig.(4) in terms of the one-particle Green's functions and the two-particle vertex in the superconductive state. It is at once obvious that the self-energy itself has a gap 3Δ. This implies that in the passage to superconductivity, not only does the quasi-particle acquire a minimum energy Δ, it looses its decay processes and therefore becomes sharp until its energy reaches 3Δ where it can again decay and acquire a width. This means for $T \ll T_c$, for any k such that $E_k = \sqrt{\varepsilon_k^2 + \Delta^2}$ is between Δ and 3Δ, the one particle spectra has a peak near E_k and a broad hump beginning at 3Δ. Such a predicted spectra is shown in Fig.(5) taken from Ref.8. The clear verification of this prediction by angle resolved photoemission spectroscopy requires a net average resolution better than Δ. This is not available yet. There are already interesting experimental results, however, which do show a two-humped structure.

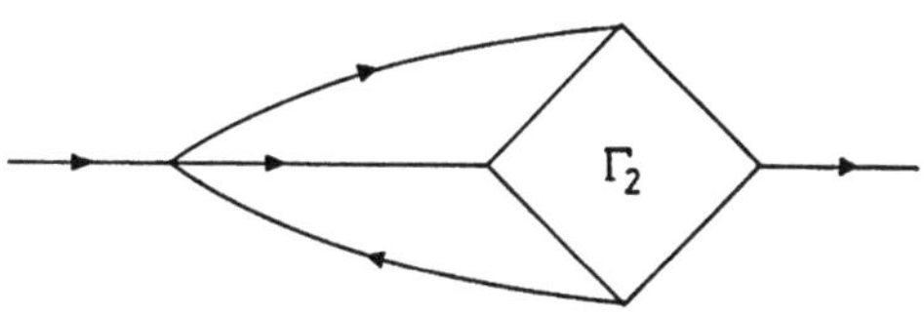

Fig. (4): Exact representation of the one-particle self-energy in terms of the two-particle vertex.

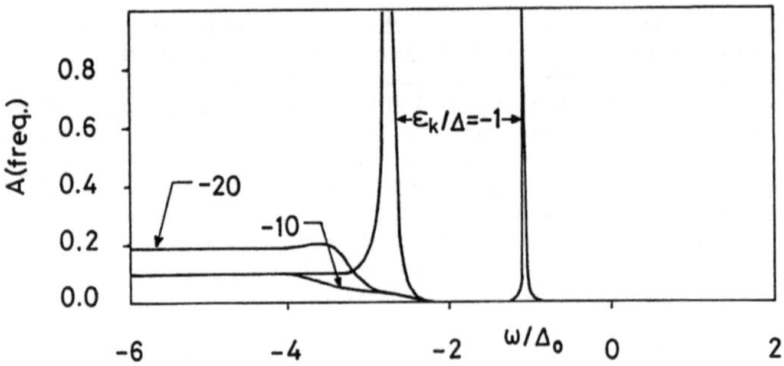

Fig. (5): Predicted one-particle spectral weight for a fixed q as a function of ω for $T \ll T_c$.

4. Unresolved Problems

As I mentioned earlier, a principal problem remaining is the form of the vertex Γ and T_c. I wish here to propose a simple extension of the phenomenology to the superconductive state for Γ in the particle-hole channel. It is possible to show from Fig.(2), which gives the exact expression for the polarizability, that one way to get Eq.(1) for the normal state is to have $\Gamma(\omega)$ of the same form. Below T_c, a new scale $\Delta(T)$ should enter. I propose that at the very least

$$\mathrm{Im}\Gamma(\omega) \begin{array}{l} \sim \omega/\sqrt{T^2+\Delta^2(T)}, \quad \omega \ll \max(\Delta,T) \\ \sim \text{constant}, \quad \omega \gg \max(\Delta,T) \end{array} \qquad (5)$$

This should soften the strong feed back effect due to the reduction of the inelastic scattering which leads to the first order transition in the calculations mentioned. It is also possible that $\mathrm{Im}\Gamma(\omega)$ itself will develop a gap below T_c. In any case, there should be a reduction in $\mathrm{Im}\Gamma(\omega)$ near T_c of the form — $\Delta^2(T)/T_c^2$.

Another problem that needs looking into carefully is optical absorption and Raman scattering below T_c. As noted,[8,14] BCS type theory with pairing (and normal) self-energy from electronic excitations would lead to a gap 4Δ in such experiments. Impurities will produce a threshold at 2Δ. The important thing to investigate is the role of mutual quasi-particle interactions (i.e., theory beyond BCS) in causing absorption below 4Δ as well as the role of the periodic lattice on it.

5, Acknowledgments:

All the work reviewed here was done in collaboration with Y.Kuroda and P.B.Littlewood.

References

1. See, for example the articles by B.Batlogg and N.P.Ong in this book.
2. C.M.Varma, P.B.Littlewood, S.Schmitt-Rink, E.Abrahams and A.E.Ruckenstein, Phys.Rev.Lett. 63, 1996 (1989).
3. G.Kotliar et al., Europhysics Letters (1991).

4. C.G.Olson et al., Phys.Rev. **B42**, 381 (1990).
5. S.M.Holden et al., Phys.Rev.Lett. **66**, 82 (1991).
6. A.E.Ruckenstein and C.M.Varma, Proceedings of M^2S-III Conference at Kanazawa, to be published in Physica (1991).
7. Y.Kuroda and C.M.Varma, Phys.Rev. **B42**, 8619 (1990).
8. P.B.Littlewood nd C.M.Varma, Journal Appl.Phys. **69**, 4979 (1991).
9. E.J.Nicol, J.P.Carbotte and T.Timusk, Phys.Rev. **B43**, 473 (1991), Solid State Comm. **76**, 937 (1990). E.J.Nicol, J.P.Carbotte, Phys.Rev. **B43**, 1158 (1991) and preprint.
10. M.Nuss et al., Phys.Rev.Lett. **66**, 3305 (1991).
11. K.Holczer et al., Phys.Rev.Lett. **67**, 152 (1991).
12. D.A.Bonn et al., (preprint).
13. D.S.Dessau et al., Phys.Rev.Lett. **66**, 2160 (1991).
14. J.Orenstein, S.Schmitt-Rink, and A.E.Ruckenstein, to be published in the Proceedings of the International Winterschool on Electronic Properties of Superconductors, Kirchberg, Austria, February (1990).

Antiferromagnetic Fluctuations in High-T_c Oxides

T. Moriya

Department of Physics, Faculty of Science and Technology,
Science University of Tokyo, 2641 Yamazaki, Noda 278, Japan

Anomalous physical properties of high T_c oxides above T_c are explained in terms of antiferromagnetic fluctuations in two dimensional itinerant electron systems as described by the self-consistent renormalization theory of spin fluctuations. Discussions are given for the applicability of the model and approach to high T_c oxides as doped Mott insulators.

1. Introduction

The most striking features of high T_c oxides are their anomalous physical properties above T_c. For example the electrical resistivity is linear in temperature in a wide range of temperature above T_c, the infrared conductivity deviates from the Drude form, showing the relaxation rate proportional to the frequency, and the nuclear spin-lattice relaxation rate shows anomalous temperature dependence different from that in normal metals, etc.

In order to explain these properties various models and approaches have been proposed. It is now generally believed that the high T_c oxides are doped Mott insulators, including charge transfer insulators, and the electrons responsible for the conduction are mostly confined in each of the CuO_2 planes of the crystal, forming two dimensional systems.

In this note we first discuss how the above-mentioned anomalous physical properties are consistently explained in terms of antiferromagnetic fluctuations in two dimensional itinerant electron systems as described by the self-consistent renormalization(SCR) theory of spin fluctuations, an established approach around the critical phase boundary for the magnetic long range order [1]. Next we discuss on the possible physical reasons for the successful application of this model and approach to high T_c oxides. For this purpose we start from the discussion on the Mott insulators and Mott transitions and then proceed to the doped Mott insulators. The relation between the present approach and the approaches from the strong correlation limit will also be discussed.

Springer Series in Solid-State Sciences, Vol. 106 Physics of High-Temperature Superconductors
Eds.: S. Maekawa and M. Sato

2. Anomalous physical properties above T_c

We first show in the left hand side columns of Table 1 the predictions of the SCR theory on the anomalous temperature dependences of various physical quantities due to spin fluctuations around the magnetic phase boundary in three dimensional (3D) itinerant electron systems. These anomalies are predicted in the shaded area in Fig.1 and all of them have been confirmed by experimental investigations in 1970's and early 1980's. For details we refer to [2].

The extension of the SCR theory to two-dimensional (2D) itinerant electron systems gives the predictions shown in the middle columns of Table 1. The right hand side column gives the behaviors of the normal Fermi liquids.

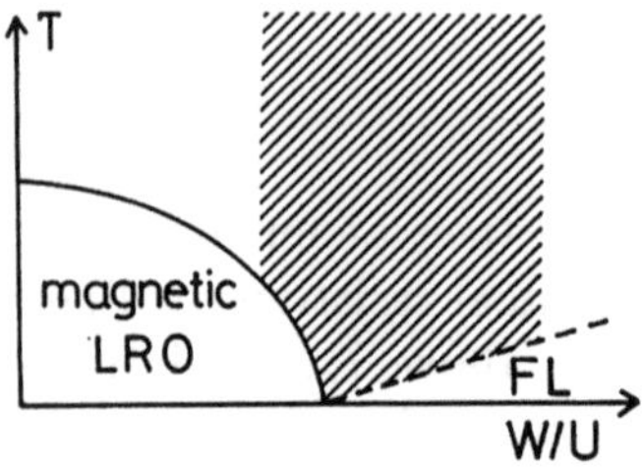

Fig.1. Phase diagram of an interacting itinerant electron system. W/U symbolizes the ratio of kinetic energy and magnetic interaction. LRO and FL stand for long range order and Fermi liquid, respectively. In the shaded area anomalous temperature dependences of physical quantities due to spin fluctuations are predicted (see Table 1).

Table1. Anomalous temperature dependences of physical quantities around the critical phase boundary for itinerant ferro- and antiferromagnetism predicted by the SCR theory. χ_Q: magnetic susceptibility, R: electrical resistivity, T_1: nuclear spin-lattice relaxation time, C_m: magnetic contribution to the specific heat, C.W. (Curie Weiss): T+θ.

	3-dimensional		2-dimensional		normal Fermi liquid
	ferro(Q=0)	a.f.	ferro(Q=0)	a.f.	
χ_Q^{-1}	$T^{4/3} \rightarrow$ C.W.	$T^{3/2} \rightarrow$ C.W.	$T^{4/5} \rightarrow$C.W.	$T + \theta$	a+bT^2
R	$T^{5/3}$	$T^{3/2}$	$T^{4/3}$	T	T^2
T_1^{-1}	$T\chi$	$T\chi_Q^{1/2}$	$T\chi^{3/2}$	$T\chi_Q$	T
C_m/T	$-\ln\vert 1-\alpha_s\vert$		$\vert 1-\alpha_s\vert^{-1/2}$	$-\ln\vert 1-\alpha_s\vert$	

It is striking to find that the anomalous behaviors due to the antiferromagnetic fluctuations in 2D systems are just what have been observed in high T_c materials. Furthermore, the systematic comparisons between the theory and experiment have been made on the quantitative level. In other words the SCR theory contains a smaller number of parameters than the number of experimental methods to determine them. Thus the parameter-free test of the theory has been made successfully for some of the physical quantities, for example, optical conductivities [1].

3. Mott insulator and Mott transition

Let us now consider a Hubbard model with a half-filled band (transfer integral is not limited between the nearest neighbors only). A general phase diagram in the $(t/U) - T$ plane is sketched in Fig.2. It is well-known that for the Mott insulators the approach from the strong correlation limit, i.e. an expansion in terms of t/U leading to the Heisenberg model or local moment picture, is established [3]. At the same time it should be kept in mind that the antiferromagnetic ground state for small t/U is well described also in terms of the Bloch electrons within the Hartree-Fock (HF) approximation [2,4]. Furthermore in this case the spin wave excitations from the ground state are correctly described by the random phase approximation (RPA) [2].

As t/U increases beyond a certain value the Mott transition to metallic phase takes place. For the ground state this generally is a transition from an antiferromagnetic insulator phase to an antiferromagnetic metal phase. With still increasing t/U, we encounter a phase boundary between antiferromagnetic and paramagnetic metals. Around this phase boundary the SCR theory is established in all the temperature ranges of interest. In the ground state the HF-RPA result is renormalized by the effect of zero-point spin fluctuations but qualitative nature of the ground

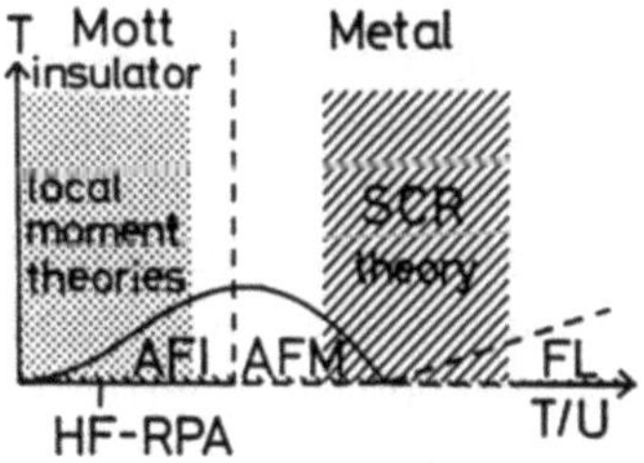

Fig.2. Phase diagram of an interacting electron system with a half-filled band, showing the Mott transition. The ranges of applicability of valid approaches are indicated. AFI and AFM stand for antiferromagnetic insulator and metal, respectively and FL for Fermi liquid.

state is maintained [5]. Thus the ground state is described in terms of the Bloch electrons within the Hartree-Fock theory with certain many body corrections in the entire range of (t/U). At finite temperatures, however, the HF-RPA theory always fails and we must use an appropriate theory of renormalized spin fluctuations in each regime; the local moment theories for small (t/U) and the SCR theory for weak itinerant ferro- and antiferromagnets as indicated in Fig.2. Thus we have established theories in the small (t/U) regime and around the critical phase boundary for magnetic long range order. In the intermediate regime we have a spin fluctuations theory, though within a static approximation, to interpolate between the two established regimes [6]. An important point is that the local moment approach from the strong correlation limit is smoothly connected with the spin fluctuation theory of weak itinerant magnetism though within a static approximation.

4. Doped Mott insulators

Now we discuss doped Mott insulators. When the Mott insulator is doped with small number of electrons or holes the Anderson localization is expected to take place because of the effect of random impurity potentials. With significant degrees of doping, however, fairly free motion of carriers are expected because of the screening effects on the impurity potentials. Thus by neglecting the random impurity potentials for brevity, we deal with the system of interacting itinerant electrons. It is now well known that the ground state of itinerant electron systems are fairly well described by the band theory particularly when the magnetic long range order is large. Even for the paramagnetic ground state of strongly correlated itinerant electron systems the band theory gives a good starting point of the theory; the Fermi surface is correctly given though the effective mass is generally enhanced significantly from the calculated value.

Since the ground state of the high T_c oxides, when the superconductivity is suppressed, is expected to be weakly or nearly antiferromagnetic, it is natural to apply the SCR theory for these systems.

It should be noted here that the above arguments apply certainly to three dimensional systems but not quite evidently to two dimensional systems of the present interest. However, the successful application of the SCR theory for 2D systems to interpret the anomalous properties of high T_c oxides seem to be a strong support for the above reasoning. It should also be noted that the Fermi liquid theory seems to apply to a 2D electron gas.

5. Superconductivity

We have seen that the anomalous physical properties including electrical conductivity in high T_c oxides above T_c are well explained in terms of antiferromagnetic spin fluctuations. This fact naturally makes us to expect that the superconductivity is also explained in terms of antiferromagnetic fluctuations. This problem is treated by using the weak coupling and strong coupling theories. The order parameter is shown to be of B_{1g} character and the critical temperature is evaluated to be of the right order of magnitude. We refer to [1] for this part of the discussion.

References

[1] T. Moriya, Y. Takahashi and K. Ueda: J. Phys. Soc. Jpn. **59** (1990) 2905; in "Physics and Chemistry of Oxide Superconductors" eds. H. Yasuoka and Y. Iye (Springer, 1991);
T. Moriya and Y. Takahashi: J. Phys. Soc. Jpn. **60** (1991) 776.

[2] See for example: T. Moriya "Spin Fluctuations in Itinerant Electron Magnetism" (Springer, 1985).

[3] P.W. Anderson: Phys. Rev. **115** (1959) 2; in "Solid State Physics" **14**, eds. F. Seitz and D. Turnbull (Academic, 1963) p.99.

[4] N.F. Mott: "Metal-Insulator Transitions" (Taylor and Francis, 1974).

[5] T. Moriya and A. Kawabata: J. Phys. Soc. Jpn. **34** (1973) 639; **35** (1973) 669.

[6] T. Moriya and Y. Takahashi: J. Phys. Soc. Jpn. **45** (1978) 397;
T. Moriya and H. Hasegawa: J. Phys. Soc. Jpn. **48** (1980) 1490.

The Physics Underlying High-T_c Superconductivity: Novel vs Exotic*

J.R. Schrieffer

Department of Physics, University of California,
Santa Barbara, CA 93106, USA

ABSTRACT: Abstract: The parent compounds of high temperature oxide superconductors are antiferromagnetic. The nature of hole excitations in the parent compounds is discussed in both the weak coupling or spin density wave (SDW) regime and the strong coupling, or t-J model limit. One finds that the hole becomes dressed by a cloud of antiferromagnetic spin fluctuations which describe the reduction of the staggered antiferromagnetic order parameter in the vicinity of the hole. In addition a twist of the spin quantization axis occurs near the hole. As the hole concentration x increases, long range antiferromagnetic order is suppressed with finite range spin fluctuations occuring in the superconducting phase. The nature of the hole-like excitations in this regime is discussed. One finds that as in the ordered antiferromagnetic regime, the one particle spectral function is dominated by two peaks centered near the SDW or Mott Hubbard upper and lower bands. In addition, there appears a third peak of fundamental importance in understanding the po sition of the chemical potential as a function of doping. This peak located between the upper and lower bands corresponds to a fully dressed quasi particle with very small weight z_k which is populated by doping and can be probed by photoemission spectroscopy. This split off band removes spectral weight from both the lower and upper bands and leads to structure of the photoemission intensity versus binding energy curve in the gap as seen experimentally. For increased doping the spin fluctuations are reduced in amplitude as the spin-spin correlation length becomes of order the atomic spacing. In this regime the Landau

---

* This presentation is based on "The Evolution of Spin Bags from Small to Large U: Implications for Spectroscopies", A.P. Kampf and J.R. Schrieffer, Proc. Argonne High T_c Conference, March, 1991.

Springer Series in Solid-State Sciences, Vol. 106 Physics of High-Temperature Superconductors
Eds.: S. Maekawa and M. Sato

quasiparticle scheme is restored. It is argued that spin bags or spin polarons evolve smoothly from the weak coupling SDW to the strong coupling large U regime.

I. INTRODUCTION

Current theoretical efforts to understand high temperature superconducting materials can be classified into two categories, namely novel and exotic. In novel approaches a pairing type condensation is assumed to exist with fermion quasi particles of charge $\pm e$ and $s = \frac{1}{2}$ being paired by an effective attractive interaction, with superconductivity already latent in a two dimensional copper oxide plane and phase fluctuations being suppressed when interplanar hopping extends the pairing correlations to three dimensions.

In exotic approaches, many if not all of these conditions are violated. For example, in the resonating valence bond and related approaches, the excitations are taken to have charge $\pm e$ and spin 0 (holon) and charge 0 and spin one-half (spinon). Thus, it is presumed that for the doped material in the absence of long range antiferromagnetic order, an injected hole spontaneously fissions into a holon and spinon, the presumed stable excitations. While at first sight such a spontaneous decay appears peculiar, similar effects actually occur in quasi one-dimensional conductors such as polyacetylene where the excitations are charged spinless solitons and neutral spin one-half solitons. The statistics of such excitations depends on the theoretical framework being used. So-called anyon statistics implies that the ground state of the system breaks time reversal symmetry as in the ground state of the quantum Hall effect where an external magnetic field B_0 is imposed. In high T_c materials spontaneously broken time reversal symmetry is assumed although the existence of a ground state having such broken symmetry has yet to be demonstrated either experimentally or theoretically starting from a credible Hamiltonian. It is well known that one can transmute the statistics of particles in two dimensions by attaching statistical flux tubes to the particles. Thus, anyons can be transformed to a bosonic or fermionic representation with a suitable interaction between the statistical flux tubes being included. Most of these exotic schemes require at least weak three dimensional coupling for superconductivity to occur.

While the two categories of theories appear to be exceedingly different, there are certain common features between them. The first such feature is the fact that the quasiparticle renormalization constant z_k is small compared to unity. That is, the amplitude for a bare hole to be inserted in a fully dressed state is small compared to one. This feature occurs not only in the two schemes mentioned above but also in the so-called marginal Fermi liquid[1] approach in which the hole self energy exhibits anomalous behavior near the chemical potential.

A second common feature is the strong correlations leading to non-Fermi liquid temperature and energy dependence in the normal state which occurs in both limiting approaches. Such anomalous normal state properties are dictated by experiment. What is less clear is which of the approaches to the superconducting phase will ultimately be correct. Recent experiments probing the broken time reversal symmetry of the superconducting phase suggest evidence both pro and con. We must await the final experimental resolution to this fascinating question.

The generic phase diagram of high temperature superconductors is shown in Figure 1a.

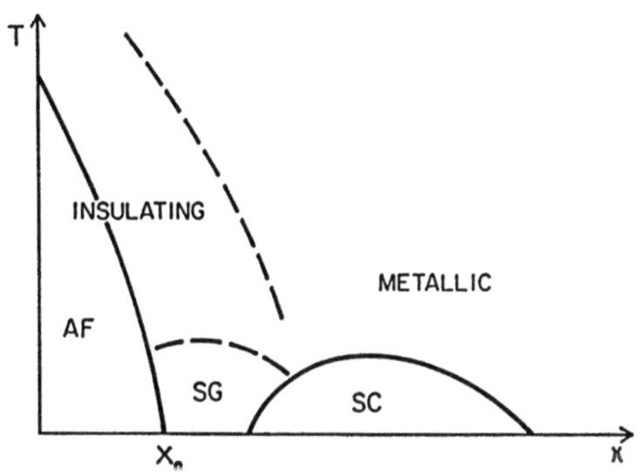

Figure 1a Phase diagram of high T_c materials of doping x.

For weak doping the system is an antiferromagnetic insulator with the Néel temperature vanishing at a concentration x_0 which is typically of order 3% holes. The smallness of x_0 indicates that the magnetic form factor of a dressed hole involves the suppression of antiferromagnetism in a significant region surrounding the hole encompassing of order 10 neighboring sites. This picture is consistent with the fact that long range antiferromagnetic order in such an anisotropic antiferromagnet continues only to small hole concentrations. For doping larger than x_0 a spin glass-type phase appears in which the dressed holes are likely pinned to charged impurities, perhaps in the fashion of weak localization. For larger doping, this glass melts and the semiconducting behavior of the spin glass converts into metallic behavior leading to superconductivity. In this regime, one is dealing with a strongly correlated system or a "strange metal". Finally, as the doping continues, the system presumably approaches Fermi liquid behavior although thermodynamically it is difficult to achieve such large dopings in a stable material.

Figure 1b shows a schematic density of states as a function of energy for each of the principal regimes–the antiferromagnetic insulator, the strongly correlated metal and the paramagnetic Fermi liquid.

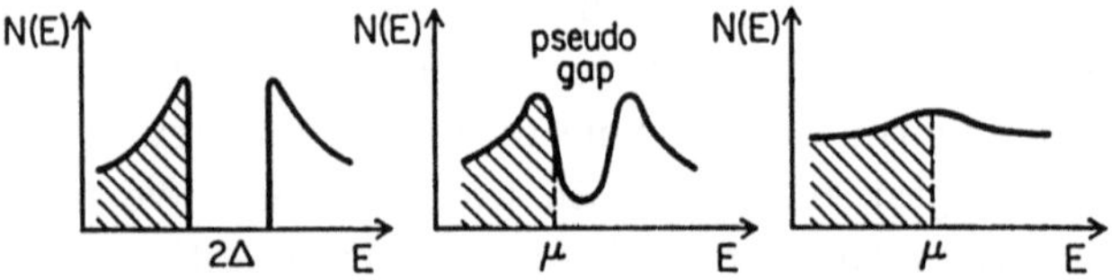

Figure 1b A schematic density of electronic states for the antiferromagnetic, pseudogap and Fermi liquid regimes.

One sees that the gap characteristic of the antiferromagnetic insulator is transformed into a pseudogap in the strange metal phase and finally a bump characteristic of a mass enhancement in the Fermi liquid regime. The principal theoretical puzzle is to understand the normal phase of this strange metal and how the peculiar excitations in this regime lead to high T_C superconductivity.

II. THE HUBBARD MODEL

The essential physics of cuprate superconductors is centered on transport in the two dimensional CuO_2 planes. As mentioned above, phase fluctuations of the superconducting order parameter are quenched by hopping along the direction c perpendicular to the ab planes. Since there are three important orbitals per copper oxides unit cell, namely the copper $d_{x^2-y^2}$ and p_σ for each of the two oxygen atoms, there are three bands important for the spectroscopic properties of these materials. However, since the lower two bands are filled and the upper band is half empty for the parent antiferromagnetic insulator, it is likely that the dominant low energy effects essential for superconductivity can be mapped into a one-band problem in which the copper d orbital and the oxygen p_σ orbitals mix to form an antibonding, partly filled band. It is this upper antibonding band which we consider in an effective two dimensional one band Hubbard model. Interband transitions are, no doubt, important for optical spectra.

The one band Hubbard model is defined by the Hamiltonian

$$H = -t \sum_{\substack{<ij> \\ \sigma}} (c_{i\sigma}^{+} c_{j\sigma} + h.c.) + U \sum_{i} n_{i\uparrow} n_{i\downarrow}, \qquad (1)$$

In Figure 2a, the mean field Néel temperature T_N is sketched as a function of the coupling constant $g \equiv U/4t$.

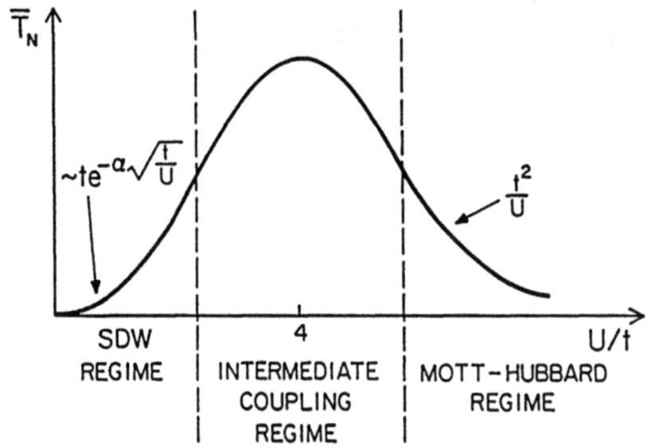

Figure 2a Mean field Néel temperature versus U/t.

The Néel temperature peaks for g of order unity, with T_N vanishing for both large and small values of g. The region g small compared to unity corresponds to the spin density wave or weak coupling behavior while, g large compared to unity corresponds to the Mott Hubbard strong coupling regime. Experimental evidence suggests that physical reality in the cuprate superconductors occurs for intermediate coupling, with g of order unity.

In Figure 2b the density of $\uparrow$ spin (dashed line) and $\downarrow$ spin (solid line) is shown for the three regimes.

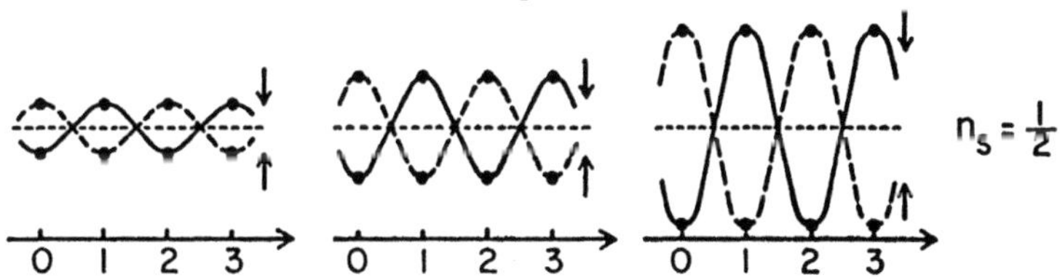

Figure 2b Density of spin up (dashed line) and spin down (solid line) electrons in the weak, intermediate and strong coupling regimes.

For g of order unity the $\uparrow$ spin tends to sit mainly on the even sites while the down spin sits primarily on the odd sites, although there are fluctuations to the opposite spin configuration. For weak coupling, scattering of electrons by the oscillating exchange potential set up by the spin density wave leads to an energy gap 2Δ. This gap grows as the amplitude of the spin density wave increases and saturates at the value U for large values of g. It is important to note that the gap essentially disappears well above the Néel temperature for weak coupling while it remains essentially U in the strong coupling regime. One can view the persistence of the gap in the spin disordered regime either as a local Coulomb correlation effect or equally well as a local exchange interaction set up by the local moments on the individual sites. These two views are related by a simple mathematical identity

$$2\,U n_{i\uparrow} n_{i\downarrow} = \frac{-U}{2}(n_{i\uparrow} - n_{i\downarrow})^2 + \frac{U}{2}(n_{i\uparrow} + n_{i\downarrow})^2. \tag{2}$$

We first consider the weak coupling approach for small doping[2]. The sublattice magnetization M_z is shown in Figure 3 as a function of g.

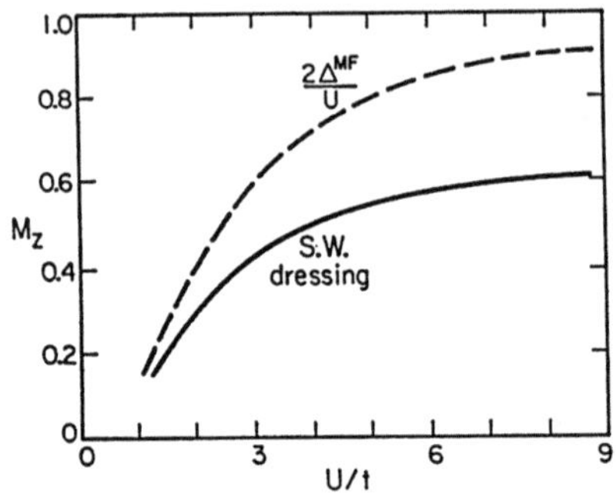

Figure 3 Sublattice magnetization as a function of U/t in the mean field approximation (dashed line) and with gaussian spin fluctuation dressing of the ground state (solid line).

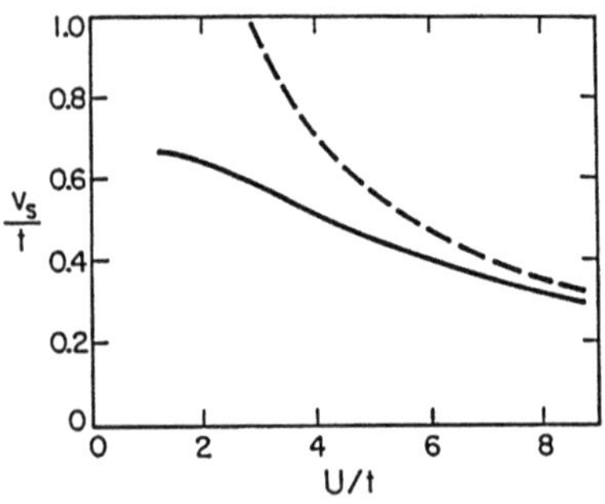

Figure 4 Spin wave velocity for harmonic spin waves about the mean field SDW state (solid line) and the result of the 2-d Heisenberg model with $J = 4t^2/U$.

The results of the mean field calculation are shown in the dashed line with the reduction of the sublattice magnetization due to Gaussian spin fluctuations as indicated by the solid line. In Figure 4, the spin wave velocity in units of ta is shown, where a is the lattice spacing.

The solid line indicates the results of mean field plus Gaussian spin wave dressing while the dashed line indicates the results of the antiferromagnetic Heisenberg model. Interestingly, the mean field result reduces to Anderson's result for the Heisenberg model in the large U limit in the absence of spin fluctuation corrections. Thus it appears that the mean field plus Gaussian spin fluctuation approximation gives a reasonable account of the undoped antiferromagnet over the entire coupling constant domain.

III. Fermion Excitations

As shown in Figure 5, for $U = 0$ the electronic spectrum is given by

$$\varepsilon_k = -2t(\cos\ k_x a + \cos\ k_y a). \tag{3}$$

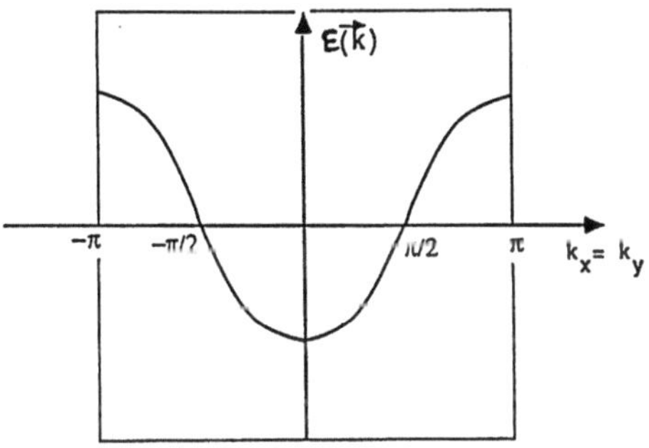

Figure 5 The Bloch energy ϵ_k for $k_x = k_y$.

In the presence of the spin density wave, exchange Bragg scattering leads to a gap 2Δ in the spectrum with the energies E_{k} being

$$E_{\mathrm{k}} = \pm\sqrt{\varepsilon_k{}^2 + \Delta^2}, \tag{4}$$

as shown in Figure 6a.

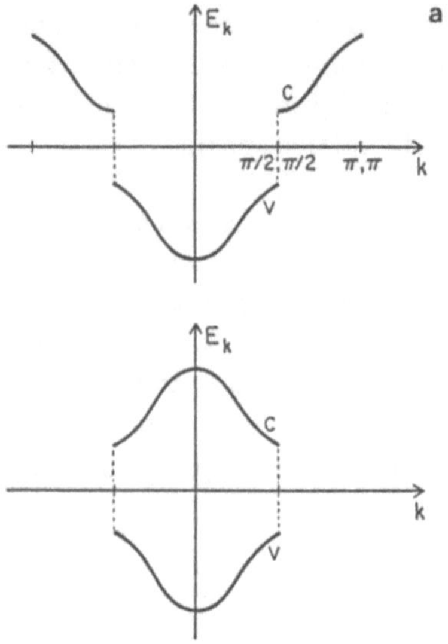

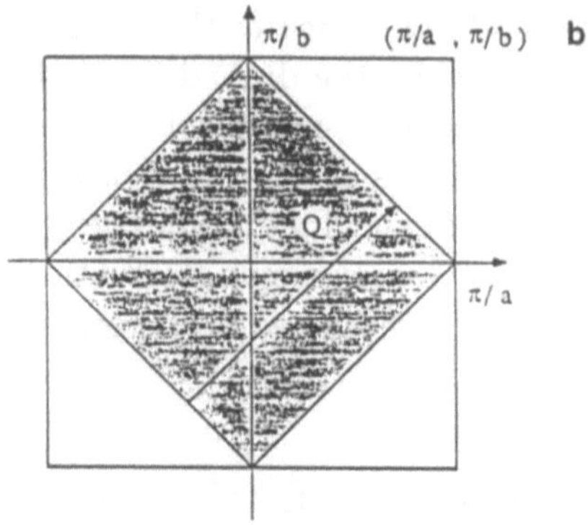

Figure 6a The mean field SDW Bloch energy ϵ_k in the full zone (top panel) and in the magnetic (reduced) zone (lower panel).

Figure 6b The full zone and the magnetic zone, illustrating the nesting vector Q.

Electronic states can be described either in the original Brillouin zone or in the magnetic zone where the upper band is folded back into this smaller zone, indicated by the shaded portion of Figure 6b.

The quasiparticle operator γ_k creating its mean field excitations is given by

$$\gamma_{ks} = u_k c_{ks} + s v_k c_{k+Q,s}, \tag{5}$$

where **Q** is the SDW wave vector $(\pi/a, \pi/a)$. In the undoped antiferromagnetic insulator, the lower band is fully occupied while the upper band is empty. In the large U limit the two bands go over to the lower and upper Mott Hubbard bands. In actuality, an added electron or hole locally weakens antiferromagnetism and leads to a dressing of the fermion with a cloud of spin fluctuations. The dressed excitation, a spin bag or one band polaron, is illustrated in Figure 7 where the staggered magnetization

$$\overrightarrow{M}_n = < \overrightarrow{S}_n > \cos \overrightarrow{Q} \cdot \overrightarrow{R}_n, \tag{6}$$

is reduced in magnitude within a region of size ξ_{SDW} defined by

$$\xi_{SDW} = \hbar v_F / \pi \Delta_{SDW}. \tag{7}$$

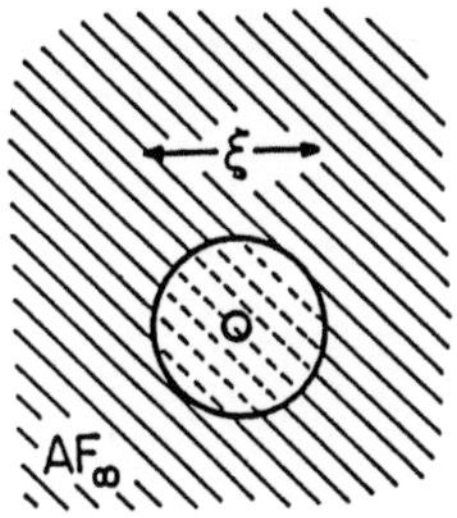

Figure 7 Reduction of the staggered magnetization within the spin bag of size ξ.

The hole self energy $\Sigma_S(\mathbf{k}, \omega)$ is shown within the one loop approximation in Figure 8.

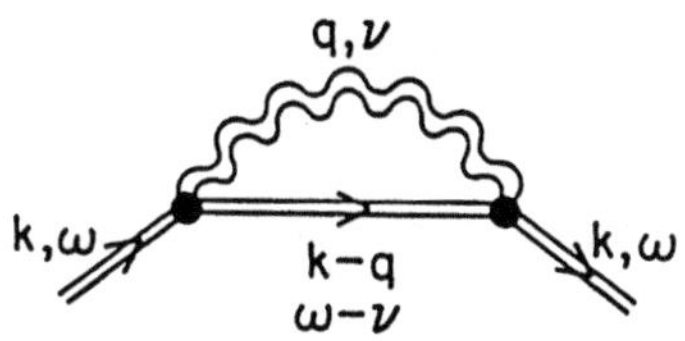

Figure 8 Hole self energy, where the fermion propagator and the vertices represent γ particles, Eqn. (5), and the boson is the RPA spin wave, both longitudinal and transverse modes.

Notice that both the fermion and boson line are excitations within the mean field SDW approximation with coherence factors entering at the vertices. The boson line is the longitudinal or transverse spin susceptibility for the SDW. χ^{zz} leads to the reduction of M_z within the bag while $\chi^{\pm}$ creates the spin twist about the bag with the spatial variation given by an inverse power law falloff. The spin bag is analogous to the polaron in a semiconductor where the band gap is locally reduced in the presence of an added electron or hole.

When two holes are added the interaction is given to leading order by the exchange of one dressed spin fluctuation as indicated in Figure 9a.

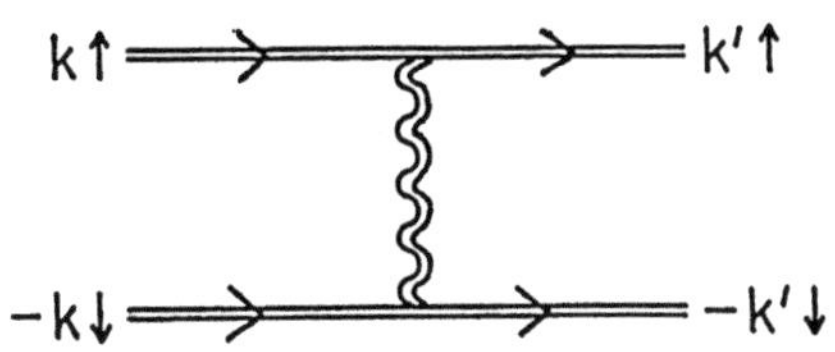

Figure 9a The pairing interaction between γ states within the one spin wave approximation.

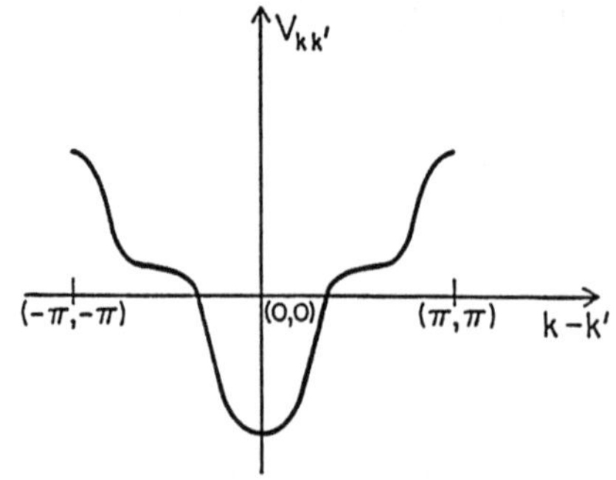

Figure 9b Pairing potential for thc graph of Figure 9a.

The pairing potential $V_{\mathbf{kk'}}$ within this approximation is shown in Figure 9b.

These result first derived by Wen, Zhang and one of the authors shows a region of strong attraction for momentum transfers $\mathbf{k} - \mathbf{k'}$ smaller than the inverse bag size. This is in sharp contrast with the Born approximation for bare particles interacting with bare phonons as shown in Figure 10 with V being repulsive for all values of $\mathbf{k} - \mathbf{k'}$. Therefore, the bag effect is essential to obtain an attractive interaction in this system.

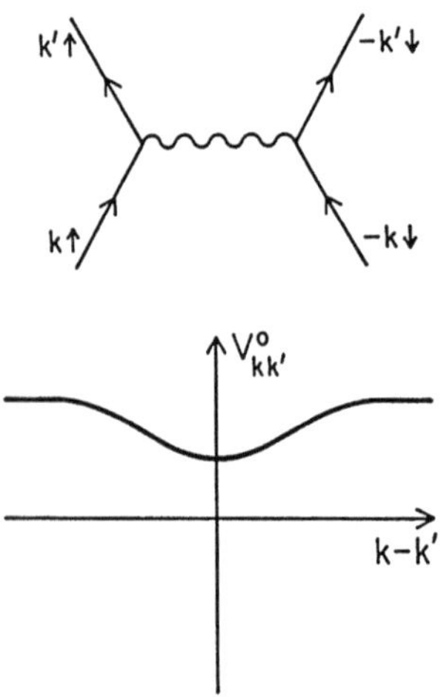

Figure 10 Pairing interaction diagram between bare Bloch electrons c_k (upper panel), illustrating that the potential is repulsive in this approximation for all $k - k'$.

IV. Intermediate Doping and Short Range Spin Correlations

We next consider the doping range x corresponding to a strongly correlated metal as shown in Figure 1. This regime can be approached from the large x regime where Fermi liquid theory presumably holds. On reducing x, antiferromagnetic spin fluctuations build up, both in amplitude and in spacial coherence with the spin-spin correlation length eventually diverging as one approaches the antiferromagnetic insulator instability.

Kampf and Schrieffer[3] have explored this regime by considering the one particle self energy $\Sigma_S(\mathbf{k},\omega)$ as illustrated in Figure 11 within the one loop approximation.

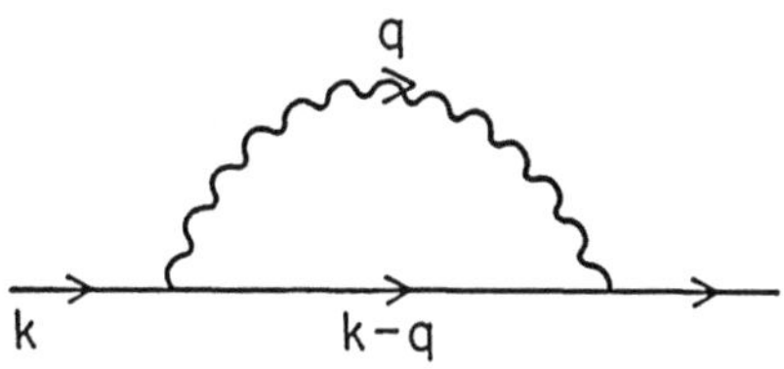

Figure 11 One loop self energy in the paramagnetic regime. The spin wave is summed over longitudinal and transverse modes.

The boson propagator was represented by the susceptibility

$$\chi(\mathbf{q},\omega) = \lambda^2 \sum_Q \frac{\Gamma}{(\mathbf{q}-\mathbf{Q})^2+\Gamma^2} F(\omega), \tag{8}$$

where Γ is the reciprocal of the spin-spin correlation length and $F(\omega)$ is given by

$$F(\omega) = \int \frac{2\nu g(\nu)\, d\nu}{\omega^2-\nu^2+i\delta}, \tag{9a}$$

$$g(\nu) = \begin{cases} \alpha \mid \nu \mid, \mid \nu \mid < \omega_0 \\ \quad 0, \quad otherwise. \end{cases} \tag{9b}$$

This form is similar to that used by Millis, Monien, and Pines[4]

$$\chi_{AF}(\mathbf{q},\omega) = \frac{\chi_{\mathbf{Q}}}{1+(\mathbf{Q}-\mathbf{q})^2\xi^2 - i\cdot\frac{\omega}{\omega_{SF}}}, \tag{10}$$

For large doping the spin-spin correlation length is very short and one obtains a self energy of the form shown in Figure 12.

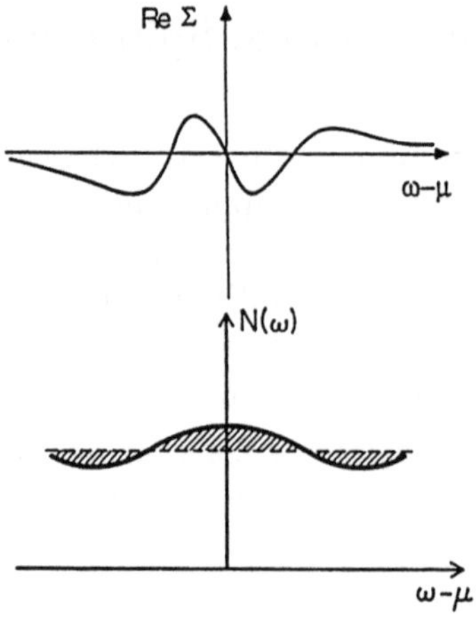

Figure 12 The real part of the electron self energy $Re\Sigma$ in the Fermi liquid regime (top panel) and the corresponding density of states enhancement at the chemical potential, corresponding to $m^t/m > 1$.

An essential point is that in this regime the slope of Σ as a function of ω is negative over a substantial region around the Fermi surface. This leads to an increase of the density of states near the Fermi surface and a depletion away from the Fermi surface.

This conventional Fermi liquid picture is drastically altered when the spin-spin correlation length becomes large compared to the interatomic spacing. In this case, the real part of Σ takes on a very different form as shown in Figure 13a.

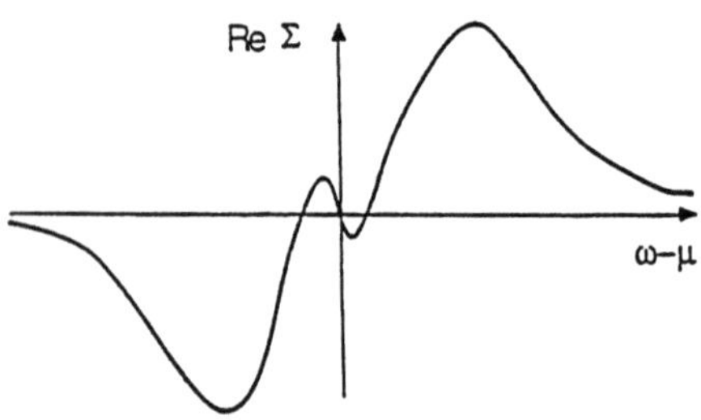

Figure 13a $Re\Sigma$ in the presence of strong antiferromagnetic spin fluctuations in the paramagnetic phase (top panel).

Except for the small wiggle near the chemical potential, the basic slope of Σ is positive rather than negative. Thus, since the real part of Σ corresponds to the level shift effect, states of positive energy are pushed

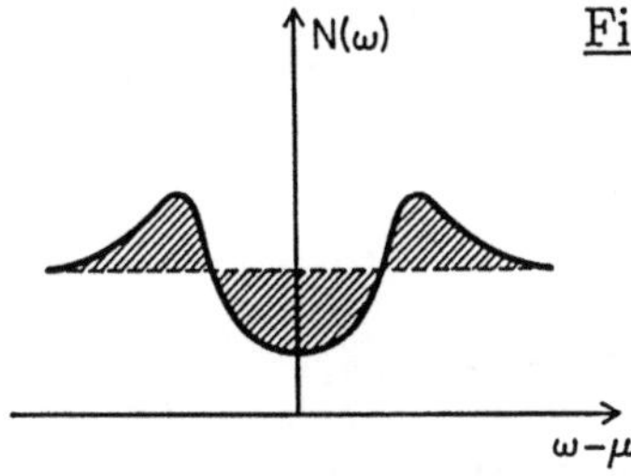

Figure 13b The formation of a pseudo gap.

up and states of negative energy are pushed down leading to a pseudo gap indicated in Figure 13b.

The physical origin of these two quite different energy dependences of Σ can be understood as follows: two possible time orderings of the vertices in Fig. 11 can occur. In Figure 14a one has the conventional polaron self energy in which an injected hole is scattered into an intermediate hole state beneath the Fermi surface, as for a hole polaron in a semiconductor.

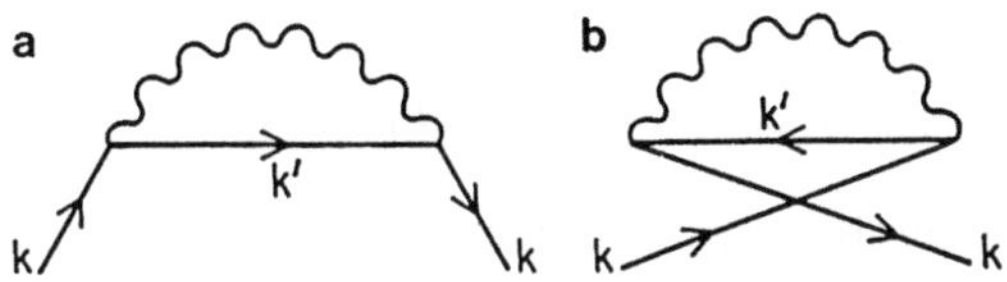

Figure 14a Polaron time ordering of the self energy graph, in which a hole occupies the intermediate state k'.

Figure 14b The exchange graph which describes the Pauli principle suppression of a spin fluctuation process which lowers the system energy in the absence of the added hole. Thus, this contribution raises the hole (and the k electron) energy and leads to a pseudo gap.

However, as shown in Figure 14b an injected hole can also be converted into an electron above the Fermi surface in the intermediate state. This second diagram physically arises from the suppression of vacuum spin fluctuations by the action of the exclusion principle effect associated with the added hole. This can best be seen as in Figure 15 where the added hole in state $\mathbf{k}$ excludes a vacuum fluctuation.

In this state a second hole, also in state k, is contained in the vacuum spin fluctuation and is forbidden by the exclusion principle. As Feynman showed, such Pauli principle violating effects can be handled by adding another diagram given by exchanging the lines of the two violating particles as is shown in Figure 15b. This extra diagram precisely cancels the

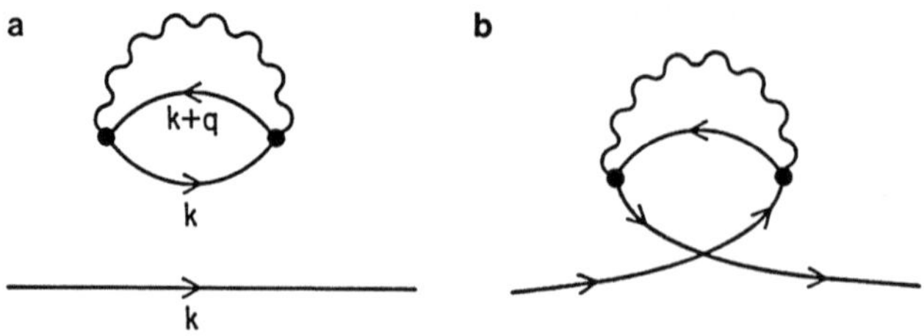

Figure 15a The vacuum fluctuation referred to in Figure 14b.

Figure 15b The Feynman exchange graph essential to preserve the Pauli principle and the linked cluster expansion.

forbidden graph, and also allows one to preserve the linked cluster expansion which is crucial to the proper volume scaling of the system energy. Since the vacuum fluctuation energy lowers the total system energy in the absence of the hole, the introduction of the hole raises the energy and therefore leads to the pseudo gap behavior characteristic of Figure 13, recalling that increasing hole energy is measured toward negative $\omega - \mu$.

V. Implications for Photoemission Spectroscopy

Photoemission and inverse photoemission spectroscopy in essence measure the one particle spectral function

$$A(\mathbf{k},\omega) = \mid \frac{1}{\pi} ImG(\mathbf{k},\omega) \mid = \mid \frac{1}{\pi} Im \frac{1}{\omega - \varepsilon_{\mathbf{k}} - \Sigma(\mathbf{k},\omega)} \mid . \quad (11)$$

Angular resolved spectra measure both the $\mathbf{k}$ and ω dependence of A, as shown in Figure 16.

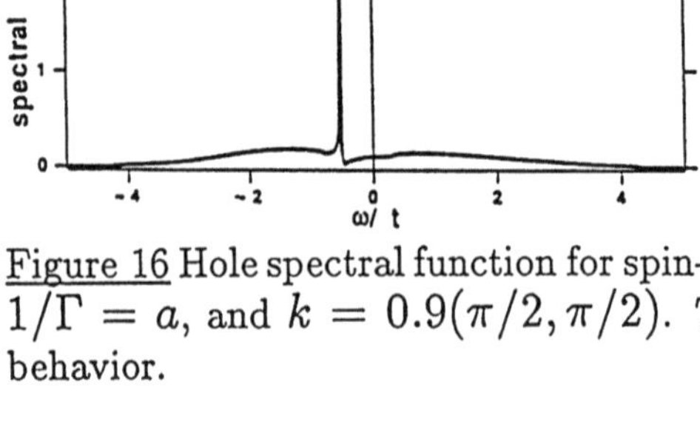

Figure 16 Hole spectral function for spin-spin correlation length $L_{SS} = 1/\Gamma = a$, and $k = 0.9(\pi/2, \pi/2)$. This illustrates Fermi liquid behavior.

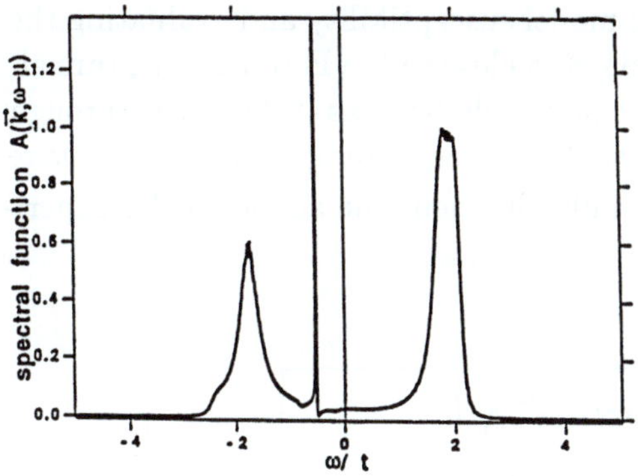

Figure 17 Hole spectral function for $L_{ss} = 20a$ and $k = 1.15(\pi/2, \pi/2)$, illustrating sharpening of the incoherent backgrounds which arise from quasi-Bragg scattering of the hole by the large amplitude slow spin fluctuations. The quasi particle amplitude z_k is small.

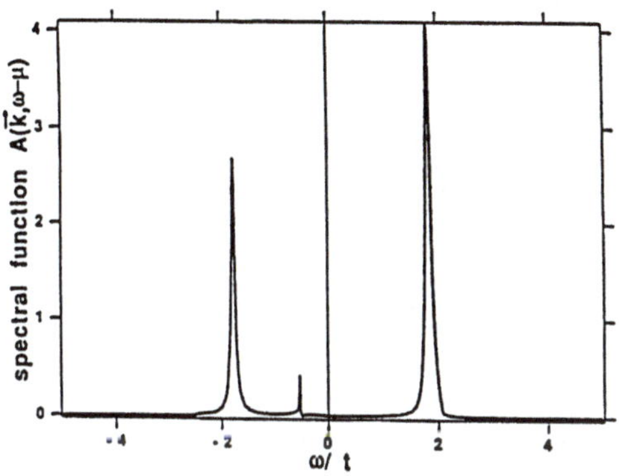

Figure 18 Hole spectral function near the SDW transition, for $L = 100a$. Clearly the quasi particle has extremely small weight and the SDW split peaks appear. The smaller of the two peaks in called the shadow band.

In the large doping limit the Landau theory presumably holds with a sharp quasiparticle peak located at the renormalized single particle energy and two incoherent peaks, one for holes and the other for electrons occurring at large energy. As the doping is reduced the spin-spin correlation length grows and the weight of the quasiparticle z_k is reduced[5], with the incoherent backgrounds becoming more sharply defined, as shown in Figure 17.

Finally, in the antiferromagnetic phase, if **k** is near the magnetic Brillouin zone boundary, A becomes two sharp peaks as seen in Figure 18.

Using the phenomenological model susceptibility and evaluating the spectral function $A(k,\omega)$ for momenta close to the Fermi momentum allows one to follow the Landau quasiparticle peak as it disperses through the Fermi level. Figure 19 shows the corresponding finite temperature photoemission spectrum, very similar in their line shape to the experimental data by Olson et al.[6]

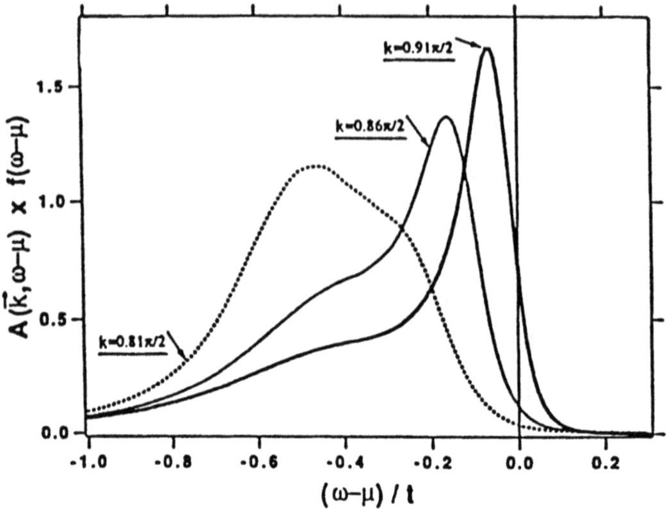

Figure 19 Photoemission spectrum for several values of k near k_F.

For each $\mathbf{k}$ an additional sharp feature appearing in the spectral function in the pseudogap regime should also be observable near momentum $\mathbf{k}+\mathbf{Q}$ where $Q = (\pm\pi, \pm\pi)$ is the antiferromagnetic wave vector. Although there is a unique quasiparticle state for each $\mathbf{k}$ in the reduced magnetic Brillouin zone, the wave function for such a state is a linear combination of Bloch waves for wave vectors $\mathbf{k}$ and $\mathbf{k}+\mathbf{Q}$. Therefore the emitted electron will carry momentum $\mathbf{k}$ as well as $\mathbf{k}+\mathbf{Q}$ even though a single unique quasihole state is created. The spectral intensity associated with the exchange Bragg scattering therefore leads to a shadow band displaced by $\mathbf{Q}$.

An interesting feature in the spectrum has been observed by Allen and coworkers[7] in $Nd_{2-x}Ce_xCuO_{4-y}$. They find that in contrast to a rigid band model, their data are consistent with a picture in which each hole splits off a state from the valence band, thereby pinning the chemical potential inside the gap. It is suggestive that this effect is the result of spin bag formation, which splits one state off from a linear combination of the valence and conduction bands, with weights favoring the band whose edge is nearer the spin bag energy. For large doping, this effect will smear into the band edge, as in heavily doped semiconductors. To treat this effect, it is essential to include in Σ the two loop crossed line diagram. In essence, one loop produces the pseudogap discussed above and the second loop accounts for the local depression of the pseudogap to form a bag. This is in analogy with the situation in the ordered antiferromagnet, where the mean field potential produces the gap and the loop accounts for the depression of the gap in the vicinity of the hole to form the bag. Model calculations of this effect are in progress[8].

VI. The Large $U/4t$ Limit

Many authors have studied both the 2d one band Hubbard model for large U and the t-J model using a variety of techniques, both analytic and numerical. The results are remarkably similar to those discussed above in the weak coupling theory. In particular, bag effects are observed including the spin twist mentioned above.

In addition, the dressed holes are observed to attract as in the weak coupling situation. The one hole spectral weight function $A(k, \omega)$ exhibits a small or vanishing z_{k_F} corresponding to a large incoherent shake-off spectrum when a hole is created. The reader is referred to the literature concerning such effects[9,10].

Concerning charge and spin separation postulated in the exotic approaches, to date no calculations on the Hubbard, t-J and related models have shown these quantum numbers to be deconfined so that the charge and spin of the excitations are very likely to be the conventional ones for a hole or an electron, namely charge $\pm e$ and spin $\frac{1}{2}$, although the issue remains open at present.

This work was supported in part by NSF grant DMR89-16582 and by the Electric Power Research Institute, contract RP8009-18.

REFERENCES

1 C.M. Varma, et al. Phys.Rev.Lett. **63**, 1996 (1989).

2. J.R. Schrieffer, X.-G. Wen, and S.-C. Zhang, Phys.Rev.B. **39**, 11663 (1989); Phys.Rev.Lett. **60**, 944 (1988).

3. A.P. Kampf and J.R. Schrieffer, Phys.Rev.B. **42**, 7969 (1990).

4. A.J. Millis, H. Monien, and D. Pines, Phys.Rev.B. **42**, 167 (1990); H. Monien, P. Monthoux, and D. Pines, Phys.Rev.B. **43**, 275 (1991).

5. A.P. Kampf and J.R. Schrieffer, Phys.Rev.B. **41**, 6399 (1990).

6. C.G. Olson, et al., Phys.Rev.B. **42**, 381 (1990).

7. J.W. Allen, et al., Phys.Rev.Lett. **64**, 595 (1990).

8. A. Kampf and J.R. Schrieffer, work in progress; see also, C. Melo and S. Doniach, Phys. Rev. **B** 41, 6633 (1990).

9. D.J. Scalapino, The Los Alamos Symposium, High Temperature Superconductivity Confr. Proc., Addison-Wesley (1990).

10. E. Dagotto, International Journal of Physics B **5**, 77 (1991).

A (Very) Brief Overview of Semionic Superconductivity

A. Zee

Institute for Theoretical Physics, University of California, Santa Barbara, CA 93106, USA and Yukawa Institute for Theoretical Physics, Kyoto University, Kyoto 606, Japan

In the very short time available I can only give the briefest of brief overviews. Fortunately, detailed reviews of semionic superconductivity are now available [1]. As you can see from my byline, I have had the pleasure of spending some time at the Yukawa Institute and enjoying the sights of Kyoto. When the organizers of this meeting heard that I was going to be in Kyoto they asked me to give a talk. I hesitated partly because so many people here have already heard me talk here in Japan and at various meetings around the world and partly because I am going to be addressing "real" physicists who know about the differences between Cu and La and such arcane acronyms as ARPES. As you all know, current research in high temperature superconductivity spans a rather broad spectrum. At the other end of the spectrum from the "real" physicists are the so-called "   " physicists [2] who write down topological field theories and claim to have solved everything. Their heroes are named Chern and Simons who, curiously enough, are not even physicists. (The latter is reportedly a wealthy business person [3].) According to their "philosophy", what's important is not the minute detail of the microscopic lattice physics, but the ***global, topological, and gauge*** structure peculiar to the two-dimensional character of the high T_c materials. The words in italics are of course some of the most fashionable words in theoretical physics today.

So, I am going to sketch a *possible* theory of superconductivity based on the speculation that the spin and charge degrees of freedom obey semion statistics. For those of you who are interested, my original research can be found in a number of publications in various journals [4–7]. I don't have time to give proper credit to everyone; suffice it to say that the pioneers in this field include Laughlin, Kalmeyer, Kivelson, Rokshar, Sethna, Dzyloshinski, Polyakov, Wiegmann, and many, many others.

Certainly one of the most amazing features of quantum physics is the existence of two types of particles, bosons and fermions. This is a totally quantum phenomenon with no classical analog at all. When we interchange two bosons we get the phase factor $+1$, and when we interchange two fermions, we get -1. What is another interesting number besides $+1$ and -1? A very interesting number is i. It would be fun to postulate that there are particles called semions such that when two semions are interchanged the quantum wave function gets a factor of i. The word "semion", being half Latin and half Greek, was coined to describe a particle half way between a boson and a fermion. More generally, as Leinaas and Myrheim [8] and later Wilczek showed [9,10], there can be particles called anyons such that when you interchange two anyons, you get any phase factor you want.

Let us emphasize immediately that this very interesting and novel phenomenon is possible only in $2+1$ dimensional spacetimes, but not in the 3+1 dimensional spacetime that we actually live in: 2+1 dimensional spacetime is really different from 3+1 dimensional spacetime. It's not just less than a good thing. In 2 dimensional space a new physical concept emerges: the concept of "going around". A particle in 2 dimensional space, that is to say a plane, can actually go around another particle. This concept of going around does not make sense in 3-dimensional space. In 3-dimensional space, there is no way you can define whether a point particle has gone around another point particle. As a noted architect once said, less is more.

We should be able to do some new physics with this new physical concept and indeed we can. We can postulate an infinite ranged phase interaction between two particles. Consider two particles such that when one particle goes around the other through an angle ϕ the quantum wave function acquires a phase $e^{i(\frac{\theta}{\pi})\phi}$. I'm free to postulate such an interaction; whether or not such an interaction actually exists in Nature is, of course, another story. Here θ is just a parameter characteristic of the particle. For different θ's we have different types of particles. Let's understand what θ means. Take the angle $\phi = \pi$ so that one particle has gone half way around the other particle. If you follow this operation by a translation, it is equivalent to interchanging the two particles. According to our definition, the wave function acquires the phase $e^{i\theta}$, and so for $\theta = 0$ we have bosons, for $\theta = \pi$, fermions. In general we can have any phase we want. (This infinite ranged phase interaction is known in the literature as fractional statistics, a potentially confusing terminology. First of all, this parameter θ can in fact be any real number, it doesn't have to be a fraction. The word statistics also confuses some people, because we ordinarily think of statistics as the representation of the permutation group. But as we can see we are dealing here with representations of a group more general than the permutation group, two interchanges is not effectively the same as no interchange.)

A very important property of anyons is that they violate time reversal and parity invariances if $\theta \neq 0$ or π. This is very easy to understand. Consider time reversal. Suppose I take a particle around another particle, then according to our formula I will get a phase $e^{i\theta}$. But now if I reverse the direction of time, so that I take the particle going the other way, then I will get $e^{-i\theta}$. These two factors are not the same unless θ is equal to 0 or π. Parity violation is also evident. Just reflect in a mirror.

Another important point is that fractional statistics is equivalent to an infinite ranged gauge interaction whose dynamics is governed by the (T and P odd) Chern-Simons term, as was noted by Wilczek and I some years ago [10]. In a nutshell, then, the idea is that the attractive gauge force between the holes would lead to pairing and superconductivity [6,11–14].

Here I would like to emphasize a generic consequence of these gauge theories of high temperature superconductivity. In these theories, there are two gauge symmetries, one associated with the gauge potential a_μ responsible for the fractional statistics, which may be denoted by $U(1)_a$, and another associated with the electromagnetic gauge potential A_μ, denoted by $U(1)_A$. As far as the gauge symmetries are concerned there are four possible states: (1) both gauge symmetries are unbroken, (2) $U(1)_a$, but not $U(1)_A$, is broken, (3) $U(1)_A$, but not $U(1)_a$, is broken, and (4) both gauge symmetries are broken. Superconductivity is obtained only when no unbroken gauge symmetry is left and thus only in state (4). Thus, generically, there may be three normal states. This point was emphasized and studied by Wen and I in the context of a generalized Landau-Ginzburg theory [7]. Note that the very general remarks given here do not depend on whether or not T and P are violated and on whether or not fractional statistics plays a role. It only depends on the appearance of another gauge field. The violation of T and P is logically independent. Recently, Sachdev [15] has studied the possible occurence of three normal states in more detail. I mention this point because, as you have heard repeatedly at this meeting, the normal state properties of the high temperature superconductors are more extraordinary than the superconducting state properties.

I find it convenient to organize this subject by asking four "big questions". First, under what circumstances does the ground state of a commonly accepted model Hamiltonian spontaneously violate T and P? As I emphasized on one of the first transparencies, fractional statistics particles violate T and P. We know, of course, that if I give you a piece of superconducting material, the underlying Coulomb interaction of the electrons certainly does not violate time reversal and parity. These invariances have to be dynamically and spontaneously violated. So the first question is whether this is possible? Second, how do we describe the dynamics of the excitations about the ground state? What are their quantum numbers and in particular what are their statistics? What is the long distance effective theory? The third question is how does a gas or liquid of fractional statistics particles behave? Is it a superfluid and does it superconduct if these particles are charged? And of course, most important is the question whether there are experimental tests that we can invent to test our scenario.

There have been several theoretical approaches towards answering this question which unfortunately I don't have time to explain in detail. All I can give you is a headline service. There are four approaches that I know of. (1) The antiferromagnet can be mapped by mathematical transformation into a problem of hard core boson gas in a magnetic field [11]. (2) We can also study the long-distance physics of an antiferromagnet as represented by a nonlinear sigma model with a topological term added [4]. (3) Wen, Wilczek and I proposed a so-called chiral spin state which is in some sense a new state of matter [5]. (4) Finally, we can also use the duality relation between the soliton and particle sectors [4]. Never mind the details of these theoretical approaches. They are however mutually consistent in determining the statistics of the excitations to be $\theta = \frac{\pi}{2n}$ where n is an integer. Take the simplest case $n = 1$: the excitations in the system are semions.

Perhaps I should mention the chiral spin liquid [5] and what the order parameter is. The order parameter is a spin triple cross product $\vec{S}_i \cdot \vec{S}_j \times \vec{S}_k$ where the $\vec{S}$'s are three neighboring spin operators in a t-J model representation, say, of the high temperature superconductor. This quantity, if it is nonzero, obviously violates T and P, because if you reverse time all the spins reverse direction and this quantity will change sign. So if you do a calculation and this quantity is not equal to zero, then it's a signal that T and P are violated. This local order parameter can be extended to a global order parameter closely related to the Wilson loop [5].

An improved treatment has been given by Fukuyama and his collaborators [16,17] based on the incommensurate flux phase [18]. For an up-to-date treatment see recent work by Zou, Levy, and Laughlin [19].

Why gauge field? I'm often asked this question by condensed matter physicists. One straightforward answer to this question has already been alluded to, that the gauge field originates in the phase fluctuation of the order parameter. A somewhat more mathematical answer is that in two dimensional systems the electromagnetic current, J_μ, by virtue of its being conserved $\partial^\mu J_\mu = 0$, can be written as a spacetime curl $J^\mu = \epsilon^{\mu\nu\lambda}\partial_\nu a_\lambda$. Clearly, a_λ is defined only up to the transformation $a_\lambda \to a_\lambda + \partial_\lambda \Lambda$. Thus, the concept of gauge potential appears naturally and almost instantly.

Fröhlich and I [20] have exploited this simple observation as a starting point for constructing effective theories of the quantum Hall fluid. Effective theories of the Hall fluid and anyon superfluid have been constructed by a number of authors [21,22]. It turned out, after various field theoretic manipulations the Lagrangian (density) for the Hall fluid is given by $\mathcal{L} = \frac{1}{4\pi}\alpha K \epsilon \partial \alpha$, and for the anyon superfluid $\mathcal{L} = \frac{1}{4\pi}\alpha \Lambda \epsilon \partial \alpha$. The long distance physics of both fluids are given by a matrix Chern-Simons theory, with the matrix Λ constructed from the matrix K. The physics of the anyon superfluid may be obtained from that of the Hall fluid in just one extra step. Here I would like to emphasize that the long distance of the Hall fluid, of the chiral spin fluid, and of the anyon superfluid may be described with closely related mathematical structures. The mathematical structures of these fluids are extraordinarily rich, with concepts from topology, algebra, and number theory all playing a role. For instance, in some cases the matrix K turns our to be related to Cartan's root matrix for the Lie algebras of the simple unitary groups [20,23].

Mathematical beauty does not necessarily have anything to do with reality, of course. However, given the close mathematical similarity between these three fluids, which we may perhaps call topological fluids [24], it is tempting to think that since the Hall fluid is physically realized the other two fluids may also be realized in some physical systems.

Let me now give some heuristic arguments on why we might expect a liquid of semions to behave like a superfluid or a superconductor. First, notice that two semions do not make a fermion. You may think that semions are half way between bosons and fermions and so if I put two semions together they might make a fermion. Well, suppose we move a bound state of two semions half way around another bound state of two semions, thus interchanging the two bound states. Then we get lots of phases. This semion gets a phase of i going around that semion but he also gets a phase from going around the other semion, and so on. Now we must do a mathematical calculation: $2 \times 2 = 4$. We get the basic phase four times. Each time a semion goes around another semion we get a phase of i. Since we get this phase 4 times we get i^4, which is equal to 1. Thus, two semions actually make a boson.

We may now give some handwaving arguments. We all know that fermions like to stay apart and bosons like to stick together. Semions, being half way between fermions and bosons, are more likely than fermions to pair. When they do pair, they form bosons, whose condensation can then lead to superfluidity and superconductivity. Just think, if you were a fermion and you want to relax and condense into the ground state, what could you do? You have basically two strategies: (1) Find another fermion and pair with him or her, or (2) turn yourself into a boson. Strategy (1) is of course what was followed in BCS superconductivity. Strategy (2) is an attractive possibility, but in the systems under discussion you can apparently make it only half way and turn yourself into a semion.

Much of the physics of anyons is contained in the following deep mathematical identity $\theta = 0 + \theta = \pi - (\pi - \theta)$. What is the mathematics trying to tell us? The first equality says that an anyon with statistics θ is effectively a boson (statistics 0) with a gauge interaction of "strength" θ; the second equality says that it is also a fermion (statistics π) with a gauge interaction of "strength" $-(\pi - \theta)$. The physics is made completely clear by looking at the 2-anyon problem. In the center of mass, the Schrödinger equation has a centrifugal potential like $(\ell + \theta/\pi)^2/r^2$ with ℓ an integer. The smallest possible value of this potential is thus $0/r^2$ for bosons, $1/r^2$ for fermions, and $1/4r^2$ for semions. Anyons are like bosons with "centrifugal repulsion" between them or fermions with a "centripetal attraction". Thus, theorists have a choice. They can begin either with a fermi gas [12–14] — certainly not a superfluid — and show that the attraction gives pairing and superfluidity, or with a boson gas [6,25] – which with a short ranged repulsion is a superfluid as was shown by Bogoliubov ages ago — and show that the repulsion does not destroy superfluidity.

People have made some progress using a mean field approximation [26]. Consider moving an anyon through a big loop enclosing an area A. Since in the mean, there are nA anyons in the loop, the wave function picks up the phase $e^{i(\theta/\pi)(2\pi)nA}$. But we know that a charged particle moving in a magnetic field b picks up a phase e^{ibA}. Thus, we can consider an approximation: the particles act as if they are moving in a mean magnetic field given by $b = 2n\theta$ and generated by all the other particles.

In the quantum Hall effect we have electrons moving in a plane in the presence of an external (real) magnetic field. Thus, in the mean field approximation the semion fluid is closely related to the Hall fluid. There is however a crucial difference between the two fluids. The hallmark of the Hall fluid is its incompressibility. The basic physics of this striking phenomenon may be understood heuristically. The electrons are forced by the external fixed magnetic field to move in circles. Even though the electrons are point particles, they have to occupy a minimal and finite amount of room. This is perhaps reminiscent of the housing situation in Japan. Because of fermi statistics, the circles can't "overlap." The combination of fermi statistics and the magnetic field makes the system incompressible. Superficially, the semion fluid and the Hall fluid sound similar. We have particles and magnetic fields (real in one case, statistical in the other). What's the difference?

The crucial difference is that the magnetic field is external and fixed in one case, but generated by and dynamically tied to the particles in the other. This makes all the difference in the world. (It is sort of like the difference between a fixed political system imposed on the people and a dynamical political system generated by the people.) There is at least no *a priori* reason for the semion fluid to be incompressible and in fact it isn't.

The essential physics behind the compressibility of the semion fluid can be captured, I believe, by a hydrodynamic approximation [6] treating the semions as bosons with gauge repulsion between them. To say a fluid is compressible is to say that it has gapless excitations, that is, excitations that cost infinitesimally little energy. Consider a long wavelength density fluctuation. Where the density is higher, the effective statistical magnetic field is also higher according to $b = 2n\theta$, and so the (Landau or Larmor) circles the particles move in are correspondingly smaller, in precisely such a way that they don't bump into each other any more than in the low density regions. As the wavelength becomes longer and longer, the energy of the density wave becomes smaller and smaller. In gauge theory language, the massless mode corresponds to the gauge boson.

Another way of approaching the physics is to calculate the effective energy density. Since in a magnetic field the Landau or Larmor energy is $\frac{1}{2}\hbar\omega_c \sim b$, the energy density is $\sim nb \propto n^2$. The quadratic dependence of the energy density on the number density is as if there is a short ranged repulsion between the bosons. We may then invoke Bogoliubov's work to say that we have a superfluid.

I am not qualified to review in detail the experimental situation regarding semionic superconductivity. Let me just say that in contrast to a number of theories of superconductivity which make more or less generic predictions, this theory goes out on a limb and says there should be T and P violation — it's certainly a falsifiable theory.

To look for T and P violation, we can study the polarization of the light reflected from a high-T_c material when the incident light is linearly polarized [7]. Thus far, three experiments have been done with two showing a rotation of the polarization axis and one showing no effect [27]. Of course there are numerous solid state effects to worry about. Also, it is important to see both T and P violation with TP conserved. A rotation of the polarization axis shows only P violation, but not T violation. Another worry is that the statistics parameter θ may alternate in sign as we go from layer to layer, in which case the effect would be washed out. An alternative experiment is to look at the precession of muons injected into the material [14]. In contrast to the polarized light experiment, the muon is a local probe. The experiment has been done and fails to show a signal at the expected level. Thus, we have now two negative experiments and two quasi-positive experiments.

To summarize let me list the current answers to the "four questions". (1) Does the ground state violate T and P? Possible (but unlikely?). (2) What are the quantum numbers of the excitations? If the answer to (1) is yes, then we have semions. (3) Is a semion fluid a superfluid? Almost certainly yes. (4) Are there experimental tests? Definitely a falsifiable theory.

The experimental situation is far from clear and we can only hope at this point. One nice thing about going around giving talks about this subject is that I get taught some relevant local sayings. At a meeting in Italy, I learned the saying "Se è rosa, fiorirà." Here in Japan, I have learned to say "Kitai, yume, kibo."

In closing, I would like to remind you of an elementary point of logic: Even if the presently known high temperature superconductors are not semion superconductors, it does not follow that semion superconductors do not exist. In particle physics, there is a rule known as Gell-Mann's rule stating that whatever is not forbidden is required. I was told by condensed physicists that they have a priority claim: they have long known this rule as Herring's rule. (And of course the humanists know it as T.H. White's rule.) Unlike particle physicists who have only one single universe to look at, condensed matter physicists have a wealth of systems to study, particularly in this era of great progress in materials science when you can almost have designer materials made to order. Let us then end with the credo of an extreme optimist. It would certainly be lots of fun if the presently known superconductors turn out to be semionic superconductors, but won't it be even more exciting if the presently known superconductors turn out not to be semionic superconductors? That would mean that there may be a new class of high temperature superconducting materials out there waiting to be discovered!

This research was supported in part by the National Science Foundation under Grant No. PHY89-04035.

References

1. I recommend, for obvious reasons, "Semionics: a theory of high temperature superconductivity," A. Zee, in *High Temperature Superconductivity*, edited by K. Bedell, D. Coffey, D.E. Meltzer, D. Pines and J.R. Schrieffer (Addison-Wesley 1990).
2. My fellow travellers on that end of the spectrum objected to my using the term "fake" physicists in some earlier talks I've given. Hence the blank here.
3. This is just unsubstantiated hearsay.
4. X.G. Wen and A. Zee, *Phys. Rev. Lett.* **61**, 1025 (1988); *Nucl. Phys.* **B316**, 641 (1989); *J. de Physique* **50**, 1623 (1989); *Phys. Rev. Lett.* **62**, 2873 (1989); **63**, 461 (1989); *Nucl. Phys.* **B326**, 619 (1989); *Phys. Rev. B* **43**, 5595 (1991).
5. X.G. Wen, F. Wilczek and A. Zee, *Phys. Rev. B* **39**, 11413 (1990).
6. X.G. Wen and A. Zee, *Phys. Rev. B* **41**, 240 (1990).
7. X.G. Wen and A. Zee, *Phys. Rev. Lett.* **62**, 1937 (1989); *Phys. Rev. B* **43**, 5595 (1991).
8. J.M. Leinaas and J. Myrheim, *Il Nuovo Cimento* **37B**, 1 (1977).
9. F. Wilczek, *Phys. Rev. Lett.* **49**, 957 (1982).
10. F. Wilczek and A. Zee, *Phys. Rev. Lett.* **51**, 2250 (1983); F. Wilczek and A. Zee, unpublished, SBITP report.
11. V. Kalmeyer and R. Laughlin, *Phys. Rev. Lett.* **59**, 2095 (1987); and R. Laughlin, *Science* **242**, 525 (1988).
12. A. Fetter, C. Hanna, and R. Laughlin, *Phys. Rev. B* **39**, 9679 (1989).
13. J. Hetrick, Y. Hosotani, and B.-H. Lee, TPI-MINN-90/40-T, and references therein.

14. Y.H. Chen, F. Wilczek, E. Witten, and B.I. Halperin, *Int. J. Mod. Phys.* **B3**, 1001 (1989).
15. S. Sachdev, Yale preprint (1991).
16. H. Fukuyama, O. Narikiyo, and K. Kuboki, *J. Phys. Soc. Jpn.* **59**, 807 (1990).
17. Y. Hasegawa, O. Narikiyo, K. Kuboki, and H. Fukuyama, *J. Phys. Soc. Jpn.* **59**, 822 (1990).
18. Y. Hasegawa, P. Lederer, T.M. Rice, and P.B. Wiegmann, *Phys. Rev. Lett.* **63**, 907 (1989).
19. Z. Zou, H. Levy, and R. Laughlin, Stanford preprint (1991).
20. J. Fröhlich and A. Zee, SBITP preprint (1991) to appear in *Nucl. Phys.*
21. S.C. Zhang, T.H. Hansson and S. Kivelson, *Phys. Rev. Lett.* **62**, 82 (1989); See also S. Girvin and A.H. McDonald, *Phys. Rev. Lett.* **58**, 1252 (1987). For a more recent formulation, see J. Fröhlich and T. Kerler, ETH-TH/90-26. X.G. Wen and Q. Niu, *Phys. Rev. B* **41**, 9377 (1990); B. Blok and X.G. Wen, *Phys. Rev.* **42**, 8133, 8145 (1990). M.P.H. Fisher and D.H. Lee, *Phys. Rev. Lett.* **63**, 903 (1989); *Phys. Rev. B* **39**, 2756 (1989).
22. X.G. Wen and A. Zee, SBITP preprint, to appear in *Phys. Rev. B.*
23. B. Blok and X.G. Wen in Ref. [21], and N. Read, *Phys. Rev. Lett.* **65**, 1502 (1990).
24. For more details, see A. Zee, in Proceedings of the Kyoto Conference in "Low dimensional field theories and condensed matter physics," (1991).
25. H. Mori, Hiroshima preprint (1990).
26. D. Arovas, R. Schrieffer, F. Wilczek, and A. Zee, *Nucl. Phys.* **B251**, 117 (1985).
27. For a review, see the talks by H.J. Weber, A. Kapitulnik, and K.B. Lyons in *Physics and Mathematics of Anyons*, edited by S.S. Chern, *et al.*, (World Scientific Publishing, 1991).

Fluctuations in Type-II Superconductors: A Dense Vortex Plasma Picture

Z. Tešanović

Department of Physics and Astronomy, The Johns Hopkins University, Baltimore, MD 21218, USA

Abstract. The critical fluctuations in type-II superconductors near $H_{c2}(T)$ exhibit certain universal features due to the constraint on the form of the order parameter imposed by the external magnetic field. We discuss the physical consequences of this constraint and outline the description of fluctuations which provides a unified picture of the behavior of type-II superconductors both above and below $H_{c2}(T)$.

1. Introduction

Most of the new superconducting materials discussed in this School are strongly type-II systems: Examples are high temperature superconductors, organic and heavy fermion superconductors, M_3C_{60}, etc. In such systems there is a wide region in the $H-T$ phase diagram where the superconducting order parameter coexists with the internal magnetic field. The superconducting transition in type-II systems is commonly described using the Abrikosov mean-field theory (MFT) [1]. In the MFT the transition occurs at $H_{c2}(T)$ or, equivalently, $T_{c2}(H)$, where $\Psi(\vec{r})$ becomes finite and its zeros (vortices) form a (Abrikosov) lattice (in this lecture I will use both $H_{c2}(T)$ and $T_{c2}(H)$ to denote the mean-field transition).

Recent excitement in the field of superconductivity has led to an intense interest in general theoretical properties of type-II systems and, in particular, the issue of how the fluctuations affect the results obtain in the MFT. The problem of fluctuations in type-II systems has been studied since early 60's and in this limited time we can mention only a fraction of the relevant contributions. One overall conclusion is that fluctuations in type-II systems at $H \neq 0$ tend to be enhanced over the zero field case. This enhancement is the manifestation of a "dimensional reduction" which takes place near $H_{c2}(T)$ [2], [3]. This dimensional reduction is of a fundamental nature and arises directly from the gauge invariance resulting in the formation of the "Cooper pair" Landau levels (see below). In the past, these fluctuations have been studied in the perturbation theory starting from far above $T_{c2}(H)$ [4], [5]. The perturbation theory suggests that the effective dimensionality of this problem is $D-2$ and leads to significant corrections to the mean-field results for specific heat, magnetization, etc. [6]. However, the perturbative expansion does not generate significant correlations between vortices, and does not show any indication of a phase transition to the Abrikosov lattice even when carried out to a very high order [4], [5]. This is in a sharp contrast to the low temperature, far below $T_{c2}(H)$, behavior where we expect the vortex lattice to be a stable state. If we start from this assumption and consider small fluctuations of vortices from their equilibrium positions we will end up with the elastic theory of the Abrikosov lattice [7]. Within the elastic theory one concludes that there must be a region of in the $H-T$ phase diagram, far from $T_{c2}(H)$, where the vortex "solid" will not melt. Eventually, as one moves toward $T_{c2}(H)$, the rms displacements become large and, relying on the Lindemann criterion for melting, one is led to conclusion that the lattice would melt and the vortex "liquid" will become a stable state before one reaches $H_{c2}(T)$ [8], [9]. This is very suggestive of the importance of fluctuations. However, this approach which exploits the Lindemann criterion does not represent the theory of melting and is unsatisfactory in principle since the elastic theory starts from the state which breaks translational symmetry by hand and clearly cannot describe the vortex "liquid" phase.

In this lecture, I will discuss these problems and outline how a theory of critical fluctuations in type-II superconductors can be devised which overcomes above difficulties by bridging the gap between the high and the low temperature behavior. Such a theory must provide a unified description of both vortex "liquid" and "solid" phases. Therefore, we have to construct a theory in which the formation of an inhomogeneous structure (the Abrikosov lattice) arises spontaneously within a translationally invariant description: We clearly want to avoid breaking the symmetry by hand, as is done in the elastic theory. Using this approach to critical fluctuations one should be able to understand both the $D-2$ dimensional perturbative regime far above $T_{c2}(H)$ and the (highly non-perturbative, as it turns out) formation of the Abrikosov lattice below $T_{c2}(H)$ as well as the crossover between the two regimes.

2. Dense Vortex Plasma

The critical fluctuations in type-II superconductors near $H_{c2}(T)$ are typically studied by considering the partition function:

$$Z = \int \mathcal{D}\Psi(\vec{r})\mathcal{D}\Psi^*(\vec{r}) \int \mathcal{D}\vec{a}(\vec{r}) \exp\{-(F_s[\Psi(\vec{r}), \vec{a}(\vec{r})] + F_m[\vec{a}(\vec{r})])/k_BT\} \quad , \tag{1}$$

where

$$F_s = \int d\vec{r}\{\alpha(T)|\Psi(\vec{r})|^2 + \frac{\beta}{2}|\Psi(\vec{r})|^4 + \frac{n}{2m_\mu^*}|[\partial_\mu + \frac{2ei}{c}(A_\mu + a_\mu)]\Psi(\vec{r})|^2\} \quad ,$$

$\vec{A} = \vec{H}$, $\mu = x, y, \zeta$, n is the electron density, and F_m is the magnetic free energy. It is customary here to neglect fluctuations in $\vec{a}$ which is justified for strongly type-II systems (*i.e.* $\kappa \gg 1$) [4], [5]. Physically, an important characteristics of a type-II superconductor in a magnetic field is the constraint placed by the field on the form of the fluctuating order parameter $\Psi(x, y, \zeta)$ (note that I am using ζ to denote the third coordinate, $\hat{\zeta} \parallel \vec{H}$). This constraint arises from the quadratic piece of the free energy in Eq. (1). The quadratic piece is diagonalized by the set of Landau orbitals corresponding to particles of charge $e^* = 2e$ (Cooper pairs) in the magnetic field H. In the vicinity of $H_{c2}(T)$ one can expand $\Psi(x, y, \zeta) = \sqrt{2\pi\ell^2} \sum_{n,m} b_{nm}(\zeta)\varphi_{nm}(z, z^*)$, where $\ell = \sqrt{c/e^*B}$, $z = (x + iy)/\ell$ and $\varphi_{nm}(z, z^*)$ are the Landau eigenfunctions in a suitable gauge. The dominant contribution comes from the lowest Landau level (LLL), $n = 0$. The fluctuations from higher Landau levels $\{b_{n\neq 0,m}\}$ have a gap and, according to conventional wisdom, can be ignored near $H_{c2}(T)$ [4]. They will "smoothly" renormalize the kernels of the Lagrangian for the LLL. Thus, these degrees of freedom of $\Psi(x, y, \zeta)$ are effectively taken out from the dynamics. The problem is now reduced to the study of the remaining fluctuations in $\{b_{n=0,m}(\zeta)\}$. This is the "dimensional reduction" mentioned above. The fluctuations have been studied in Ref. [4] by considering the perturbative expansion in the quartic term, assuming one is high above $H_{c2}(T)$. The expansion of the free energy was carried out to the sixth order in β. The Padé and Borel-Padé approximants were then used to extrapolate the series and extract the low temperature (below $H_{c2}(T)$) behavior. These authors have found no evidence of a sharp phase transition but only a gradual crossover in the mean squared value of the order parameter. The perturbation series is closely related to the $D - 2$ dimensional Ginzburg-Landau theory which is the consequence of the above dimensional reduction and the "quenching" of gradients in the LLL. Recently it was demonstrated that the series does not converge to the Abrikosov form of the free energy and that the perturbation theory cannot be used to describe the formation of the Abrikosov lattice [5].

We should not be particularly surprised at this failure of the perturbation theory to describe low temperature behavior. The solidification transition of any type is hard to analyze by perturbation theory. Often the best one can do is to selectively resum a subset of diagrams and note that they may diverge at some finite wavevector indicating the instability toward an inhomogeneous state. This basically amounts to doing the mean-field theory. Further difficulty arises from the crucial physical aspect specific to our problem. Due to the dimensional reduction described above, the upper critical dimension in the RG sense is $D_c = 6$ [10]. Thus, both 2D and 3D cases are "low-dimensional", with interactions likely to be essential and the fluctuation spectrum likely to be very non-linear and far from the "Gaussian" behavior. This is caused by the infinite degeneracy of the LLL. The quadratic part of the free energy does not contain any correlations among different configurations of $\Psi(\vec{r})$ in the xy-plane. Such correlations are only built by the quartic part. Within the pedestrian perturbation theory all diagrams are completely "flat" and do not depend on the internal index of the LLL. On the other hand, if we break the symmetry by hand, like in the MFT and the elastic theory, such correlations immediately become apparent. Consequently, the Abrikosov form of the interaction among vortices is most likely "hidden" deep in the non-perturbative sector of the field theory described by Eq. (1). This situation is reminiscent of the one encountered in the study of the quantum Hall effect where the interactions are essential in lifting the degeneracy of the LLL and where the ground state cannot be described by the perturbation theory [11].

The confinement to the LLL provides a severe restriction on the form of $\Psi(\vec{r})$. We can exploit this by noting that in study of superconductivity it is often beneficial to write $\Psi(\vec{r}) = A(\vec{r})\exp(i\chi(\vec{r}))$ and study fluctuations in amplitude and phase separately since the nature of these fluctuations can be quite different. In the critical region, however, A and χ cannot fluctuate independently. The confinement to the LLL provides a stringent constraint which couples the fluctuations in A and χ and places a restriction on the allowed fluctuations in $\Psi(\vec{r})$. In general, such a constraint is difficult to enforce in the partition function (1). In the *symmetric* gauge, however, the restriction on $\Psi(\vec{r})$ becomes particularly transparent and the constraint can be enforced exactly:

$$\Psi(\vec{r}) = \sqrt{2\pi\ell^2}\sum_{m=0}^{N} b_m(\zeta)\varphi_{0m}(z,z^*) = \Phi(\zeta)\prod_{i=1}^{N}(z - z_i(\zeta))\exp(-|z|^2/4) \quad , \tag{2}$$

where $\varphi_{0m}(z,z^*) = z^m(2\pi\ell^2 m!2^m)^{-\frac{1}{2}}\exp(-|z|^2/4)$ and $N = \Omega/2\pi\ell^2$, Ω being the area of the system in the xy-plane. Φ and $\{z_i\}$, which are in highly non-linear relation to $\{b_{0m}\}$, are now free to fluctuate independently. Thus, the continuous field $\Psi(\vec{r})$ has been expressed in terms of a discrete set of new variables. The Jacobian of the transformation (2) is given by:

$$\mathcal{J} \propto \prod_{i<j}^{N}|z_i(\zeta) - z_j(\zeta)|^2(\Phi(\zeta)\Phi^*(\zeta))^N \quad .$$

By changing variables in the functional integral from $\{b_{0,m}(\zeta)\}$ to $\Phi(\zeta)$ and $\{z_i(\zeta)\}$ I finally obtain:

$$Z \propto \frac{1}{N!}\int\prod_{i=1}^{N}\frac{dz_i(\zeta)dz_i^*(\zeta)}{2\pi}\prod_{i<j}^{N}|z_i(\zeta) - z_j(\zeta)|^2\int d\Phi(\zeta)d\Phi^*(\zeta)(\Phi(\zeta)\Phi^*(\zeta))^N$$

$$\exp\{-\frac{2\pi\ell^2 N}{k_B T}\int d\zeta[\tilde{\alpha}(T)|\Phi(\zeta)|^2\langle f^2(\zeta)\rangle + \frac{\beta}{2}|\Phi(\zeta)|^4\langle f^4(\zeta)\rangle]$$

$$-\frac{2\pi\ell^2 N}{k_B T}\int d\zeta \int (d^2r/\Omega)\frac{n}{2m_\zeta^*}|(\partial_\zeta + \frac{2ei}{c}A_\zeta)\Phi(\zeta)f(z|\{z_i(\zeta)\})|^2\} \quad , \qquad (3)$$

where $f(z|\{z_i(\zeta)\}) = \Pi_i(z - z_i(\zeta))$, $\langle f^p\rangle = \int \frac{dzdz^*}{2\pi N}\exp(-\frac{p|z|^2}{4})f^p(z|\{z_i(\zeta)\})$ and $\tilde{\alpha} \equiv \alpha(1 - H/H_{c2}(T))$.

Φ and $\{z_i\}$ are the "natural" variables of this problem representing two distinct tendencies in $\Psi(\vec{r})$ near $H_{c2}(T)$: The "global" superconducting correlations are embodied in Φ, while the "local" amplitude and phase fluctuations, coupled by the confinement to the LLL, arise from the motion of $\{z_i\}$. I now briefly summarize some of the relevant physics that will emerge from Eq. (3). Most of what is said in this summary comes from a detailed study of a (quasi) 2D case, where, by a combination of analytical and numerical methods, the physics described by Eq. (3) can be illuminated to a great extent [12]. First finding is that the behavior of fluctuations in Φ and $\{z_i\}$ in the vicinity of $T_{sc}(H)$ is quite different, indicating the "low-dimensional" character of the transition. In MFT Φ becomes finite and $\{z_i\}$ "freeze" at the same point, $T_{c2}(H)$. When the fluctuations are included, as shown below, $\{z_i\}$ still form a liquid long after the fluctuations in Φ have subsided and its value is dominated by the saddle point contribution. In fact, a growth of this saddle point value of Φ provides an effective interaction among $\{z_i\}$ which ultimately leads to their solidification at $T_{sc}(H)$. This interaction can be recovered only after the integration over $\Phi(\zeta)$ which leads to the effective many-body system of interacting particles $\{z_i(\zeta)\}$. I call this many-body system a *dense vortex plasma* (DVP). DVP contains the information about superconducting correlations when these are dominated by the vortex core overlap and provides a paradigm for study of various cooperative phenomena in type-II systems in the critical region near $H_{c2}(T)$.

Unlike the perturbative expansion or the elastic theory, DVP describes both the liquid and the solid state of dense system of vortices. These two phases of DVP are separated by the solidification transition which we identify as $T_{sc}(H)$. This is the thermodynamic superconducting transition which in the MFT occurs at $T_{c2}(H)$. Only below $T_{sc}(H)$ can the expectation value of the superconducting order parameter be finite, although its amplitude rises at $T_{c2}(H)$ and there would still be a significant amount of entropy associated with $T_{c2}(H)$. Following traditional argument, $T_{sc}(H)$ should also correspond to the resistive transition since, once DVP "freezes", small disorder will pin this vortex solid and result in the dissipationless flow for weak currents. The interaction which causes freezing of DVP is undetected in large order perturbation theory [5], as emphasized earlier, and is extracted *only* through the integration over

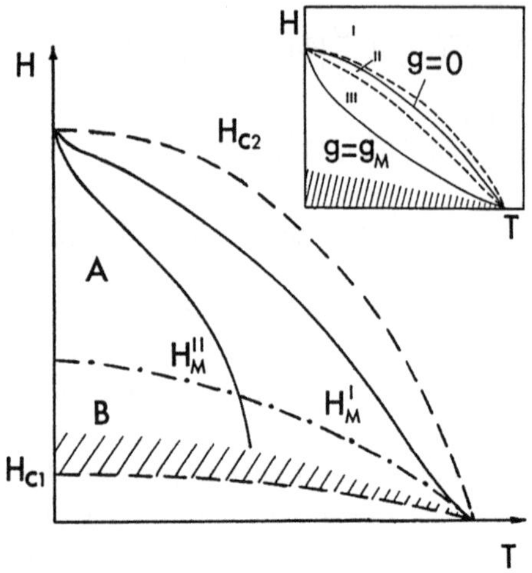

Fig. 1. A schematic phase diagram for type-II superconductors. The dense vortex regime (A) is separated from the dilute vortex regime (B) by the dashed-dotted crossover line. $H_{sc}(T)$ can either look like H_M^I or H_M^{II} depending on the strength of fluctuations (see text). In the dashed region the separation between vortices is larger than the magnetic field penetration depth. The inset shows the same phase diagram in the DVP scaling form. Regions I, II and III are defined in the text.

Φ. In performing this important integral one picks up contributions from the non-perturbative sector of (1) which assures correct behavior both above and below $H_{c2}(T)$. It takes a while for this (Abrikosov) interaction to become sufficiently strong and, in the properly scaled form, $T_{sc}(H)$ is *always far* from $T_{c2}(H)$ both in 2D and 3D. Furthermore, the constraint on the form of the order parameter, imposed by Eq. (2), results in the properties of DVP being determined by a single dimensionless parameter, g, which is a particular combination of coefficients appearing in Eq. (3). Consequently, the transition line $T_{sc}(H)$ of type-II superconductors can be inferred from the "universal" DVP scaling law $g(T,H) = g_M$, where $g_M < 0$ depends only on dimensionality (see Fig. 1).

2.1 Two-Dimensional Case

In a 2D system, the ζ-dependence is quenched when $(n/2m_\zeta^*|\tilde{\alpha}|)^{\frac{1}{2}} \equiv \xi_\zeta(T,H)$ is comparable or larger than d, where d is either the film thickness or the interlayer separation in the case of highly anisotropic layered superconductors. As a consequence, this crucial step of integration over Φ can then be performed *exactly*. This is what makes the 2D example so valuable. The result is

$$Z = [\int \prod_{i=1}^{N} \frac{dz_i dz_i^*}{2\pi} \langle f^4 \rangle^{-\frac{N+1}{2}} \prod_{i<j}^{N} |z_i - z_j|^2]^{-1} \int \prod_{i=1}^{N} \frac{dz_i dz_i^*}{2\pi} \langle f^4 \rangle^{-\frac{N+1}{2}} \prod_{i<j}^{N} |z_i - z_j|^2$$

$$\times \exp[\frac{1}{2} N V^2 - \frac{1}{2} N V \sqrt{V^2+2} - N \sinh^{-1}(V/\sqrt{2})] \quad , \qquad (4)$$

where $V(\{z_i\}) \equiv g\langle f^2 \rangle / \sqrt{\langle f^4 \rangle}$ and $g = \tilde{\alpha}\sqrt{2\pi\ell^2 d/2\beta k_B T}$.

Eq. (4) describes a classical incompressible many-body system of particles $\{z_i\}$: a 2D DVP. The incompressibility is assured by the purely entropic long range interactions arising from the Jacobian $\mathcal{J}$. The basic dynamical interaction among $\{z_i\}$, $V(\{z_i\})$, is long ranged and of purely geometric nature. It contains two-, three-, four-, etc. body forces reflecting the dense vortex limit (unlike the dilute limit, the forces cannot be reduced to two-body terms only). It also exhibits scale invariance. This implies that Z depends only on the dimensionless coupling constant g and is independent of density (It is therefore useful to normalize Z to $Z(g=0)$, as is done in Eq. (4).). We recognize the variance $\frac{\langle f^2 \rangle}{\sqrt{\langle f^4 \rangle}}$ as a square root of an inverse of Abrikosov's β_A except that now $\{z_i\}$ do not form a lattice as in Ref. [1] but are allowed to move all over the xy-plane. $\frac{\langle f^2 \rangle}{\sqrt{\langle f^4 \rangle}}$ varies between 0, for "exploded" or "imploded" zeros, and 0.928, for triangular lattice (note that values larger than 0.928 are not possible for a finite density of zeros).

There are three physical regimes of this DVP as depicted in Fig. 1. For $g \gg \sqrt{2}$ one is in the regime I far above $T_{c2}(H)$, with interactions being very weak. The influence of the quartic term in Eq. (3) is basically perturbative. The effective interaction in the exponent of Eq. (4) goes as $-N\ell n(\sqrt{2}V) + N\mathcal{O}(g^0)$ in this region. The higher order terms coincide with the perturbative expansion [5]. Thus, in this regime, the perturbation theory can be used to extract the contribution of fluctuations to various physical quantities. As one approaches the MFT transition ($T_{c2}(H)$ or $g = 0$) from above one enters the regime II where, for $-\sqrt{2} \ll g \ll \sqrt{2}$, the effective interaction goes as $-\sqrt{2}NV + N\mathcal{O}(g^2)$. Thus, the interaction of $\{z_i\}$ sets in, although it is far too feeble to produce solidification. Here DVP is still deep into the liquid state. In this regime perturbation theory should start running into trouble as its large order behavior starts being dominated by an unphysical saddle point [5]. Finally, for $g \ll -\sqrt{2}$, the effective interaction is $NV^2 + N\ell n(\sqrt{2}V) + N\mathcal{O}(g^0)$. Two leading terms are contributions from the saddle point approximation to the integral over Φ indicating that in the regime III the fluctuations in Φ have largely subsided. As $|g|$ increases, this saddle point value of Φ continues to grow leading to enhanced interaction ($\propto V^2$) among $\{z_i\}$. These correlations among $\{z_i\}$ show that this problem is not $D-2$ dimensional, as suggested by the perturbation theory in the region I. Note that the leading term in the interaction becomes exactly the Abrikosov mean-field free energy if $\{z_i\}$ are frozen into a triangular lattice. DVP in 2D, therefore, correctly reproduces all the relevant fluctuation regions, from the perturbative regime I to the highly non-perturbative regime III, in which the leading interaction is of the Abrikosov form. By a detailed study of DVP in Eq. (4) one should be able to understand the nature of critical fluctuations near $H_{c2}(T)$ in quasi 2D superconductors.

Eventually, deeply in the region III, the interactions become strong enough to cause solidification. The phase transition in DVP could be of the continuous Kosterlitz-Thouless-Berezinskii-Halperin-Nelson-Young (KTBHNY)-type or a weak first order since the forces are long ranged. The transition occurs at $g = g_M$, where g_M is a pure number. The Monte Carlo simulation of DVP gives $g_M \simeq -8$ [12]. The size of g_M testifies to the deep non-perturbative character of the transition. $g = g_M$ is, in fact, the implicit equation for $T_{sc}(H)$, and, in this "universal" scaling form, is the same for all quasi 2D type-II superconductors. Thus, while in the actual $H - T$ phase diagram the position of $T_{sc}(H)$ relative to $T_{c2}(H)$ may vary considerably from one superconductor to another, depending on material parameters, in this proper scaling form it always occurs at the same place, and, with $|g_M| \simeq 8$, it is very far from $T_{c2}(H)$. This is illustrated in Fig. 1. The above theoretical results for $T_{sc}(H)$ agree very well with the recent experimental data on the thin films of Nb_3Ge [13].

2.2 Three-Dimensional Case

In 3D Z again depends on a single scaling parameter. The appropriate coupling constant is $g' = \tilde{\alpha}\sqrt{2\pi\ell^2\xi_\zeta/2\beta k_B T}$, with d replaced by $\xi_\zeta(T,H)$ in comparison to the 2D case. Unfortunately, in 3D the integration over Φ cannot be carried out exactly, due to ζ-dependence, and one has to rely on approximate calculations in deriving the effective DVP. Motivated by a 2D example one still expects that the situation is similar and qualitatively described by Fig. 1. Thus, $g' = g'_M$ should still be deep in the region III and we may use the saddle point approximation to perform the integral over Φ. Of course, one may be more pragmatic and simply use the freezing (melting) condition $g'(T,H) = g'_M$ to fit an experimentally determined $T_{sc}(H)$. This one-parameter fit gives g'_M, which is the same for all 3D superconductors. From the recent data on YBCO [14] one finds $g'_M \simeq -3.5$ which is smaller than g_M as expected (as the dimension increases toward D_c, $T_{sc}(H)$ should move closer to $T_{c2}(H)$) and, most likely, in the region III. Hence, the saddle point evaluation of the Φ integral in Eq. (3) should lead to a DVP in the regime appropriate to the vicinity of $T_{sc}(H)$. Here I follow Ref. [15]. The saddle point form of $\Phi(\zeta)$ is determined from:

$$\tilde{\alpha}\Phi(\zeta)\langle f^2(\zeta)\rangle + \beta|\Phi(\zeta)|^2\Phi(\zeta)\langle f^4(\zeta)\rangle = \frac{n}{2m_\zeta^*}\langle f^*(\zeta)\partial_\zeta^2(\Phi(\zeta)f(\zeta))\rangle \quad . \quad (5)$$

ζ-dependence again makes this equation untractable for an arbitrary arrangement of $\{z_i(\zeta)\}$. However, if the ζ-derivatives were absent, the solution could be easily found. Introducing parameter $\epsilon \ll 1$ to accompany the ζ-derivative terms in the kinetic energy (3) we can find the solution of Eq. (5) as an expansion in ϵ, and then set $\epsilon = 1$. In practice, this is probably feasible only in the lowest order. Using this saddle point approximation and setting $\epsilon = 1$ we obtain a 3D DVP:

$$Z = [\int \mathcal{D}[z_i(\zeta)]\langle f^4\rangle^{-\frac{N+1}{2}} \prod_{i<j}^{N} |z_i - z_j|^2]^{-1} \int \mathcal{D}[z_i(\zeta)]\langle f^4\rangle^{-\frac{N+1}{2}} \prod_{i<j}^{N} |z_i - z_j|^2$$

$$\times \exp\{-\int_0^{d/\xi_\zeta} d\zeta [\frac{1}{2}g'^2 \sum_{i,j} \frac{dz_i^*}{d\zeta} M_{ij}(\zeta)\frac{dz_j}{d\zeta} - NV^2 - N\ell n(\sqrt{2}|V|)]\} \quad , \quad (6)$$

where

$$M_{ij}(\zeta) = 4N\frac{\langle f^2\rangle}{\langle f^4\rangle}\langle\frac{\partial f^*}{\partial z_i^*}\frac{\partial f}{\partial z_j}\rangle \quad ,$$

V is defined below Eq. (4) and only the leading powers in g' have been retained. In deriving (6) I have assumed that $\frac{\partial_\zeta f}{f} \gg \sqrt{\frac{\langle f^4\rangle}{\langle f^2\rangle}}\partial_\zeta(\sqrt{\frac{\langle f^2\rangle}{\langle f^4\rangle}})$. This should be a reasonable approximation near $T_{sc}(H)$. The general form of the "kinetic" energy is more complicated but it is derived straightforwardly.

The above form of DVP should provide a qualitatively reasonable description of the superconducting fluctuations in region III, near $T_{sc}(H)$. Eq. (6) represents a "quantum" version of DVP, if we interpret $\{z_i(\zeta)\}$ as the world-lines of particles located at $\{z_i\}$ [16], [9]. The "kinetic" energy term is quite unusual, with velocity dependent forces. The ground state of this quantum DVP (appropriate for $d/\xi_\zeta \to \infty$) could be either a liquid or a solid depending on the size of g'. The superconducting transition corresponds to the transition between the quantum liquid and the quantum Wigner crystal in the DVP as g'^2 increases, which is equivalent to a decrease in the "effective" Planck constant. The transition will most likely be first order, as anticipated in [10]. The important new result is that the "lateral" correlations (V^2 term in (6)) which are essential for description of the transition are now included on equal footing with the fluctuations along $\vec{H}$, considered in Ref. [10]. The calculation of this quantum melting process, which can be carried out using the variational approach [17], would yield a value for g'_M.

3. Conclusions

DVP description of critical fluctuations near $H_{c2}(T)$ presented here enables one to study various superconducting properties of type-II systems in a well-behaved theoretical framework. This is particularly true of a quasi 2D case where the DVP many-body problem (Eq. (4)) can likely be solved using available numerical methods. Identification of the melting (freezing) point of the DVP with the thermodynamic superconducting transition has led to the transition line $T_{sc}(H)$ in 2D and 3D, obtained from a simple scaling law. This DVP scaling should be a universal characteristic of critical behavior in type-II superconductors.

It is important to stress here that the DVP description holds only near $H_{c2}(T)$ and cannot be extended to very low fields. There the vortices (and antivortices) are far apart and one must consider the familiar opposite limit of a *dilute vortex plasma* (dVP). In dVP the vortices interact through pairwise logarithmic interactions generated by long range phase correlations. The amplitude fluctuations are suppressed and this is the essential difference from the DVP description in the critical region where amplitude fluctuations are crucial. Properties of dVP have already been discussed extensively in the literature: Refs. [18] and [19] provide analysis of the 2D case, while the 3D situation is considered in Refs. [16], [9], [20] and [21]. The crossover between DVP and dVP is controlled by the contribution of higher Landau levels in $\Psi(\vec{r})$. Near $H_{c2}(T)$ (region A in Fig. 1) the contribution from higher Landau levels is very small and they do not significantly affect the dynamics. Near $H_{c1}(T)$ (region B), on the other hand, a large contribution from higher Landau levels is necessary to construct $\Psi(\vec{r})$ which has a uniform amplitude in the large regions outside vortex cores. The crossover region is denoted by the dashed-dotted line in Fig. 1.

We can illustrate this crossover on the example of $T_{sc}(H)$ in 2D. The crossover will always take place at sufficiently low fields, where $T_{sc}(H)$ drops far below $T_{c2}(H)$ [12]. $T_{sc}(H)$ is found from $g^2(T,H) = g_M^2$. We can rewrite this as $e(t,h) = \theta_M th$, where $t \equiv \frac{T}{T_{c0}}$, $h \equiv \frac{H}{H_{c2}(0)}$, $e(t,h)$ is the BCS condensation energy normalized to its value e_c at $T = H = 0$, and $\theta_M = g_M^2 T_{c0}/2\pi\xi_0^2 d e_c$. Clearly, θ_M is the parameter which determines the proximity to the mean-field transition. Since $\theta_M \sim g_M^2/(k_F^2\xi_0 d)$, for most superconductors $\theta_M \ll 1$ and $T_{sc}(H)$ will be near $T_{c2}(H)$. However, for $1-t \ll \theta_M$, $T_{sc}(H)$ deviates substantially from $T_{c2}(H)$ and the corrections to DVP will have to be taken into account. Thus, at sufficiently low fields the transition line will drop far below H_{cross} and start approaching $H_{c1}(T)$ and one always has to abandon the DVP description in favor of the dVP one. Typically, this will be true only at very low fields, far smaller than

$H_{c2}(0)$. In thin films, for example, as one reduces the thickness, $H_{sc}(T)$ would evolve from H_M^I to H_M^{II} reflecting the decrease in θ_M.

Acknowledgements

I thank L. Boulaevskii, H. Fukuyama, D. R. Nelson, M. Rasolt, H. J. Schulz, O. T. Valls, and L. Xing for useful discussions and comments.

References

[1] A. A. Abrikosov, *Zh. Eksp. Teor. Fiz.* **32**, 1442 (1957).
[2] K. Maki, *J. Low Temp. Phys.* **1**, 513 (1969).
[3] P. A. Lee and S. R. Shenoy, *Phys. Rev. Lett.* **28**, 1025 (1972).
[4] G. J. Ruggeri and D. J. Thouless, *J. Phys.* **F6**, 2063 (1976).
[5] E. Brézin, A. Fujita, and S. Hikami, *Phys. Rev. Lett.* **65**, 1949 (1990).
[6] R. Ikeda *et al.*, *J. Phys. Soc. Jpn.* **58**, 1906 (1989).
[7] E. H. Brandt, *Int. J. Mod. Phys.* **B5**, 751 (1991) and references therein.
[8] A. Houghton, R. A. Pelcovits, and A. Sudbo, *Phys. Rev.* **B40**, 6763 (1989).
[9] D. R. Nelson and H. S. Seung, *Phys. Rev.* **B39**, 9153 (1989).
[10] E. Brézin, D. R. Nelson, and A. Thiaville, *Phys. Rev.* **B31**, 7124 (1985).
[11] For study of formation of a CDW of electrons in the LLL see, for example, H. Fukuyama, P. M. Platzman, and P. W. Anderson, *Phys. Rev.* **B19**, 5211 (1979).
[12] Zlatko Tešanović and L. Xing, *Phys. Rev. Lett.* **67**, XXXX (1991).
[13] P. Berghuis, A. L. F. van der Slot and P. H. Kes, *Phys. Rev. Lett.* **65**, 2583 (1990).
[14] P. L. Gammel, L. F. Schneemeyer, and D. J. Bishop, *Phys. Rev. Lett.* **66**, 953 (1991).
[15] Zlatko Tešanović, unpublished.
[16] D. R. Nelson, *Phys. Rev. Lett.* **60**, 1973 (1988).
[17] L. Xing and Zlatko Tešanović, *Phys. Rev. Lett.* **65**, 794 (1990).
[18] B. A. Huberman and S. Doniach, *Phys. Rev. Lett.* **43**, 950 (1979).
[19] D. S. Fisher, *Phys. Rev.* **B22**, 1190 (1980).
[20] M. V. Feigel'man, V. B. Geshkenbein, and V. M. Vinokur, *Pis'ma v Zh. Eksp. Teor. Fiz.* **52**, 546 (1990).
[21] D. S. Fisher, M. P. A. Fisher, and D. A. Huse, *Phys. Rev.* **43**, 130 (1991).

Part III

Experiment

Material Design of High-T_c Oxides

Y. Tokura

Department of Physics, University of Tokyo, Tokyo 113, Japan

Abstract. Structures of existing cuprate superconductors can be viewed as alternating stacks of conducting CuO_2-sheets and insulating ionic layers (block layers). Using the concept of the block layer, we argue the classification and material design of CuO_2-layer based superconductors. In particular, dependence of carrier-doping characteristics on the Cu-O network and bond length is discussed based on the observed variation of some electronic and vibrational parameters.

1. Introduction

Structures of the high T_c copper oxide compounds are characterized by two-dimensional (2D) Cu-O networks, which consist of CuO_4 squares sharing their corners. Depending on the number (=0,1,2) of the apical oxygen(s) on each copper site, the Cu-O sheets show some variations of two-dimensional (2D) networks such as Cu-O squares, pyramids and octahedra, as shown in Fig.1. The figure shows further variations of CuO_2-sheets accompanying halogen ions at the apex position of Cu.

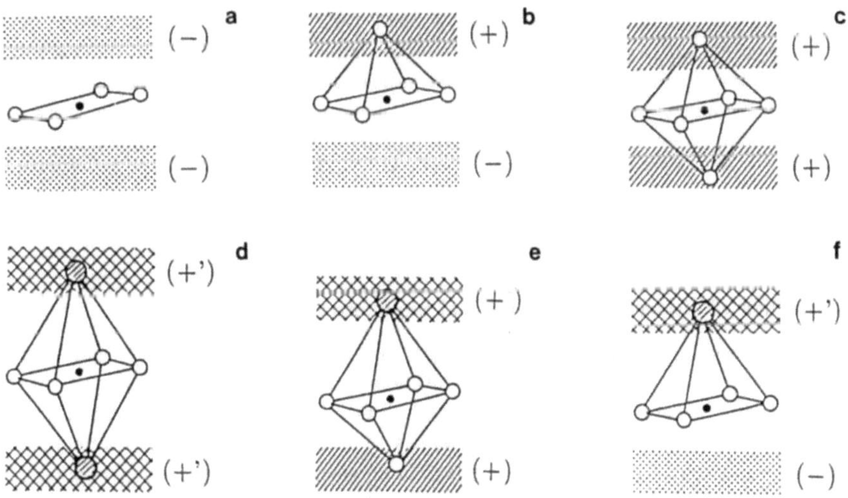

Fig.1: Basic Cu-O and Cu-O-halogen patterns. The notations, +,- and +', represent the attributes of the block layers sandwiching the CuO_2-sheet (see section **3** in the text).

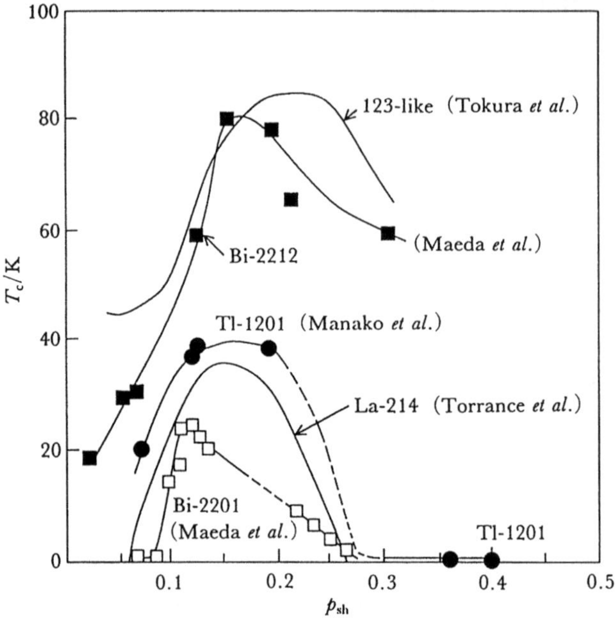

Fig.2: Dependence of T_c on $[CuO]^p$ charge in the sheets for several compounds.

It is known at present that the high-T_c superconductivity is observed to arise when these Cu-O planes are doped with an appropriate density of charge-carriers. Amount of chemically doped carriers, which may be measured as the effective $[CuO]^{+p}$ charge (or the nominal copper valence 2+p), is one of the most important factors which govern T_c [1,2]. We show in Fig.2 the correlation between T_c and $[CuO]^p$ charge in the Cu-O sheets for several CuO_2-layered superconductors. As clearly seen in this figure, the high-T_c superconductivity emerges in the narrow window of the carrier concentration. This indicates that we need to finely control the $[CuO]^p$ charge to produce the high-T_c superconductivity. We present here a comprehensive view on structures of layered copper oxide compounds. In particular, we pay special attention to the Cu-O network- and bond length-dependent variations of carrier-doping characteristics and several important electronic/vibrational parameters.

2. Variation of electronic and vibrational parameters

Before proceeding to the material design, we should note the fact that some important physical parameters in the CuO_2-sheets are quite sensitive to the Cu-O coordination, network and bond length. In Fig.3, we have summarized experimentally observed variation in energies of some characteristic parameters in insulating analogs of cuprate superconductors[3]; electronic gap (E_g), superexchange interaction between the Cu-spins (J_s), and in-plane Cu-O stretching phonon ($\hbar\omega_s$). To represent the material dependence of these quantities, we have adopted the one-dimensional axis on which the in-plane Cu-O bond length (d) is plotted for each compound. (Note that the both axes in Fig.3 are logarithmic in scale.)

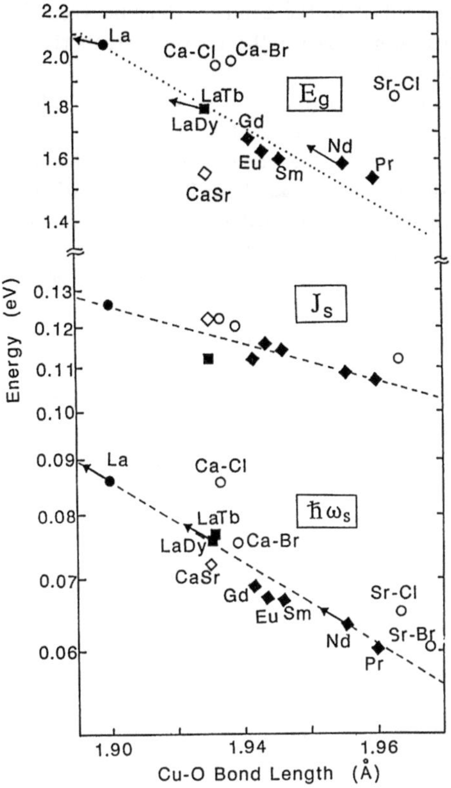

Fig.3: Variation of electronic and vibrational parameters in various CuO_2-layered compounds with T- (oxygen- or halogen-apex), T^*-, and T'-phase structures, plotted against the in-plane Cu-O bond length (d); the electronic gap (E_g), superexchange energy (J_s) and in-plane Cu-O stretching phonon mode ($\hbar\omega_s$). The labels attached to the data points represent the compounds; for "La" for La_2CuO_4, "CaSr" for $(Ca,Sr)CuO_2$, "Ca-Cl" for $Ca_2CuO_2Cl_2$, and etc. Arrows indicate the changes of E_g and ω_s against d under hydrostatic pressures. A dotted line for E_g is merely a guide to eye.

Among those quantities, the E_g-value shows a large variation depending on the compounds; from 2.0_5eV in La_2CuO_4 down to 1.5_3eV in Pr_2CuO_4. The electronic gap in the insulating compound is governed by the charge transfer (CT) process from O 2pσ to Cu $3d_{x^2-y^2}$ (upper Hubbard) state. Such a critical dependence of E_g on the lattice structure can be explained by the Madelung type contribution to the CT type gap[4]. The arrows attached to the data points of the three compounds indicate an increasing rate of E_g against the shrinkage of d under hydrostatic pressure. This shows that the E_g-value depends not only on the d-value but on the whole lattice structure, or a little more specifically on the coordination around the Cu-sites, reflecting the non-local nature of Madelung type interaction.

In contrast with the case of E_g, the variation of J_s, which was estimated from the Raman scattering for spin-pair excitations, is more gentle. As discussed in previous studies[4-6], this feature can be ascribed to the compensation between the two competing quantities, t_{pd} (electron transfer between the nearest-neighboring p and d state) and $E_g^{(0)}$ (bare CT gap when $t_{pd} \to 0$).

The in-plane Cu-O stretching mode, whose energy ($\hbar\omega_s$) is also shown in Fig.3, is the highest-lying phonon mode (with pseudo-E_u symmetry at $\mathbf{k}=0$) in this class of compounds. As pointed out by Tajima et al.[7], the mode frequency shows a significant variation, scaling approximately such that $\omega_s \propto d^{-7}$. In fact, application of hydrostatic pressure shifts the mode frequency[3] as indicated by arrows in Fig.3, which is along the dashed line representing the above relation. The in-plane Cu-O bond length reflects the valence of the in-plane oxygen. Therefore, the above observed steep change of mode frequency ω_s with d indicates that this oxygen vibration strongly couples with the valence state in the CuO_2-sheets.

To sum up, the bond covalency increases in the CuO_2-sheets, as d increases. This may be contrary to the simplest intuition, but obviously reflect the variation of E_g which is sensitive to the Cu-O coordination and overwhelms the change of t_{pd}. The relative bond covalency can be measured by variation of d or $\hbar\omega_s$.

3. Concept of block layer

The physico-chemical properties of the CuO_2-layered compounds are quite sensitive to the Cu-O coordination and bond length, as exemplified above. To describe such structural characteristics with sufficient accuracy but preferably in a simple-minded way, let us first introduce the simplified structural rule for constructing CuO_2-layered superconductors. In the present classification scheme[8], we view that every high T_c compound possesses an alternating one-to-one stack of CuO_2-sheets and "block layers" along the normal to the basal plane, as schematically shown in Fig.4. Here, the block layer (BL) stands for an intervening atomic layer which is composed of metal ions (mostly alkaline-earth and/or rare-earth ions) and oxygens or halogens. As a result, each CuO_2 sheet is isolated from each other and shows no direct connection via the apical oxygens.

Figure 5 shows the structures of the two simple compounds, $La_{2-x}Sr_xCuO_4$ (so-called T-phase) and $Nd_{2-x}Ce_xCuO_4$ (so-called T'-phase): There exist La_2O_2 layers with the rock salt-type structure and Nd_2O_2 layers with the fluorite-type structure, respectively, which sandwich the CuO_2-sheets. Hereafter, the La_2O_2 (or similar rock salt-type) and Nd_2O_2 (or similar fluorite-type) layers will be referred as **L**- and **G**-

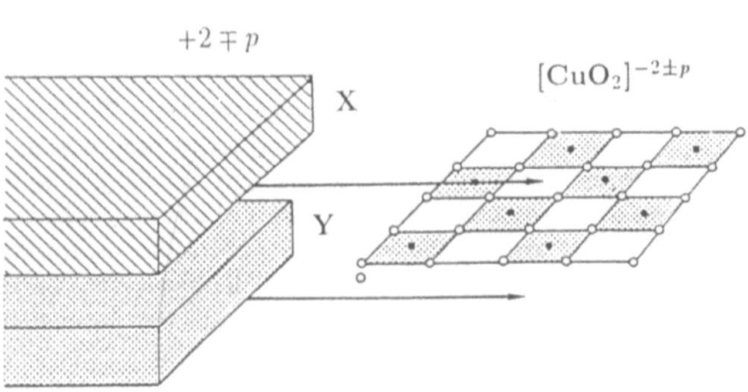

Fig.4: Concept of block layer (BL).

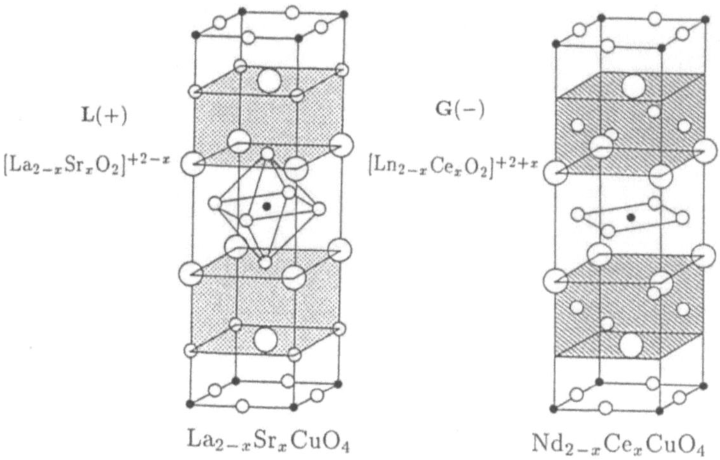

Fig.5: Block layers in T-phase and T'-phase compounds.

layer. By using the concept of the BL, the structures of these compounds can be expressed as sequences of -**L**-[CuO_2] -**L**-[CuO_2] - and -**G**-[CuO_2] -**G**-[CuO_2] -. Within the **L**-layer, the O atom positions above (or below) the Cu-site of the CuO_2-sheet as an apex, while in the **G**-layer the O atom is deviated from the apical position. Here, we denote the BL which contains the apical oxygen for the CuO_2-sheet as the + type and the BL with no apical oxygen as the - type. According to this definition, the **L**-layer is of the + type, while the **G**-layer of the - type.

One of the important roles of the BL is to supply the Cu-O planes with charge-carriers. From the viewpoint of ionic crystals, an insulating CuO_2-sheet consists of Cu^{2+} and O^{2-} and hence is charged -2 ($[CuO_2]^{2-}$). To maintain the charge neutrality in whole crystal, the average charge of the BL in an insulating compound must be +2. In fact, the charge of the **L**- and **G**-layer is +2 without Sr^{2+}-or Ce^{4+}-substitution. However, when the charge of the BL is reduced (by -q) or increased (by +q) from +2 with Sr^{2+}- or Ce^{4+}-substitution, the excess charge of +q or -q enters the Cu-O plane per Cu. In $La_{2-x}Sr_xCuO_4$, the charge of the **L**-layer is 2-x, then the effective $[CuO]^P$ charge equals x. Similarly, the charge of the **G**-layer is reduced to 2-x in $Nd_{2-x}Ce_xCuO_4$ and hence the $[CuO]^P$ charge is (-x). The former process corresponds to hole-doping, while the latter to electron-doping. Thus, the BL plays a role of controlling carrier-concentration in Cu-O planes.

We show in Fig.6 the BL structures in three 80-100K class superconductors, (a) $Bi_2Sr_2CaCu_2O_{8+y}$ (Bi-2212 phase) and (b) $Pb_2Sr_2(Ca,Y)Cu_3O_{8+y}$ (Pb-3212 phase), (c) $YBa_2Cu_3O_{6+y}$ (123 or Cu-1212 phase). In the Bi-2212 phase, a pair of CuO_2-sheets shows a face-to-face stack via an intervening Ca□ layer, □ being an oxygen vacancy. From the viewpoint of the alternating stacking of the BL and CuO_2-sheet, we may consider that the Bi-2212 compound is composed of two kinds of BLs; $\mathbf{B_2}$-layer=[(SrO)$(BiO)_2$(SrO)] and **V**-layer=Ca□. Then, the 2212-phase can be represented as a doubly periodical sequence of -$\mathbf{B_2}$-[CuO_2] -**V**-[CuO_2] - or equivalently as the abbreviated formula ($\mathbf{B_2}$,**V**). Note that charge of the $\mathbf{B_2}$-layer and **V**-layer in the nominal (y=0) composition of the Bi-2212 phase is both +2, assuming that the valency of Bi is +3. The hole-type carriers in this compound are

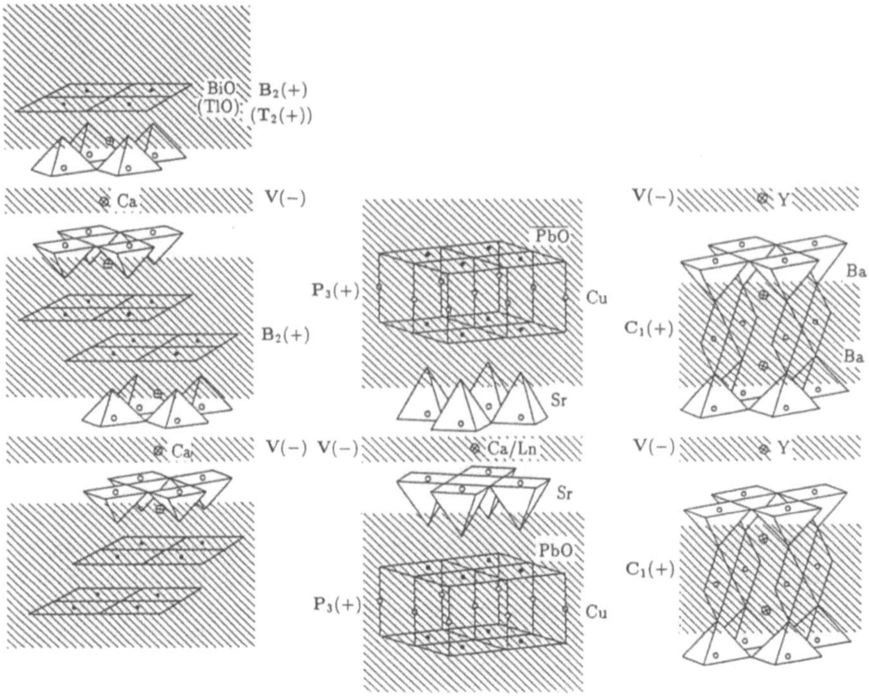

(a) $Bi_2Sr_2CaCu_2O_8$ ($Tl_2Ba_2CaCu_2O_8$) (b) $Pb_2Sr_2(Y,Ca)Cu_3O_8$ (c) $YBa_2Cu_3O_7$

Fig.6: Block layers in (a) Bi-2212, (b)Pb-3212 and (c) 123 (or Cu-1212) compounds.

introduced by charge reduction of the $\mathbf{B_2}$-layer. It appears that the BiO plane contains excess oxygens (y) which reduce the charge of the $\mathbf{B_2}$-layer and hence introduce the positive charge in the Cu-O plane. Usually, the as-prepared sample possesses the excess $[CuO]^p$ charge ($p\sim0.2$) due to the oxygen nonstoichiometry and readily shows 80K-class superconductivity.

A similar procedure for extraction of BLs is possible for the Pb-3212 phase (Fig.6(b)). In this case, we can extract a new + type BL, that is $\mathbf{P_3}$-layer=[(SrO)(PbO)Cu(PbO)(SrO)]. Another BL in the Pb-3212 structure is a modification of **V**-layer; **V**=[(Ca,Y)□]. In the parent (**V**=[Y]) compound $Pb_2Sr_2YCu_3O_8$, the arrangement of the positively charged BLs and negatively charged CuO_2-sheets is such as -$\mathbf{P_3}^{1+}$-$[CuO_2]^{2-}$-$\mathbf{V}^{3+}$-$[CuO_2]^{2-}$-. Contrary to the previous examples, charges of the BLs show disproportion into +3 and +1, but as an average the whole charge neutrality is maintained. To dope this structure with hole-type carriers, the Y^{3+}-sites are partially substituted with Ca^{2+} (i.e. $Y_{1-x}Ca_x$) to reduce the charge of the **V**-layer from +3 to +3-x, and then the positive charge of +x/2 is introduced into Cu-O planes per [Cu-O] unit.

In the case of the 123 (or Cu-1212) structure (Fig.6(c)), there is another Cu-O network, CuO_y, in between the two CuO_2-sheets. This Cu-O chain unit associated with two [BaO] layers on the both sides can be considered to be a sort of the BL as

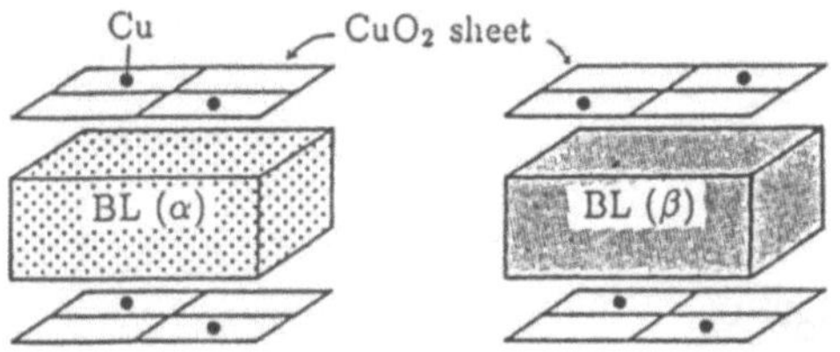

Fig.7: Phase relation between the two CuO_2 sheets adjacent to α and β type BL.

shown in Fig.6(c) and we name this the C_1-layer. Following the concept of the BL, we consider the y=0 compound ($Y_1Ba_2Cu_3O_6$) as a parent compound of the 123 phase. At y=0, the valence of Cu in the C_1-layer is +1 since it shows the two-oxygen coordination, and hence the charge of the **V**-layer [(BaO)Cu(BaO)] is +1. Charges of the BLs in this insulating compound are disproportionate, such as (V^{+3}, C_1^{+1}), as in the Pb-3212 phase. As is well-known, hole-doping in the 123 phase is conducted by oxygenation of the C_1-layer. However, nominal Cu valence in the C_1-layer varies with the oxygen content y and also depends on detailed ordered (or disordered) structures of Cu-O networks within the C_1-layer in a complicated manner.

We can use the above procedure in other CuO_2-layered compounds to extract BLs. These BL components are characterized by the values of the BL charge (Q) and its controllable range (by means of elemental substitutions) as well as by the "in-plane bond length" d_s and "thickness" w. Here, the quantity d_s is defined as a half of the unit cell parameter of the single BL structure and hence nearly equals the in-plane Cu-O bond length. The "thickness" w is defined as the repeated unit length of the single BL structure along the normal to the sheet, i.e. the thickness of the unit of one CuO_2 plus one BL.

Here, it should be mentioned about another attribute of the BL. There are different situations concerning the relative position of the two CuO_2 sheets below and above a specific BL (Fig.7); whether the two sheets on the both sides of the BL shows the Cu-Cu correspondence along the normal to the plane or otherwise the "phase shift" by (**a**+**b**)/2 in the lateral direction. (Here, let **a** and **b** be the two-dimensional unit cell vectors for the pseudo-tetragonal CuO_2 sheet.) Hereafter, we refer the former and latter type of BL as the α-type and β-type, respectively. For example, the BLs, C_1, P_s and **V**, are of the α-type, while the **L**-, **G**- and B_2-layers are of the β-type (see Fig.5 and 6). Using these BLs and their attributes, we can describe all the existing CuO_2-layered compounds including the high-T_c superconductors and further predict structures, lattice parameters and ways of carrier doping for hitherto unknown compounds.

4. Carrier-doping characteristics

Six basic patterns of 2D Cu-O networks shown in Fig.1 can be formed by combination of two of the +, - and +' type BLs. Here, the notation +' represents the BL which supplies an apical halogen (e.g. Cl^- and Br^-). The Cu-O networks are entirely governed by the combination of BLs positioned above and below the CuO_2-sheet. Up to now, however, the combination of the same sign of BLs, such as (+,+),

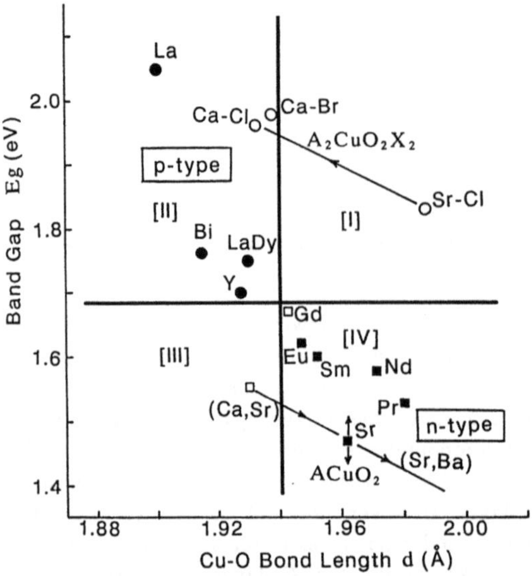

Fig.8: Empirical diagram for the carrier-doping characteristics; the data are plotted in (E_g, d) space. Closed circles and rectangulars represent the compound which can doped with hole and electron type carriers, respectively, and superconduct.

(-,-) and (+',+'), are only realized in the single-BL structures. Most of existing double-BL structures are composed of the pyramidal CuO_2-sheets with the (+,-) combination.

The important point to be stressed here is that the type of charge carriers (hole-like or electron-like) depends on the pattern of Cu-O networks: The CuO_2-sheets with (a) octahedral and (b) pyramidal networks sustain hole-like (p-type) carriers, while those in (c) squaric sheets are electron-like (n-type) ones. Further, it has been empirically known that the hole-type carriers cannot be introduced to the squaric sheets and the electron-type carriers neither to octahedral nor pyramidal sheets mostly due to the crystallographical instability or carrier-compensating oxygen nonstoichiometry.

To elucidate such structure-sensitivity of the doping-characteristics, we may have to pay attention to the critical variation of electronic structures in the parent compounds. In Fig.8 we present a little more elaborate diagram of the empirical rule for the carrier-doping in the CuO_2-sheets, in which the value of the band gap (E_g) in the insulating parent compounds are plotted against the in-plane bond length d (or half of the a-axis length). The CT gap is quite sensitive to the surrounding circumstance around the Cu-site and hence critically reflect the Madelung type interaction, as discussed in section 2. On the other hand, the degree of hybridization between the Cu d and O p states, or equivalently the covalency of the in-plane Cu-O bond, is accurately reflected in variation of d. The E_g-d space shown in Fig.8 is divided into quadrants according to relative difficulty of p or n type doping. Empirically, the compounds locating on the left-top part of the present diagram characterized by large E_g and small d is easy to be doped with hole-type carrier. In fact, all the hoe-doped (p-type) superconductors position in the second qaudrant (II) of this diagram. By contrast, the compounds on the right-bottom part with small E_g and large d_s is relatively easily doped with electron-type carriers. For example, the T'-phase compound (or the **G**-layer compound) positioning on the fourth quadrant (IV) show the n-type behavior, yet the Gd_2CuO_4 with relatively long bond length is difficult at present to be doped with sufficient carriers and to be made superconducting. All the compounds locating in the first or third quadrant (I or III) are very hard to be doped with carriers.

Let us exemplify two more prototypical cases; the so-called infinite layer compound with the **V**-layer , $ACuO_2$ (A=Ca,Sr and Ba), and the K_2NiF_4-type compound with $\mathbf{H_1}$-layer, $A_2CuO_2X_2$ (A=Ca and Sr; X=Cl and Br). These compounds have been difficult to be hardly doped with charge-carriers; For example, the conventional **V**-layer compound, $Ca_{0.84}Sr_{0.16}CuO_2$, is located in the third quadrant (III) of the diagram shown in Fig.8. To dope this structure with appreciable amounts of n-type carriers, the bond length should be elongated so that the bond covalency may be increased. Very recently, such an attempt has fruited to realize new electron-doped superconductors with $T_c{\sim}40K$[9]; $SrCuO_2$ doped with Nd or Pr. The new compound $Sr_{1-x}Nd_xCuO_2$ prepared at high pressures shows the so-called infinite layer structure, or the **V**-layer compound, with an elongated bond length (~ 1.92A) and hence falls in the fourth quadrant (IV), as shown in Fig.8, where the electron-doping by increasing the BL charge of the **V**-layer may be easier.

On the other hand, the K_2NiF_4 compounds with apical halogens, $A_2CuO_2X_2$ (A=Ca,Sr; X=Cl,Br), are ranging on the upper half of the diagram shown in Fig.8 and hence likely to be doped with hole-type carriers. So far, several attempts to introduce the hole-type carriers into $Sr_2CuO_2Cl_2$, for example by substitution of Sr-sites with Na, have not been successful as yet. The Cu-O bond length of $Sr_2CuO_2Cl_2$ may be too long for hole-doping and we may have to concentrate efforts on Ca-substituted analogs locating in the second quadrant (II).

5. Classification of double-BL structures - method of BL matrix-

Hereafter, let us discuss the double BL structures with pyramidal Cu-O sheets (Fig.1 (b)) which are formed by a combination of + and - type BLs. Most of the existing superconducting copper oxide compounds show the double-BL structures with such a (+,-) combination of BLs. Using the attributes of the BLs and assigning the structures to the pseudo-tetragonal lattice, lattice structures can be predicted for the double-BL compounds according to the following rule: The lattice constants are approximately given such that $a=d_s^{(1)}+d_s^{(2)}$, and that $c=w^{(1)}+w^{(2)}$ for the (α,α) or (β,β) combination and $c=2(w^{(1)}+w^{(2)})$ for the (α,β) combination. Here, $d_s^{(i)}$ and $w^{(i)}$ (i=1,2) are the "in-plane Cu-O bond-length" and "thickness" of the composing BLs, as defined in the section **3**.

To classify the double-BL structures, the representation by the BL matrix is useful as shown in Table 1: In the row and column, we list prototypical - and + type BLs, respectively. In this representation, the compounds are expressed as "matrix elements", whose representative stoichiometries are indicated in the table together with crude values of maximum T_c's so far observed for superconducting compounds. As seen in this BL matrix, many of the pyramidal sheet compounds are readily synthesized using either - type BL of **V** or **G**. (For complete references,

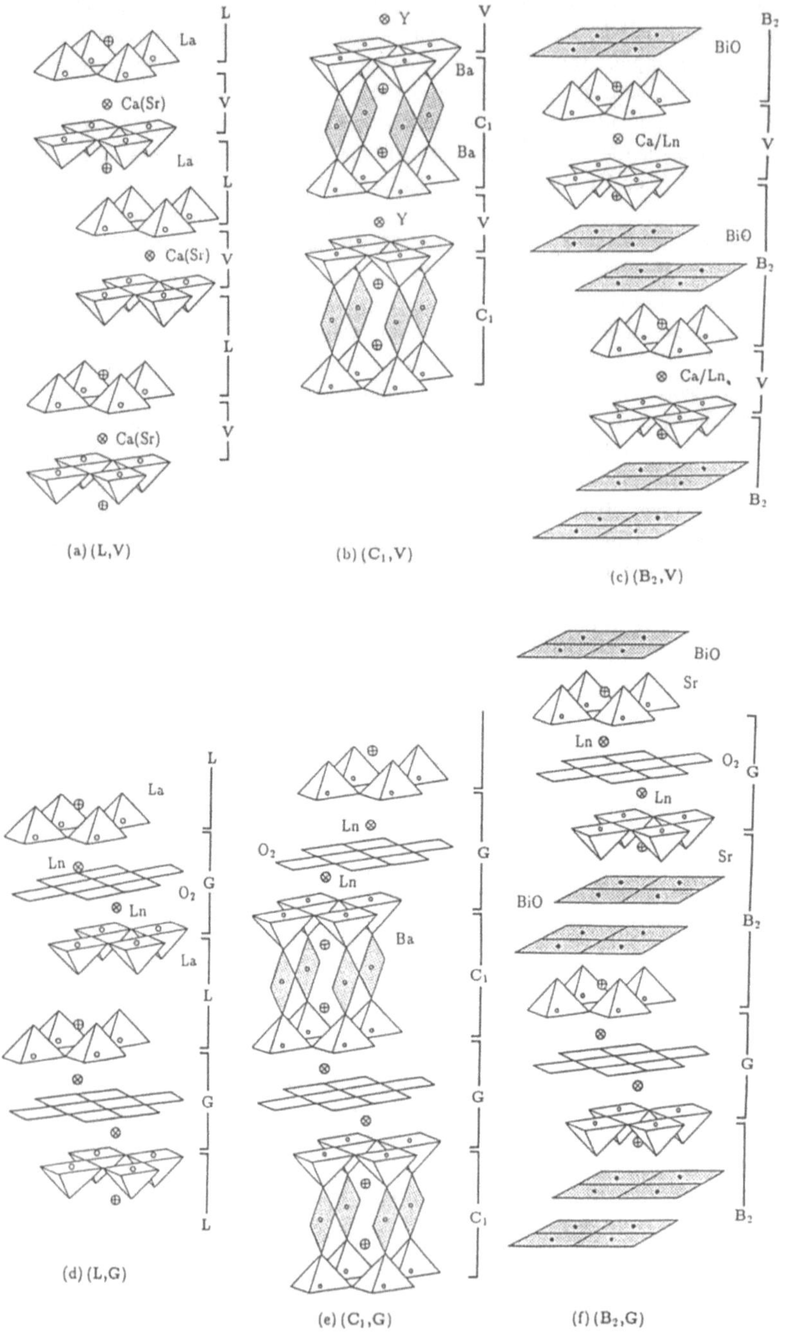

Fig.9: Corresponding **V**-based (a,b,c) and **G**-based (d,e,f) double BL structures; (a) 2126 or double-layered perovskite, (b) 123 (or Cu-1212), (c) Bi-2212, (d)T^*, (e) 223 (or Cu-1222), and (f) Bi-2222 phases.

see ref.[8].) We would like to emphasize that some of the **G**-based compounds such as (**L**,**G**), ($\mathbf{B_2}$,**G**) and ($\mathbf{T_2}$,**G**), have been successfully synthesized following the strategy described by this BL matrix.

In Fig.9, we show some examples for the structures of the corresponding **V**- and **G**-based structures which have the common + type BLs. Fig.9 (a) and (d) shows the (**L**,**V**) and (**L**,**G**) structures which have the **L**-layer as a common + type layer. The (**L**,**V**) structure is the so-called 2126 structure with a typical stoichiometry of $(La,Sr)_2CaCu_2O_6$, which has also become a member of High T_c compounds[10]. In the compound (d) (**L**,**G**) (so-called T^* phase), the - type BL unit is replaced with the **G**-layer[11]. Similar substitutions of the oxygen-vacant **V**-layers with the fluorite-like **G**-layers are possible also in other structures as exemplified for the so-called 223 (or Cu-1222) compound[12], (e) ($\mathbf{C_1}$,**G**), transformed from the 123 (or Cu-1212) compound, (b) ($\mathbf{C_1}$,**V**), and the Bi- or Tl-2222 compound [13], (f) ($\mathbf{B_2}$,**G**) or ($\mathbf{T_2}$,**G**), from the Bi- or Tl-2212 compound, (c) ($\mathbf{B_2}$,**V**) or ($\mathbf{T_2}$,**V**).

In Fig.10, we plot the structural data on (a) a-axis and (b) c-axis lengths (with the a x a pseudo-tetragonal setting) for double-BL structures against the calculated values by the above-mentioned method with the d_s and w values of the constituent BLs. Most of the data fall quite well on the straight broken lines representing that $a_{obs.}=a_{cal.}$ and $c_{obs.}=c_{cal.}$. The result shows that the choice of the BL governs the lattice parameters in the double-BL structures. In particular, the **G**-layer based compounds, e.g. the (**L**,**G**) and ($\mathbf{B_2}$,**G**), can show a large variation of the a-axis length (or the in-plane Cu-O bond length), which may enable us to investigate experimentally the effect of the Cu-O deformation on the superconducitng properties, or simulate the high pressure effect.

Another important quantity to characterize the double-BL compound is the averaged BL charge Q_{av}, which equals $(Q^{(1)}+Q^{(2)})/2$. The $[CuO]^p$ charge is then given as $+2-Q_{av}$. For the p-type compounds described by the (+,-) BL-matrix, the averaged charge of the two BLs must be smaller than +2. In particular, to realize the hole-conducting (p-type) superconductors, the averaged BL charge should be adjusted to be 2-q, where q approximately ranges form 0.05 to 0.25, as shown in Fig.1.

In Table I, we have also included two more BLs of the - type; $\mathbf{H_1}=[(Pb,Sr)_2Cl)$ with the CsCl-type structure [14] and $\mathbf{G_3}=[(Ln,Ce)_3O_2]$ with the accumulated fluorite structure[15]. Up to now, the (+,-) compounds with these - type BLs have not been made superconducting, yet we may have new superconductors by further adjusting the bond length and increasing carrier density, according to the strategy described above.

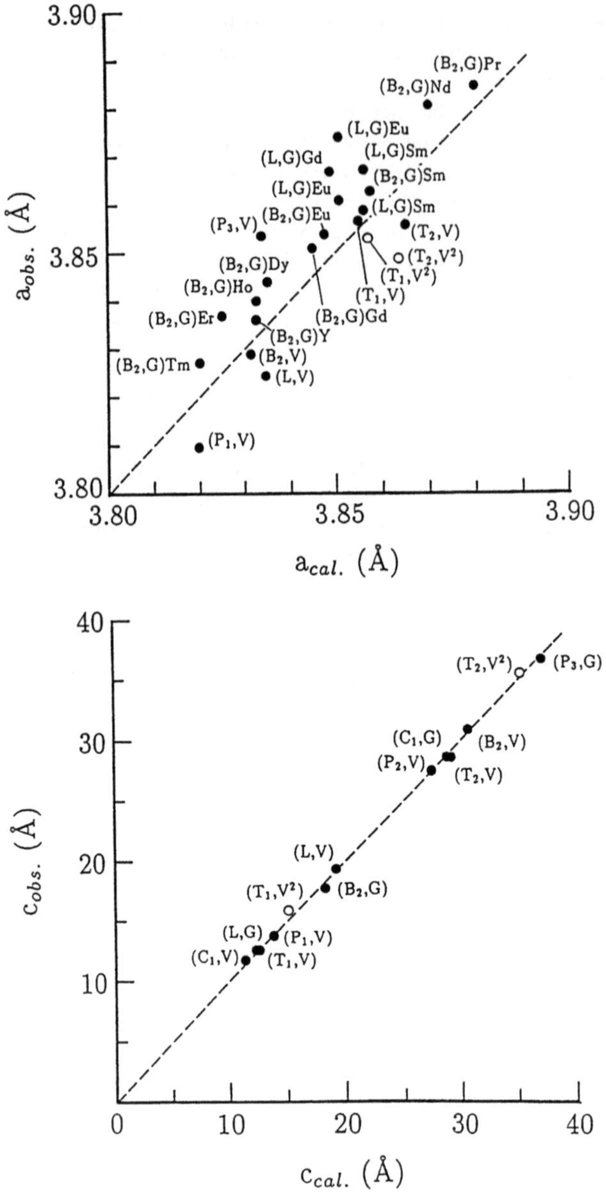

Fig.10: Comparison of the calculated and observed lattice parameters, a and c, for existing double-BL (closed circles) and triple-BL (open circles) structures. Compounds are indicated by the abbreviated BL-formulas (see Table I) and the attached label of a rare-earth element means that the constituent **G**-layer is composed of that element.

Table I. BL-matrix for the (+,-) double-BL compounds; ◎:superconducting, ○:synthesized but not superconducting as yet, ×:not synthesized as yet.

−type BL / +type BL	**V** (Ca,Y) $Q=2\sim3$	**G** $(Ln,Ce)_2O_2$ $Q=2+\delta$	$\mathbf{H_1}$ $(Sr,Pb)_2Cl$ $Q=3$	$\mathbf{G_3}$ $(Ce,Ln)_3O_4$ $Q=3\pm\delta$?
L $(La,Sr)_2O_2$ $Q=2-\delta$	◎~ 60 K $(La,Sr)_2CaCu_2O_6$	◎~ 40 K $(La,Ln,Sr)_2CuO_4$	× $(La,Sr)_2SrPbCu_2O_8Cl$	× $(La,Sr)_2(Ce,Ln)_3Cu_2O_{10}$
$\mathbf{C_1}$ Ba_2CuO_{3-y} $Q=1\pm\delta$	◎~ 90 K $YBa_2Cu_3O_7$	◎~ 40 K $(Ln,Ce)_2(La,Ba)_2Cu_3O_9$	× $Ba_2SrPbCu_3O_7Cl$	○ $(Ce,Ln)_3Sr_2Cu_3O_{11}$
$\mathbf{C_2}$ $Ba_2Cu_2O_4$ $Q=1-\delta$	◎~ 90 K $YBa_2Cu_4O_8$	× $(Ln,Ce)_2Ba_2Cu_4O_{10}$	× $Ba_2SrPbCu_4O_8Cl$	× $(Ce,Ln)_3Ba_2Cu_4O_{12}$
$\mathbf{T_1}$ $TlBa_2O_3$ $Q=1+\delta$	◎~ 100 K $TlBa_2CaCu_2O_7$	○ $TlBa_2(Ln,Ce)_2Cu_2O_9$	× $TlBa_3PbCu_2O_8Cl$	○ $TlBa_2(Ln,Ce)_3Cu_2O_{11}$
$\mathbf{T_2}$ $Tl_2Ba_2O_4$ $Q=2-\delta$	◎~ 110 K $Tl_2Ba_2CaCu_2O_8$	○ $Tl_2Ba_2(Ln,Ce)_2Cu_2O_{10}$	× $Tl_2Ba_3PbCu_2O_9Cl$	× $Tl_2Ba_2(Ln,Ce)_3Cu_2O_{13}$
$\mathbf{B_2}$ $Bi_2Sr_2O_{4+y}$ $Q=2-\delta$	◎~ 90 K $Bi_2Sr_2CaCu_2O_8$	◎~ 40 K $Bi_2Sr_2(Ln,Ce)_2Cu_2O_{10}$	× $Bi_2Sr_3PbCu_2O_9Cl$	× $Bi_2Sr_2(Ln,Ce)_3Cu_2O_{13}$
$\mathbf{P_1}$ $(Pb,Cu)Sr_2O_x$ $Q=1+\delta$	◎~ 70 K $(Pb,Cu)Sr_2(Ca,Y)Cu_2O_y$	◎~ 30 K $(Pb,Cu)Sr_2(Ln,Ce)_2Cu_2O$	× $(Pb,Cu)Sr_3PbCu_2O_yCl$	○ $(Pb,Cu)Sr_2(Ce,Ln)_3Cu_2O_y$
$\mathbf{P_2}$ $PbSrBaCuO_x$ $Q=1+\delta$	◎~ 50 K $PbSrBa(Ca,Y)Cu_3O_y$	○ $PbSrBa(Ln,Ce)_2Cu_3O_y$	× $Pb_2BaSr_2Cu_3O_yCl$	× $PbSrBa(Ce,Ln)_3Cu_3O_y$
$\mathbf{P_3}$ $Pb_2Ba_2CuO_4$ $Q=1+\delta$	◎~ 80 K $Pb_2Ba_2(Ca,Y)Cu_3O_8$	○ $Pb_2Sr_2(Ln,Ce)_2Cu_3O_{10}$	○ $Pb_3Sr_3Cu_3O_8Cl$	○ $Pb_2Sr_2(Ce,Ln)_3Cu_3O_{20}$
$\mathbf{M_1}$ $TaBa_2O_4$ $Q=1$	○ $TaBa_2YCu_2O_8$	× $TaBa_2(Ln,Ce)_2Cu_2O_{10}$	× $TaBa_3PbCu_2O_8Cl$	× $TaBa_2(Ce,Ln)_3Cu_2O_{12}$

6. Design of n-type superconductors

As an application of the method of the BL matrix to material design, we present here possible structures and stoichiometries for electron-conducting (n-type) CuO_2-layered compounds. Until very recently, the superconductors with electrons as the charge-carriers have been limited to the so-called T'-phase structure (or **G** structure), which is in contrast with a variety of p-type superconductors. The single-BL compounds with the **V**-layers, $(Sr_{1-x}Nd_x)CuO_2$, has been recently made superconducting (T_c 40K) by increasing the BL charge[9]. However, the compound could be synthesized only under high pressures (~ 2GPa). We need to have a further variety of the n-type superconductors on which we can scrutinize electronic properties to make detailed comparison with those of hole-doped superconductors.

As described in the previous sections, whether the CuO_2-sheets is of the p or n type largely depends on the local coordination around the Cu-site and relative difficulty of carrier-doping also depends on the Cu-O bond length of the parent compound. To realize the n-type superconductors, we have to fabricate the CuO_2-sheets without apical oxygen and with relatively large Cu-O bond length. Therefore, we should use BLs except for the + type ones, i.e. the - type BLs or possibly the +' type BL with apical halogens, and adjust the composition of such a BL to have relatively large Cu-O bond length.

The double-BL structures without apical oxygens are represented by the BL matrix composed of - or +' type (Table II). For example, two **G**-based structures and their representative stoichiometries are shown in Fig.11, in which the electron-

Table II. BL-matrix for the (-,-) or (-,+') compounds; ◎:superconducting, ○:synthesized but not superconducting as yet, ×:not synthesized as yet.

	V (Ca,Y) $Q=2\sim3$	**G** $(Ln,Ce)_2O_2$ $Q=2+\delta$	$\mathbf{H_1}$ $(Sr,Pb)_2Cl$ $Q=3$	$\mathbf{G_3}$ $(Ce,Ln)_3O_4$ $Q=3\pm\delta$?
V $Q=2\sim3$	◎~ 40 K $(Sr,Ln)CuO_2$	× $(Ln,Ce)_2CaCu_2O_6$	×	× $(Ce,Ln)_3CaCu_2O_8$
G $Q=2+\delta$		◎~ 30 K $(Ln,Ce)_2CuO_4$	×	× $(Ln,Ce)_5Cu_2O_{10}$
$\mathbf{H_1}$ $Q=3$			×	×
$\mathbf{G_3}$ $Q=3\pm\delta$?				× $(Ce,Ln)_3CuO_6$

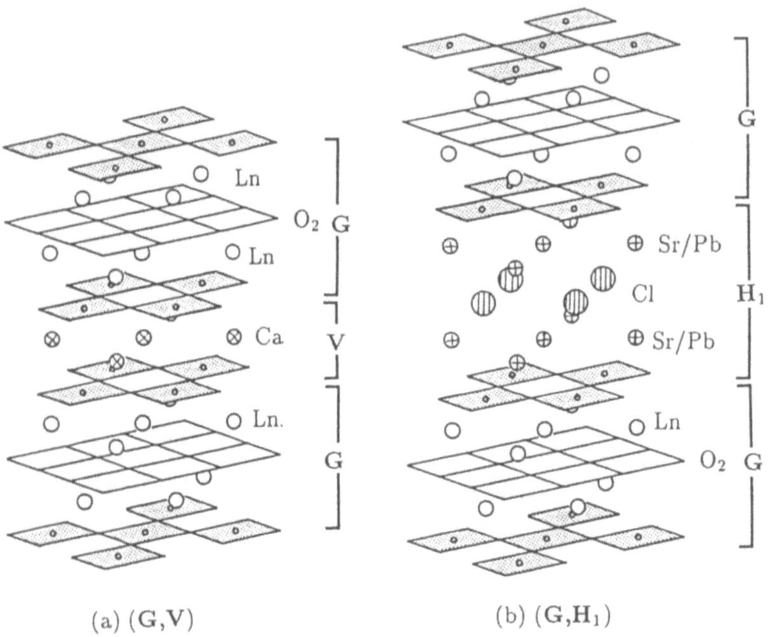

Fig.11: Structures of possible n-type copper oxide superconductors.

doping might be relatively easy by adjusting the charge of the **G**-layer with Ce-doping. The point is to control the averaged BL charge to be slightly higher than +2, i.e. +2+q (0 ⟨q ⟨0.3). The most promising one is the **(G,V)** structure (Fig.8(a)), e.g. $Ca(Ln_{1-x}Ce_x)_2Cu_2O_6$, where the averaged BL charge ([CuO] charge) is +2+x (-x). On the other hand, the charge of the **G**-layer should be rather decreased in the $(\mathbf{G},\mathbf{H_1})$ compound (Fig.8(b)), e.g. $(Sr,Pb)_2(Ln_{1-x}Ca_x)_2Cu_2O_6Cl$, since the charge of the $\mathbf{H_1}$-layer is +3 and seems to be difficult to control further. Most of the double-BL structures with the (-,-) combination have not been synthesized as yet to our knowledge. This is because the BLs with the same sign tend to form the solid solution with each other and is difficult to order sequentially along the normal to the plane, in particular by thermal processing at high temperatures. Therefore, the n-type (-,-) double-BL structures may be difficult to synthesize by conventional preparation methods such as solid state reaction of powder stuff. One of the possible breakthroughs may be achieved by the epitaxial growth of the BL structure.

7. Summary

We have argued the carrier-doping characteristics which largely depend on the Madelung type contribution and the Cu-O bond covalency. These two factors seem to manifest themselves in the CT gap energy (E_g) of the parent compound and the in-plane Cu-O bond length (d). This enables us to make the empirical E_g-d (or almost equivalently, coordination vs. bond length) diagram for carrier-doping characteristics in the Cu-O sheets.

The useful classification scheme for the layered cuprate superconductors was presented with use of the concept of block layer (BL), which can describe structures of all the existing superconducting cuprate compounds without exception. In principle, we can arbitrarily design multi-BL structures even with two or more patterns of Cu-O networks[8]. Furthermore, we can predict essential Cu-O bond length parameters as well as appropriate ways of p or n type carrier-doping. In the near feature, techniques of the epitaxial growth of the BL unit may be further developed to fabricate new superconducting structures.

Acknowledgement

The author is grateful to T.Arima for his collaboration.

References

[1] J.B.Torrance et al., Phys.Rev.Lett. **61**, 1127 (1988); J.B.Torrance, A.Bezinge, A.I.Nazzal and S.S.P.Parkin: Physica C **162-164**, 291 (1989).
[2] H.Takagi, S.Uchida and Y.Tokura, Phys.Rev.Lett. **62**, 1197 (1989).
[3] Y.Tokura et al., to be published.
[4] Y.Ohta, T.Tohyama and S.Maekawa, Phys.Rev.Lett. **66**, 1228 (1991).
[5] Y.Tokura et al., Phys.Rev.B **41**, 11657 (1990).
[6] S.L.Cooper et al., Phys.Rev.B **42**, 11605 (1990).
[7] S.Tajima et al., Phys.Rev.B **43**, 10496 (1991).
[8] Y.Tokura and T.Arima, Jpn.J.Appl.Phys.**29**, 2388 (1990).
[9] M.G.Smith et al., Nature, **351**, 549 (1991).
[10] R.J.Cava et al.: Nature, **345**, 602 (1990).
[11] J.Akimitsu et al., Jpn.J.Appl.Phys. **17**, L1859 (1988); Y.Tokura et al.,Phys.Rev.B**40**, 2568 (1989).
[12] H.Sawa et al., J.Phys.Soc.Jpn. **58**, 2252 (1989).
[13] Y.Tokura et al., Nature, **342**, 890 (1989).
[14] R.J.Cava et al., Physica C**167**, 67 (1990).
[15] T.Wada et al., Physica C**171**, 344 (1990).

High Pressure Synthesis and Characterization of Single Crystalline $YBa_2Cu_4O_8 (T_c = 80\,K)$ and $Y_2Ba_4Cu_7O_{15\pm x}$ $(14\,K \leq T_c \leq 94\,K)$

E. Kaldis, J. Karpinski, and S. Rusiecki

Laboratorium für Festkörperforschung, ETH-Hönggerberg, CH-8093 Zürich, Switzerland

Abstract. The merits of the high oxygen pressure synthesis of the YBaCuO oxides and the advantages of a two—chamber autoclave are briefly reviewed. The P—T—x and the flux P—T phase diagrams are presented for the first time. The P—T—x diagram shows the P—T cuts across the compositions 123, 123.5 and 124; the 123 is not stable in the low temperature region and the 123.5 must be cooled through the stability range of 124, with the corresponding implications. The flux P—T diagram shows the fields of crystallisation under high oxygen pressure of the above three HT_C—phases. Crystallization experiments give some first results about the complex growth mechanism of 124 and 123.5, in which also the vapour phase seems to be important.

A brief survey of the physical properties of the double chain compounds is also given : STM investigations show that micron range areas of as grown surfaces are atomically smooth with perfect singular steps with unit cell height. Although the T_C of the 124 crystals is 7—9 K lower than that of the ceramic (81 K) — probably due to traces of aluminum — the diamagnetic transition of single crystals is very steep (ΔT=2 K). Due to the thermodynamic stability of the double chains, 124 and 123.5 do not have a—b twins and, therefore, are ideal for measurements of anisotropy. Transport and optical properties show an anisotropy of 3—5 . Electronic Raman scattering in single crystals of 124 shows the existence of two gaps, attributed to the chains and the plains. Also the critical fields H_{C1} and H_{C2} of 124 have been measured and found approx. 3 times lower than those of 123 . This means that the coherence length is also longer in 124 .

1. Introduction

Among the few existing strategies for the discovery of new compounds with similar physical properties to an existing one, the most difficult but also powerful strategy is the exploration of the phase boundaries. To this end, one has to study the thermodynamic (or kinetic !) phase diagram and define the region of stability of the parent compound. Careful investigation of the phase boundaries may bring two important results:

a) The variation of the physical properties of a compound with the concentration of nonstoichiometric defects, which may be statistically distributed or ordered in the form of superstructures. In this way, the dependence of the physical properties on carriers concentration or slight lattice distortions influencing the band structure can be investigated. This is tool number one in the materials research of solid state physics, unfortunately too difficult for most laboratories to apply.

b) The discovery of new neighbouring phases which may have similar or even more interesting physical properties.

Both these investigations lead to a better understanding of the physical properties and avoid their misinterpretation.

Experimentally, an appreciable effort is necessary. The three dimensional P—T—x phase diagram has to be measured up to high pressures and temperatures. The construction of such diagrams, is very difficult experimentally.

The oxide superconductors and particularly the 123 phase, with its feebly bound oxygen of the chains, seemed to us from the beginning as a classical example for the application of this thermodynamic exploration strategy. Our original idea was to suppress decomposition by applying during melting a partial pressure of oxygen larger than the decomposition pressure. Application of 3000 bar did not achieve this goal, but showed for the first time that the single chains of the 123 lattice, with their unusual square planar coordination, are unstable at high oxygen pressures and that double chains are favoured. This led to the discovery of the new phase $Y_2Ba_4Cu_7O_{15}$ (247 or 123.5) and to the first bulk synthesis of $YBa_2Cu_4O_8$ (124) [1,2].

With progressing characterization, the sensitivity of the $Y_2Ba_4Cu_{6+n}O_{14+n}$ family of HTc—superconductors to nonstoichiometry and the corresponding structural changes became visible, so that we devoted our research to three directions:

—thermodynamic investigations; the refinement of the phase diagram of the YBaCuO system at high oxygen pressures and particularly the study of the phase boundaries,

—structural investigations; the study of the bond lengths and lattice distortions as a function of the nonstoichiometry and hydrostatic pressure.

—Crystal growth under high oxygen pressures of 124 and 123.5 and the characterization of their properties.

In this paper we review some aspects of the first and the last topic. Other aspects have been reviewed recently [1,2].

2. Pressure as a Tool of Synthesis of HT_C—Superconductors

As mentioned above, the application of high oxygen pressure for the synthesis of many HT_C—superconductors is a sine qua non, in view of the high decomposition pressures of these compounds at moderate temperatures. As we have stressed in the past, the influence of gas pressure on the thermodynamic equilibria is twofold [3].

—As a reactive component of the chemical equilibrium. Additional reactivity comes from the increased fugacity of real gases at pressures above one kilobar and particularly at medium temperatures. A good example is nitrogen which at 10 kbar and 700 K has a fugacity corresponding to that of an ideal gas, at 10^3 kbar [4].

—Due to the volume contraction effect resulting from the hydrostatic pressure. Change of solubilities, of defect equilibria and of diffusion coefficients are expected. An example is the change of solubility of aluminum in silicon, found by Japanese scientists [5]. Under 54 kbar the solubility is nine times higher than under normal pressure.

With increasing pressure, the volume of the sample decreases due to technical limitations and the experimental difficulties increase steeply. Gas pressures for the synthesis of solids reach presently 25–30 kbar. For pure oxygen, only our own investigations went up to 3 kbar [6]. To increase this pressure, a series of experiments with mixtures of oxygen and argon is necessary, in order to find the compatibility range for a given autoclave. Exceeding this pressure could possibly lead to uncontrolled burning.

Hydrostatic pressures can go to much higher values, usually above 50 kbar, but in laboratory equipment the sample volume is only $\lesssim 1$ cm^3. Thus, it is difficult to grow even tiny crystals and perform an extensive physical characterization of the samples.

2.1 Two–chamber Autoclave for Pure Oxygen Atmosphere (3000 bar)

Figure 1, shows schematically the design of our two–chamber autoklave for pure oxygen pressures up to 3000 bar, and temperatures up to 1700 ^0C. This is an internally heated autoclave with water cooled walls. Some details have been given in the past [3]. Here, we point out some of the advantages of this construction design.

The second (internal) chamber is an alumina crucible (20 cm long, 12mm int. diam.), with an isothermal volume of up to 10 cm^3, which is fed directly from a reservoir of high purity oxygen (5N). The opening of the crucible is sealed with a silver gasket. The walls of this crucible are supported by the argon atmosphere filling the outer (furnace) chamber of the autoclave. Automatic regulation of the argon pressure leads to pressure differences across the crucible walls of less than 1–5 bar in 3000. Thus, the brittleness of the crucible walls is not important.

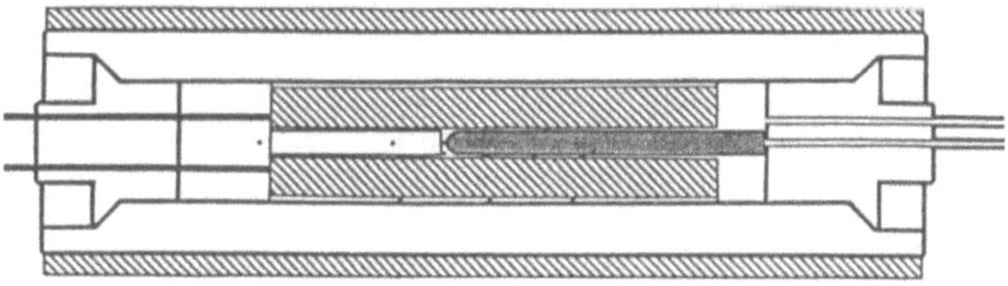

Figure 1. Schematic drawing of the novel two–chamber autoclave for high oxygen pressures (3000 bar). The alumina crucible is gray, black dots indicate the position of thermocouples.

An important safety aspect in high pressure work is to keep the volume of the free gas under high pressure as small as possible, to decrease the energy released by an explosion. This is achieved in our autoclave with alumina rods — fitting exactly the internal diameter of the alumina crucible — which are filling the nonisothermal parts of the crucible. Thus, the free volume of oxygen in the high temperature zone of the autoclave is only 1–2cm^3 under 3000 bar. Therefore, in the case of a leak – which has never happened up to now — only 3–6 liters of hot oxygen will be released.

An additional advantage of the small residual free volume is that it enables an accurate measurement of the pressure changes inside the crucible, a very important aspect for the measurements of the phase diagrams.

Using externally heated autoclaves [7] pressure ($P \lesssim 200$ bar), temperature ($T \lesssim 1000$ ^{0}C),and the lifetime, are much more limited. Thus, with such a system no single crystals of 124 or 123.5 could be grown up to now.

3. First results of the P–T–x Phase Diagram up to 3000 bar

For reasons of simplification, the phase diagram of the Y–Ba–Cu–O system is generally considered as a pseudobinary system where for a given fixed ratio of the metal contents only the oxygen stoichiometry and therefore its decomposition pressure P_{O_2} is changing with temperature. This metal composition dependence of the high pressure phase diagram has been unfortunately neglected, even contradicted in the past [8]. Here, we show in a rather schematic way, the stability regions of the pure 124 and 247 phases as determined by x–rays in rapidly cooled samples. A detailed discussion will be given elsewhere [9]. Considering the vast amount of work necessary for the high pressure experiments, the controlled cooling, the characterization of mixtures of phases, and the great difficulties associated with establishing equilibrium inspite of the sluggish kinetics, we believe that these first results are only a first approximation to the P–T–x diagram, which we hope to improve in the following years. For comparison, the "simple" T–x diagram at normal pressure is not yet quite clear after 5 years of work from approx. 50 laboratories.

Figure 2 shows three composition cuts of the three dimensional P–T–x phase diagram, corresponding to the stoichiometries x= 123, 123.5(247), and 124. Three other intermediate compositions have been also measured and will be published soon [9]. The diagrams of Figure 2 have been measured up to 2500 bar [14], but due to lack of space we can show here only up to 300 bar. In the high temperature part, the boundary of the appearence of the melt (L) is visible. The hatched two–phase region 124–247 is seen for the compositions 124 and 123.5(247). Also the 123 stability range is shaded, to indicate that due to the very slow kinetics, its boundaries – particularly those in the x=123 composition – cannot be considered as final. The white 124 stability range for x=124 is the single phase region where only 124 appears. The same is true for the 123.5 (247) phase at x=123.5 .

With increasing CuO content, the pressure and temperature ranges in which 123 is stable, decrease strongly from $T \gtrsim 760^0$C and $P \lesssim 200$ bar at

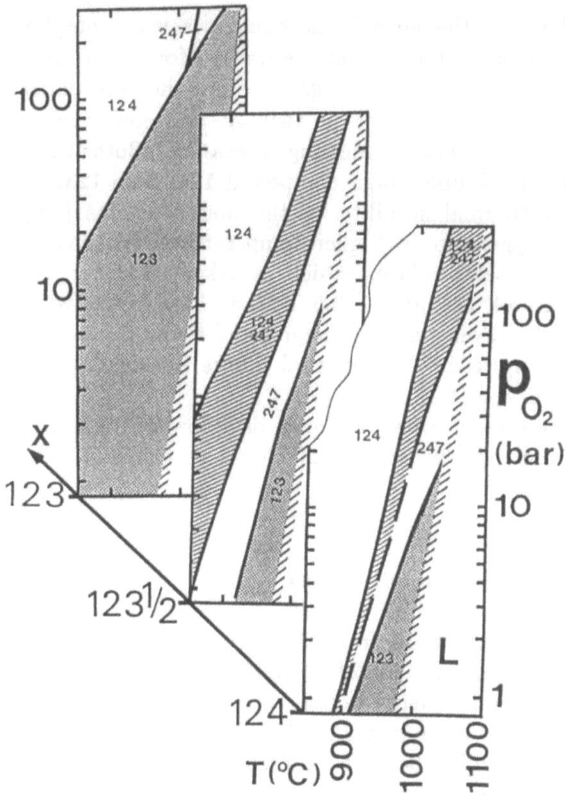

Figure 2. Schematic P—T—x phase diagram of the [Y—Ba]—CuO system.

x=123, to T≥910°C and P≤15 bar at x=124. We point out, that as mentioned already in the past [2], at lower temperatures 123 is not thermodynamically stable. Thus at the interpretation of physical measurements one should keep in mind *the state of metastability* of the samples.

Figure 2 makes also clear the difficulties which arise in the preparation of single phase 123.5. The cooling path has to cross the stability range both of the two—phase region 124—247 and 124. Thus, quenching is necessary [1].

The P—T—x phase diagram shows also clearly that *the dominating phase is 124 and not 123.*

4. Crystal Growth of 124 and 123.5 under High Oxygen Pressures

Presently, the importance of single crystals of the double chain compounds is great, because many interesting properties are not yet measured. Unfortunately, more than in the case of 123, the crystal growth of 124 and

123.5 is extremely difficult because the growth mechanism is very complex and a new high pressure phase diagram is necessary for each flux composition. On the other hand, crystals of 123 grown from the flux under normal pressure, have the disadvantage of a–b twining and low oxygen content in the as grown state, making annealing necessary. Both these disadvantages do not occur in the double chain compound 124. Also 123.5 is not twined. Due to the high thermal stability of the double chains [2,6], these compounds keep their oxygen up to higher temperatures, so that no ortho→ tetrag phase transition takes place, which is responsible for the twining [2]. Therefore, crystals of 124 and 123.5 do not have a–b twins. Relatively thick platelets, however, show frequently twining in the c–direction, resulting from the rotation of structural blocks by 90^0, leading to an exchange of the a– and b–axis.

The twin free single crystals of these compounds allowed the direct measurement of the a–b anisotropy by means of electrical transport and optical measurements [10] (see Sect.6).

4.1 Phase Diagram for Crystal Growth from the Flux

For the crystal growth of these compounds from the flux, knowledge of the P–T phase diagram of the system (Flux–124, 123.5) is necessary up to approx. 2000 bar oxygen, which is different from the diagrams of Figure 2 . The reason is that the flux consists of an excess of BaO and CuO ($BaCuO_2$), so that the concentration relationships become different. This leads to the formation of a eutectic temperature, lower than the peritectic melting point [11]. However, for the growth of 124 and 123.5, the flux method has to be used under high oxygen pressure. Our measurements have shown that the pressure dependence of the eutectic and peritectic equilibria is quite different (Figure 3).

Thus, the eutectic temperature increases very rapidly with the oxygen pressure, from $\cong$ 900 ^{0}C at P_{O_2}=1 bar to 1100 ^{0}C at P_{O_2}=1000 bar. In the same pressure interval, the peritectic equilibrium is shifted only from 1010 to 1120 ^{0}C. Therefore, although under normal pressure a ΔT(per–eut)=110 ^{0}C is available for crystal growth, only a ΔT(per–eut)=20 ^{0}C remains at P_{O_2}=1000 bar. A further implication is that at P>1800–2000 no YBaCuO phases crystallize from the flux, an indication that the peritectic and eutectic lines are crossing and the phase relationships change. Figure 3 shows the pressure ranges in which crystals of the 124, 247 and 123 phases can be grown. They correspond to steps of a P–x type diagram which has been published elsewhere [1,12].

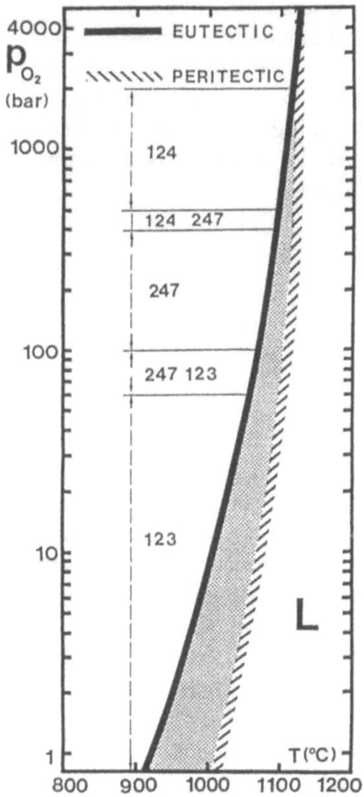

Figure 3. P–T phase diagram at high oxygen pressures of the system (Flux+Solute) with total metal composition $YBa_3Cu_{7.5}O_x$.

4.2 Crystal Growth Mechanism

The disadvantages expected from the narrow field of crystallization are :

—The adjustement of the absolute temperature becomes more crucial.

—The crystallized mass should decrease due to the small ΔT, and the crystals should become smaller because the driving force (deviation from equilibrium) is smaller. However, the size–growth time relationship was found to be dependent on the kinetics of formation of the flux.

In a typical experiment the autoclave is charged with a mixture of phases (flux+solute) with the total composition $YBa_3Cu_{7.5}O_x$ (4 pellets, $\cong$ 10 g). The heating schedule consists of the actual heating (t_h), dwell at the highest temperature (t_d), and cooling (t_c) periods. We have used cooling rates 1–30 $^0C/h$, and t_d's between 0 and 36 h. Experiments with $t_d=0$ do not show any crystal growth, and the shape of the pellets does not change, indicating that the flux did not have the time to form the eutectic. For $t_d=12$–24 h a meniscus of liquid is formed at the lower part of each pellet. Many crystals are growing out of it into the surrounding atmosphere (Figure 4); and for $t_c \geq 36$ h much more flux is formed which covers almost all crystals. As no specific solvent for the flux is existing, these crystals cannot

Figure 4.
Filamentary growth of flat 124 needles, out of the flux–meniscus forming at the vertical side of a pellet (view from above)

be recovered from the solidified flux. The fact that, as Figure 4 shows, the crystals grow out of the flux, is an indication that the vapour phase is involved in the growth process, so that a vapour–liquid–solid (VLS) mechanism could be possible. Also, at the walls of the crucible many small needles appear near the charge, indicating the involvement of the vapour phase. Further support comes from the filamentary habit (flat needles with longer b–axis) of most crystals.

Unfortunately, the chemical composition of the vapour phase is very difficult to measure under high oxygen pressure.

According to the above, following conclusions can be made :

–The crystals which can be recovered from the solidified flux, are those who grow into the vapour phase. They do not grow during cooling, but at the isothermal period of the experiment. In this case, the driving force for crystallization is due to concentration gradients along the appearing phases, with the vapour possibly contributing to a VLS mechanism.

–Thus, the driving force is much larger for 124 and 123.5 than for 123, and several ($\approx$ 1–5) mm can grow in the optimal time of 24 h. This corresponds to growth rates 1.0 – 6.0 x 10^{-6} cm/sec which lie between typical and strongly enhanced values for vapour growth. Flux growth rates are generally smaller.

–The long time necessary for the formation of the eutectic melt shows the existence of the severe kinetic hinderance even at 1100 ^{0}C. This behaviour is similar to our earlier observations in the decomposition of 123 and the synthesis of the 124 and 123.5 phases [2].

5. Step Structure of as Grown Surfaces by STM

Recently, a first investigation of the as grown surfaces of 124 and 123.5 single crystals has been performed [13]. The main characteristic are atomically flat surfaces on micron scale, with singular steps, without any

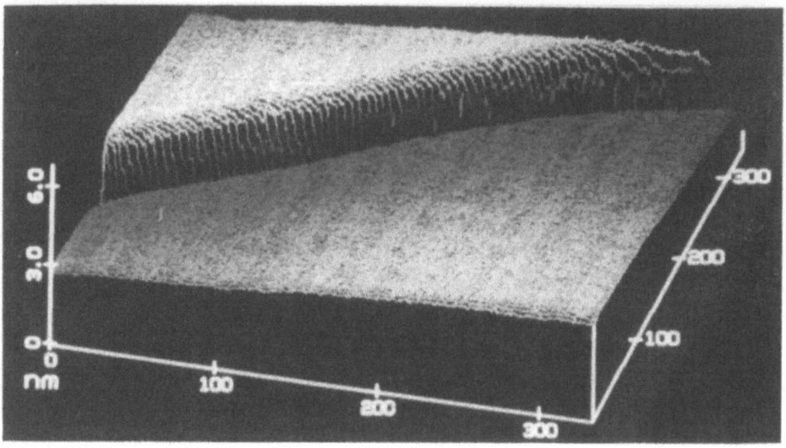

Figure 5. STM picture of the surface of an 124 single crystal, after [13].

roughness. The height of the steps is monatomic, in agreement with our X–ray data [2]. Figure 5 shows such a surface of an 124 crystal, with step height 2.7 nm.

The radius of curvature of the steps is very large, so that most of them are seen as straight. Only, near step–pinning centers much smaller radii appear.

In 123.5 with its complex unit cell [2], steps with height less than the unit cell (equal to a subunit of the structure) may be formed under the influence of the electric field of the tip.

The surfaces of most crystals do not show any rests of liquid phase, thus supporting the assumption of growth from the vapour. However, crystals have been also found with quenched traces of liquid near the step edge.

Very interesting is that, contrary to 124 and 123.5, the as grown surfaces of thick 123 crystals do not have singular habit but show a certain roughness. This can be explained by the fact that according to our phase diagram [14] the 123 phase is not thermodynamically stable.

6. Survey of Characteristic Physical Properties

The transition temperature of 124 single crystals is with T_C=72–74 K, 7–5 K lower to that of the ceramic material. As the chemistry of preparation of the crystals is different due to the presence of the flux and vapour phases, it is possible that a small amount of aluminium is present in the single crystals. This can be estimated to <0.1% , which is the limit of detection by EDAX. On the other hand, the transition width remains very narrow (down to ΔT=2 K), an indication for high crystal perfection (Figure 6).

Due to the thermodynamic stability of the double Cu–O chains with square planar coordination, no orthorhombic to tetragonal phase transition takes place [2]. Due to this reason the 124 and 123.5 crystals do not have

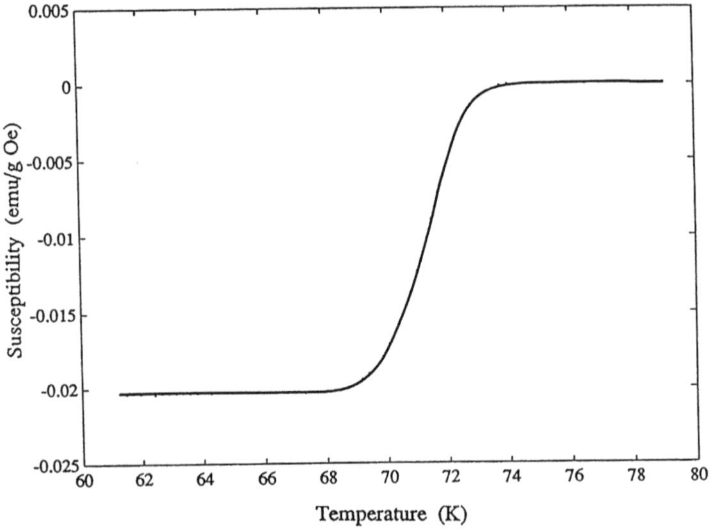

Figure 6. D.c. susceptibility of a 124 crystal, after [19].

a–b twins, in contrast to the 123 phase. These phases are, therefore, ideal for the measurement of *the a–b anisotropy.*

Measurements of both reflectivity and electrical resistivity show clearly an anisotropy of a factor ≈3–5. Very characteristic for 124 is the *nonlinear* temperature dependence of the resistivity. Further, the linear part of the resistivity does not extrapolate to zero but to 36 K [10].

Electronic Raman scattering shows in 124 single crystals that two energy gaps $2\Delta_1/kT_c=2.3$ and $2\Delta_2/kT_c=6.8$ are opening in the superconducting region [15]. This is confirmed idependently by anomalies of phonon frequencies, line widths and Raman intensities, thus showing an appreciable electron–phonon coupling. This is in agreement with band structure calculations showing that in 124 four bands cross the Fermi level, two arising from the planes and two from the chains. One would then expect that in the superconducting state energy gaps will open in all four sheets of the Fermi–surface.

Polarized IR–reflectivity spectra in 123.5 [16] indicate also the existence of a gap at $2\Delta/kT_c=3.6$ for crystals with $T_c=70K$ and $2\Delta/kT_c= 4.4$ for a crystal with low oxygen content ($T_c=30$ K).

Measurements of magnetization as a function of the magnetic field were performed to determine the lower critical field [17]. It was shown that $H_{c1}(T=0)=44\pm1$ Oe parallel to the a,b plane, and $H_{c1}(T=0)=197\pm8$ Oe parallel to the c–axis. Therefore, the c/a,b anisotropy is 4–5, as compared to 3–4 for 123, the uncertainty coming mainly from the errors in the demagnetization factor. The values of the first critical field H_{c1} for 124 are approximately three times lower than those for 123. This can be attributed to the expansion of the CuO_2–Y–CuO_2 sandwiches, which is almost 0.2nm (17%) longer in 124 than in 123.

A study of the magnetoresistive transition [18] gave first estimated values for the upper critical magnetic field H_{c2}. Depending on the criterion used, these values were found 3–5 times lower than those of 123. This would also mean that the coherence length is 3–5 times larger in 124 than in 123.5. The c–a,b anisotropy of Hc_{c2} was with 10 slightly larger than that of 123.

References

1. E.Kaldis, J.Karpinski, S.Rusiecki, B.Bucher, K.Conder, and E.Jilek, M^2S–HTSC III, Kanazawa; Physica C 1991
2. E.Kaldis, and J.Karpinski, Eur.J.Solid St.Chemistry, 27(1990)143
3. J.Karpinski and E.Kaldis, J.Cr.Growth 79(1986)477
4. J.Karpinski and S.Porowski, J.Cr.Growth 66(1984)11
5. H.Mii, M.Senoo, and I.Fujishiro, Jap.J.Appl.Phys.15(1976)77
6. J.Karpinski, E.Kaldis, Nature 331(1988)242; J.Karpinski, E.Kaldis, S.Rusiecki, B.Bucher, E.Jilek, Nature 336(1988)660
7. Morris Research Inc., 1862 Euclid Avenue #240, Berkeley, CA 94709, USA ; information sheet on autoclave
8. M.R.Chandrachood, D.E.Morris and A.P.B.Sinha, Physica C 171 (1990)187
9. J.Karpinski et al (to be published)
10. J.Schoenes, E.Kaldis, and J.Karpinski, J.Less–Comm. Metals 164–165 (1990) 50
11. K.Dembinski, M.Gervais, P.Odier, and J.P.Coutures, J.Less–Common Metals 164–165(1990)177
12. J.Karpinski, S.Rusiecki, E.Kaldis, and E.Jilek, J.Less–Comm. Metals 164–165(1990)3
13. H.P.Lang, J.P.Ramseyer, D.Brodbeck, T.Frey, J.Karpinski, E.Kaldis, Th.Wolf, Proceedings STM '91 Int.Conference, Interlaken, August 1991; also H.P.Lang et al, to be published
14. J.Karpinski, S.Rusiecki, E.Kaldis, E.Jilek, Proceed. HIP–3 Conference, Osaka, June 1991,Elsevier, December 1991.
15. E.T.Heyen, M.Cardona, J.Karpinski, E.Kaldis, and S.Rusiecki, M^2S–HTSC III, Kanazawa, Juli 1991, Physica C
16. B.Bucher, J.Karpinski, E.Kaldis, and P.Wachter, Phys.Rev.B (submitted)
17. J.C.Martinez, J.J.Prejean, J.Karpinski, E.Kaldis, and P.Bordet, Solid State Comm. 75(1990)315
18. J.C.Martinez, O.Laborde, J.J.Prejean, C.Chappert, J.P.Renard, J.Karpinski, and E.Kaldis, Europhys.Letters 14(1991)693
19. D.Zech, J.Karpinski, H.Keller, E.Kaldis,and W.Kundig (unpublished)

Physical Properties of Cuprate Superconductors: An Introduction

B. Batlogg

AT&T Bell Laboratories, 600 Mountain Ave.,
Murray Hill, NJ 07974, USA

Abstract. An introductory overview is given over experiments in the field of cuprate superconductivity. The selection focuses on basic crystal-chemical aspects, spin dynamics and charge response in the normal state, and essential microscopic aspects of the superconducting state, including length scales, gap spectroscopies and order parameter symmetry.

1. Introduction

Five years have passed since J. Georg Bednorz and K. Alex Müller discovered cuprate superconductivity, [1] and by the end of this year, the number of publications in this field will have exceeded the 20,000 mark. Keeping track in detail of all the various measurements on all different compound families is a major task, since many early results are superseded by results on samples with improved quality. This lecture is an attempt to critically focus on basic crystal-chemical concepts, normal state charge and spin dynamics, and the superconducting state. A number of summary articles provide an excellent overview of activities in subfields and they also give a more extensive list of references. [2-10]

The field of high temperature superconductivity is now characterized by advances through careful, quantitative work on well characterized high quality materials. With the advances in single crystal growth, new insight is gained, and it is gratifying to find detailed quantitative agreement between results from different research groups. An example of this is the field of nuclear magnetic resonance studies. Thus, it is hoped that the results discussed here may not be subject to significant changes in the near future.

2. Building Blocks and Charge Transfer

Despite the many structural families known by now, all the cuprate superconductors have common crystal-chemical properties. [11-13] Most useful to describe the general aspects is the concept of charge transfer between "charge reservoir" layers and the electronically active CuO_2 layers. This simplifying scheme, presented already four years ago (at the Interlaken meeting), has found extensive support. The "charge reservoir" includes e.g. the Bi-O, Tl-O, or Pb-Cu-O layers, or the Cu-O chains in $Ba_2YCu_3O_7$. The details of charge transfer and the resulting changes in the superconducting properties have been studied in great detail in $Ba_2YCu_3O_x$. [14] The

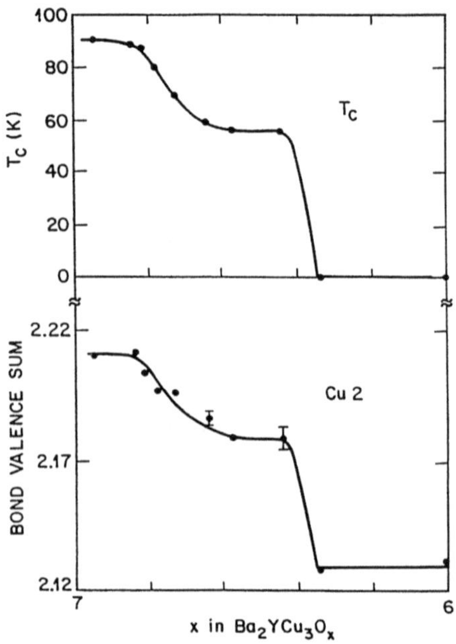

Fig. 1. The concept of charge transfer from the CuO_2 layers to the "charge reservoirs" can be demonstrated best in $Ba_2YCu_3O_x$. As function of oxygen content x, the position of the apical oxygen shifts with respect to the CuO_2 plane, resulting in a change of the bond valence sum of the Cu atom in the CuO_2 plane. The concomitant change of T_c is shown in the lower panel.

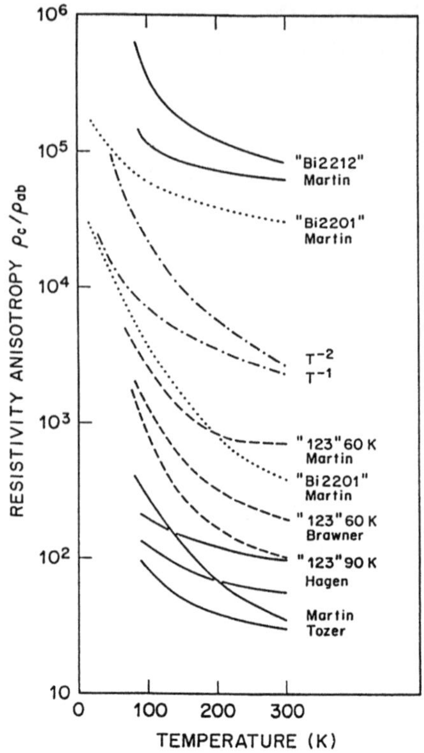

Fig. 2. Examples of resistivity anisotropies in cuprate crystals. The resistivity perpendicular to the planes can be larger than the parallel resistivity by up to 5 orders of magnitude.

position of the apical oxygen, which connects the CuO_2 layers with the Cu-O chains, sensitively reflects the amount of charge transferred in or out of the CuO_2 layers as function of the oxygen content. The close connection between T_c and the amount of charge transfer is shown in Fig. 1. From the detailed structural studies, the various bond lengths have been determined and the bond valence sum of Cu in the CuO_2 layer has been calculated. (Incidentally, the dependence of T_c on the bond valence sum is essentially the same as the dependence of T_c on the hole concentration derived from transport or chemical measurements.)

The structural parameters used in Fig. 1 were derived from neutron diffraction experiments and therefore are averaged over rather long length scales (of order 10^2 Å). In more recent experiments, different probes have been employed to study the charge redistribution in $Ba_2YCu_3O_x$ as function of x on a more local scale. These include measurements of the crystal-electric field strength experienced by the 4f electrons of various Rare Earth ions on the Y site, [15] and the occupation changes of various orbitals through x-ray absorption spectroscopy. [16]

Whenever the carrier concentration is changed by chemical means, such as heterovalent substitution or total oxygen content variation, it is useful to keep in mind that the modification of the electronic structure will have a strong local component due to the impurity potential of the charge donor or acceptor. Therefore, the total "hole concentration" deduced e.g. from the chemical formula, is not enough to determine the electronic properties. Equally important also is the local geometric arrangement. The clearest possible example is found in $Ba_2YCu_3O_{6.45}$, where room temperature annealing (at constant oxygen content) of the Cu-O chain fragments led to remarkable changes of the superconducting properties. [17] Therefore, the "electronic granularity" associated with heterovalent substitution or oxygen non-stoichiometry might make it rather difficult to map out he generic phase diagram of layered cuprates as function of carrier concentration. Furthermore, it is useful to keep in mind that various experimental techniques probe the structural and electronic properties on different length scales, and therefore what appears homogeneous in one experiment might appear inhomogeneous in another.

3. The General Phase Diagram

High temperature superconductivity occurs when the carrier concentration is reduced (or increased) by $\sim 0.1-0.2$ units per CuO_2, e.g. in $La_{1.85}Sr_{0.15}CuO_4$ or in $Nd_{1.85}Ce_{0.15}CuO_4$. [18] The parent compounds, such as La_2CuO_4 or $Ba_2YCu_3O_6$, are charge-transfer insulators. The changes of transport and magnetic properties upon introduction of charge carriers have been studied extensively. Much similarity of the physical properties is found among the various structural families for the composition with optimized superconducting properties. Therefore, in the following discussions we will concentrate mainly on these compositions. It is worth noting that superconducting properties, such as the specific heat anomaly at T_c and the Meissner signal, deteriorate rapidly away from these optimal compositions. The microscopic origin of this effect is the subject of current research.

4. Charge Response

4.1 DC Transport

The layered structure of the cuprate superconductors results in a highly anisotropic electrical conductivity with different characteristic temperature dependences parallel and perpendicular to the CuO_2 layers. The magnitude of the anisotropy depends on the "charge reservoir" layers and is smallest for $Ba_2YCu_3O_7$ and can reach values $> 10^5$ in the Bi-compounds. Although it is technically difficult to measure very large anisotropies, several examples are shown in Fig. 2. Perpendicular to the CuO_2 layers the resistivities ρ_c decrease with increasing temperature, and follow a power-law like temperature dependence T^{-n}, with n ranging from 1-2, and not a thermally activated behavior. The perpendicular resistivity in $Ba_2YCu_3O_7$ has been found to be either "metallic" ($d\rho/dT > 0$) or "semiconducting " ($d\rho/dT < 0$), and no definitive agreement has been reached on the intrinsic character. In any case, ρ_c is much larger than estimates based on minimum metallic conductivity arguments (in 3 dim). (Perhaps $Ba_2YCu_3O_7$ is an exception).

The in-plane resistivities ρ_{ab} are surprisingly similar in many optimized compounds and follow quite closely a linear temperature dependence over large temperature ranges, and extrapolate to small values at $T=0$. In Fig. 3 some examples from single

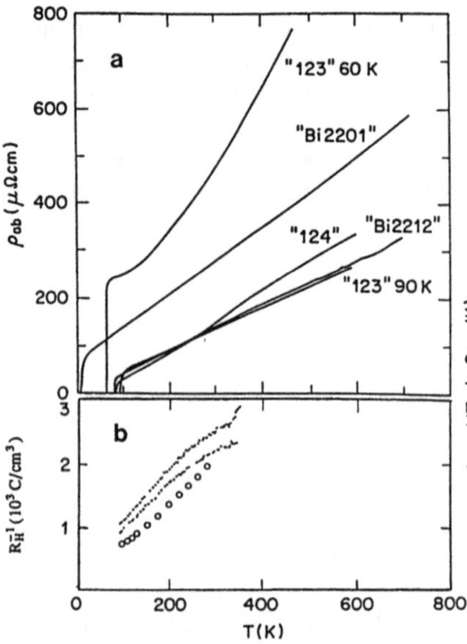

Fig. 3a. Selected in-plane resistivities ρ_{ab} of metallic cuprates with composition near the optimum. The near-linearity of ρ_{ab} over a wide temperature range has become a hallmark of these compounds. Pronounced non-linear temperature dependences have been associated with conduction outside the CuO_2 planes.

Fig. 3b. The inverse of the Hall coefficient measured on three different $Ba_2YCu_3O_7$ crystals. The strong temperature dependence is also seen in other cuprates and appears to be most pronounced when T_c is optimized.

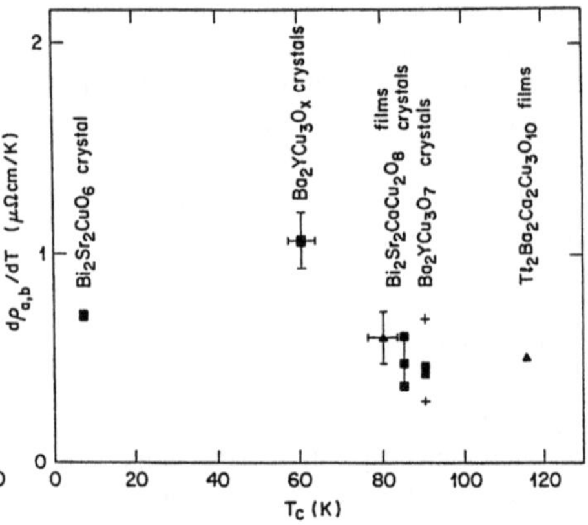

Fig. 4. Temperature coefficient $d\rho_{ab}/dT$ of the in-plane resistivity ρ_{ab}. In single domain crystals of Bi2212 and $Ba_2YCu_3O_7$, an anisotropy of ρ_{ab} of up to a factor of 2 has been found.

crystal studies are shown. (Some compounds show distinct non-linear T-dependences of ρ_{ab} which has been associated with conduction in secondary building blocks such as the double-Cu-O chains in $YBa_2Cu_4O_8$).

Within the CuO_2 layers the resistivity can be anisotropic up to factors of 2, e.g. in single domain samples of $Ba_2YCu_3O_7$ and in Bi 2212. It remains to be clarified how much of this in-plane anisotropy is due to the anisotropy of the CuO_2 layer itself (orthorhombic structure) or due to conduction by the Cu-O chains or the Bi-O layers, respectively.

The in-plane resistivities ρ_{ab} have a remarkable similar temperature coefficient $d\rho/dT$ of typically $0.3-0.8\mu\Omega$ cm. No correlation exists between $d\rho_{ab}/dT$ and the transition temperature. (Fig. 4) From the absence of saturation at highest temperature and the value of $d\rho/dT$, together with the plasma frequency, the coupling strength λ had been estimated to be in the range of 0.2-0.4. This will be contrasted later to the large gap-to-T_c ratios and the substantial renormalization of quasiparticle dispersion near E_F suggested by photoemission spectroscopy. (Refs. for this chapter: [19-32])

4.2 Optical Conductivity $\sigma(\omega,T)$

A close connection exists between the dc resistivity and the optically probed charge response $\sigma(\omega,T)$. The most extensive set of data is available for $YBa_2Cu_3O_x$ which has been studied as function of oxygen concentration and more recently on untwinned samples.[33-40] Although substantial qualitative agreement is found among the various results, some quantitative discrepancies still remain to be resolved. In the "optimized" compositions, $\sigma(\omega,T)$ deviates from a simple Drude behavior, i.e. a single scattering rate τ_0^{-1}. With increasing frequency, the measured conductivity drops more slowly than in a simple Drude model. A description of $\sigma(\omega,T)$ up to photon frequencies of several tenths of an eV in terms of a frequency and temperature dependent scattering rate reveals the peculiar nature of the charge dynamics in the cuprates: the relaxation rate $\hbar/\tau^*(\omega,T)$ is given by the greater of kT or $\hbar\omega$. A possible form would be $(\hbar/\tau^*)^2=(\alpha k_B T)^2+(\beta\hbar\omega)^2$ with α and β of order unity). The difference between the simple Drude conductivity and the cuprate results, highlighted in Fig. 5, has also been interpreted as due to a superposition of two different conductivity processes, one of them being a Drude-type process at lowest energies, the other one at finite energies (of a few tenths of an eV), termed "mid-1R absorption." It has been pointed out that these two components can be discerned more easily in samples with less than optimal composition. [34]

4.3 Electronic Raman Scattering

The observed $\sigma(\omega,T)$ has been taken as evidence for charge carriers interacting with excitations that exist over a wide energy range. Excitations of this kind may have been observed in Raman scattering. In addition to the sharp phonon lines, [41], a broad continuum of electronic excitations has been measured. [42] The response function describing this spectrum, reproduced in Fig. 6 for two different temperatures, increases linearly with energy up to $\hbar\omega \simeq kT$ and then remains

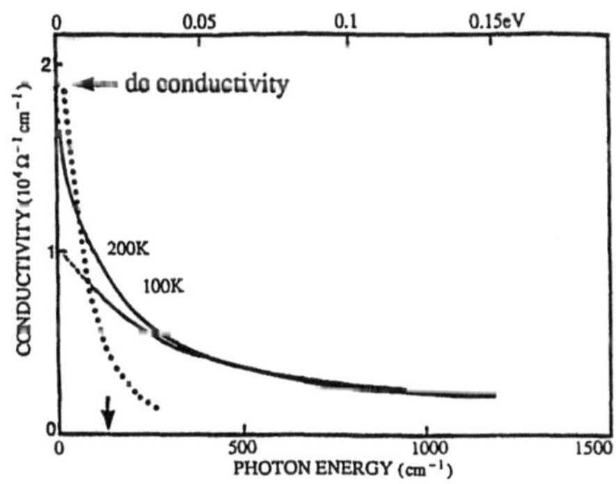

Fig. 5. Frequency dependence of the conductivity $\sigma(\omega,T)$ in $Ba_2YCu_3O_7$ at two different temperatures in the normal state. The dotted line is a simple Drude model with a constant scattering rate $1/\tau_0$ of order $k_B T/\hbar$. The measured $\sigma(\omega,T)$ deviates from this model, particularly at energies $\hbar\omega >> kT$. The arrow indicates the energy equivalent of 200K.

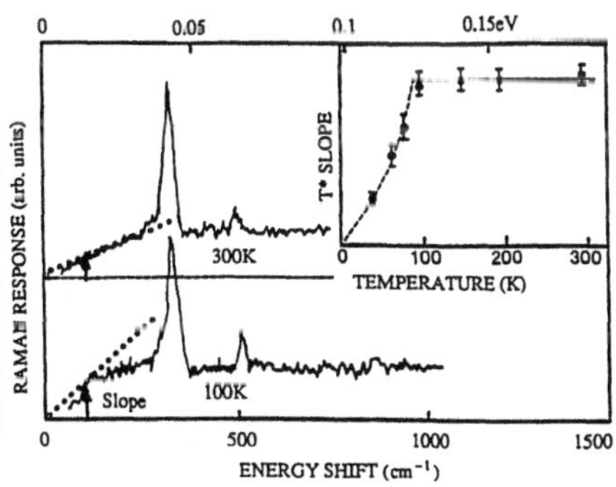

Fig. 6. Imaginary part of the Raman response function in $Ba_2YCu_3O_7$ at two temperatures. In addition to the sharp phonon lines, a broad electronic background is observed, extending to high energies. At low energies, this electronic scattering increases nearly linearly with a slope proportional to 1/T in the normal state. (See insert).

approximately constant up to the experimental limit. The slope of the linear increase is proportional to 1/T, and thus, the product of the slope times T is constant (for $T > T_c$) as shown in the insert to Fig. 6. [43] The nature of the excitations giving rise to this characteristic response, and the mechanism that couples light to them, is not yet clear. The frequency dependent charge response and the inelastic light scattering experiments suggest that there is no scale for the low-energy excitations other than the temperature itself. [44]

When cuprates are "overdoped", e.g. in $La_{2-x}Sr_xCuO_4$ with x>0.25, superconductivity disappears [45,46], the resistivity increases faster than linearly with temperature and its value is smaller than in the optimized samples [45-47]. It is interesting to note that recent optical conductivities in the overdoped region reveal again a close relationship between $\rho(T)$ and $\tau^*(\omega)$, where $\tau^*(\omega)$ is the generalized Drude relaxation parameter. It was found that $1/\tau^*(\omega)$ up to 0.4eV increases faster than linearly with ω, but is smaller than in the optimized compounds. [48]

4.4 Hall Effect

The Hall effect is similarly characteristic of the cuprates as the resistivity is. (A separate lecture at this school deals with the Hall effect in more detail; see [32]). Instead of being nearly constant above a fraction of the Debey temperature ($0.2-0.5\theta_D, \theta_D \sim 400-450K$), the Hall coefficient in the cuprates is strongly temperature dependent in the entire range of measurement. An explanation in terms of two-band conduction would require an unlikely combination of T-dependent carrier concentrations and mobilities. A possible reason for the observed characteristic $\rho(T)$ and $R_H(T)$ behavior has been suggested, which involves different scattering processes for the two transport phenomena ($\tau_{Hall}^{-1} \sim T^2, \tau_\rho^{-1} \sim T$ and $\tau_H^{-1} << \tau_\rho^{-1}$). [32] An example of $R_H^{-1}(H)$ is shown in Fig. 3b.

5. Magnetism

The nonmetallic parent compounds are magnetic insulators which are well understood by now, and can be described as spin 1/2, 2-dimensional square lattice Heisenberg systems [49, 50]. The coupling is strongly antiferromagnetic with exchange parameters in the range from 100-130 meV. In search for quantum effects in this $S = 1/2$ system, one has found only small renormalization of the numerical parameters. However, no quantum frustration or qualitatively new phenomena in the dynamic properties could be discovered. When carriers are introduced, the long range magnetic order is rapidly destroyed. The static susceptibility $\chi_0(T)$ varies gradually in the metallic regime as function of carrier concentration: $\chi(T)$ is essentially T-independent at (or near) the optimal composition, and decreases with decreasing T in sub-optimized compositions. Examples are given in Fig. 7 for $(La,Sr)_2CuO_4$ and $Ba_2YCu_3O_x$. [51] By now the s-shaped $\chi(T)$ curves for non-optimized superconductors are well documented, and are recognized to be generic to the cuprates.

The spin susceptibility χ^s can be derived from the total measured χ by correcting for core diamagnetism χ_c, Van-Vleck paramagnetism χ_{vv} associated with the d-orbitals, and by taking into account the anisotropy of χ_{vv}, [52]. Information on

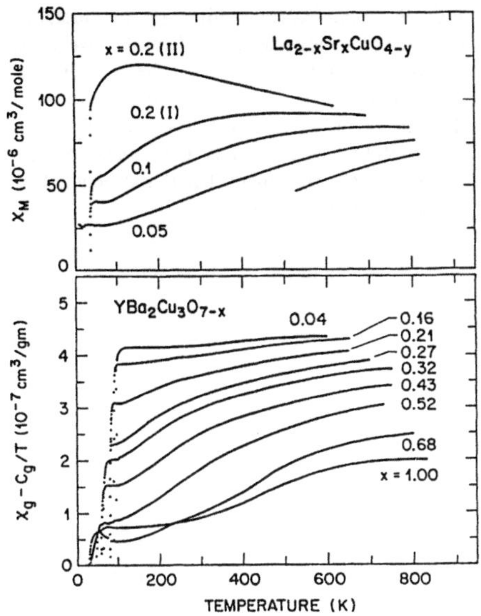

Fig. 7. The static susceptibility changes systematically with total (nominal) carrier concentration. $\chi(T)$ is nearly constant for the compositions with maximal T_c, and decreases at lower temperature in sub-optimized compounds.

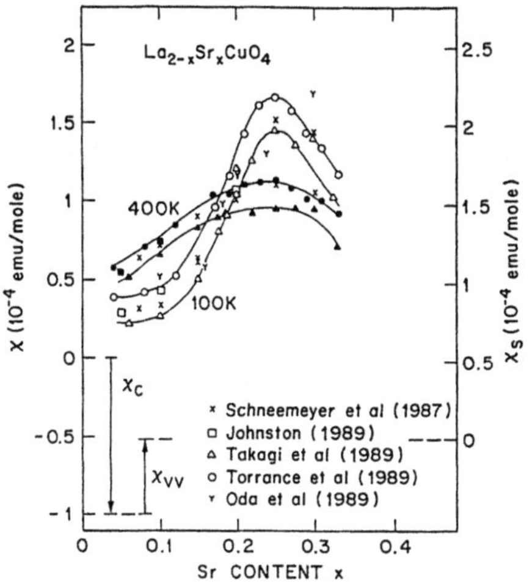

Fig. 8. Summary of susceptibility data on $La_{2-x}Sr_xCuO_4$ powders at 100K and 400K. The total measured susceptibility χ is corrected for core diamagnetism χ_c and Van Vleck paramagnetism χ_{vv} to obtain the spin susceptibility χ_s.

χ_{vv} comes from NMR studies [53, 54]. Fig. 8 is a summary of susceptibility data on $(La,Sr)_2CuO_4$ from several research groups, indicating both the substantial agreement and the magnitude of the χ_{vv} and χ_c corrections. [29, 46, 51, 55, 56] Similar comparisons for other compound families, together with single crystal work, lead to the conclusion that χ^s is essentially determined only by the CuO_2-layer arrangement. Within experimental uncertainty χ^s is the same for a specific CuO_2 layer arrangement, independent of the other building blocks in the unit cell. [52] For instance, χ^s for the single-layer compounds is $\sim 1.2-1.5\ 10^{-4}$emu/mole, and $\sim 2.3-2.5\ 10^{-4}$ emu/mole for the double-pyramidal layers. (Fig. 9) Antiferromagnetic coupling between closely spaced double layers may cause χ^s to be somewhat smaller than twice χ^s in a single layer compound.

Local probes of the susceptibility, such as NMR Knight shifts measured at the Cu or O sites, are in substantial agreement with the static measurements of χ^s, and suggest that a single-component spin-system description is appropriate. [53, 54, 57, 58]

Inelastic neutron scattering, which requires large crystals, and NMR relaxation rate measurements, which can be done on powders, are techniques to study the generalized magnetic response $\chi(q,\omega)$. The relaxation rates of the Cu and O nuclei are different and follow different T-dependences. This can be understood because the form factor for the two sites enhances different parts of q-space. One concludes that the susceptibility in the $\omega \to 0$ limit (of the NMR experiment) has a distinct q-dependence and peaks near the Brillouin zone corner, $q=(\pi/a_0,\pi/a_0)$, indicating the existence of antiferromagnetic fluctuations. The study of $\chi(q,\omega)$ at higher energies by

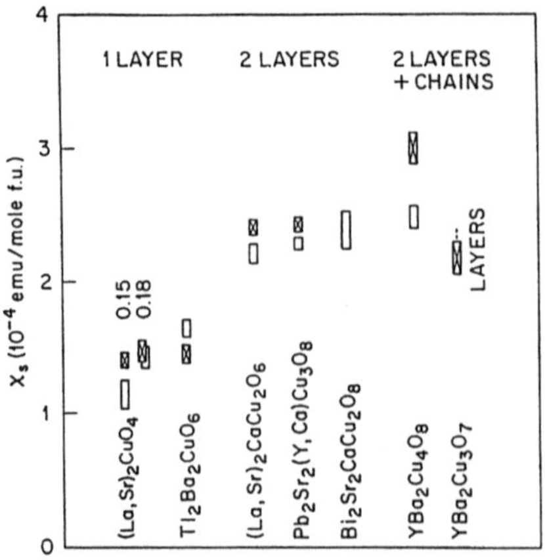

Fig. 9. Spin susceptibility χ_s for metallic cuprates of various structural families. Near the composition for optimal T_c, the temperature dependence of χ_s is weak (open symbols: T=100K, crossed symbols: T=400K). The spin susceptibility is essentially determined by the CuO_2 layer arrangement, but is independent of other structural aspects or T_c.

neutron scattering requires large and homogeneous crystals, which are rather difficult to grow. Crystals with optimal carrier concentration are scarce and thus a comprehensive picture of $\chi(\omega,q)$ is emerging only slowly. So far, these measurements have revealed maxima of $\chi(q,\omega)$ near the zone-corner. Interestingly, the antiferromagnetic fluctuations appear to be commensurate, i.e. at $(\pi/a_0,\pi/a_0)$, in $Ba_2YCu_3O_x$ $(x\simeq6.9)$ but incommensurate in sub-optimized $(La,Sr,Ba)_2CuO_4$. [49, 59, 60] The local magnetic response $\chi_0(\omega)$ in metallic $(La,Sr,Ba)_2CuO_4$ has a characteristic energy dependence. The imaginary part $\chi''_0(\omega)$, an odd function of ω, increases with ω up to a frequency $\hbar\omega\approx kT$ and remains constant at higher energies $\hbar\omega>>kT$. This suggests that $\chi''_0(\omega,T)$ is a "universal" function of ω/T. [61, 62] The similarity between this magnetic response and the Raman response is remarkable, although the latter one is known not to be due to spin fluctuations. [63]

There exists an interesting difference in $\chi(T)$ between optimized T_c and reduced T_c materials. While $\chi(T)$ is essentially T-independent for the former, $\chi(T)$ for the reduced T_c metals suggests a reduction in the spin excitation spectrum at low T, but well above T_c. This is seen both in the static $\chi(T)$, in NMR and neutron scattering studies, and also in a less well-known technique that measures the line width of intra-4f transitions of rare earth ions. The width is a measure of the local susceptibility $\chi_0(\omega_0)$ at the particular energy $\hbar\omega_0$ of the 4f-excitation ($\hbar\omega_0=12$meV in a particular experiment [64]). All these probes consistently find a gradual decrease of χ well above T_c without a pronounced (or even noticeable) change at T_c.

6. Electronic States Near E_F

The Cu-O bond involves Cu 3d and O2p orbitals and is both of ionic and covalent character. It is the near degeneracy in energy of the p and d orbitals, and therefore this particular combination of elements in the square-planar arrangement that determines the electronic structure of cuprates. (In the $BaBiO_3$-based superconductors, a similar degeneracy is found between the 6s level of Bi and the 2p states of oxygen). The states at E_F have both p and d character, and have been studied in detail using various electron spectroscopies.[65-70]

Although it was not clear initially whether or not these correlated systems have a Fermi surface, several types of experiments suggest indeed the existence of a Fermi surface. More precisely, angle resolved photoemission and inverse photoemission spectroscopy mapped out the E(k) relations and found crossings of the energy bands at points in k-space which agree with band structure calculations. [71-73] The quasiparticle lifetime may be derived from the line shapes as function of $|E-E_F|$. It is not possible, so far, to decide unambiguously whether the quasiparticle lifetime grows linearly or quadratically with $|E-E_F|$. While the latter would be consistent with an ordinary Landau Fermi liquid description, the former would suggest an unconventional normal state. The "dispersion" of the quasiparticle state E(k) near E_F as derived from the experimental spectra is considerably smaller than that which is derived from band structure calculations. It is noteworthy that this mass enhancement in Bi2212 ($m^*/m_b \approx 2$[74], $3-5$[72]) is observed over a much wider energy range than the phonon frequencies, therefore ruling out electron-phonon interaction as the main cause of the suggested mass enhancement. Positron annihilation measurements, probing the momentum distribution of electrons at the site of annihilation, are also in agreement with band structure calculations.[75-77]

The electronic density of states within an energy window of several times kT can be derived from two experimental sources: the electronic specific heat γT and the spin susceptibility χ_s. The electronic density of states is rather low in the cuprates and therefore it is an experimental challenge to measure the electronic contribution to the specific heat. All the estimates suffer from quite large uncertainties. [78-80] For $La_{1.85}Sr_{0.15}CuO_4$, $\gamma \approx 5-6 mJ/moleK^2$ and for $Ba_2YCu_3O_7$, γ's range from 14 to 25 $mJ/moleK^2$ (and higher). The spin susceptibility, discussed above, ranges up to $\sim 1.5\ 10^{-4}$ emu/moleCu. A comparison of the thus derived density of states is given in Fig. 10. Here the values of χ_s have been converted into γ through $\gamma/\chi_s = \pi^2 k^2/3\mu_B^2$. The units of N(0) are states /eV-Cu (both spin states). Apparently, neither γ nor χ_s are much enhanced over the band structure value. The experimental scatter precludes to deduce precise value of the enhancement factor λ.

A useful way to compare superconductors is to plot T_c vs. the measured density of states, expressed in the γ term. (Fig. 11) In such a plot, the high T_c superconductors stand out in that their T_c is high when compared to other materials with comparable γ values. This indicates, independent of any microscopic model, a high characteristic energy of the pairing mediating excitations. Not all oxide superconductors fall in the thus defined category of "high T_c/γ." What seems to be necessary is that the states at the Fermi level have both oxygen p and metal d(or s)

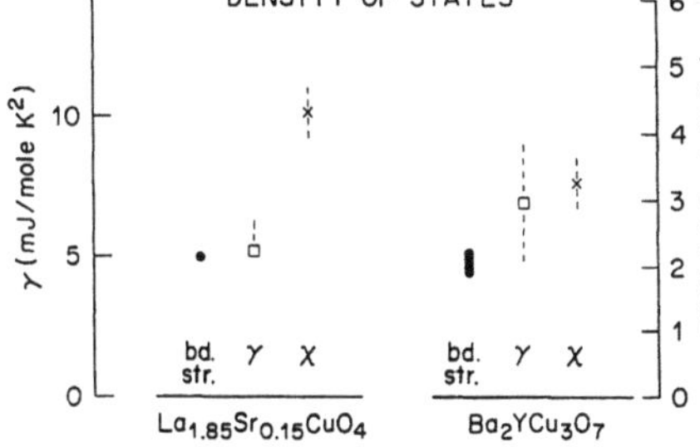

Fig. 10. Density of states in $La_{1.85}Sr_{0.15}CuO_4$ and $Ba_2YCu_3O_7$. The experimental values are deduced from the Sommerfeld constant γ (= lines-in-T specific heat) and the spin susceptibility χ_s, with an estimate of the uncertainties indicated by the broken lines. The band structure calculation results are shown by the closed symbols. (The units for N(0) refer to both spin directions. For easy comparison, the $Ba_2YCu_3O_7$ values are given per one Cu).

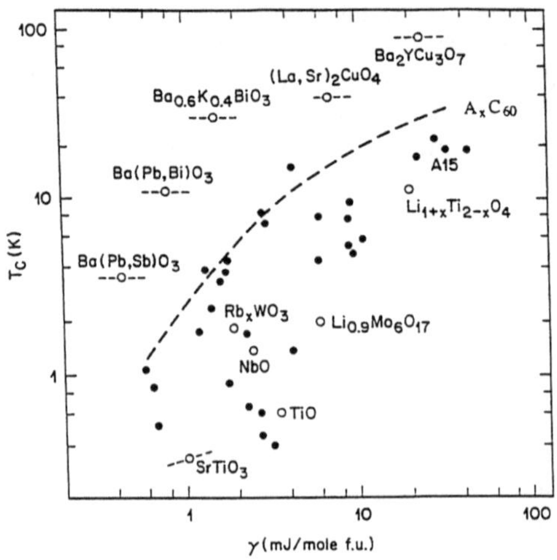

Fig. 11. An updated version of the $\gamma - T_c$ plot. While the Cu-O and Bi-O superconductors have high T_c's compared to other materials with the same density of states at $E_F(\propto\gamma)$, other superconducting oxides are not so anomalous. A preliminary estimate for superconducting fullerenes A_xC_{60} places them in the high γ range near the A15's.

character. A preliminary estimate of γ for the superconducting fullerenes places them in the high-density of states upper right corner, near the A15 compounds.

7. The Superconducting State

The superconducting state involves pairing of quasiparticle ($e^*=2e$). [ac-Josephson effect, flux jump measurements, flux quantum ϕ_0 through decoration experiments, Little-Parks oscillations]. The Andreev reflections between cuprate superconductors and a normal metal suggests that quasiparticles with opposite momentum form the pairs. [81-85]

The anisotropy of the structure, with conducting CuO_2 layers separated by the charge reservoir layers, is reflected in a pronounced anisotropy of the electronic properties both in the normal and in the superconducting state. The microscopic length scales, such as coherence length ξ and magnetic penetration length λ are different parallel and perpendicular to the CuO_2 layers. $Ba_2YCu_3O_7$ is the least anisotropic case and from several different experiments the following parameters have been determined: $\xi_{ab}=14\pm2\,\text{Å}$, $\xi_c\,1.5-3\,\text{Å}$, $\lambda_{ab}=1400\,\text{Å}$, $\lambda_c\sim7000\,\text{Å}$. Larger anisotropies are measured in other compounds, which then act in many respects as stacks of weakly coupled two-dimensional superconductors. Space limitations do not permit a discussion of the many interesting aspects of magnetic flux dynamics in the cuprates. The extreme electronic anisotropy is at the core of these phenomena.

The short coherence length in the cuprates follows immediately from the high T_c (and the large gap) and the rather slow Fermi velocity of several times 10^7 cm/sec [73]. The long penetration length λ reflects the low carrier concentration, and thus the low plasma frequency.

Some consequences of a short coherence length are noteworthy. First it means that the number of superconducting pairs per coherence volume is by many orders of magnitude smaller than in usual metallic superconductors, and actually this small number (of order 5 pairs per $\xi_{ab}^2\xi_c$) makes one wonder why the fluctuation effects are not more pronounced, and why most aspects of superconductivity are close to mean-field like. Secondly, a short coherence length makes it more challenging to produce devices that are based on tunneling and proximity effects. Excellent control of materials composition on an atomic length scale is required, or one can take advantage of short-length perturbation (for instance, grain boundaries) to fabricate junctions. Third, the upper critical field H_{c2} is very large since $H_{c2}=\phi_0/2\pi\xi^2$, and this is a necessary condition to produce high field magnets from the cuprates.

7.1 Gap Spectroscopies

Many types of experiments have been performed on cuprates in an attempt to shed some light on the symmetry and magnitude of the gap parameter $\Delta(k,T)$. A variety of spectroscopies suggest gap values well in excess of the weak coupling BCS value of $3.5kT_c$. A short comment on the individual techniques follows.

Electron tunneling was extremely successful in traditional superconductors, yet it relies on well-defined tunneling barriers and no degradation of the superconductor at the interface. In the cuprates all of the reported spectra deviate from the classical tunneling curves by showing broadened structures, finite conductance (of variable magnitude) in the "gap" region, or multiple gap structures. Despite all these shortcomings, one might define a characteristic energy and associate it with Δ. The values of $2\Delta/kT_c$ range from ~3.5 to ~6 and a summary of several experiments is shown in Fig. 12 [86-88]. When electrons are injected parallel to the CuO_2 planes, Δ tends to be larger than for tunneling perpendicular to the planes. Whether or not this indicates intrinsic gap anisotropy remains to be investigated. The T-dependence of features associated with Δ is not clear because some studies suggest a decrease as T_c is approached, while others find only a gradual filling of the gap below Δ.

Optical spectroscopy, mainly on $Ba_2YCu_3O_7$, reveals an absorption that rises steeply between ~400cm^{-1} and 800cm^{-1}. While the experimental results are in essential agreement, no consensus has been reached on the interpretation. [33, 34, 37-40, 89] It seems to be inappropriate to identify this edge with a conventional BCS gap since it is not sensitive to the transition temperature T_c in crystals with reduced T_c, nor is there a noticeable shift in energy as the temperature approaches T_c.

Phonon renormalization also gives an estimate of Δ. The width and frequency of the q=0 phonon lines, measured in Raman spectroscopy, change below T_c when the phonon frequency is comparable to Δ. In $Ba_2YCu_3O_7$ this method gives a $2\Delta/kT_c$

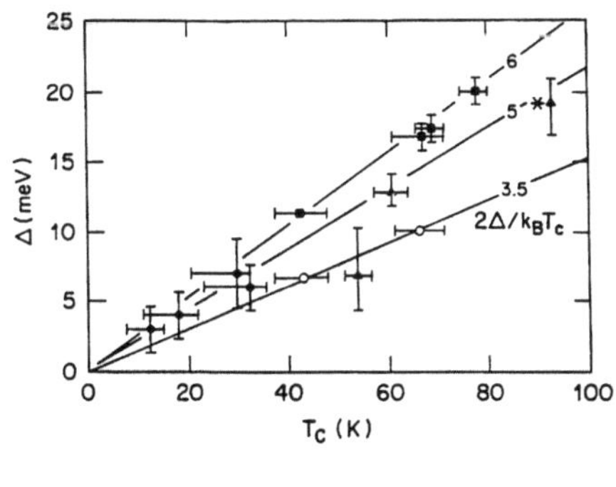

Fig. 12. Characteristic energies Δ from tunneling spectroscopy on cuprates with different T_c's. For tunneling parallel to the CuO_2 layers, Δ tends to be larger than for perpendicular tunneling.

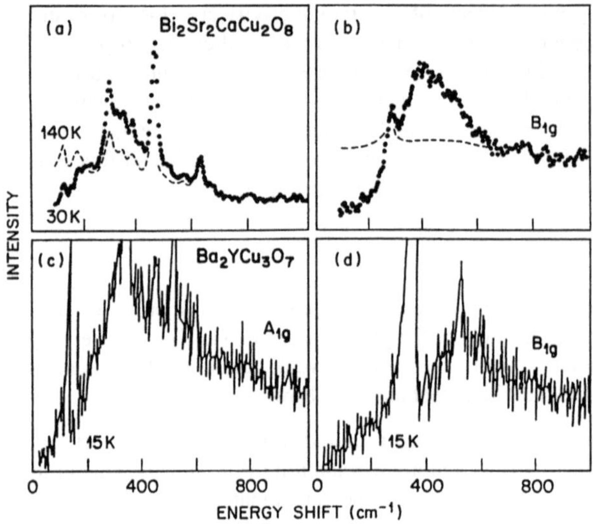

Fig. 13. Raman scattering intensity for $Ba_2YCu_3O_7$ and Bi2212 above and below T_c. The broad background scattering is of electronic origin and becomes redistributed below T_c. The position of the peaks depends on the scattering geometry (A_{1g},B_{1g}). Unlike traditional superconductors, no sharp cut-off is observed below the peak energy.

very close to 5. It is not evident yet which portion of k-space is probed. The electron-phonon coupling strength for these $q \approx 0$ phonon modes is found to be very small ($\sim 10^{-2}$). [90, 91]

The redistribution below T_c of the broad *electronic excitation spectrum* mentioned above has also been associated with the gap formation below T_c. In contrast to the formation of a well-defined gap below 2Δ in ordinary superconductors, the electronic spectrum is not completely suppressed in the cuprates. An early example is shown in Fig. 13 for $Ba_2YCu_3O_7$ and Bi2212. The spectrum below T_c peaks at energies between ~5.5 and 8 times kT_c, and this value depends on the scattering geometry. [42, 92-94]

Andreev reflection between $Ba_2YCu_3O_{6.9}$ and a gold tip resulted in a conductance enhancement by the expected amount and could be modeled by a gap of 25 meV, which is equivalent to $2\Delta/kT_c = 6.4$. [85]

Photo-emission spectroscopy techniques have been improved so much that they could be employed to even measure the superconducting gap. Both angle integrated and angle resolved spectra indicated gaps in the range of 6-8 kT_c. It has been pointed out [71, 95], however, that the finding of the same values in the two types of measurements may not be consistent. [96]

Despite all the imperfections of the data and limitations of the various techniques, they all point to gap values in cuprates that are significantly larger than the weak-coupling BCS value of $3.5kT_c$, typically in the range from $5-8kT_c$. An average value of 5 ± 1 may be the best estimate at this point.

7.2 Strong Coupling Effects?

The coupling strength between quasiparticles and excitations that give rise to the normal state resistivity has been estimated to be small, of order 0.2-0.4 [97], and our comparison of experimentally estimated density of states with band structure calculations also indicates only a modest enhancement, perhaps a λ of order <1. In contrast to this, the specific heat anomaly at T_c in optimized $Ba_2YCu_3O_7$ and in $La_{1.85}Sr_{0.15}CuO_4$ ($\Delta C/\gamma T \approx 2.5-4$; vs. 1.43 in weak coupling BCS) [78-80], the precipitous drop of the nuclear spin relaxation rate below T_c [98], the increase of the superfluid density ρ_s below T_c, and the various gap spectroscopies mentioned above indicate "strong coupling" effects. Relevant to resolving this apparent puzzle is the observation of *strong pair-breaking*. From $\rho(T)$ and $\rho(H)$ measurements in the normal state, the phase-breaking time has been estimated, and it is found to be essentially the same as the normal state transport scattering time. Like $\rho(T)$, the phase-breaking rate also increases linearly with T. Therefore, there appears to be a close connection between (if not an identity of) the scattering mechanisms that give rise to the characteristic $\rho(T)$ behavior and that induce phase-breaking. The large gap-to-T_c ratios are often interpreted as due to strong coupling in cuprates, but the observed large pair-breaking effects need to be considered as well.

The recently discovered "coherence" peak in the high-frequency conductivity below T_c [101, 102] has been proposed to be *not* a result of type II coherence factors, but due to a rapidly changing inelastic scattering rate below T_c. Otherwise, a similar peak would be expected in the nuclear relaxation rate below T_c (Hebel-Slichter peak), which is not observed in cuprates.[98]

7.3 Superfluid Density

In the absence of a detailed spectroscopic measurement of $\Delta(k,T)$, the superfluid density ρ_s provides more global information on $\Delta(k,T)$. Particularly, one would expect to learn from the low temperature variation of ρ_s whether or not the gap function is finite everywhere on the Fermi surface. In heavy fermion superconductors, the gap function vanishes along lines or at points which lead to power-law variation of many physical properties at $T \ll T_c$, which can be experimentally distinguished from an exponential variation. Measurements related to ρ_s include penetration length $\lambda(T)$, lower critical field $H_{c1}(T)$ and the spin susceptibility $\chi_s(T)$. In Fig. 14 $\lambda(T)$ and $H_{c1}(T)$ of $Ba_2YCu_3O_7$ are plotted on a normalized scale to be proportional to ρ_s. {103-106] Near T_c, ρ_s increases faster than the weak coupling BCS curve in agreement with a specific heat anomaly $\Delta C/\gamma T_c$ larger than 1.43. In the low T region only one measurement of $\lambda(T)$, based on a microstrip transmission line technique, is sensitive enough to address the "power-law vs. exponential" question. According to this measurement, there is indeed a true gap of at least 2.5kT in the excitation spectrum. Earlier studies of $\lambda(T)$ found indication of a T^2 contribution to $\lambda(T)$, but a more recent systematic investigation on thin films found a decrease of the T^2 term as the sample quality improves, suggesting a non-intrinsic origin of this contribution. [107]

Fig. 15 summarizes the NMR Knight shift ^sK results for Cu and O sites in the CuO_2 planes of $Ba_2YCu_3O_7$. [108, 109] The data from the two research groups agree remarkably well in that $^sK(T)$, which is proportional to the susceptibility, drops more rapidly below T_c than expected from weak coupling BCS. The temperature dependence of ^sK in Fig. 5 seems to be consistent with a gap 2Δ larger than $3.5kT_c$. [110]

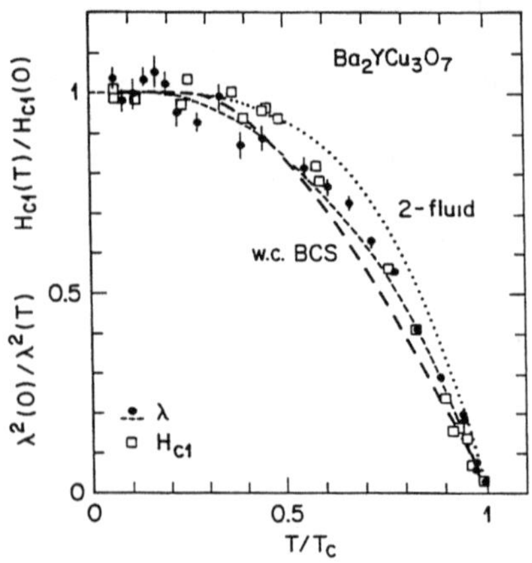

Fig. 14. Normalized $H_{c1}(T)$ and $\lambda^{-2}(T)$ results for $Ba_2YCu_3O_7$, reflecting the T-dependence of the superfluid density ρ_s, compared with ρ_s from weak coupling BCS and 2-fluid models. The two examples for $\lambda^{-2}(T)$ are from μSR (dots) and strip-line resonator measurements (short dashed), respectively.

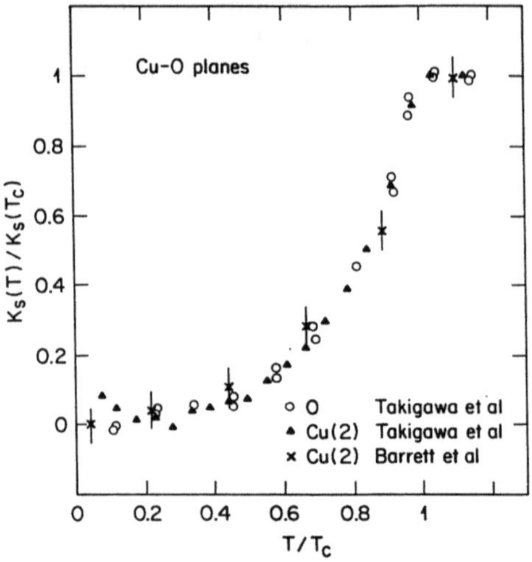

Fig. 15. The spin susceptibility χ_s from NMR Knight shift measurements on the Cu and O sites in the CuO_2 layers of $Ba_2YCu_3O_7$.

While the results discussed so far may be interpreted in support of a conventional (perhaps s-wave) order parameter, a new set of NMR experiments might reflect a non-conventional order parameter. In the normal state, the ratio of Cu nuclear relaxation rates with the external magnetic field parallel and perpendicular to the planes is temperature independent dependent. This ratio, however, changes drastically below T_c and becomes temperature dependent. [111, 112] The relaxation rate of the O nuclei shows no such anomaly. This intriguing behavior is observed not only in $Ba_2YCu_3O_7$, but also in $YBa_2Cu_4O_8$. [113]

7.4 Isotope Effect

Ever since the first reports of very small or vanishing isotope effects in $Ba_2YCu_3O_7$ and $La_{1.85}Sr_{0.15}CuO_4$ [114, 115], this type of study has attracted a great deal of interest. (A comprehensive overview can be found in Ref. 116). The essential finding of the many experiments is that the isotope effect is indeed very small (or zero) for all cuprate families when the chemical composition is adjusted to optimize T_c. This includes even the n-type compound $Nd_{1.85}Ce_{0.15}CuO_4$ with no apical oxygen. [117] Thus, the isotope effect α is not related to the absolute value of T_c, and particularly α is not intrinsically larger for optimized superconductors with relatively low T_c. When a given compound's composition is modified to reduce T_c, however, then α can be large and can even exceed 1/2. While this has been interpreted as a sign for phonon mediated pairing, one has to keep in mind an important materials aspect. Concomitant to the reduction of T_c occurs also a reduction of (1) the Meissner signal and (2) in the cases where specific heat has been measured in the comparable system, also a reduction of the specific heat anomaly at T_c. This is strong evidence that changes of T_c are accompanied by changes of the superconducting volume fraction. An interpretation of the large "isotope effects" in terms of a simplistic model that considers only changes of the phonon frequencies alone is not sufficient. Complicated materials issues such as phase homogeneity on a microscopic scale have to be addressed also.

Concluding Remarks

Since this lecture was intended as an introduction to the wide variety of experiments on high T_c cuprates, we have only mentioned those types of studies that may contribute to the understanding of some central questions. The selection of topics is limited and the list of references cannot be exhaustive for obvious practical reasons.

Acknowledgement

I would like to gratefully acknowledge the many collaborations and stimulating discussions with many colleagues at AT&T Bell Laboratories and elsewhere, that make such an overview lecture possible. I thank in particular R. J. Cava for continuous collaboration and L. W. Rupp, Jr. for expert technical support.

References

1. J. G. Bednorz and K. A. Müller, *Z. Phys.* **B64,** 189 (1986).
2. "Physical Properties of High Temperature Superconductors Vol. I," Donald M. Ginsberg, (ed.) World Scientific Publishing Co., (1989); Vol. II (1990).
3. "Mechanisms of High Temperature Superconductivity," H. Kamimura and A. Oshiyama (eds.), Springer Series in Materials Science, Vol. 11, Springer-Verlag Berlin Heidelberg, 1989.
4. IBM J. Res. Devel. **33,** No. 3, 197-404, (1989).
5. "Strong Correlation and Superconductivity," H. Fukuyama, S. Maekawa and A. P. Malozemoff (eds.), Springer Series in Solid State Sciences, Vol 89, Springer-Verlag Berlin Heidelberg, 1989.
6. "High Temperature Superconductivity," Proc. of Los Alamos Symposium-1989, K. S. Bedell, D. Coffey, D. E. Meltzer, D. Pines and J. R. Schrieffer, eds., Addison-Wesley Publishing Co., (1990).
7. "Electronic Properties of High-T_c Superconductors and Related Compounds," H. Kuzmany, M. Mehring and J. Fink (eds.), Springer Series in Solid-State Sciences, Vol. 99; Springer Verlag Berlin, Heidelberg 1990.
8. *Physics Today,* Special Issue: High-Temperature Superconductivity, **44** /6, June 1991.
9. B. Batlogg, Proc. of Los Alamos Symposium-1989, Ref. 6, p. 37-82; and Review Modern Physics (1991); *Physica B* **169,** 7 (1991) [Proc. LT 19]; *Physics Today,* **44,** 44 (1991).
10. K. Levin, Ju H. Kim, J. P. Lu and Qimiao Si, *Physica C* **175,** 450 (1991).

11. R. J. Cava, *Scientific American,* p. 42, August 1990.

12. A. W. Sleight, *Physics Today,* **66,** 24, June 1991.

13. J. D. Jorgensen, *Physics Today,* **66,** 34, June 1991.

14. R. J. Cava, A. W. Hewat, E. A. Hewat, B. Batlogg, M. Marezio, K. M. Rabe, J. J. Krajewski, W. F. Peck, Jr., L. W. Rupp, Jr., *Physica C* **165,** 419 (1990).

15. A. Furrer et al., *Eur. J. Solid State Inorg. Chem.* **28,** 627 (1991), and Proc. Kanazawa Conf. (paper 2A-09).

16. A. Fontaine, H. Tolentino, F. Baudelet, T. Gourieux, G. Krill, J. Y. Henry and J. Rossat-Mignod, Proc. Kanazawa Conf., (paper 5B-17).

17. B. W. Veal, A. P. Paulikas, H. You, H. Shi, Y. Fang, J. W. Downey, *Phys. Rev.* **B42,** 6305 (1990).

18. Y. Tokura, H. Takagi and S. Uchida, *Nature* **337,** 345 (1989).

19. M. Gurvitch and A. P. Fiory, *Phys. Rev. Lett.* **59,** 1337 (1988).

20. S. Martin, A. T. Fiory, R. M. Fleming, L. F. Schneemeyer and J. V. Waszczak, *Phys. Rev. Lett.* **60,** 2194 (1988).

21. S. W. Tozer, A. W. Kleinasser, T. Penney, D. Kaiser and F. Holtzberg, *Phys. Rev. Lett.* **59,** 1768 (1987).

22. S. J. Hagen, T. W. Jing, Z. Z. Wang, J. Horvath and N. P. Ong, *Phys. Rev.* **B37,** 7928 (1988).

23. T. Penney, S. von Molnar, D. Kaiser, F. Holtzberg, A. M. Kleinasser, *Phys. Rev.* **B38,** 2918 (1988).

24. D. A. Brawner, Z. Z. Wang and N. P. Ong, *Phys. Rev.* **B40,** 9329 (1989).

25. B. Bucher, J. Karpinski, E. Kaldis and P. Wachter, *Physica* **C167,** 324 (1990).

26. U. Welp, S. Fleshler, W. K. Kwok, J. Downey, Y. Fang, G. W. Crabtree, J. Z. Lin (preprint).

27. T. A. Friedmann, M. W. Rabin, J. Giapintzakis, J. P. Rice and D. M. Ginsberg, *Phys. Rev.* **B42,** 6217 (1990).

28. T. R. Chien, D. A. Brawner, Z. Z. Wang, N. P. Ong, *Phys. Rev.* **B43,** 6242 (1991).

29. H. Takagi, Y. Tokura and S. Uchida, in Ref. 3, p. 238.

30. Y. Iye, in Ref. 3, p. 263.

31. A. Zettl, A. Behrooz, G. Briceno, W. N. Craeger, M. F. Crommie, S. Hoen and P. Pinsukanjana in Ref. 3, p. 249.

32. For an updated review see N. P. Ong (this volume); and in Ref. 2, p. 459.

33. T. Timusk and D. B. Tanner, in Ref. 2, Vol. I, p. 339.

34. J. Orenstein, G. A. Thomas, A. J. Millis, S. L. Cooper, D. H. Rapkine, T. Timusk, L. F. Schneemeyer and J. V. Waszczak, (Phys. Rev. **B42,** 6342, 1990)).

35. M. P. Petrov, A. I. Grachev, M. V. Krasin'kova, A. A. Nechitailov, V. V. Poborchii, S. I. Shagin and S. V. Miridonov, Pis'ma Zh. Eksp. Teor. Fiz. **50,** 25 (1989) *[JETP Lett.* **50,** 29 (1989)].

36. B. Koch, H. P. Geserich, and TH. Wolf, *Solid State Commun.* **71,** 495 (1989).

37. Z. Schlesinger, R. T. Collins, F. Holtzberg, C. Feild, S. H. Blanton, U. Welp, G. W. Crabtree, Y. Fang and J. Z. Liu, *Phys. Rev. Lett.* **65,** 801 (1990).

38. J. Schützmann, W. Ose, J. Keller, K. F. Renk, B. Roas, L. Schultz, G. Saemann-Ischenko, *Europhys. Lett.* **8,** 679 (1989).

39. K. Kamaras, S. L. Herr, C. D. Porter, N. Tache, D. B. Tanner, S. Etamad, T. Venkatesan, E. Chase, A. Juan, X-D. Wu, M. S. Hedge, B. Dutta, *Phys. Rev. Lett.* **64,** 84 (1990).

40. B. Koch, M. Dürrler, H. P. Geserich, Th. Wolf, G. Roth and G. Zachmann, in Ref. 7, p. 290.

41. C. Thomsen and M. Cardona, in Ref. 2, Vol. I, p. 411.

42. S. L. Cooper and M. V. Klein, Comments on Solid State Physics (1990) and references therein.

43. F. Slakey, M. V. Klein, J. P. Rice and D. M. Ginsberg, *Phys. Rev.* **B43,** 3764 (1991).

44. C. M. Varma, P. B. Littlewood, S. Schmitt-Rink, E. Abrahams and A. E. Ruckenstein, *Phys. Rev. Lett.* **26,** 1996 (1989).

45. R. B. van Dover, R. J. Cava, B. Batlogg and E. A. Rietman, *Phys. Rev.* **B35,** 5337 (1987).

46. J. B. Torrance, A. Bezinge, A. I. Nazzal, T. Ch. Huang, S. S. P. Parkin, D. T. Keane, S. J. La Placa, P. M. Horn and G. A. Held, *Phys. Rev.* **B40** /13, (1989). See also M. Suzuki, *Phys. Rev.* **B39,** 2312 (1989).

47. Y. Kubo, Y. Shimakawa, T. Manakako, T. Satoh, S. Iijima, T. Ichiashi and H. Igarashi, *Physica C* **164-166,** 991 (1989).

48. S. Uchida (this volume).

49. R. J. Birgeneau and G. Shirane, in Ref. 2, Vol. I, p. 151.

50. S. Chakravarty, in Ref. 6, p. 136.

51. D. C. Johnston, *Phys. Rev. Lett.* **60,** 762 (1988).

52. B. Batlogg, R. J. Cava, S-W. Cheong, L. W. Rupp, Jr., J. J. Krajewski, W. F. Peck, Jr. and A. S. Cooper (to be published).

53. R. E. Walstedt and W. W. Warren, *Science,* **248,** 1082 (1990).

54. C. H. Pennington and C. P. Slichter, in Ref. 2, Vol. II, p. 269.

55. L. F. Schneemeyer, J. V. Waszczak, E. A. Rietman and R. J. Cava, *Phys. Rev.* **B35,** 8421 (1987).

56. M. Oda, T. Ohguro, N. Yamada and M. Ido, *J. Phys. Soc. Jpn.* **58,** 1137 (1989).

57. M. Takigawa in "Dynamics of Magnetic Fluctuations in High Temperature Superconductors," eds. G. Reiter, P. Horsch and G. Psaltakis; Plenum Publishing Co.

58. H. Alloul, *Physica* **B169,** (1991).

59. J. Rossat-Mignod, L. P. Regnault, C. Vettier, P. Burlet, J. Y. Henry and G. Lapertot, *Physica* **B169,** Proc. Kanazawa Conf.

60. S-W. Cheong, G. Aeppli, T. E. Mason, H. Mook, S. M. Hayden, P. C. Canfield, Z. Fisk, K. N. Clausen, J. L. Martinez, *Phys. Rev. Lett.* **67** (1991).

61. S. M. Hayden, G. Aeppli, H. Mook, D. Rytz, M. F. Hundley and Z. Fisk, *Phys. Rev. Lett.* **66,** 821 (1991).

62. B. Keimer, R. J. Birgeneau, A. Cassanho, Y. Endo, R. W. Erwin, M. A. Kastner and G. Shirane (preprint).

63. M. Klein, Proc. Kanazawa Conf.

64. R. Osborn and E. A. Goremychkin (preprint), and Proc. Kanazawa Conf.

65. H. M. Meyer III and J. H. Weaver in *Physical Properties HTSC, Vol. II, p. 369.*

66. J. Fink, N. Nücker, H. Romberg, M. Alexander, P. Adelmann, R. Claessen, G. Mante, T. Buslaps, S. Harm, R. Manzke and M. Skibowski, Proc. of a NATO Science Forum '90, "Highlights in the Eighties and Future Prospects in Condensed Matter Physics," Biarritz, France 1990.

67. J. W. Allen and C. G. Olson, *MRS Bull.* **XV** /6, 34 (1990).

68. H.Eskes, H. Tjeng and G. A. Sawatzky, in Ref. 3, p. 20.

69. A. Fujimori, in Ref. 3, p. 176; and this volume.

70. N. Nücker (this volume).

71. C. G. Olson, R. Liu, A-B. Yang, D. W. Lynch, A. J. Arko, R. S. List, B. W. Veal, Y. C. Chang, P. Z. Jiang and A. P. Paulikas, *Science* **245,** 731 (1989).

72. R. Manzke, T. Burlaps, R. Claessen, M. Skibowski and J. Fink, *Physica C,* **162-164,** 1381 (1989);

73. For a summary of band structure calculations see W. E. Pickett, *Rev. Mod. Phys.,* **61,** 433 (1989).

74. C. G. Olson, R. Liu, D. W. Lynch, R. S. List, A. J. Arko, B. W. Veal, Y. C. Chang, P. Z. Jiang and A. P. Paulikas, *Phys. Rev.* **B42,** 381 (1990).

75. H. Haghighi, J. H. Kaiser, S. Rayner, R. N. West, J. Z. Liu, R. Shelton, R. H. Howell, F. Solal and M. J. Fluss, *Phys. Rev. Lett.* **67,** 382 (1991).

76. L. P. Chan, D. R. Harshman, K. G. Lynn, S. Massidda, D. B. Mitzi, *Phys. Rev. Lett.* **67,** 1350 (1991).

77. P. E. Mijnarends et al. *Physica C* **176,** 113 (1991).

78. A. Junod, in Ref. 2, p. 13.

79. N. E. Phillips, R. A. Fisher and J. E. Gordon, in "Progress in Low-Temperature Physics 13," eds. D. Brewer, Elsevier, The Netherlands (1990).

80. J. W. Loram and K. A. Mirza; in Ref. 7, p. 92; and *Physica C* **153-155,** 1020 (1988).

81. C. E. Gough, M. S. Colclough, E. M. Forgan, R. G. Jordan, M. N. Keene, C. M. Muirhead, A. I. M. Rae, N. Thomas, J. S. Abell and S. Sutton, *Nature* **326,** 855 (1987).

82. P. L. Gammel, D. J. Bishop, G. J. Dolan, J. R. Kwo, C. A. Murray, L. F. Schneemeyer and J. V. Waszczak, *Phys. Rev. Lett.* **59,** 2592 (1987).

83. P. L. Gammel, P. A. Polakos, C. E. Rice, L. R. Harriot, D. J. Bishop, *Phys. Rev.* B, 1990.

84. P. J. M. van Bentum, H. F. C. Hoevers, H. van Kempen, L. E. C. van de Leemput, M. J. M. F. de Nivelle, L. W. M. Schreurs, R. J. M. Smokers and P. A. A. Teunissen *Physica C* **153-155,** 1718 (1988); also *Phys. Rev.* **B36,** 843 (1987).

85. K. E. Gray, *Mod. Phys. Lett.* **B2,** 1125 (1988).

86. For a review see J. Geerk, G. Linker, O. Meyer, Q. Li, R-L. Wang and X. X. Xi. *Physica C* **162-164,** 837 (1989).

87. J. M. Valles and R. C. Dynes, *MRS Bulletin* XV, 44 (June 1990).

88. J. S. Tsai, I. Takeuchi, J. Fujitu, T. Yoshitake, S. Miura, S. Tanaka, T. Terashima, Y. Bando, K. Iijima and K. Yamamoto, *Physica C* **153-155,** 1385 (1988), and in Ref. 3, p. 229.

89. D. van der Marel, H. U. Habermeier, D. Heitmann, W. König and A. Wittlin, *Physica C,* **176,** 1 (1991); *Phys. Rev.* **B43,** 8606 (1991).

90. B. Friedl, C.Thomsen and M. Cardona, *Phys. Rev. Lett.* **65,** 915 (1990).

91. C. Thomsen, M. Cardona, B. Friedl, C. O. Rodriguez, I. I. Mazin and O. K. Andersen, *Solid State Commun.* **75,** 219 (1990).

92. A. Yamanaka, T. Kimura, F. Minami, K. Inove and S. Takegawa, *Jpn. J. Appl. Phys.* **27,** L359 (1988), and *Physica C* **162-164,** 1099 (1989).

93. M. V. Klein, *Physica C* **162-164,** 1701 (1989).

94. R. Hackl, W. Gläser, P. Müller, D. Einzel and K. Andres, *Phys. Rev.* **B38,** 7133 (1988). and *Physica C* **162-164,** 1241 (1989).

95. J. M. Imer, F. Patthey, B. Dardel, W-D. Schneider, Y. Baer, Y. Petroff and A. Zettl, *Phys. Rev. Lett.* **62,** 336 (1989).

96. Y. Petroff, in Ref. 7, p. 114.

97. M. Gurvitch and A. T. Fiory, *Phys. Rev. Lett.* **59,** 1337 (1987).

98. For a comparison of various compounds see Y. Kitaoka, K. Ishida, K. Fujiwara, K. Asayama, H. Katayama-Yoshida, Y. Okabe and T. Takahashi in Ref. 4, p. 277.

99. A. G. Aronov, S. Hikami and A. I. Larkin, *Phys. Rev. Lett.* **62,** 965 (1989).

100. K. Semba, T. Ishii and A. Matsuda, *Phys. Rev. Lett.* **67,** 769 (1991).

101. M. C. Nuss, P. M. Mankiewich, M. L. O'Malley, E. H. Westerwick, P. Littlewood, *Phys. Rev. Lett.* **66,** 3305 (1991).

102. K. Holczer, L. Forro, L. Mihaly and G. Grüner, *Phys. Rev. Lett.* **67,** 152 (1991).

103. S. Sridhar, D-H. Wu and W. Kennedy, *Phys. Rev. Lett.* **63,** 1873 (1989).

104. D. R. Harshman, G. Aeppli, E. J. Ansaldo, B. Batlogg, J. H. Brewer, J. F. Carolan, R. J. Cava, M. Celio, A. C. D. Chaklader, W. N. Hardy, S. R. Kreitzman, G. M. Luke, D. R. Noakes, M. Senba, *Phys. Rev.* **B36,** 2386 (1987).

105. R. F. Kiefl, T. M. Riseman, G. Aeppli, E. J. Ansaldo, J. F. Carolan, R. J. Cava, W. N. Hardy, D. R. Harshman, N. Kaplan, J. R. Kempten, S. R. Kreitzman, G. M. Luke, B. X. Yang, D. L. Williams, *Physica C* **153-155,** 757 (1988).

106. S. Anlage, B. W. Langley, G. Deutscher, R. W. Simon, J. M. Murduck, C-B. Eom, M. R. Beasley, T. H. Geballe, in Proc. 19th Int. Conf. on Low Temperature Phys. (LT 19), Vol. III, *Physica B,* (1991), and preprint.

107. A. Hebard, private communication.

108. M. Takigawa, P. C. Hammel, R. H. Heffner and Z. Fisk, *Phys. Rev.* **B39,** 7371 (1989).

109. S. E. Barrett, D. J. Durand, C. H. Pennington, C. P. Slichter, T. A. Friedmann, J. P. Rice and D. M. Ginsberg, *Phys. Rev.* (1990).

110. D. Pines, *Physica* **B163,** 78 (1990).

111. M. Takigawa, J. L. Smith, W. L. Hults, *Phys. Rev.* B (1991).

112. C. P. Slichter et al., Proc. Kanazawa Conf.

113. H. Zimmermann, M. Mali, M. Bankay and D. Brinkmann, Proc. Kanazawa Conf.

114. B. Batlogg, R. J. Cava, A. Jayaraman, R. B. van Dover, D. W. Murphy, L. W. Rupp, Jr., H. S. Chen, A. E. White, K. T. Short, A. M. Mujsce, E. A. Rietman and G. A. Kourouklis, *Phys. Rev. Lett.* **58,** 2333 (1987).

115. L. C. Bourne, M. F. Crommie, A. Zettl, H-C. zur Loye, S. W. Keller, K. L. Leary, A. M. Stacy, K. J. Chang, M. L. Cohen, D. E. Morris, *Phys. Rev. Lett.* **58,** 2337 (1987).

116. R. B. Schwarz, P. J. Yvon, D. Coffey; in Vol. 8 of "Studies of High Temperature Superconductors," A. Narlikar (ed.), Nova Science Publ., NY (1991).

117. B. Batlogg, S-W. Cheong, H. Takagi, G. A.Thomas, S. L. Cooper, L. W. Rupp, Jr., D. H. Rapkine and A. S. Cooper, Proc. Kanazawa Conf. (1991).

Successive Structural Transitions of $La_{2-x}Ba_xCuO_4$

M. Sato

Department of Physics, Faculty of Science, Nagoya University,
Furo-cho, Chikusa-ku, Nagoya 464-01, Japan

Abstract. $La_{2-x}Ba_xCuO_4$ with x~1/8, undergoes successive structural transitions with decreasing temperature as, tetragonal (I4/mmm) → orthorhombic (Cmca) → tetragonal ($P4_2/ncm$). Although both transitions are due to the condensation of the CuO_6 rotations at (π/a, π/a, 0) in the reciprocal space (a:lattice constant), only the second one has large effects on various physical properties, including the almost complete suppression of the superconducting transition temperature T_c. A number of the experimental results accumulated by many groups indicate that the conduction electrons are involved in the driving mechanism of the second transition, which seems to present clues how to describe the strongly correlated electron system in high-T_c oxides.

1. Introduction

As was first reported by Moodenbaugh et al.[1], superconductivity of $La_{2-x}Ba_xCuO_4$ is almost completely suppressed in the x-region around 1/8, where in addition to the well known structural transition at $T=T_0$ from the high temperature tetragonal (HTT) to low temperature orthorhombic (LTO) phase, another transition from LTO to low temperature tetragonal (LTT) phase has been revealed to occur at T_1 by Axe et al.[2] At T_0, one of the two degenerate modes of the CuO_6 rotations about the $[110]_t$ and $[1\bar{1}0]_t$ axes condenses, where the suffix t indicates that the indices refer to the HTT unit cell. At T_1, the condensation of the two rotational modes with equal amplitudes takes place and the rotational axis changes abruptly to $[100]_t$ with no significant change in the rotational angle of the octahedra (see Fig. 1).

The interesting point of these results is that although both transitions are due to the condensations of the similar rotational modes at $(\pi/a, \pi/a, 0)_t$ in the reciprocal space (a: the lattice parameter), only the second one has serious effects on the superconductivity. It is one of the very rare examples which directly correlate with the occurrence of the superconductivity and the exploration of the origin(s) of these transitions

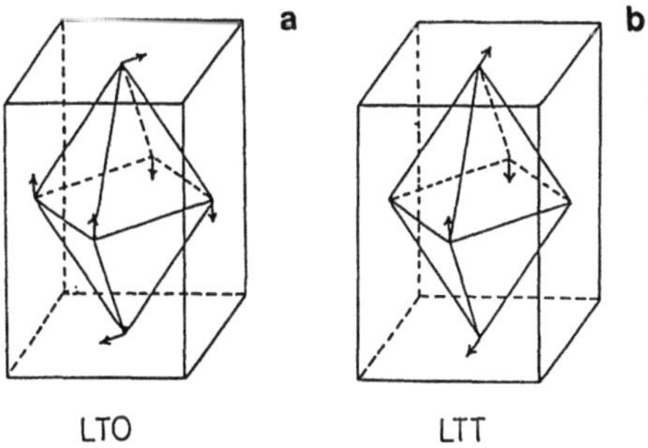

Fig. 1 (a) Rotation of the CuO_6 octahedra which condenses at the HTT→LTO transition. (b) Rotation of the CuO_6 octahedra in the LTT phase. Two inequivalent oxygen sites appear in the CuO_2 plane.

may present a firm basis to construct a realistic model to describe the highly correlated electron system of high-T_c oxides.

In this article, various experimental results which have been accumulated by many groups are overviewed and essential problems which remain to be explained are pointed out.

2. Experimental Results

Besides the almost complete depression of T_c of $La_{2-x}Ba_xCuO_4$ around x=1/8, Moodenbaugh et al. reported the resistivity anomaly at $T_1 \sim 60K$,[1] which was later shown by Axe et al.[2] to be due to the LTO→LTT transition. They carried out the Rietveld refinement of their neutron diffraction data and pointed out that the LTT phase has a structure with the space group $P4_2/ncm$, which can be derived, as mentioned above, by considering the simultaneous condensation of the two degenerate rotational modes. After these works, significant anomalies on various physical properties have been revealed by the detailed studies carried out by Sera et al.[3] Especially in the transport properties such as resistivity ρ, thermoelectric power S and Hall coefficient R_H and the magnetic susceptibility χ, they observed rather drastic anomalies in all these quantities at the transition temperature T_1. As an example of their observation of the anomalies, the thermoelectric power S are plotted as a function of T in Fig. 2 for various values of x, where steep drop of S can be observed for the samples with x~1/8 at $T=T_1 \sim 60K$. The phase boundary determined from the anomalies is shown in Fig. 3 by open circles with the T_c vs. x curve reported in ref. 1. In the figure, the data points determined by Kumagai et al.[4] from the anomalies of the thermal dilation coefficient β are also shown by closed circles.

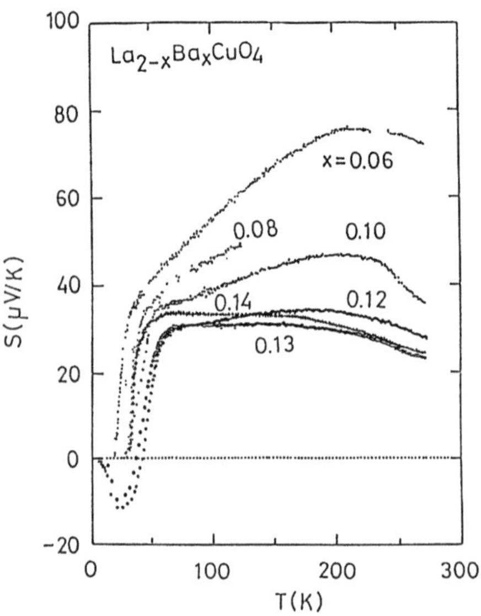

Fig. 2 Temperature dependence of the thermoelectric power S of $La_{2-x}Ba_xCuO_4$ with various x values.

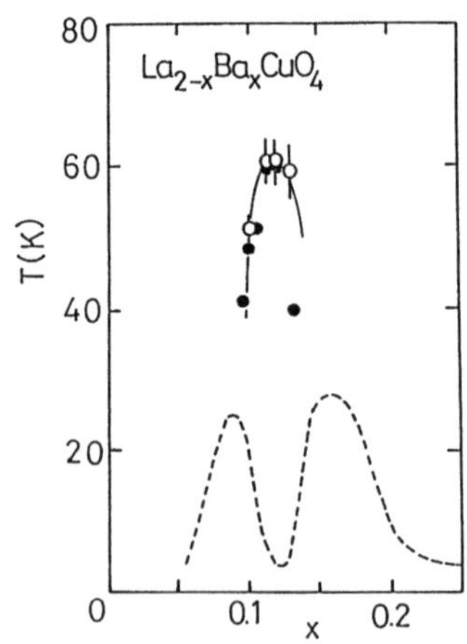

Fig. 3 Phase boundary between the LTO-LTT structures. Open (closed) circles are obtained from the results in ref. 3 (4). The dotted line shows the superconducting transition temperature reported in ref. 1.

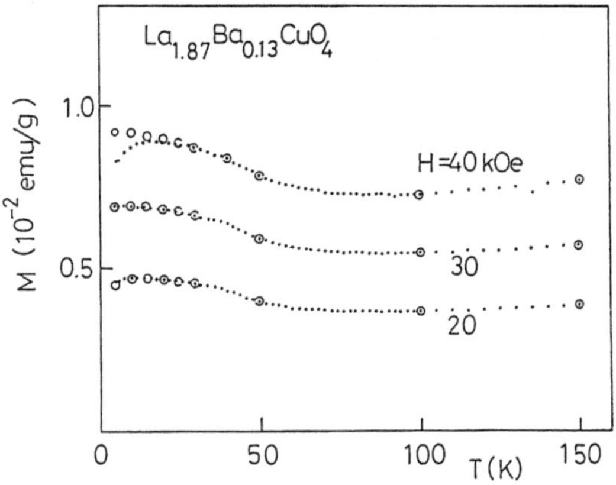

Fig. 4 Magnetic moment of $La_{1.87}Ba_{0.13}CuO_4$ is plotted as a function of T for various applied field. See text for details.

Studies by using the x-ray diffraction method for various x values have also been carried out by Axe et al.[5] and Suzuki et al.[6] Their phase boundaries are slightly broader than those of refs. 3 and 4. The discrepancy may be attributed to the difference of the specimens. For example, the anomalies observed in the ρ-T curves are very sharp, while the x-ray measurements observed rather broad transition widths.

The existence of the significant anomalies in various physical quantities suggests the large change of the electronic state of $La_{2-x}Ba_xCuO_4$ at the LTO→LTT transition, possibly due to the coupling of the Fermi surface with the lattice distortion. Therefore, it has a clear contrast with the case of La_2CoO_4, which is insulating or does not have a Fermi surface. As the experimental evidences for the reduction of the conducting carrier (hole) number p, there can be listed several data. First of all, the resistivity increases at the LTO→LTT transition (It is still difficult to see if $1/\rho$ remains finite or not as $T\to 0$, because of the filamentary superconductivity which remains within samples). The low temperature specific heat coefficient γ has been found to have a clear minimum as a function of x at $x\sim 1/8$.[7] The increase in the thermal conductivity κ[8,9] at T_1 with decreasing T can also be considered as another evidence for the reduction of p: Because in high-T_c oxides, the phonon system has a dominant contribution to κ, the decrease in the number of holes which act as the scatterers of the phonons, leads to the increase in κ. Details are in ref. 8. Here, it may be better to notice that the Hall coefficient R_H decreases as T decreases below T_1.[3] However, it cannot simply be considered as an evidence for the p-increase, because there are many ambiguous factors to correctly know the behavior of R_H.

The decrease in the carrier number p induces the change of the magnetic properties. As was first reported by Sera et al.[3], the magnetic susceptibility χ increases as a result of the LTO→LTT transition. Recently, the measurement was extended to the lower temperatures,[10] the result of which is shown in Fig. 4, where dots (open circles) are the results measured with increasing (decreasing T). First, a sintered sample of $La_{1.87}Ba_{0.13}CuO_4$ was cooled in zero field and H=40kOe was applied. Then, the magnetization M was measured with varying T up and down and H was changed to 30kOe at 5K. Again, M was measured in the same way as before and H was changed to 20kOe. Very small hysterisis effect certifies that the Meissner diamagnetism from the filamentary superconductivity which often remains in samples, does not disturb the result.

At low temperatures (≲30K), the static magnetic field was detected in the LTT phase by μSR.[11] However, the neutron powder diffraction experiment carried out at JRR-3 in JAERI has not detected any evidence for the 3-dimensional spin ordering.[10] If any ordering exists, the ordered moment is

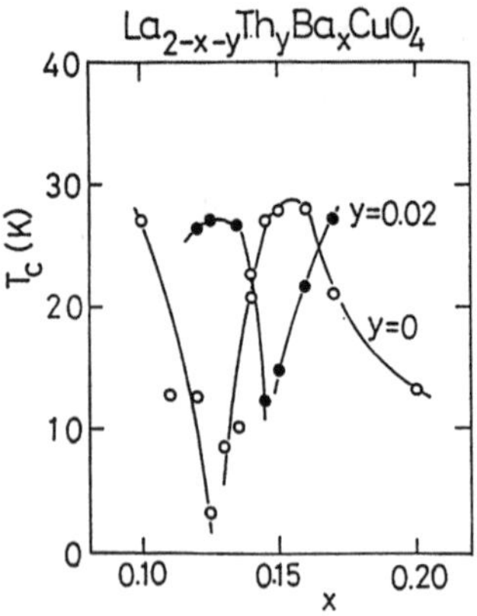

Fig. 5 T_c vs. x curve of $La_{2-x-y}Th_yB_xCuO_4$. The result indicates that the p-hole value which is determined as (x-y) is an important parameter for the occurrence of the superconductivity or the LTO→LTT transition.

significantly smaller than $1\mu_B$/Cu. Then, it is plausible that as a result of the reduction of the carrier number, there develops only a short range order within the CuO_2 plane, which is difficult to detect by neutron diffraction experiment. Due to the effect of the Dzialoshinsky-Moriya type interaction, χ increases with the development of the short range spin ordering, as discussed by Thio et al.[12] for the case of La_2CuO_4. The M-H curves measured at temperatures $2K \lesssim T < T_1$, do not exhibit, up to 42kOe, a metamagnetic behavior observed in the antiferromagnetic La_2CuO_4.[12, 13] This is also consistent with the lack of the long range order, although it should be noted that the possibility of the spin structure which does not exhibit the metamagnetic behavior has also been pointed out.[14]

At low temperatures, the system seems to become mictomagnetic, where the static field is expected and as really observed by μSR.[11]

So far, various experimental results which indicate that the coupling of the conduction electrons with the lattice distortion is relevant to the realization of the LTO→LTT transition, has been shown. The coupling, however, may not be very strong, because in the LTO phase one of the rotational modes of CuO_6 octahedra remains soft and therefore, it is relatively easy for the system to transform to the LTT phase, where both rotational modes are in the condensed state.

Here, it may be better to mention the experimental results which have ruled out the atomic ordering mechanism of the transition. The possibility of the atomic ordering was considered because the transition is confined to rather narrow region of x around the simple rational number 1/8 (see Fig. 3). Moreover, in $La_{2-x}Sr_xCuO_4$, in which the ionic radius of the doped atoms is closer to that of the host atom La than in $La_{2-x}Ba_xCuO_4$, the LTO→LTT transition has not definitively been observed (although the T_c vs. x curve of $La_{2-x}Sr_xCuO_4$ has a slight dip-like structure at $x\sim1/8$[16], and the ultrasonic study carried out by Fukase's group[17] observed the small anomaly below 10K). On the contrary, the existence of the similar successive transition in undoped La_2CoO_4 is one of the counter examples against the atomic ordering model. Moreover, a doping effect study carried out by Shamoto et al.[18] on $La_{2-x-y}Ba_xCu_{1-y}Li_yO_4$ has revealed that the LTO→LTT transition always takes place at the hole concentration p which corresponds to $x\sim1/8$ of $La_{2-x}Ba_xCuO_4$, irrespective of the values of the Ba concentration. Similar experiments have also been carried out by many other groups.[19] Among these results, the beautiful data taken by Maeno et al. for $La_{2-x-y}Th_yBa_xCuO_4$ with Th^{4+} are shown in Fig. 5, where it is clearly indicated that the occurrence of the LTO→LTT transition is not determined by the Ba-concentration but by the carrier concentration p.

Another thing which should be noted here is that the occurrence of the LTO→LTT transition, depends on the ratio of the ionic radii of doped and host atoms, possibly through the change in the elastic energy of the corresponding distortion. Pressure effects were studied by Oda et al.[20] and Murayama et al.[21] and the superconducting transition temperature of $La_{2-x}Ba_xCuO_4$ was found to depend on the applied pressure very strongly at x~1/8, which may indicate the sensitiveness of the LTO→LTT transition to the elastic energy.

3. Discussion

Now, the most interesting subject to be studied is why the LTO→LTT transition takes place at the fixed carrier concentration p~1/8Cu or why the Fermi surface couple with the lattice distortion only in the narrow region of x≃1/8. Even if the coupling of the conduction electrons with the lattice distortion is not very strong as mentioned above, the sharp x-dependence is the difinite experimental evidence which presents a characteristic of the electronic structure of the system. Only a small number of theoretical works, to the author's knowledge, exist on this problem. Barisic and Zelenko[22] considered a crystal field induced electron-phonon interaction, in which the charge fluctuations between the neighboring oxygen sites have an essential role in realizing the transition, which introduces the two inequivalent oxygen sites in the CuO_2 plane (see Fig. 1). On the other hand, based on their local density functional band calculations, Pickett et al.[23] proposed a band Jahn-Teller like mechanism of the transition, where the electronic density of states at the Fermi level decreases down to about 1/2 as a result of the distortion. These theories use the existence of the van Hove singularity in the density of states, at which E_F locates when x~1/8. However, in the x-dependence of the magnetic susceptibility, there does not exist any indication of the van Hove singularity even at temperatures just above T_1, which makes the special meaning of the x value of 1/8 rather obscure.

The strong x-dependence of the oxygen isotope effect parameter α ($\equiv \delta \ln T_c/\delta \ln M$, M: oxygen isotope mass) has been reported for $La_{2-x}Ba_xCuO_4$ and $La_{2-x}Sr_xCuO_4$ by Crawford et al.[24] (Fig. 6). The data seem to have a possible correlation with the LTO→LTT transition. To understand the whole behavior of α, various complicated effects of the transition on T_c or on α should carefully be taken into account. For example, the existence of the soft mode in the LTO phase seems to be very important for the possible isotope mass dependence of the anharmonic effect.

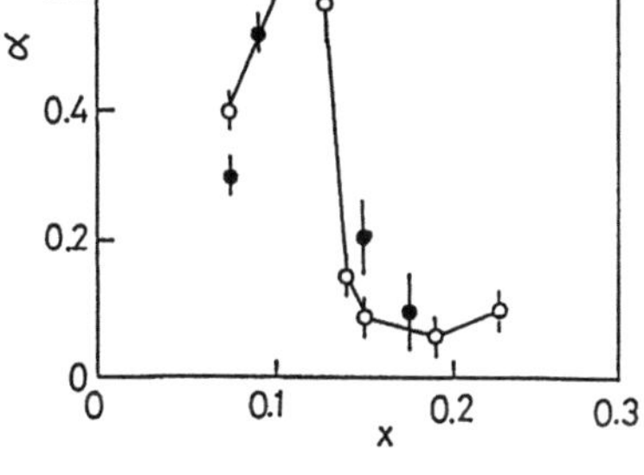

Fig. 6 Isotope effect parameter $\alpha \equiv \delta \ln T_c/\delta \ln M$ is plotted as a function of x for $La_{2-x}M_xCuO_4$ (M=Ba and Sr).

As is well known, the electrons in high-T_c oxides have very large intraalomic Coulomb repulsion energy U. Then, the simple band picture has encountered a quite hard difficulty in describing the behaviors of the electron system and an alternative approach which treats the non Fermi liquid like electron system has been explored by a number of researchers. The present problem on the successive transitions gives a quite clear opportunity for the theorists who intend to make their electronic model to be more realistic, to check if they can predict the anomalous behaviors mentioned in this chapter.

4. Concluding Remarks

In the present article, various experimental results on the LTO→LTT transition are reviewed. They indicate that in case of $La_{2-x}Ba_xCuO_4$, the coupling of the conduction electrons with the lattice distortion is relevant to the transition. It is still unclear why the transition is confined to the very narrow region of $x \sim 1/8$. The explanation of this problem may present a definite clue for a construction of a model to describe the electron systems of high-T_c oxides. Experimentally, the relationship between the occurrence of the second transition and the ratio of the ionic radius of the doped atoms which are usually divalent and that of the host atoms is interesting for the future studies.

Acknowledgement. The author thanks a number of researchers who sent him their preprints and reprints on the successive transitions of $La_{2-x}Ba_xCuO_4$

References

[1] A. R. Moodenbaugh, Y. Xu, M. Suenaga, T. J. Folkerts and R. N. Shelton, Phys. Rev. B38 4596 (1988)
[2] J. D. Axe, D. E. Cox, K. Mohanty, H. Moudden, A. R. Moodenbaugh, Y. Xu and T. Thurston, IBM J. Res. Dev. 33 382 (1989)
[3] M. Sera, Y. Ando, S. Kondoh, K. Fukuda, M. Sato, I. Watanabe, S. Makashima and K. Kumagai, Solid State Commun. 69 851 (1989)
[4] K. Kumagai, H. Matoba, N. Wada, M. Okaji and K. Nara, J. Phys, Soc. Jpn 60 1448 (1991)
[5] J. D. Axe, A. H. Moudden, D. Hohlwein, D. E. Cox, K. M. Mohanty, A. R. Moodenbaugh and Youwen Xu, Phys. Rev. Letters 62 2751 (1991)
[6] T. Suzuki, T. Fujita, J. Phys. Soc. Jpn 58 1883 (1989)
[7] Y. Okajima, K. Yamaya, N. Yanada, M. Oda and M. Ido, Solid State Commun. 74 767 (1990)
[8] M. Sera, S. Shamoto and M. Sato, Solid State Commun. 74 951 (1990)
[9] T. Uchida et al., private communication
[10] M. Sato, M. Sera, S. Shamoto, S. Yamagata and K. Kakurai, Proc. Int. Conf. M^2S Kanazawa submitted
[11] S. Uchida et al., and K. Kumagai et al. private communication
[12] T. Thio, T. R. Thurston, N. W. Preyer, P. J. Picone, M. A. Kastner, H. P. Jenssen, D. R. Gabbe, C. Y. Chen, R. J. Birgeneau and A. Aharony, Phys. Rev. B38 905 (1988)
[13] M. Sato, S. Hosoya, K. Fukuda, M. Sera, M. Onoda, S. Shamoto, K. Oka and H. Unoki, Physica 148B 363 (1987)
[14] Y. Koshibae, Y. Ohta and S. Maekawa, Proc. Int. Conf. M^2S Kanazawa
[15] T. R. Thurston, R. J. Birgeneau, N. W. Preyer, J. D. Axe, P. Boni, G. Shirane, M. Sato, K. Fukuda and S. Shamoto, Phys. Rev. B39 4327 (1989)
[16] H. Takagi, Y. Tokura and S. Uchida, Proc. 2nd NEC Symp., Hakone, October 1988 edited by H. kamimura and A. Oshiyama p238
[17] T. Fukase, Y. Koike, T. Goto, T. Nomoto, Y. Iwabuchi and T. Hanaguri, Research Report on Mechanism of Superconductivity p108 February 1990
[18] S. Shamoto, M. Sera and M. Sato, Solid State Commun. 76 923 (1990)

[19] For example , Y. Maeno et al. in preparation and Y. Koike T. Kawaguchi, N. Watanabe, T. Noji and Y. Saito, Solid State Commun. to be published
[20] M. Oda and M. Ido, private communications.
[21] C. Murayama, T. Tamegai, Y. Iye, N. Mōri, I. Oguro, S. Yomo, H. Takagi, S. Uchida and Y. Tokura, Physica B169 639 (1991)
[22] S. Barisic and J. Zelenko, Solid State commun. 74 367 (1990)
[23] W. E. Pickett, R. E. Cohen and H. Krakauer, Proc. Int. Conf. on Manifestations of El-ph Interaction in Cu-O Superconductors, edited by R. Baguero and J. Carbotte Oaxtapec, Mexico, December 1990 World Scientific 1991
[24] M. K. Crawford, W. E. Farneth, E. M. McCarron, R. L. Harlow and A. H. Moudden, Science 250 1390 (1990)

The Normal-State Transport Properties of the High-Temperature Cuprate Superconductors

N.P. Ong

Joseph Henry Laboratories of Physics, Princeton University,
Princeton, NJ 08544, USA

Abstract. A discussion of the normal-state Hall effect and magnetoresistance is given. Recent results on the temperature dependence of the Hall current in YBCO implies that the Hall damping rate $1/\tau_H$ of charge carriers varies as T^2, in contrast with the linear-T dependence in the transport scattering rate. The effect of impurities on $1/\tau_H$ is discussed. Magnetoresistance provides a measurement of the carrier dephasing rate $1/\tau_\phi$. In Bi 2201, $1/\tau_\phi$ is found to vary as $T^{1/3}$, implying greatly enhanced level widths compared with k_BT. New observations of spin-dependent magnetoresistance in longitudinal field are discussed.

1. Introduction

Transport measurements have an important role to play in our efforts to understand the normal-state properties of the high T_c cuprates, because they couple directly to the charge carriers. The linear temperature (T) dependence of the in-plane resistivity ρ_a is anomalous when compared with conventional metals, as emphasized by Anderson.[1] The out-of-plane resistivity ρ_c is non-metallic in most of the cuprates. Other transport coefficients are equally puzzling. Recent progress in understanding the anomalous temperature dependence [2] of the Hall coefficient R_H in YBCO ($YBa_2Cu_3O_7$) will be reported. Magnetoresistance studies to probe the quantum corrections to the normal state conductivity have not been widely performed, because superconductiving fluctuations intervene at high temperatures. Some recent results[3] on non-superconducting $Bi_2Sr_2CuO_6$ (Bi 2201) will be presented. In addition to the usual orbital effects arising from coherent back-scattering, there are novel spin dependent effects observed in a longitudinal magnetic field.

2. Normal-state Hall current

For perspective, let us look at the T dependence of R_H in the transition metals. In Cu and Ag, R_H increases only when T decreases below ~100 K (R_H is T independent above 100 K).[4] It peaks near 30 K, at a value that depends on sample purity as measured by the RRR and then rapidly decreases. In Ca, Mg and W, the same pattern is observed.[5] R_H is almost T independent until T decreases below a value $\zeta\Theta_D$ where ζ is a fraction between 0.2 and 0.4, and Θ_D is the Debye temperature. Ziman[6] explains the T variation of R_H in terms of a temperature-dependent *anisotropy* of the transport relaxation time $\tau_{tr}(\mathbf{k})$. R_H is strongly inluenced by the anisotropy of the mean-free-path $\mathbf{l}_\mathbf{k} = \mathbf{v}_\mathbf{k}\tau_{tr}$ over the Fermi Surface (FS) where $\mathbf{v}_\mathbf{k}$ is the velocity in state $\mathbf{k}$. At low temperatures, the anisotropy of τ_{tr} changes significantly with T as phonon scattering increasingly weights regions of the FS with small calipers, so that R_H becomes strongly dependent. At high temperatures, however, the phonon distribution is classical and τ_{tr} is isotropic over the FS. The anisotropy in $\mathbf{l}_\mathbf{k}$ is determined solely by that of $\mathbf{v}_\mathbf{k}$, so that R_H is independent of T. From experiment, the crossover occurs at $\zeta\Theta_D$.

In the hole-type cuprates, R_H varies (roughly like $1/T$ + const.) up to room temperature (See Ref. 2 for a review). In YBCO this continues above Θ_D (~430 K). This anomalous temperature dependence has now been observed in almost all the high T_c cuprates, and should be regarded as an important property of the holes in the CuO_2 plane. In $La_{2-x}Sr_xCuO_4$ (214), the combined data of several groups [7,8,9] clearly show a strongly temperature-dependent R_H when x falls between 0.14 and 0.18. When x is less than 0.14, the temperature dependence goes away. In $YBa_2Cu_3O_7$ (YBCO), a strongly temperature dependent R_H is well-established in the 90 K phase[10]. When doped with Pr[11], Ni or Co[8,12], the slope of n_H vs. T is observed to be rapidly suppressed ($n_H = 1/R_He$). Clayhold *et al* showed that R_H is strongly T dependent in ceramic samples of $Bi_2Sr_2Ca_2Cu_3O_{10}$ (Bi 2223) and $Tl_2Ba_2Ca_2Cu_3O_{10}$ (Tl 2223)[13]. In $Tl_2Ba_2CaCu_2O_8$ (Tl 2212) and $Tl_2Ba_2CuO_6$ (Tl 2201), Shimakawa *et al* [14] recently reported that T_c is a strong function of the oxygen content. They also demonstrated that as T_c is varied in this way, the Hall slope changes as well, in a way similar to that in doped YBCO. Recent work on $Bi_2Sr_2CaCu_2O_8$ (Bi 2212) shows that the R_H vs. T profile approaches the behavior in YBCO as the crystals are improved.[15] The single-plane Bi 2201 continues to show an almost T-independent R_H down to 1 K.[15] However, in this compound, T_c is ~6 K or less.

Recently, T.R. Chien *et al* [16] reported measurements of the Hall conductivity σ_H in single-crystal YBCO over a wide temperature range. They find that, whereas ρ_{xy} itself is not strictly linear in $1/T$, σ_H varies like $1/T^3$ rather accurately between 100 and 370 K (**Fig. 1**). This surprisingly simple power-law behavior implies that the anomalous behavior in YBCO originates from the Hall current, rather than the Hall resistivity $\rho_{xy} = \sigma_H/\sigma_a^2$, which involves both σ_H and the in-plane conductivity σ_a.

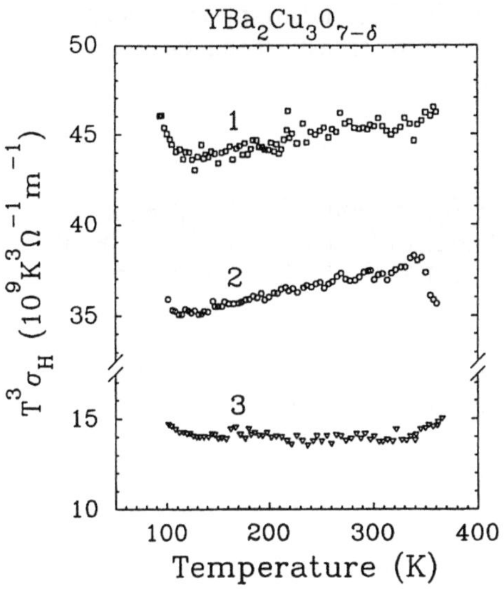

Figure 1 Plot of the product $T^3\sigma_H$ versus T in three YBCO single crystals, all with $T_c = 91$ K. The field (8T) is applied parallel to **c**. The results indicate that σ_H is very nearly proportional to $1/T^3$, instead of $1/T^2$. [Chien, Brawner, Wang and Ong[16]]

3. Hall angle relaxation

I will sketch a way to analyze the temperature dependence of σ_H. In the usual Boltzmann equation approach, the Hall current may be calculated as a two-step process (I take the electric field $\mathbf{E}$ //$\mathbf{x}$ and $\mathbf{B}$//$\mathbf{z}$). First, the FS is displaced along $\mathbf{E}$, so that the electron distribution is given, as usual, by $\delta f_E = e(-\partial f/\partial \varepsilon_k)\mathbf{E}.\mathbf{v}_k \tau_{tr}$. Then each state $\mathbf{k}$ is accelerated by the field $\mathbf{B}$ through the Hall angle θ_H so that the distribution is further altered by $\delta f_B = e\tau_H \mathbf{v}_k \times \mathbf{B}.\nabla \delta f_E$. The Hall current J_H, equal to the sum of $e\delta f_B v_y$ over the FS, involves the two time scales τ_{tr} and τ_H describing the relaxation of the longitudinal current and the Hall acceleration, respectively. In the usual quasi-particle picture of metals, the two times are the same, at least for temperatures above $\zeta\Theta_D$. However, the data in YBCO strongly suggest that they may, in fact, be very different. To bring this out, we compute the ratio $\sigma_a/\sigma_H = \cot\theta_H$ which involves only the parameter τ_H. Thus, the reciprocal of the Hall angle should reflect only the scattering rate of the Hall response (Anderson[17]). Cot θ_H, plotted against T^2, falls on a straight line passing close to the origin, viz.

$$\cot\theta_H = \alpha T^2 + C \qquad (1)$$

where $\alpha = 5.1 \times 10^{-3}$ K^{-2} (C is shown later to be proportional to disorder). Equation 1 suggests that the Hall scattering rate varies as T^2 as opposed to the linear T dependence of the transport scattering rate $\hbar/\tau_{tr} = \eta k_B T$, where $\eta \approx 2$. If we write

$$\hbar/\tau_H = k_B T^2/\Theta_H, \tag{2}$$

the value of α fixes the energy scale Θ_H at about $19(m^*/m_0)$ K, a surprsingly low energy scale. (In Anderson's theory Θ_H should be of the order of the exchange energy $J \sim 1{,}400$ K.) Unlike the transport scattering rate which varies linearly with T, the Hall angle relaxation rate varies as the square of T. This difference results in the $1/T^3$ dependence observed for σ_H and the approximately $1/T$ variation in ρ_{xy}. The two scattering rates become equal at the temperature $\eta\Theta_H \approx 300$ K if $m^*/m_0 \approx 8$. Thus, it appears that above 300 K, the Hall angle relaxation rate becomes increasingly larger than $1/\tau_{tr}$, and the observed R_H is steadily *suppressed* below its intrinsic value. This thermal Hall suppression extends beyond Θ_D.

A further test for the unusual temperature dependence in Eq. 1 derives from the effect of impurtiy scattering on the Hall angle. It is known that both Zn and Ni substitute for Cu in the planes of YBCO, and strongly scatter the holes. Recently, Chien *et al* [18] have measured how ρ_{xx} and ρ_{xy} vary with Zn and Ni content in single crystals of $YBa_2Cu_{3-x}M_xO_7$ (M = Ni or Zn) in the range $0 < x < 0.1$. I briefly summarize the results. With increasing x, the derivative dn_H/dT is

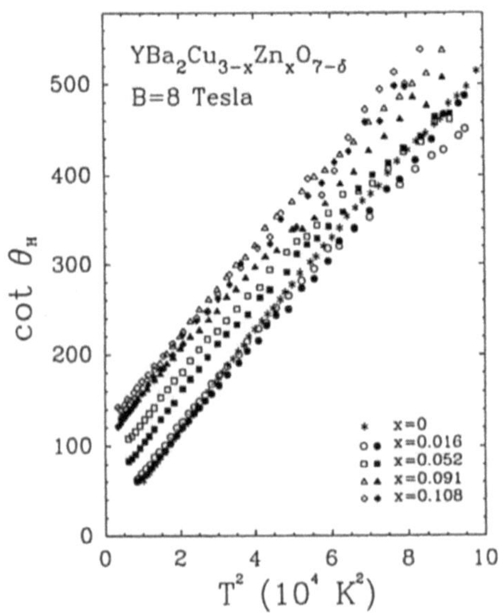

Figure 2 Variation of cot θ_H with temperature for single crystal YBCO doped with Zn. Cot θ_H remains linear in T^2, but the intercept C increases linearly with x. [Chien, Wang and Ong.18]

gradually suppressed in agreement with earlier work on ceramic samples. However, if we plot cot θ_H vs. T^2, an interesting pattern emerges (**Fig. 2**). Data from all samples are consistent with Eq. 1. Chien *et al* find that the slope α is unaffected by x. Further, C varies with x as βx, where β is independent of T. These relationships impose mathematical constraints on θ_H which all but rule out conventional two-band models [consider $\partial f/\partial T^2$ and $\partial f/\partial x$, where cot $\theta_H = f(x,T)$]. The data strongly support the existence of a single band of charge carriers whose Hall scattering rate $\hbar/\tau_H$ is proportional to $\alpha T^2 + \beta x$.

4. Magnetoresistance and quantum corrections

I turn next to magnetoresistance effects. In conventional metals quantum-interference effects (weak localization) and interaction effects introduce quantum corrections to the low temperature conductivity.[19-21] In two-dimensional (2D) systems these corrections are best observed in the low-temperature magnetoresistance (MR). Important quantities such as the Thouless length and dephasing rate of the carriers may be inferred from such studies. The application of these techniques to the high T_c cuprates has been hampered by the intervention of superconductivity at relatively high temperatures. The problem is exacerbated by superconducting fluctuations [Aslamazov-Larkin (AL) and Maki-Thompson (MT) terms] which may extend 20 K above the observed T_c.[22] The suppression of these terms by a magnetic field leads to positive MR contributions which dominate the much smaller normal-state corrections that are of interest. Before describing the studies of Jing *et al* [3] on non-superconducting Bi 2201 crystals, let us review the underlying physics. [19-21]

Consider a 2D sample in which the scattering potential scattering time τ is much shorter than the inelastic scattering time τ_{in}. The electron undergoes many elastic scattering off impurities before it encounters an inelastic processe (scattering off a phonon or another electron). Depending on the energy exchanged, it may take several inelastic events to disrupt the electron's phase memory. The average duration when the phase memory remains coherent is called the dephasing time τ_ϕ. Typically, we have $\tau_\phi > \tau_{in} >> \tau$. The amplitude for an electron to diffuse in this way between two points, 1 and 2, is the sum of contributions over all Feynman paths linking 1 and 2. Those paths that self-intersect at one or more points are subject to the interesting quantum interference effect that leads to weak localization. A wave packet that diffuses, say clockwise, around the closed loop is the time-reversed partner of the wave that diffuses counter-clockwise around the same loop (i.e. the two waves scatter off the same set of impurities but in opposite sequence). It is clear that the two waves interfere constructively at the intersection, and the amplitude for locating the electron at the intersection is enhanced by the interference. Thus, every closed path traversed by an electron in the time τ_ϕ introduces a negative contribution to the current. The dominant correction arises from the largest closed loop that fits into a phase coherent region. The linear size of this loop is the (temperature dependent) dephasing length $l_\phi = \sqrt{(D\tau_\phi)}$, where D is the diffusion constant. A field **B** applied normal to the sample shifts the phase of the two waves by equal and opposite amounts. At sufficiently high fields, the phase shift destroys the constructive interference, removing the weak localization. This occurs when 1/4 of a flux quantum $\phi_0 = h/e$ is inserted into the largest loop. Thus, the characteristic field B_ϕ at which this occurs is given by $B_\phi = \hbar/(4eD\tau_\phi)$. By measuring the negative MR observed with **B** normal to the plane, we may determine the important time τ_ϕ. The calculation [19-21] shows that the MR is given by

$$\Delta\sigma = (\alpha e^2/2\pi^2\hbar)\,[\psi(1/2 + 1/\Omega\tau) - \psi(1/2 + 1/\Omega\tau_\phi) - \ln(\tau_\phi/\tau)] \qquad (3)$$

where ψ is the digamma function, α is a constant, and $\Omega = \hbar/4eHD$.

5. Magnetoresistance in Bi 2201

Jing *et al* selected crystals of Bi 2201 in which no superconducting fluctuations (AL-MT terms) are observed down to 0.4 K.[3] These would swamp the effects we are searching for. (By comparing with the MR in other crystals in which T_c falls in the range 0.3 to 0.1 K, we estimate that T_c in these crystals is lower than 50 mK.) The resisivity in these crystals shows a linear-T profile between 20 and 300 K. Unlike YBCO, however, there is a large intercept (ρ_{a0} = 140 μΩcm) when the ρ_a line is extrapolated to $T = 0$, indicating significant in-plane disorder, perhaps caused by cation vacancies. Despite the large residual ρ_{a0}, the samples are far from the Ioffe-Regel limit. The metallic parameter $k_F l_0$ equals 23. (If we regard the system as 2D CuO_2 planes separated by 12.3 Å, this corresponds to a sheet resistance of 1,140 Ω/□.) Near 20 K, localization causes an upturn in ρ_a. However, the localization corrections remain small (<0.1 of ρ_{a0} at 0.5 K) so that the perturbation expansion underlying Eq. 3 is applicable.

The MR in Bi 2201 shows a very rich behavior as a function of T and field orientation.We first describe the transverse MR results. There is a remarkable anisotropy. At 4.2 K, the transverse MR is negative and logarithmic at high fields whereas the longitudinal MR is positive and approximately quadratic.[3] At each temperature, we subtract the longitudinal part from the transverse to obtain the orbital contribution alone (we assume that the spin dependent part is isotropic). The orbital contribution is then fitted to Eq. 3 and the potential scattering time τ and dephasing time τ_ϕ extracted. We find that Eq. 3 in fact provides a very good fit to the subtracted data in the temperature range 0.4 to 20 K (**Fig. 3**). The most surprising finding from these fits is that the Thouless length is unusually short (~130 Å at

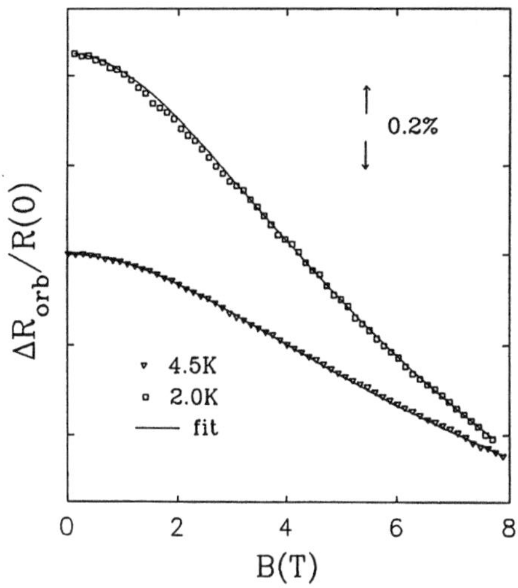

Figure 3 Fits of the orbital component of the transverse MR to the weak localization expression (Eq. 3) in Bi 2201 single crystal. The fit parameters at 2.0 K (4.5 K) are $\alpha = 2.15$ (2.36), $l_0 = 65$ Å (58 Å), $l_\phi = 102$ Å (87 Å). [Jing *et al.* 3]

0.5 K). The elastic mean free path $l_0 = v_F \tau$ comes out to be 70 Å from the fit to Eq. 3, which is somewhat longer than that deduced from the Hall effect (40 Å). These numbers indicate that the holes in Bi 2201 interact strongly even below 1 K. Since the phonon population is negligible, this interaction must arise from strong Coulomb interaction.

In conventional 2D metals, the dephasing rate is observed to vary linearly with T and to be of the order predicted by theory[21], viz.

$$\hbar/\tau_\phi = (k_B T / k_F l_0) \ln (k_F l_0 / 2). \qquad (4)$$

From our data, the dephasing rate is not only much larger than this prediction, it also shows an anomalously slow T dependence. **Figure 4** shows that τ_ϕ varies as $T^{-1/3}$. This fractional power implies that the quasi-particle widths are much larger than $k_B T$. (At 1 K, we estimate that the ratio $\hbar/(\tau_\phi k_B T)$ is of the order of 100.) Such large values clearly invalidate the whole quasi-particle concept in these systems. We return to this question below. A power-law different from 1 implies energy scale. Writing $\hbar/\tau_\phi$ in the form

$$\hbar/\tau_\phi = k_B T \, (T_\phi / T)^{2/3} \qquad (5)$$

we find that the energy scale $T_\phi = 680$ K. This is approximately half the size of the exchange energy J (1,400 K) between Cu spins in the plane.

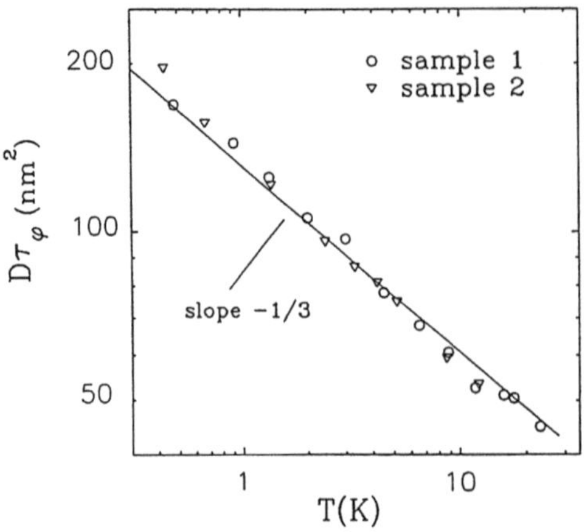

Figure 4 Temperature dependence of the dephasing time plotted as (dephasing length)^2 vs. T. The slope of -1/3 implies is very different from that observed in conventional metals. [Jing et al.[23]]

In contrast to the transverse MR, the longitudinal MR (**B**//**I**) displays novel features that cannot be readily compared with existing theories.[23] Very little is understood at present about the MR in this orientation, and the discussion here is mostly descriptive. First, we make some general remarks on the longitudinal MR. At all temperatures it is positive and of the order of 0.1 % in a field of 8 T. However, the T dependence is not monotonic. There appear to be different contributions that become important at various temperatures.

In describing the results in more detail, it is convenient to discuss the temperature region above and below 20 K separately. As we cool down from 20 K, the longitudinal MR (LMR) remains approximately quadratic in B but increases in overall magnitude. Normally, one expects the longitudinal MR to arise from coupling of the spins to **B**. However, the Zeeman field H_z scale, given by $g\mu_B H_z = k_B T$, is not observed at any temperature. At sufficiently low T (1.5 K), a new contribution to the LMR becomes evident. At low fields, the LMR is negative. The negative contribution continues to increase with decreasing T, but it is dominated by the positive term at fields exceeding ~2 Teslas, so that the LMR curve displays a broad minimum near 2 T.

We next turn to the regime above 20 K. Instead of decreasing with increasing T, the LMR signal (say at 8 T) now increases with T until it peaks near 40 K.[23] Above 40 K, it again decreases until it drops below our resolution near 80 K. The whole temperature range above 20 K, where ρ_a is linear in T is most unusual for several reasons. First, the MR is almost completely dominated by spin-dependent contributions. With increasing T the MR becomes increasingly isotropic, i.e. the transverse

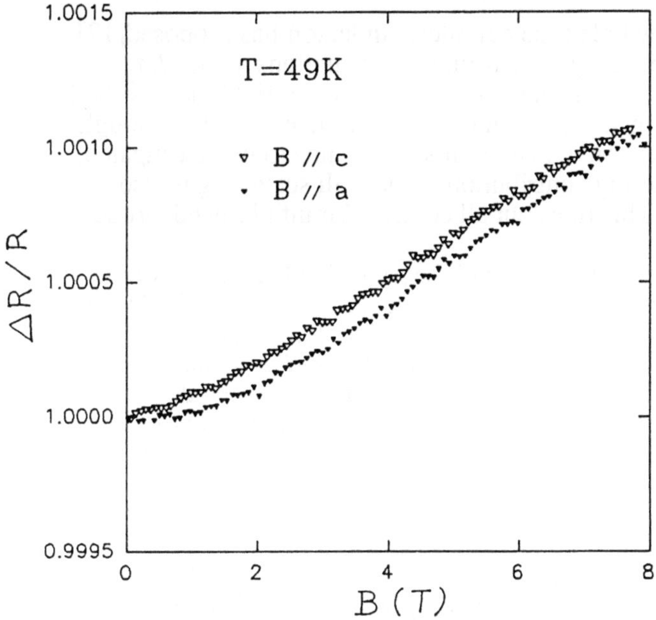

Figure 5 The fractional change in resistivity ρ_a with field for **H//I** and **H//c**. With increasing temperature above 20 K, both orientations yield increasingly similar MR. [Jing et al.[23]]

and longitudinal MR become less and less distinguishable (**Fig. 5**). A positive and isotropic MR appears to be novel. We are not aware of any mechanism that can explain this behavior. Secondly, the overall magnitude peaks at 40 K, as described above. Although this suggests some sort of magnetic ordering, there is no evidence from susceptibility thus far that such ordering occurs near 40 K.

6. Conclusions

In the cuprates, the emerging picture is that on the underdoped side ($x < 0.1$), the carriers are a few spinless holes moving in a spin-1/2 background with short range antiferromagnetic correlation. On the over-doped side ($x > 0.3$), however, *all* the Cu $3d^9$ electrons appear to participate in transport and a Fermi Surface is clearly observed. Within the superconducting window ($0.1 < x < 0.25$), we do not seem to have a model with which we can calculate transport quantities. Clearly, many questions remain open. The non-conventional nature of the cuprates is apparent from the various lifetimes. The transport scattering rate $1/\tau_{tr}$ is known from ρ_a and reflectivity to vary as $2\,k_BT$. However, this is distinct from Hall angle relaxation rate $1/\tau_H$ which varies like T^2. The rapid increase of the ratio τ_{tr}/τ_H dramatically suppresses R_H from its Drude value. This Hall suppression seems to be ubiquitous in the hole type cuprates. Anderson has proposed[17] that the T^2 scattering may arise from spin excitations alone ($1/\tau_{tr}$ involves both charge and spin excitations). Gauge field models[24,25] also predict a strong suppression of R_H with increasing T, although quantitative comparison with experiment remains to be investigated. The impurity-scattering contribution to the Hall scattering is also especially simple. This rules out all contrived multiple-band Drude models.

A central tenet of Fermi liquid theory is that the quasi-particle width becomes very sharp as the Fermi level is approached. (The width $\hbar/\tau_{in}$ decreases as T^2 in clean systems, and as T for disordered 2D metals at low T.) In Bi 2201, we find that $\hbar/\tau_\phi k_B T$ is so large that it makes the quasi-particle picture invalid. Interestingly, the dephasing rate smoothly interpolates to the transport scattering rate above 20 K when the resistivity becomes linear in T. Above 20 K, the lifetime is determined by the unknown thermal scattering responsible for the linear ρ_a. As the lifetime lengthens with decreasing T, the carriers abruptly change their behavior near 20 K. Localization corrections become important and h/τ_ϕ goes over to a $T^{1/3}$ variation. The familiar impurity-scattering regime, in which ρ_a is T independent, does not exist in the cuprates. The picture at low T is one of diffusing holes strongly interacting with each other through excitations of the spin background (recall the energy scale T_ϕ in Eq. 5).

Lastly, interaction between the holes via the background spins appears to be a dominant scattering mechanism in the linear resistivity regime above 20 K. The existence of a positive, *isotropic* MR that extends to high tempertures appears to be another generic property of the cuprates. This normal-state property should be carefully distinguished from superconducting fluctuation contributions. [22]

I wish to thank my collaborators T.W.Jing, T.R. Chien, Z.Z. Wang, D.A. Brawner, T.V. Ramakrishnan, J.M. Tarascon and K. Remschnig for permission to cite unpublished work. Useful discussions with P.W. Anderson, P.A. Lee, E. Abrahams and P. Wiegmann are gratefully acknowledged. The research is supported by the U.S. Office of Naval Research and by the Seaver Institute.

References

1. P.W. Anderson in *Frontiers and Borderlines in Many Particle Systems,* edited by R.A. Broglia and J.R. Schrieffer (North Holland, Amsterdam 1988).
2. For a review of the Hall effect see N.P.Ong in *Physical Properties of the High Temperature Superconductors*, edited by D.M. Ginsberg (World Scientific, Singapore 1990), vol. 2, p. 459.
3. T.W. Jing, N.P.Ong, T.V. Ramakrishnan, J.M. Tarascon and K. Remschnig, to be published.

4. J. S. Dugdale and L.D. Furth, J. Phys. C **2**, 1272 (1969).
5. C.M. Hurd *et al.*, Phys. Rev. B 14, 395 (1976); S.P. McAlister and C.M. Hurd, Phys. Rev. B **15**, 561 (1977).
6. J. Ziman, Phys. Rev. **121**, 1320 (1961).
7. H. Takagi *et al.*, Phys. Rev. B **40**, 2254 (1989).
8. J. Clayhold, N.P. Ong, Z.Z. Wang, J.M. Tarascon, and P. Barboux, Phys. Rev. B **39**, 7324 (1989).
9. Minoru Suzuki, Phys. Rev. B **39**, 2312 (1989).
10. Z.Z. Wang *et al.*, Phys. Rev. B **36**, 7222 (1987); P. Chaudari *et al.*, Phys. Rev. B **36**, 8903 (1987); T. Penney *et al.*, Phys. Rev. B **38**, 2918 (1988).
11. A. Matsuda *et al.*, **38**, 2910 (1988).
12. J. Clayhold, S.J. Hagen, N.P. Ong, J.M. Tarascon and P. Barboux, Phys. Rev. B **39**, 7320 (1989).
13. J. Clayhold, N.P. Ong, P.H. Hor and C.W. Chu, Phys. Rev. B **38**, 7016 (1988).
14. Y. Shimakawa, Y. Kubo, T. Manako, and H. Igarashi, Phys. Rev. B **40**, 11400 (1989).
15. L. Forro, D. Mandrus, C. Kendziora, L. Mihaly, and R. Reeder, Phys. Rev. B **42**, 8704 (1990).
16. T.R. Chien, D.A. Brawner, Z.Z. Wang, and N.P.Ong, Phys. Rev. B **43**, 6242 (1991).
17. P. W. Anderson, to be published.
18. T.R. Chien, Z.Z. Wang and N.P. Ong, to be published.
19. P.A. Lee and T.V. Ramakrishnan, Rev. Mod. Phys. **57**, 287 (1985).
20. Gerd Bergmann, Phys. Rept. **107**, 1 (1984).
21. B.L. Altshuler, A.G. Aronov, and D.E. Khmel'nitskii, J. Phys. C **15**, 7367 (1982).
22. A.G. Aronov, S. Hikami and A.I. Larkin, Phys. Rev. Lett. **62**, 965 (1989).
23. T.W. Jing, N.P. Ong, J.M. Tarascon and K. Remschnig, to be published.
24. Naoto Nagaosa and Patrick A. Lee, Phys. Rev. Lett. **64**, 2450 (1990); Phys. Rev. B **43**, 1219 (1991).
25. L.B. Ioffe and P.B. Wiegmann, Phys. Rev. B **43**, 1219 (1991).

Optical Properties of High-Temperature Superconductors

S. Uchida

Faculty of Engineering, University of Tokyo, Yayoi, Tokyo, Japan

ABSTRACT: The optical spectra of the undoped CuO_2 planes in the parent compounds of high-Tc superconductors are dominated by the electronic excitations across the Cu3d-O2p charge-transfer gap. Doping carriers, either electrons or holes, into the CuO_2 plane drastically changes the spectrum by transferring the spectral weight of the change-transfer excitation to the lower energy excitations. This gives rise to a unique electronic structure of high-Tc superconductors.

I. INTRODUCTION

The perovskite-related copper oxides, $La_{2-x}Sr_xCuO_4$, $YBa_2Cu_3O_{6+x}$, $Nd_{2-x}Ce_xCuO_4$, etc., show unprecedently high T_c superconductivity, and pose challenging problems on solid state physicists. It has been dogmatically believed after very intensive experimental efforts to search for high Tc in other systems that the high-Tc superconductivity occurs exclusively in Cu exides [1].

However, the mechanism of high-Tc is not yet understood. In the earlier days it was said that the development of high-Tc superconductivity is connected with the search for new, non-phonon, superconductivity mechanisms [2]. The most realizable one was thought to be so-called exciton mechanism where the attraction between the conduction electrons, which is needed for the appearance of superconductivity, can have its cause in other (bound) electrons in the metal and/or some substances, molecules or dielectrics, in contact with it [3-5].

In $YBa_2Cu_3O_7$ the Little's idea appeared to be realized with metallic CuO_2 planes in contact with "insulating" CuO chains where the excitons are associated with the electronic transitions between bonding and antibonding bands in the chain. However, it has become clear by the discoveries of many other high-Tc superconductors that the chains are not required, nor is the presence of alternate metallic CuO_2 planes and insulating layers such as La_2O_2 and Nd_2O_2 layers. In the high-Tc superconductors both conduction carriers and other low-energy excitations, if any, are confined within the CuO_2 planes.

One of the energy scales which characterize the electronic structure of the CuO_2 plane is the Coulomb repulsive energy U on the Cu site which is estimated to be very large U$\sim$6-8eV from the photoemission spectra [6]. This value is larger than the bandwidth ($\sim$4eV calculated from the band theory, which indicates that the band theory is inadequate to describe the electronic state in the CuO_2 plane. The "undoped" CuO_2 plane, is indeed an insulator in contrast to a metal with half-filled conduction band in the band theory [7]. Another quantity is the charge-transfer (CT) energy Δ between O2p and Cu3d states which defines an energy gap in the particle-hole excitation spectrum of the CuO_2 plane. Varma et al. [8] pointed out that, as one moves to the right of the 3d transition metal series in the periodic table, the ionization level E_d of the transition metal moves down and at Cu it is closest to, or almost coincident with the oxygen affinity level E_p. This is unusual and another example is found only in $BaBiO_3$ which is the parent compound of $BaBi_{1-x}Pb_xO_3$and $Ba_{1-x}K_xBiO_3$. A consequence of$E_d \simeq E_p$ is a small Δ [9] and large superexchange interaction $J \sim t^4/U\Delta^2$ between spin magnetic moments on the Cu atoms [10].

The present lecture focuses on the optical spectra of high-Tc superconductors in their normal state and shows that the spectra of parent materials, that is, of the undoped CuO_2 planes are predominated by electronic excitations above the Cu3d-O2p CT energy gap. The highlight is how the CT excitation spectra are affected by carrier doping. This is demonstrated by the experimental results on the two prototypical systems P-type $La_{2-x}Sr_xCuO_4$ and N-type $Nd_{2-x}Ce_xCuO_4$. They indicate that new states are created to fill in the CT gap by a transfer of spectral weight from the states involved in the CT excitations [11].

It is also interesting to compare the spectra of the copper oxides with those of non-Cu oxide systems $La_{2-x}Sr_xNiO_4$ and $Ba_{1-x}K_xBiO_3$. In the latter system relatively high Tc(~30K) was discovered which has been thought to be outside the range for electron-phonon interaction [12,13]. Both undergo an insulator-metal transition at x $\simeq$0.65 and 0.35, respectively. Apparently there is no question of magnetic correlations in thissystem different from copper oxide systems. From the optical spectra it is inferred that the metallic state is realized by the (partical) destruction of the change-density-waves in the parent compound $BaBiO_3$ [14,15].

II. OPTICAL SPECTRA OF UNDOPED COPPER OXIDES

The optical spectra shown here were measured mostly by our group at University of Tokyo on single crystals of $La_{2-x}Sr_xCuO_4$ and $Nd_{2-x}Ce_xCuO_4$ covering a wide compositional range. The samples of $Ba_{1-x}K_xBiO_3$ are epitaxial films sputtered on $SrTiO_3$ substrate [15]. In most cases the optical spectra cover a wide energy region from far infrared to vacuum ultraviolet 0.004-40eV which guarantees the accuracy of the Kramers-Kronig transformation to obtain various optical response functions from the measured reflectivity spectrum.

2.1. The parent compounds La_2CuO_4 and Nd_2CuO_4

The reffectivity spectra of La_2CuO_4 and Nd_2CuO_4 are shown in Fig.1 together with each crystal structure (schematic) in the inset. The light polarization is parallel to the CuO_2 planes ($E \perp c$). The spectra of the two compounds are similar and have common features; three dominant reflectivity edges at around 2, 13 and 30eV. The highest energy edge originates from the excitations involving all the valence electrons and the reflectivity band starting from ~4eV and ending at the second edge corresponds to the interband excitations from the O2p valence bands to the conduction bands mainly composed of 5d/4f orbitals of La or Nd [9].

The spectra below 4eV are characteristic of insulators, with sharp optical phonon peaks below 0.1eV and featureless reflectivity up to 1eV or higher. At 1.5eV for Nd_2CuO_4 and at 2.0eV for La_2CuO_4 a prominent reflectivity peak is seen. Figure 2 demonstrates that this peak dose not appear when the polarization is perpendicular to the CuO_2 planes. Hence, this peak corresponds to the excitations in the CuO_2 plane which set up a threshold for electronic excitations or an optical energy gap.

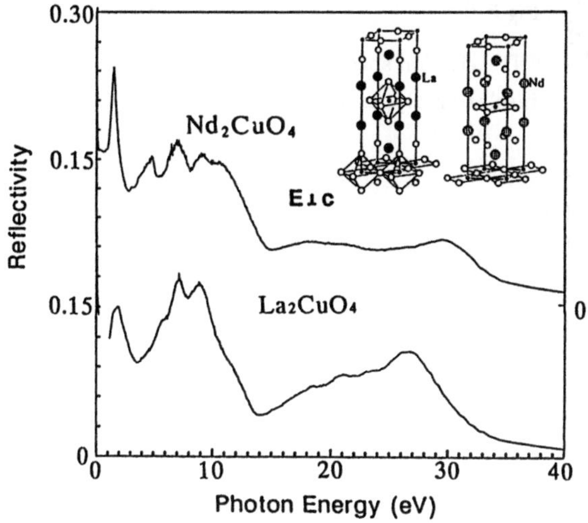

1. Reflectivity spectra of the two prototypical parent compounds of the high-Tc superconductors, La_2CuO_4 and Nd_2CuO_4, covering a wide energy range. Inset shows crystal structures of the two compounds.

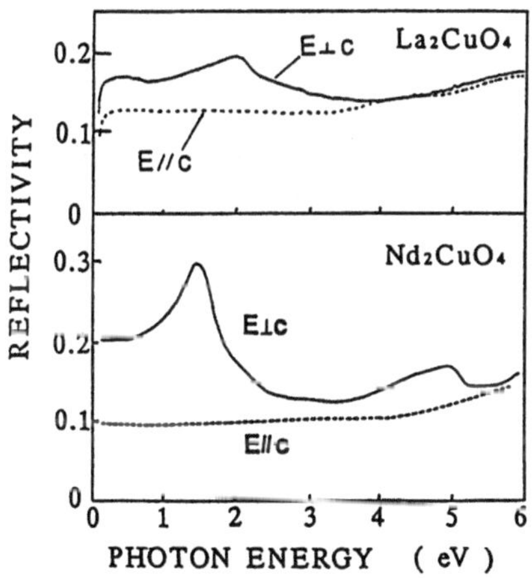

2. Reflectivity spectra of La_2CuO_4 and Nd_2CuO_4 for two different polarization, E|| c and E⊥c, where c stands for the direction perpendicular to the CuO_2 planes.

2.2. Various insulating copper oxides

In order to clarify the origin of the low-energy electronic excitation at 1.5-2.0eV, the reflectivity spectra of various insulating cuprates have been measured[16]. The spectra are shown in Fig.3 where the T*-phase $(La,Dy)_2CuO_4$ is a P-type superconductor (Tc∼30K) when doped with holes but $Sr_2CuO_2Cl_2$, $CaCuO_2$ and Ca_2CuO_3 are yet non-superconducting owing to the difficulty in doping carriers. $Sr_2CuO_2Cl_2$ is isostructural to La_2CuO_4 but two apical oxygens surrounding Cu are replaced by Cl. $CaCuO_2$ is composed of alternating CuO_2 and Ca layers and thought to be a parent material of Bi- or Tl- based cuprate super conductors with Tc exceeding 100K [17]. The Cu atoms form a

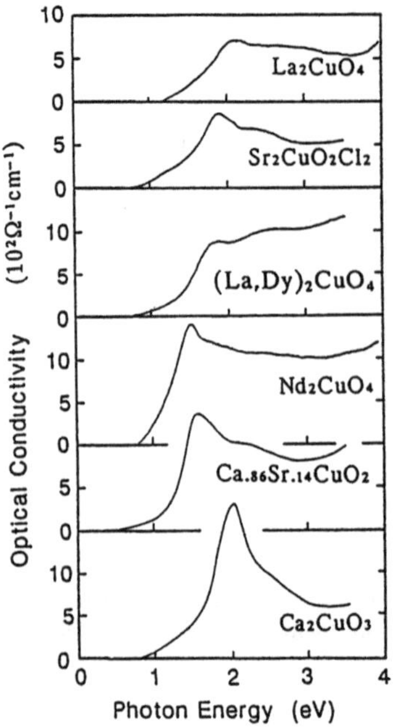

Fig.3

CT excitation spectra of various insulating copper oxides [16].

square lattice also in Ca_2CuO_3 but the oxygens intervene only in one direction (b-axis), forming CuO chains as in $YBa_2Cu_3O_7$.

Obviously the spectra of all these compounds are similar and dominated by a pronounced excitonic peak at 1.5-2.0eV, indicating that the origin of the peak can be ascribed to the excitation in the CuO_2 plane, the common structural element of these compounds. The fact that a similar peak is seen in Ca_2CuO_3 only when the polarization is parallel to the CuO chains suggests that the peak corresponds to the electronic transitions across a gap created between Cu and O states.

The results of the band calculations for these compounds indicate metallic state with half-filled $Cu3d_{x^2-y^2}/O2p$ antibonding conduction band [18-20], so that the low-energy spectra should be dominated by intraband contribution (Drude term) centered at ω=0. Therefore, the observed spectra, characteristic of insultors, give an evidence that the band theory is inadequate to describe the electronic state of the "undoped" copper oxides. The insulating electronic structure is realized when the on-Cu site Coulomb repulsive energy U is large enough to exceed the bandwidth of the conduction band. In this case the Cu3d states, specifically $3d_{x^2-y^2}$ orbital, which were strongly hybridzed with O2p orbitals in the band theory, are split into lower (full) and upper (empty) Hubband bands with a filled O2p band located in between [21]. In this energy scheme the O2p → Cu3d (upper Hubbard) change-transfer is a fundamental electronic excitation with a CT gap Δ. The prominent peak observed at 1.5-2.0 eV in every parent compond and other insulating cuprates should correspond to this CT excitations and thus these compounds are CT insulators as suggested by the photoemission spectroscopy [6].

2.3. Dependence of the CT gap on the Cu-O bond distance

Cooper et al. [22] and Arima et al. [23] measured the energy gap Δ and the exchange interaction J in a series of T'-phase Ln_2CuO_4 (Ln=Pr, Nd, Sm, Eu and Gd). It was suggested that the energy gap Δ changes predominantly with the in-plane Cu-O distance, a as

$\Delta \sim a^{-\alpha}(\alpha$=6-7). The scaling can be extended to other cuprates with various structural circumstances outside the planes, e.g., La_2CuO_4 with two apical oxygens, T*- compounds and $YBa_2Cu_3O_6$ with pyramidal oxgen configuration, although a small deviation from this scaling exists. Therefore, it appears that the dominant contribution to Δ throughout the cuprate series is very sensitive to the change in the in-plane Cu-O distance.

Phenomenologically the CT gap may be expressed as, according to Cooper et al. [22], $\Delta=\Delta_o+M/a+\delta\text{-}t$, where Δ_0 represents the energy difference between Cu3d and O2p determined by electronic ionization and affinity energies, presumably independent of a. The second term represents Madelung or electrostatic Coulombic energy associated with the CuO_2 planes and is expected to have a positive coefficient in the cuprate. $\delta \sim t^2/\Delta$ and t represent the influence of hybridization on Δ with transfer integral t; the former does energy level shift (level repulsion) due to Cu3d-O2p hybridization and the latter does hybridization bandwidth. While the last term decreases Δ as a decreases, the δ-term is expected to increase with decreasing a. Cooper et al. concluded from the observed increase in Δ with decreasing a that changes in t are outweighed by Madelung and level repulsion in the cuprate systems. This seems different from many other systems where changes in t are expected to dominate Δ.

III.OPTICAL SPECTRA OF DOPED COMPOUNDS

3.1. $La_{2-x}Sr_xCuO_4$ and $Nd_{2-x}Ce_xCuO_4$

Carrier doping is made by partial substitution of La^{3+} with Sr^{2+} (hole doping) and Nd^{3+} with Ce^{4+} or reduction of O content (electron doping) in the parent CT insulator La_2CuO_4 or Nd_2CuO_4. Doping does not cause any significant change in the optical spectra in the energy region higher than 5 eV. By contrast in the low energy region the spectrum exhibits drastic changes. The reflectivity spectra are shown in Figs. 4(a) and (b) for various compositions x. In both P-type and N-type cases the CT peak of the parent compound is rapidly suppressed, instead a reflectivity edges showing up at around 1.0 eV even for very light doping levels as seen for x=0.02 in the P-type case and for reduced Nd_2CuO_{4-y} ($y\neq 0$). Note that the edge associated with the CT excitation stays almost at the same energy , so that the new edge at ~1.0eV corresponds to different excitations. This edge does not shift appreciably with increasing dopant concentration, either.

One can learn an implication of the observed spectral change on doping from the optical conductivity spectrum $\sigma(\omega)$ which is obtained by the Kramers-Kronig transformation of the reflectivity spectrum. The $\sigma(\omega)$ spectra demonstrate that the conductivity rapidly decreases in the energy region above 1.5eV in the case of $La_{2-x}Sr_xCuO_4$ (above 1.1eV in $Nd_{2-x}Ce_xCuO_4$), while it increases below 1.5eV (1.1eV). The CT excitation dominates the higher energy region and the lower energy region involves a free carrier contribution, since metallic conduction is observed at room temperatures even for slightly doped compounds [24]. However, it cannot be concluded that the free carrier contribution extends up to 1.5eV(1.1eV). Apparently, for small x, $\sigma(\omega)$ is composed of two contributions below 1.5eV(1.1eV), a narrow band peaked at ω=0 and a broad band centered in the mid-infrared region. Similar infrared conductivity was previously reported for $YBa_2Cu_3O_{6+x}$ [25,26].

One would notice that the energy 1.5eV in $La_{2-x}Sr_xCuO_4$ or 1.1eV in $Nd_{2-x}Ce_xCuO_4$ is approximately an isosbestic point in the $\sigma(\omega)$ spectra. This indicates that the lower energy conductivity is transferred from the higher energy region on doping. The implication of these spectral changes would be that the states, O2p and Cu3d, initially separated by a CT energy gap, are redistributed by doping, or equivalently new states are reconstructed by collapsing the CT excitation.

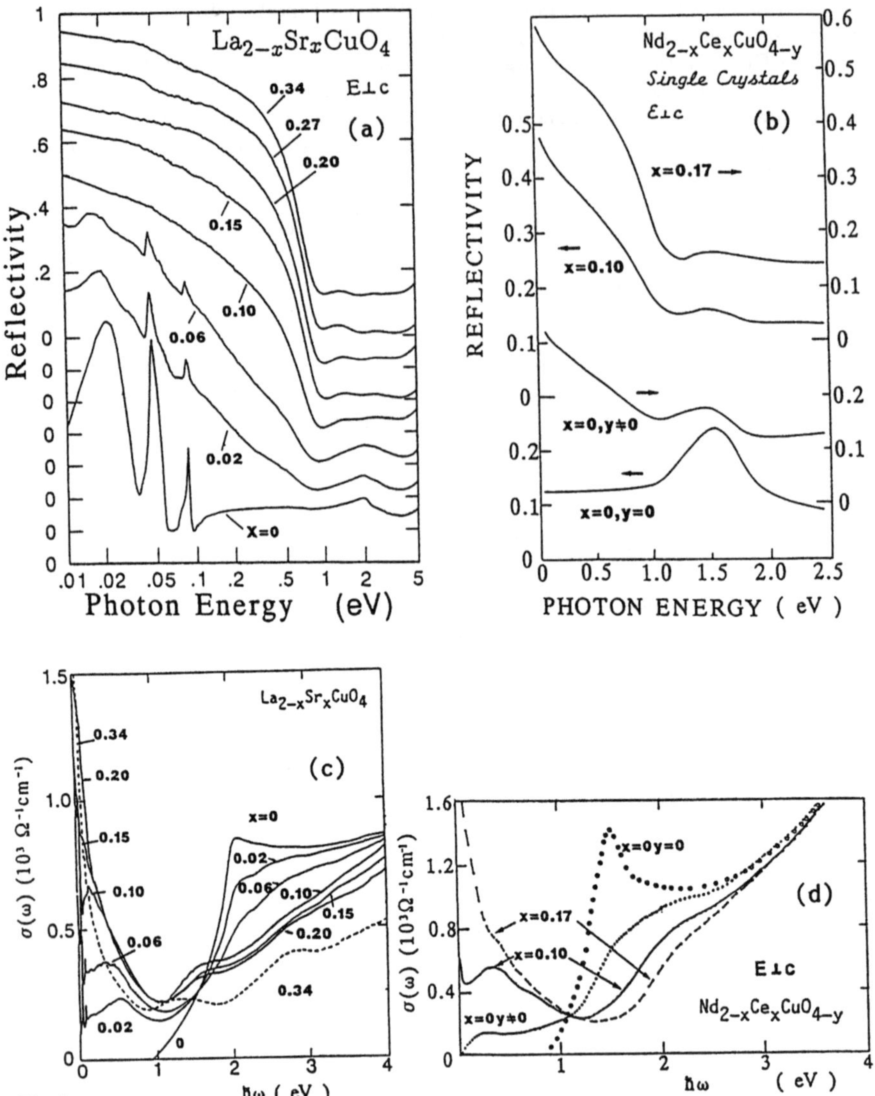

Fig.4

Reflectivity spectra of (a) $La_{2-x}Sr_xCuO_4$ and (b) $Nd_{2-x}Ce_xCuO_4$, and their K-K transformed optical conductivity spectra, (c) and (d), respectively.

3.2. Comparison with other spectroscopic experiments

The formation of the "mid-gap" states on doping has also been suggested by electon-energy-loss-spectra (EELS) [27] as well as by x-ray absorphon spectra (XAS) [28] from the core O1s level. In particular, the result of EELS measured by Romberg et al. at Karlsruhe on a series of $La_{2-x}Sr_xCuO_4$ has clearly demonstrated that upon doping the conduction band states in La_2CuO_4 are reduced and states are formed in the gap with the Fermi level crossing a band of these new states. On the other hand, unoccupied states in the gap are not clearly seen in the EELS for N-type doped systems [29]. This is in contrast

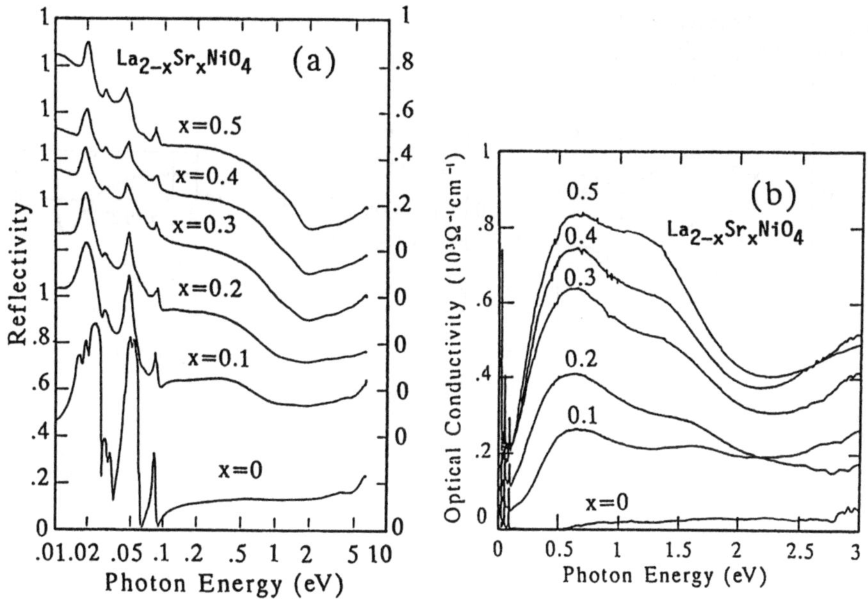

Fig.5 Reflectivity and optical conductivity spectra of $La_{2-x}Sr_xNiO_4$ ($0 \leq x \leq 0.5$).

to the optical spectra which suggest electron-hole symmetry as shown in Fig.5. However, the recent photoemission data [30,31] postulate that there is no shift of the Fermi level when going from P- to N-type doping. In addition Allen et al. [31] have shown using resonant photomission spectroscopy that for both electron and hole doping the states near the Fermi level are generated by a transfer from both conduction and valence band states of the parent insulators, consistent with the present optical data.

3.3. Comparison with other systems

Similar O1s XAS spectra were observed in doped NiO, $Li_xNi_{1-x}O$ [32]. The doped holes are located primarily in the O2p states but they are bound to the Li^{1+} impurities forming states in the O2p states in the NiO CT gap (Δ ~4eV). These states are localized "impurity" states so that the system does not exhibit an insulator-metal transition.

Some perovskite-related 3d transition-metal oxides exhibit insulator-metal transitions on doping, $La_{2-x}Sr_xNiO_4$ at x~1.0 [33], $La_{1-x}Sr_xVO_3$ at x~0.23 [34] etc. However, the situation is quite different from $La_{2-x}Sr_xCuO_4$ in that substantial dopant concentration is necessary to realize metallic states and that the CT energy gap is larger, 4eV in La_2NiO_4 and NiO. Although the apparent spectral change is similar as illustrated in Fig.5 for $La_{2-x}Sr_xNiO_4$ with a reflectivity edge appearing at about 1eV on doping [35], the transfer of the spectral weight to the low energy region is much slower than that in the case of $La_{2-x}Sr_xCuO_4$. Possibly the mid-gap states in the non-Cu oxides are more localized in the vicinity of the dopants atoms because of larger CT energy gap or smaller p-d hybridization and of smaller quantum fluctuation of larger spin quantum number of the non-Cu transition metal.

$BaBiO_3$, the parent compound of superconducting $BaBi_{1-x}Pb_xO_3$ (BPB) and $Ba_{1-x}K_xBiO_3$ (BKB), is an insulator with a gap of ~2eV [36]. In the band picture it is a metal with a half-filled conduction band derived from Bi6s-O2p antibonding orbitals which is a similar situation to La_2CuO_4 [36]. The insulating gap is created by a splitting

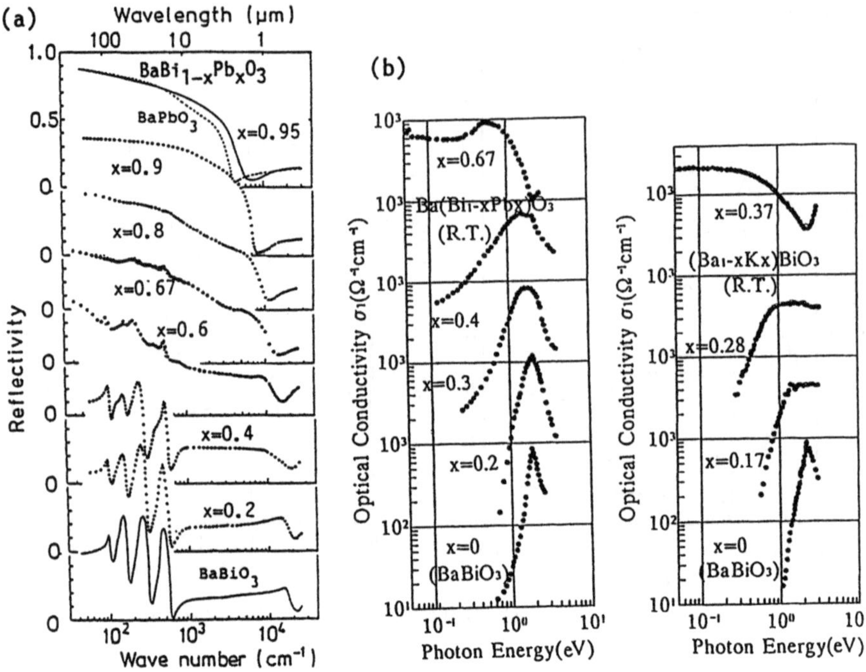

Fig.6 (a) Reflectivity spectra of $BaPb_{1-x}Bi_xO_3$ [14].
(b) Optical conductivity spectra of $BaBi_{1-x}Pb_xO_3$ and $Ba_{1-x}K_xBiO_3$ [15].

of this conduction band due to change disprortionation, $2Bi^{4+} \rightarrow Bi^{3+}+Bi^{5+}$, coupled with the oxygen breathing-mode distortion surrounding Bi atoms. As illustrated in Figs.6, the optical spectrum of $BaBiO_3$ is dominated by an absorption peak at ~2eV and anassociated reflectivity edge at ~3eV. This absorption corresponds to the electronic transition across an optical gap, ~2eV, and is intepreted in the local picture as charge-transfer excitation from Bi^{3+} to Bi^{5+} sites.

Metallic state is realized either by the substitution of Bi with Pb or by the substitution of Ba with K. The K-substitution formally decreases the average valence of Bi, so it seems that it would create a hole in the lower split-off band of $BaBiO_3$. However, unlike the cuprate systems BKB remains insulting up to x~0.35(the insulting state of BPB persists up to x ~ 0.65). Fig.6(a) shows that the reflectivity edge at ~3eV in $BaBiO_3$ moves to lower energies with increase of Pb composition, which is an indication of decrease in the optical energy gap. The decrease in the gap is more clearly seen in the optical conductivity spectra of BKB (Fig.6(b)). Different from the cuprates, this type of "CT" excitation dominates up to very large x until it is replaced by the metallic Drude-type contribution — it persists even in the matellic phase ($x > 0.65$) in BPB. Therefore, the insulator-metal transition in these systems takes place when the energy gap of decreases and finally closes.

IV. SUMMARY

In the CuO_2 plane various interaction processes appear to be competing. It is not yet understood which processes lead to high-Tc superconductivity. In the undoped CuO_2 plane the Cu atom is in the Cu^{2+} ionic state, having a localized hole with spin magnetic moment (S=1/2) due to very large onsite Coulomb energy. The Cu spins are ordered antiferromagnetically due to large superexchange coupling. The Cu^{2+}state is also characterized by the CT, Cu^{2+}-O^{2-} $\rightarrow$$Cu^{1+}$-$O^{1-}$, excitations as demonstrated in this paper. Depending on which interaction processes, spin or charge, are stressed, various theoretical models have been proposed so far. One school of views is to note that the Cu spins will persist even in the doped CuO_2 planes and spin fluctuations will play an essential role in high-Tc superconductivity. Another view is to ignore the Cu spins in the moderately doped CuO_2 plane and to stress the CT excitations—superconductivity near the CT instability ($\Delta \sim 0$).

The optical study has revealed that the doping creates new states in the CT gap region of the undoped CuO_2 plane. The observed optical absorptions in the low-energy region ($\hbar\omega$ <1eV) correspond to the electronic excitations within a band of these states. The new states should involve both Cu3d and O2p states transferred from the conduction and valence bands of the parent CT insulator, so that the Fermi level E_F would be located in the middle of the original CT gap as suggested by the photoemission studies on both P-type and N-type doped materials [30]. This implies that the doping does not reduce the magnitude of the CT gap but collapses the valence and conduction band states involved in the CT excitations in the undoped CuO_2 plane. Further, the states near E_F are highly itinerant as indicated by the transport properties [24], suggesting that the originally localized Cu3d states become somewhat itinerant in the doped CuO_2 plane. Actually, these states seem to form a Fermi liquid in the heavily doped region where superconductivity disappears. It is a subtle question whether the Cu spins and/or CT excitations persist even in the superconductivity phase of the doped CuO_2 plane. This, together with a question on the crucial interaction processes occurring in the new states which make the doped CuO_2 planes to superconduct, remain open.

ACKNOWLEDGEMENT

I would like to thank T.Ido, H.Eisaki, S.Tajima, T.Arima and Y.Tokura for collaboration in the optical studies of high-Tc superconductors.

REFERENCES

1. In the Proceedings of International Conference on Materials and Mechanisms of High-Temperature Superconductors, Interlaken,1989, PhysicaC153-155; Stanford, 1989, Physica C162-164
2. "High Temperature Superconductivity", Eds.,V.L.Ginzburg andD.A.Kirzhnits (Plerum,1982).
3. W.A.Little, Phys.Rev. A134, 1416(1964).
4. V.L.Ginzburg, Sov.Phys. JETP 20, 1549(1965)
5. D.A.Allender, J.Bray and J.Bardeen, Phys.Rev. B7, 1020 (1973); ibid. B8, 4433 (1973).
6. A.Fujimori, E.Takayama-Muromachi,Y.Uchida, and B.Okai, Phys.Rev. B35, 8814 (1987).
7. For example, L.F.Mattheiss, Phys.Rev.Lett.58, 1028(1987).
8. C.M.Varma, S.Schmitt-Rink and E.Abrahams, Solid State Commun. 62, 681(1987).
9. S.Tajima et al., J. Opt. Soc. Am. B6, 475(1989).
10. K.B.Lyons, P.A.Fleury, L.F.Schneemeyer and J.V.Waszczak, Phys. Rev.Lett. 60, 732(1988).
11. S.Uchida et al., Phys. Rev. B43, 7942(1991)
12. L.F.Mattheiss, E.M.Gyorgy and D.W.Johnson Jr., Phys.Rev. B37, 3745(1988).
13. R.J.Cava et al., Nature 332, 814(1988).
14. S.Tajima et al., Phys.Rev.B32, 6302(1985).

15. H.Sato, H.Takagi, S.Tajima and S.Uchida, Nature 338, 241(1989).
16. Y.Tokura et al., Phys.Rev.B41, 11657(1990).
17. T.Siegrist et al., Nature334, 231(1988).
18. G.Y.Guo, Z.Szotek and W.M.Temmerman, Physica C162-164, 1351(1989).
19. D.Singh,W.E.Pickett and H.Krakauer,Physica C162-164, 1431(1989); L.F.Mattheiss and D.R.Hamann, Phys.Rev.B40, 2217(1989).
20. L.F.Mattheiss, Phys.Rev.B42,354(1990).
21. V.Emery, Phys.Rev.Lett.58, 2794(1987).
22. S.L.Cooper et al.,Phys.Rev.B42, 10785(1990).
23. T.Arima et al., to appear in Phys.Rev.B.(in press)
24. H.Takagi et al., Phys.Rev.B40,2254(1989).
25. G.A.Thomas et al., Phys.Rev.Lett. 61, 1313(1988).
26. J.Orenstein et al., Phys.Rev. B42,6342(1990).
27. H.Romberg et al., Phys. Rev. B42, 8768(1990).
28. T.Takahashi et al., to appear in Proc. 3rd International Symposium on Superconductivity, 1990 Sendai, Japan.
29. M.Alexander et al., Phys. Rev. B43, 333(1991).
30. J.W.Allen et al., Phys. Rev. Lett. 64, 595(1990).
31. H.Namatame et al., Phys. Rev. B41, 7205(1990).
32. P.Kuiper et al., Phys. Rev. Lett. 62, 221(1989).
33. M.Khairy, P.Odier and J.Choisnet, J.de Physiqye 47,C1-831(1986).
34. P.Dougier and P.Hagenmuller, J.Solid State Chem.15,158(1975).
35. X.-X.Bi et al., Phys. Rev. B42, 4756(1990).
36. L.F.Mattheiss and D.R.Hamann, Phys. Rev. B28, 4227(1983).

An Introduction to Infrared Properties of High-T_c Superconductors

Z. Schlesinger[1], *L.D. Rotter*[1], *R.T. Collins*[1], *F. Holtzberg*[1], *C. Feild*[1], *U. Welp*[2], *G.W. Crabtree*[2], *J.Z. Liu*[2;*], *Y. Fang*[2], *and K.G. Vandervoort*[2]

[1]IBM T.J. Watson Research Center, P.O. Box 218, Yorktown Heights, NY 10598, USA

[2]Materials Science Division, Argonne National Lab., Argonne, IL 60439, USA

*Present address: Physics Department, University of California, Davis, CA 95616, USA

Abstract. The essential element in the high T_c cuprate superconductors appears to be the CuO_2 planes which, of course must reside in an electropositive environment (Tokura). Recognizing this, and focussing only on the properties of the CuO_2 planes, as distinct from that of their environment, allows a relatively simple picture of the phenomenology of infrared properties to be obtained. In particular one finds that the normal state conductivity drops unusually slowly as a function of frequency, more like $1/\omega^{0.8}$ than the conventional $1/\omega^2$; and that it can be described in terms of a scattering rate that is linear in frequency. This linear frequency dependence appears to be fundamentally related to the linear temperature dependence of the resisitivity, and may also be related to a linear photoemission linewidth and a broad electronic background signal observed in Raman experiments. In the superconducting state a characteristic energy scale of $\sim$500 cm^{-1} ($2\Delta/kT_c \simeq 8$) develops, and can be reasonably identified as a pair-excitation threshold (superconducting gap) at low temperature. Aspects of this feature which are similar to and different from expectations based on conventional BCS theory are discussed. In particular, one finds that the temperature dependence is quite unusual.

Introduction Infrared and optical spectroscopy has been used to study elemental metals(1), transition metal compounds(2), heavy fermions(3), and superconductors(4-8). Infrared covers the range between microwave and visible, i.e, from about 0.0005 eV to 2 eV. The visible range extends from about 2 to 3 eV. (Roughly 1 cm^{-1}=1.44 K=0.124 meV= 30 GHz.) These are important energy scales for solid state physics. Typically, thresholds for interband absorption in semiconductors and metals lie in the near-infrared (1-2 eV), the visible, or the near UV. More importantly, the characteristic energy scales for most of the exotic phenomena of solid state physics, e.g., charge density waves, superconductivity, magnetism and heavy Fermion physics, tend to lie in the infrared range, which in units of temperature is about 5 to 20,000 K. Here we will review some of the properties of conventional metals and superconductors, and compare

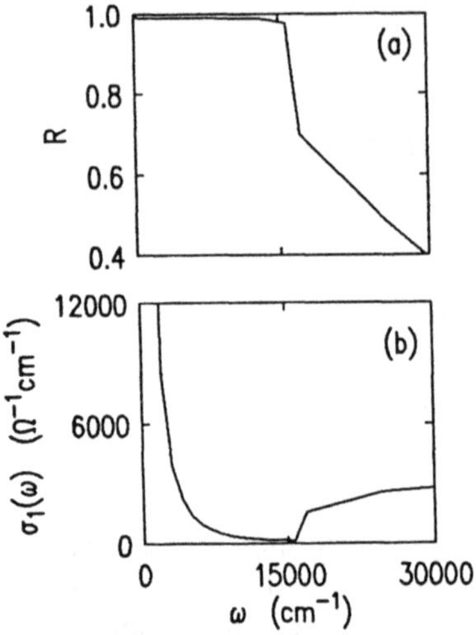

FIGURE 1

Reflectivity and conductivity (real part) vs frequency for Cu at room temperature (ref 1).

them with what we presently know of the cuprates. we will concentrate on the nature of the cuprates doped near the optimum concentration for high T_c superconductivity, where the properties seem to be most simple.

In a free-electron metal the infrared conductivity is given by(9),

$$\sigma(\omega) = \sigma_0(1 - i\omega\tau)^{-1} \quad (1)$$

where σ_0 is the d.c. conductivity. In general the real and imaginary parts of $\sigma(\omega)$ are related by Kramers-Kronig (causality)(1), and the physics in the conductivity is most readily accessed by examining $\sigma_1(\omega)$. For the free electron metal $\sigma_1(\omega)$ has a zero frequency intercept of σ_0, and a characteristic width given by $1/\tau$. Above this characteristic frequency $\sigma_1(\omega)$ falls like $1/\omega^2$. This is the essential nature of a free-electron conductivity(9).

The left half of fig 1, i.e. for $\omega \lesssim 15{,}000$ cm^{-1}, shows the reflectivity and $\sigma_1(\omega)$ for a free-electron metal in the infrared region. The reflectivity is very high and the real part of the conductivity falls like $1/\omega^2$. The conductivity has a zero frequency intercept of about 350,000 Ω^{-1}cm^{-1}, (which corresponds to $\rho_{dc} \simeq 3$ $\mu\Omega -$ cm). The parameters used here, i.e. σ_0 and $1/\tau$(300 cm^{-1}), are roughly appropriate for Cu at room temperature. This low frequency ($\omega \lesssim 15{,}000$ cm^{-1}) part of the conductivity is called the *intraband* or Drude part of the conductivity.

At higher frequency in figure 1 we show the reflectivity and conductivity of Cu (from ref 1) up to 30,000 cm^{-1} (3.7 eV). Here we begin to see the 2nd part of the conductivity, the *interband* term, which begins at a threshold energy of about 16,000 cm^{-1} (2 eV). This interband term is associated with

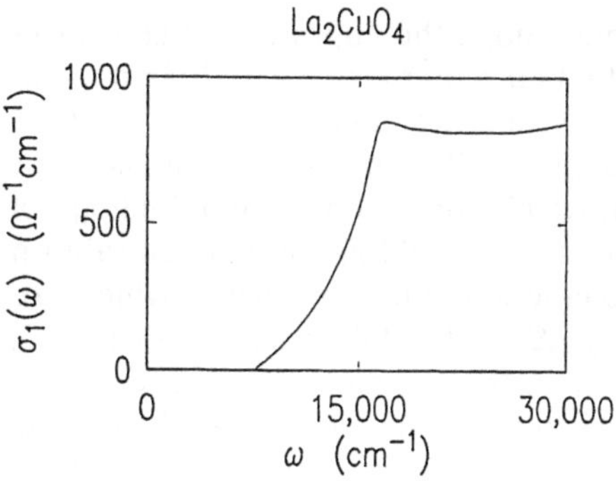

FIGURE 2

Conductivity vs. frequency for the cuprate insulator La_2CuO_4 (from Uchida et al.(10)).

transitions from occupied d-bands which lie about 2 eV below the Fermi level, to the unoccupied states (s-band) just above the Fermi level. (Note that copper is a "simple" metal, with filled d-bands, and a single electron in the 4s level (9).) It is this 2 eV transition which gives Cu its characteristic color, since it causes the reflectivity to drop significantly between the red-yellow part of the visible spectrum and the green-blue-violet part (fig 1a).

In fig 1b one can see the two characteristic parts of the conductivity for Cu. The intraband term at low energy, which corresponds to excitations within the partially occupied s-band at the Fermi level, and the interband contributions which begin at about 16,000 cm^{-1}. For the intraband term, the characteristic frequency scale is the scattering rate, $1/\tau$, above which the $\sigma_1(\omega)$ falls like $1/\omega^2$ through the intraband range (which extends up to $\sim$16,000 cm^{-1} for Cu). The high value of the reflectivity in this range, and the $1/\omega^2$ frequency dependence of $\sigma_1(\omega)$ are intimately related.

By coincidence, the insulating (undoped) cuprates also have an interband threshold of about 2 eV as shown in figure 2, which shows up clearly particularly in clean insulators such as Nd_2CuO_4 and La_2CuO_4 as studied by Uchida et al.(10). This threshold is due to transitions between a filled O_{2p}-like band and an empty Cu_{3d} band, which lies above the O_{2p} band due to the strong on-site coulomb repulsion for the Cu in the d^9 configuration (large Hubbard U). This is called a charge transfer excitation since it involves the transfer of an electron from the O site to the Cu site. Although the above description provides a useful starting point, it is also somewhat over simplified. Actually in the cuprates there appears to be very strong hybridization of the Cu and O bands, and the Zhang-Rice singlet, for example, may provide the proper starting point for a detailed understanding. The effects of strong hybridization may be particularly important for understanding the evolution of $\sigma(\omega)$ with doping, as discussed for example by Sawatzky at the Kanazawa conference(11).

To obtain superconductors, one must dope the cuprate insulator away from half-filling. Typically one finds that for either electron or hole doping, optimum superconductivity, i.e high Meissner fraction and high T_c, occurs somewhere in the vicinity of 0.2 doping. (That is, an excess hole or electron concentration of 0.2 per Cu, which correspond to band-fillings of 0.4 and 0.6, respectively). In this paper, we would like to concentrate on the nature of the conductivity for hole-doped superconductors which are close to the optimum range. For $La_{2-x}Sr_xCuO_4$, this means $0.25 \gtrsim x \gtrsim .15$; for $YBa_2Cu_3O_{7-y}$, $T_c \gtrsim 50$ K, etc.. Within this range we find several universal aspects of the data, which are phenomenologically fairly simple. At lower doping it is possible that strong localization effects may become operative, and that the data may therefore become more complicated and more variable, both from system to system and from sample to sample.

Normal state The normal state reflectivity and $\sigma_1(\omega)$ (at $T \simeq 100$ K) for a typical cuprate superconductor are shown in figure 3. Although an interband threshold for the cuprates occurs near 2 eV, the metallic cuprates do not have a yellowish color like Cu. This is because the plasma edge occurs below the visible range for the cuprates. As discussed in reference 1, the plasma edge is the frequency at which $\varepsilon_1(\omega)$ crosses from

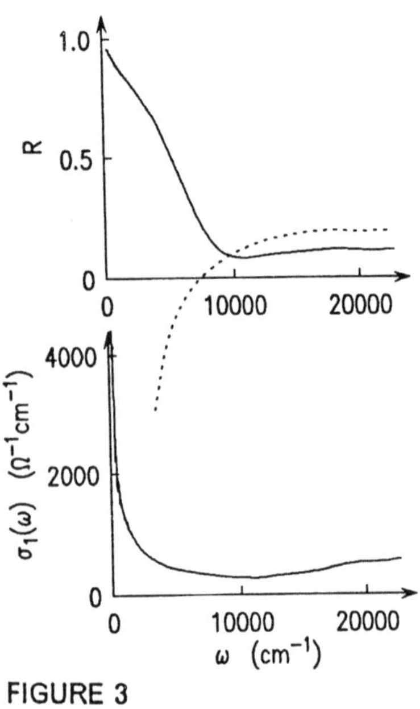

FIGURE 3

Reflectivity, conductivity and $\varepsilon_1(\omega)$ for a typical cuprate superconductor. The dotted curve is $\varepsilon_1(\omega)$, which crosses zero at about 7500 cm^{-1} and has a high frequency value of about 3.5 at 20,000 cm^{-1} (no scale is shown for clarity), (from ref 15).

negative to positive, and the reflectivity drops as a consequence. In the simplest case, this frequency is proportional to the square root of the carrier density. For the metallic cuprates it is about 8,000 cm^{-1} (1.0 eV), and thus the reflectivity is low throughout the visible range ($\sim$16,000-24,000 cm^{-1}). This is why the cuprates have a dark appearance. The dotted curve in fig 3 is $\varepsilon_1(\omega)$, which crosses zero at about 7500 cm^{-1} and has a value of about 3.5 near $\omega = 20{,}000\ cm^{-1}$.

The curves shown in fig 3 represent the reflectivity, conductivity and $\varepsilon_1(\omega)$ associated with the CuO_2 planes in $YBa_2Cu_3O_7$ ($T_c \simeq 92$ K). Very similar spectra are obtained for $Bi_2Sr_2CaCu_2O_{8-y}$ crystals with $T_c \simeq 92$ K (Terasaki et al.,12) and $La_{1.85}Sr_{.15}CuO_4$ ($T_c \sim 35$ K, Uchida et al.,10). The special problem of separating chain and plane contributions to $\sigma(\omega)$ in $YBa_2Cu_3O_{7-y}$ is discussed elsewhere(13-16). The important thing here is that the spectra shown in figure 3 are typical of layered cuprates doped close to the optimum composition range for high T_c superconductivity (see ref. 14).

Unlike the ordinary metal (fig 1) the cuprate (fig 3) does not have a clean interband threshold. $\sigma_1(\omega)$ falls for $\omega \lesssim 10{,}000\ cm^{-1}$, and it rises for $10{,}000 \lesssim \omega \lesssim 20{,}000\ cm^{-1}$, but there is no clear threshold energy for the interband transitions. The physics of the interband transitions in cuprates is not elementary. At half-filling the $\sim$2 eV transition is a transition between strongly hybridized bands where the final state is the empty upper Hubbard band. Electron-electron interactions play a crucial role here. How this feature evolves with doping, i.e., away from half-filling, is a very difficult problem (see, e.g., reference 11), thus we do not know exactly what the interband contribution to $\sigma_1(\omega)$ is for the metallic cuprates.

Here we would like to focus on a different aspect of the infrared properties, which is the behavior at *low frequency* for cuprates which are doped close to the optimal range for high T_c. Here one can make much progress in understanding the *phenomenology* with one simplifying assumption: that below some characteristic frequency, the conductivity is predominantly of intraband origin. We don't need to know exactly what that frequency is, only that such a frequency exists, and that it is significantly larger than $kT/\hbar$. I believe that that frequency actually lies somewhere between about 5000 and 10,000 cm^{-1}, however, for our purposes it does not have to be that high, and could be as low as 1000 cm^{-1} without changing the discussions which follow.

With this assumption one can conclude that the intraband conductivity is very unusual. In particular, it does not drop like $1/\omega^2$ over any frequency range, as in a conventional metal, but instead falls like $1/\omega^n$ with $n \simeq 0.8$. This is why the reflectivity looks so different from that of the Cu (fig 1) in the infrared range. The Cu reflectivity stays very high because $\sigma_1(\omega)$ drops like $1/\omega^2$ up to $\sim$15,000 cm^{-1}. The cuprate reflectivity, on the other hand, (fig 3) does not stay high because $\sigma_1(\omega)$ drops so much more slowly. There is "extra" conductivity in the infrared range, and this leads to extra absorption, hence lower reflectivity.

An alternative point of view is to assume that the intraband contribution to $\sigma(\omega)$ is ordinary, i.e., that it falls like $1/\omega^2$. One then describe the difference between an assumed $1/\omega^2$ Drude term and the actual conductivity as interband contributions to $\sigma(\omega)$, modelled as Lorentz oscillators and associated with low energy electronic excitations(17). As shown in figure 4a, one can obtain a good fit to the normal state conductivity with this approach. Following Kamaras et al.(17), we use 3 Lorentz oscillators and a Drude term (11 parameters), with the lowest finite frequency oscillator at about 300 cm^{-1} ($\sim$40 meV). Earlier it has been reported that collapsing the Drude term to a Dirac-δ function while leaving the other oscillators unchanged provides a good fit to the superconducting state conductivity. This, however, is not the case, as shown in fig 4b, and as implied in the Erratum to ref 17.

Returning to the first approach, in which one assumes that all the low frequency conductivity (below say 2000 cm^{-1}, for example) is intraband, one can follow the examples of references 2 and 3, and represent $\sigma(\omega)$ in terms of a frequency dependent scattering rate and mass. Formally this involves a generalization of the usual Drude formula (Eq. 1) which allows τ to be frequency dependent, as described in references (2,3,18,19). Obtaining τ (and m^*) from $\sigma_1(\omega)$ and $\sigma_2(\omega)$ is a straight forward problem with two equations and two unknowns, which is solved at each frequency to obtain τ and m^*. The fact that $\sigma_1(\omega)$ and $\sigma_2(\omega)$ obey a causality relation means that τ and m^* will also. Carrying out this procedure for the data shown in figure 3 (and using also $\sigma_2(\omega)$ which is not shown), one obtains the scattering rate and mass enhancement as shown in fig 4. Renormalized frequency dependent scattering rates are shown for $YBa_2Cu_3O_7$ (T_c=92 K, ref 15), $Bi_2Sr_2CaCu_2O_{8-y}$ (T_c=92 K, ref 12), and $La_{1.85}Sr_{.15}CuO_4$ (T_c=35 K, ref 10). These scattering rates are unusual in several related ways. First, they are very frequency dependent; second, they are large in the sense they are comparable to $\hbar\omega$, and, third, they are linear in ω.

From a phenomenological point of view, one should notice that the linearity of the scattering rate vs ω is fundamentally related to a linear dependence of scattering on T(20-23). (For each of the compounds in fig 4. the resistivity increases linearly with T, and it can be shown that this temperature dependence is associated with the scattering rate and mass, rather than the carrier density). The relationship between frequency and temperature dependence can be embodied in the crossover formula(22),

$$\hbar/\tau = a((\pi kT)^2 + (\hbar\omega)^2)^{1/2} \tag{2}$$

where self-energy considerations lead to the π prefactor for the kT term. A linear scattering rate with ω is also closely related to to a quasi-particle lifetime that is linear in energy, which has been discussed from photoemission data, and is also probably related to a broad background signal of electronic origin seen in Raman scattering(22).

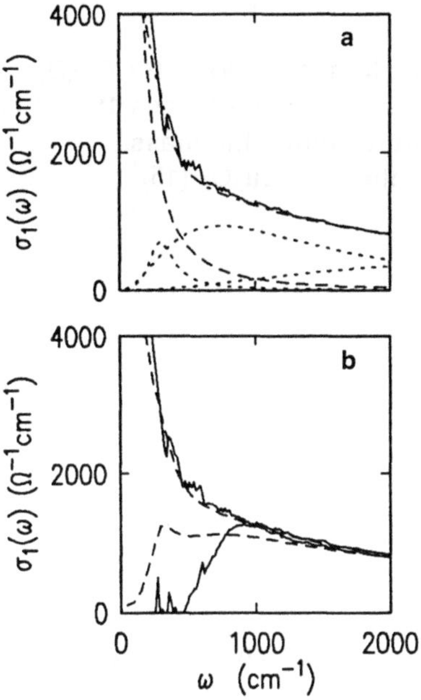

FIGURE 4

Part (a) shows a fit (dot-dashed) to the normal state (solid) using 3 Lorentz oscillators (shown individually as dotted curves) and a Drude term (11 parameters) (after Karamas et al.(17)). Part (b) shows the effect of collapsing the Drude term to a Dirac δ-function (dashed curve), which does *not* provide a good fit to the superconducting state conductivity (solid).

The simple way to think about frequency dependent scattering, is to envision a charged fermion (electron) scattering from a Bose excitation. The classic example of this is the electron-phonon interaction (Holtstein scattering). There the description is fairly straightforward because the charge carriers (electrons), and the excitations that scatter them (phonons) are clearly distinct before the interaction is "turned on". Let us imagine an electron (near the Fermi level) interacting with a single Einstein phonon with an energy $\hbar\omega_0$. Scattering is a real process in which an electron looses energy and a phonon is created with that energy, thus when the energy of the electron is less than $\hbar\omega_0$ a phonon cannot be emitted and there is no scattering. When the energy of the electron is greater than $\hbar\omega_0$, however, the electron can scatter inelastically by emitting a phonon. One can say that this inelastic scattering channel becomes open for $E \gtrsim \hbar\omega_0$. The scattering rate therefore exhibits a step-like increase at $\hbar\omega_0$ associated with the opening of the scattering channel. (Actually one probes the conductivity (and hence the scattering rate) using electromagnetic radiation. In the Holstein process an electron near the Fermi surface simultaneously absorbs a photon and emits a phonon. This

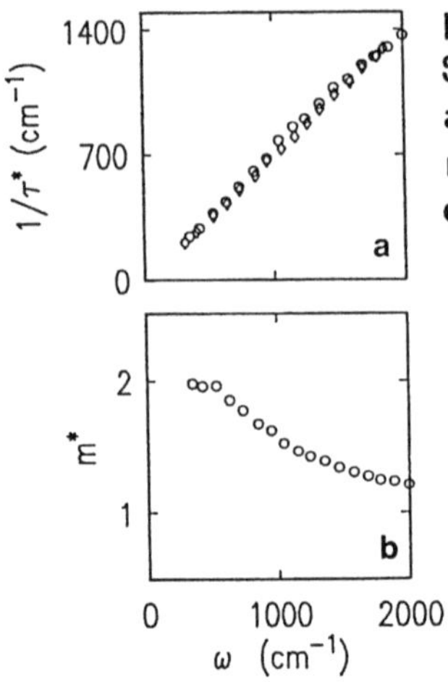

FIGURE 5
Scattering rate vs frequency for $YBa_2Cu_3O_7$ and $La_{1.85}Sr_{.15}CuO_4$ (from references 15 and 10 respectively). Part (b) shows the mass enhancement for the $YBa_2Cu_3O_7$ (15).

can occur only when the energy of the photon is greater than that of the phonon, thus there is an onset of absorption at $\hbar\omega_0$.)

Mass enhancement, on the other hand, is associated with virtual phonon emission, thus the mass is enhanced at low energy, below the scattering threshold. One can say that at low energy the electron drags a phonon cloud along with it, which makes it heavy, while at higher energy its shed this phonon cloud, as the phonons cannot follow the faster motion.

Within this context, the uniformly increasing scattering rates of fig 5 suggest an interaction with a very broad spectrum of excitations, i.e., extending from about $kT/\hbar$ to $\sim$1000 cm^{-1}. Often one thinks of these excitations as related to spin degrees of freedom in the cuprates. Although one can fit the normal state conductivity at a particular temperature reasonably well using the "phonon" based formalism, attempts to fit the data at different temperatures, particularly below T_c, fail in detail. This is true even if one allows the scattering spectrum to be strongly temperature dependent. The reason for these failures may be related to the possibility that the carriers, and the excitations that scatter them are not so clearly distinct in the cuprates. In this case one has a more subtle problem than the electron-phonon interaction, and new theoretical approaches may be desirable.

The frequency dependent mass parameter shown in fig 5b represents the enhancement of the mass over its "band", or high frequency, value due to the interactions that give the scattering rate its frequency dependence. We don't know from these measurements what that high frequency mass is; possibly one can get that from band theory, but we can measure how much it is enhanced in the infrared range. The value of the enhancement is neither extremely large, as one would expect in the polaron regime, nor extremely small. From infrared one infers an enhancement

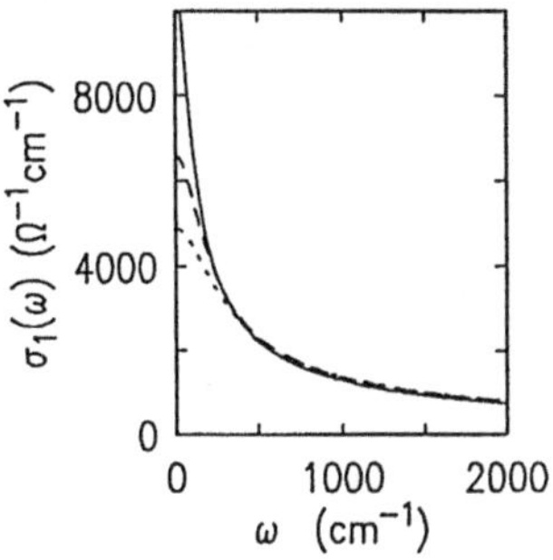

FIGURE 6
Schematic summary of the normal state conductivity in a cuprate superconductor. The dotted curve shows the conductivity at a high temperature (~300 K); the dashed curve shows $\sigma_1(\omega)$ at an intermediate temperature (200 K); and the solid curve shows $\sigma_1(\omega)$ just above T_c (~100 K). Note the normal state temperature dependence is strong only for $\omega \lesssim 3kT/\hbar$

of 2 to 4. Photoemission measurments of band dispersion seem to also yield very similar values for mass enhancement.

One can summarize the phenomenology of the frequency and temperature dependence in the normal state with the picture shown in figure 6. Near d.c., $\sigma_{1n}(\omega)$ falls like 1/T with increasing T (we know this from the d.c. resistivity). In the frequency domain, for $\omega \gtrsim 2kT$ $\sigma_{1n}(\omega)$ falls like $1/\omega^n$ with $n \simeq 0.8$. One thus finds that in the normal state conductivity there is no characteristic energy scale at low frequency (as $1/\tau$ provides for a conventional superconductor), other than kT.

Superconducting state Let us now consider the changes that occur in the infrared properties below T_c. Traditionally these changes, and the energy gap, have been studied by examining reflectivity, transmission and conductivity characteristics(4-8). In figure 7, we compare the evidence for an energy gap in measurements of reflectivity ratios of elemental superconductors, V and Nb(6), the cubic bismuthate superconductor, $Ba_{0.6}K_{0.4}BiO_3$(24), and the cuprate, $YBa_2Cu_3O_7$(25). These figures show the ratio of the reflectivity in the superconducting state to that in the normal state. In each the reflectivity is enhanced in the superconducting state below a characteristic energy. This characteristic energy has been associated with an energy gap in the elemental and bismuthate superconductors, and the gap value inferred has been $2\Delta/kT_c \simeq 3.5$-4, consistent with BCS/Eliashberg expectations. For the cuprate, $YBa_2Cu_3O_7$, one observes a similar enhancement, however, the implied characteristic energy scale is $2\Delta/kT_c \simeq 8$. The conductivity ratios shown in fig 7, compare the best infrared evidence for an energy gap in Pb(7), with the analogous measurements from (untwinned) $YBa_2Cu_3O_7$ crystals(15,26). Both show

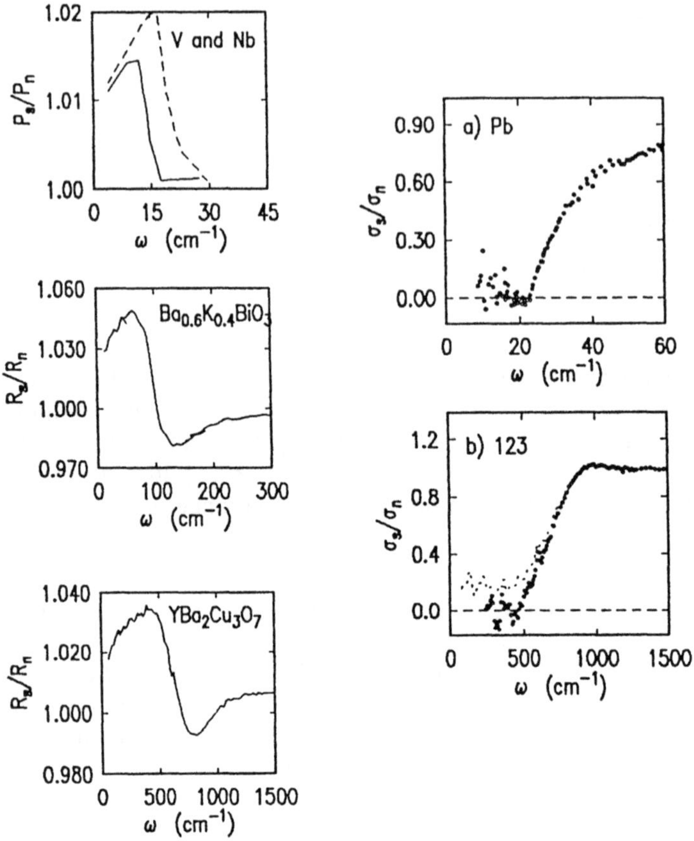

FIGURE 7

Gap inferences from ratios in conventional and cuprate superconductors. The curves on the left show reflectivity enhancements due to superconductivity for elemental(6), cubic bismuthate(24), and layered cuprate superconductors(25), as labelled. The curves on the right show conductivity ratios from Pb (7) and $YBa_2Cu_3O_7$. The latter are based on both reflectivity(15) and absortivity(26) studies of untwinned crystals.

a very similar threshold at a particular characteristic energy ($\sim$22 cm^{-1} for Pb, $\sim$500 cm^{-1} for $YBa_2Cu_3O_7$). No other characteristic energy scale is evident for $YBa_2Cu_3O_7$ (or Pb) in these data.

Previously Thomas et al.(28) and Shutzman, Renk and coworkers(29) have reported evidence for a lower energy scale, which in the former case was associated with an energy gap of conventional size (3.5 kT_c). In the later case, Renk et al.(30) have done very careful follow-up studies, and have recently reported that they find " *evidence for a superconducting gap at $2\Delta/kT_c \simeq 7$, ... no other feature ... at lower frequency, as suggested earlier*". More detailed discussion of the "gap controversy" is presented

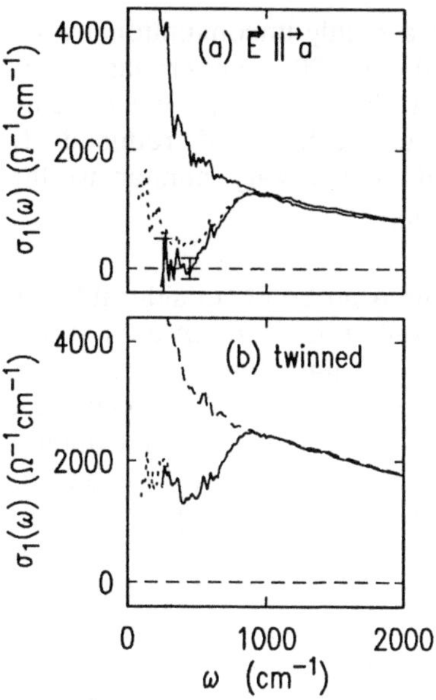

FIGURE 8

Conductivity in the normal and superconducting states for untwinned and twinnned $YBa_2Cu_3O_7$ (from references 15, 26, 18, 27). Part (a) shows measurements of only the CuO_2 plane response, which are performed by polarizing along the $\hat{a}$ direction in untwinned crystals. Below 500 cm^{-1} the conductivity may be as low as 0, or possibly as high as the dotted curve. This is a measure of the present level of experimental uncertainty, which we expect to be reduced as surface quality is improved (see ref 14 for discussion). Part (b) shows the results of the same measurement on twinned crystals. The considerable difference between the spectra in (a) and (b) is due to a chains contribution to $\sigma(\omega)$, which cannot be avoided with the twinned samples(15,16).

in the Kanazawa proceedings. Overall it appears that there is a single (large) characteristic energy scale in the in-plane response of the CuO_2 planes, and no solid evidence for any smaller gap, except when chains or perpendicular response(31) are considered.

The issue is thus not "how big is the gap?", but rather, " what is the nature of the 500 cm^{-1} feature? In what ways is it like an ordinary superconducting gap, and in what ways is it different?". Obviously the data in fig 8 is quite reminiscent of the gap in Pb. On the other hand, as discussed in references 15, 18, 31-33, the temperature dependence is very different from that of a BCS gap, which would collapse to zero as $T \rightarrow T_c$. In particular, we have found that there is no evidence at all for gap collapse-- the gap seems to be just as big at T_c as it is at low T. Also there appears to be some persistence of the gap feature into the normal state, suggesting very large fluctuation effects or a highly unconventional transition. These effect are fairly modest in the $T_c \simeq 90$ K superconductors, where a remnant of the gap persists about 10 K above T_c; but become more pronounced in the CuO_2 plane conductivity of reduced T_c $YBa_2Cu_3O_{7-y}$. In this aspect, the IR data are very similar to the NMR/NQR data, as explored in reference 16.

Conclusions Traditionally infrared has been a probe of considerable importance in the study of fundamental electronic properties of both superconductors and metals in the normal state. At half-filling the insulating cuprates show a "charge transfer" gap of about 1.5-2 eV (12,000-16,000 cm^{-1}). At low doping the conductivity at low frequency can exhibit a maximum at a frequency between about 1000 to 10,000 cm^{-1} depend on the details of the sample, doping, etc. This regime, which precedes the occurrence of highest T_c superconductivity, may be the most complicated since band structure aspects, correlation effects, and strong localization effects may all be competing here. At moderate doping ($x \sim 0.2$), where the highest T_c's and Meissner fractions occur, the phenomenology of the cuprates seems simpler. This is the regime on which we have concentrated in this paper. Here the conductivity has a reasonably universal form, dropping like $1/\omega^n$ in the normal state ($n \simeq 0.8$), and developing a 500 cm^{-1} threshold at low temperature. Very similar behavior is observed in $Bi_2Sr_2CaCu_2O_{8-y}$, $La_{1.85}Sr_{.15}CuO_4$ and $YBa_2Cu_3O_7$ (for untwinned samples). In many ways the spectral shape of the 500 cm^{-1} threshold looks like that of an ordinary (s-wave?) superconducting gap, however, the temperature dependence near T_c is unusual, and there is evidence against gap collapse. Remaining controversies in the IR field largely revolve around the interpretation, modelling or language used to describe the data, rather than the data itself.

Acknowledgemnts: The authors acknowledge valuable discussions with P. W. Anderson, E. Abrahams, S. L. Cooper H. Fukuyama, K. Kitazawa M. V. Klein, P. A. Lee, P. B. Littlewood, A. J. Millis, T. M. Rice, D. J. Scalapino, M. Takigawa, I. Terasaki, S. Uchida, C. M. Varma,

References

1. F. Wooten, *Optical Properties of Solids*, (Academic Press, New York, 1972).
2. J. W. Allen and J. C. Mikkelsen, Phys. Rev. B **15**, 2952 (1977).
3. see e.g., B. C. Webb, A. J. Sievers and T. Mahalisin, Phys. Rev. Lett. **57**, 1951 (1986).
4. R. E. Glover and M. Tinkham, Phys. Rev.104, 844 (1956), Phys. Rev.108, 243 (1957)

5. D. M. Ginsberg and M. Tinkham, Phys. Rev.118, 990 (1960)

6. P. L. Richards and M. Tinkham, Phys. Rev.119, 575 (1960)

7. L. H. Palmer and M. Tinkham, Phys. Rev.165, 588 (1968)

8. R. R. Joyce and P. L. Richards, Phys. Rev. Lett. 24, 1007 (1970)

9. N. W. Ashcroft and N. D. Mermin, *Solid State Physics*, (Holt, Rinehart and Winston, New York, 1976).

10. S. Uchida, T. Ido, H. Takagi, T. Arima, Y Tokura and S. Tajima, Phys. Rev. B**43**, 7942 (1991), and Kanazawa proceedings

11. G. W. Sawatzky, Kanazawa proceedings

12. I. Terasaki, T. Nakahashi, S. Takebayashi, A. Maeda and K. Uchinokura, Physica C165 152 (1990), and Kanazawa proceedings

13. B. Koch, H. P. Geserich and TH. Wolf, Solid State Commun.71, 495 (1989)

14. Z. Schlesinger at al., Kanazawa conference proceedings, 1991

15. Z. Schlesinger, R. T. Collins, F. Holtzberg, C. Feild, U. Welp, G. W. Crabtree, Y. Fang and J. Z. Liu, Phys. Rev. Lett.**65**, 801 (1990)

16. L. D. Rotter, Z. Schlesinger, R. T. Collins, F. Holtzberg, C. Feild, U. Welp, G. W. Crabtree, Y. Fang, K. G. Vandervoort, S. Fleshler and J. Z. Liu, submitted to Phys. Rev. Lett.

17. K. Kamaras, S. L. Herr, C. D. Porter, N Tache, D. B. Tanner, S. Etemad, T. Venkatesan, E. Chase, A. Inam, X. D. Wu, M. S. Hegde, and B. Dutta, a) Phys. Rev. Lett. **64**, 84 (1990); b) Erratum, ibid, 1962 (1990)

18. Z. Schlesinger, R. T. Collins, F. Holtzberg, C. Feild, G. Koren and A. Gupta, Phys. Rev. B**41**, 11237 (1990)

19. J. Orenstein, G. A. Thomas, A. J. Millis, S. L. Cooper, D. H. Rapkine, T. Timusk, L. F. Schneemeyer and J. V. Waszczak, Phys. Rev. B**42**, 6342 (1990)

20. P. W. Anderson, in Strong Correlation and Superconductivily, edited by H. Fukuyama, S. Maekawa and A. Malozamoff (Springer-Verlag, Berlin, 1989), and Kanazawa proceedings (1991)

21. R. T. Collins, Z. Schlesinger, F. Holtzberg, P. Chaudari, and C. Field, Phys. Rev.B. 39, 6571 (1989).

22. M. V. Klein, in Strong Correlation and Superconductivity, edited by H. Fukuyama, S. Maekawa and A. Malozamoff (Springer-Verlag, Berlin, 1989)

23. C. M. Varma, P. B. Littlewood, S. Schmitt-Rink, E. Abrahams and A. E. Ruckenstein, Phys. Rev. Lett. 63, 1996 (1989);

24. Z. Schlesinger, R. T. Collins, J. A. Calise, D. J. Hinks, A. W. Mitchell, Y. Zheng, B. Dabrowski, N. E. Bickers, and D. J. Scalapino, Phys. Rev. B40, 6862 (1989)

25. Z. Schlesinger, R. T. Collins, D. L. Kaiser, and F. Holtzberg, Phys. Rev. Lett. **59**, 1958 (1987);

26. T. Pham, M. W. Lee, D. H. Drew, U. Welp and Y. Fang, preprint

27. T. Pham, H. D. Drew, S. H. Mosely and J. Z. Liu, Phys. Rev. B**41**, 11681 (1990).

28. G. A. Thomas, J. Orenstein, D. H. Rapkine, M. Capizzi, A. J. Millis, R. N. Bhatt, L. F. Schneemeyer and J. V. Waszczak, Phys. Rev. Lett. **61**, 1313 (1988).

29. J. Schutzmann, W. Ose, J. Keller, K. F. Renk, B. Roas, L. Schultz and G. Saemann-Ischenko, Europhysics Lett. 8, 679 (1989)

30. K. F. Renk et al., preprint

31. R. T. Collins, Z. Schlesinger, F. Holtzberg and C. Field, Phys. Rev. Lett. 63, 422 (1989).

32. D. van der Marel, M. Bauer, E. H. Brandt, H.-U. Habermeier, W. Koenig, and A. Wittlin, Phys. Rev. B**43**, 8606 (1991)

33. R. T. Collins, Z. Schlesinger, F. Holtzberg, C. Feild, U. Welp, G. W. Crabtree, J. Z. Liu and Y. Fang, Phys. Rev. B**43**, 8701 (1991)

Electronic Structure Studies of Cuprate Superconductors by High-Energy Spectroscopy

N. Nücker

Kernforschungszentrum Karlsruhe, INFP, Postfacn 3640,
W-7500 Karlsruhe, Fed. Rep. of Germany

Abstract. For the understanding of superconductivity in the high temperature superconductors (HTSC) the knowledge of the electronic states near the Fermi level is of primary interest. High-energy spectroscopies are a powerful tool for the investigation of high-energy parameters which determine the low energy properties of these compounds. In this lecture high-energy spectroscopy investigations on the electronic structure of HTSC will be presented. The pecularities of these spectroscopies combined with the complexity of HTSC require very careful preparation of HTSC samples and critical interpretation of results.

1. Introduction

Immediately after the discovery of high T_c superconductors (HTSC) by Bednortz and Müller [1] many ideas and speculations on the mechanism of high T_c superconductivity were put forward. These were refined or replaced by new theories when more precise knowledge on the electronic states became available from a whealth of experimental and theoretical investigations. At this time, however, there is no general accepted theory describing the electronic structure of HTSC. For the understanding of superconductivity the knowledge of electronic states is important, in particular those near the Fermi level (E_F). The first insight came from band structure calculations [2], which, however, were unable to predict the nonmetallic antiferromagnetic ground state of undoped La_2CuO_4. From the sign of the Hall signal [3] and more directly from the observation of unoccupled oxygen states using electron energy - loss spectroscopy (EELS) [4] the nature of the charge carriers in $La_{2-x}Sr_xCuO_4$ was found to be holes residing mainly on oxygen sites. This excluded all theories based on trivalent Cu as for example the original proposal by Bednorz and Müller [1]. The observed splitting of the Cu $2p_{3/2}$ X-ray photoemission spectrum (XPS) [5] into a main peak and satellites showed the importance of onsite correlation effects as had to be expected for a 3d element. These onsite correlations were not taken into account in the early band structure calculations (BSC) using the local density approximation. Undoped La_2CuO_4 is, because of correlation effects, not a metal with a half-filled Cu $3d_{x^2-y^2}$ - O 2p band as predicted by BSC but a

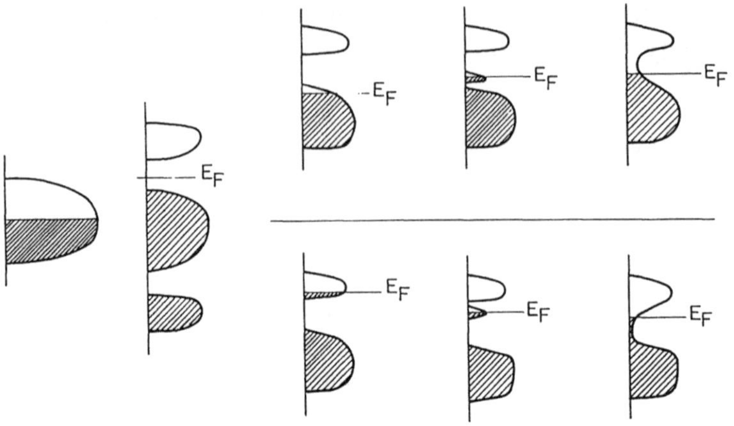

Figure1: Schematic band structure of high T_c superconductors;
left: from band structure calculations in the local density approximation
second: charge transfer insulator right: possible changes on p-type (upper)
or n-type (lower) doping (see text).

charge transfer insulator. The transition to the metallic state on p-type or n-type doping may result from a shift of the Fermi level (see Fig. 1) in rigid bands, by impurity bands [6], or by a continuous change of electronic states leading to a complete filling of the gap.

2. Spectroscopy of occupied states

For the investigation of states below and above the Fermi level several spectroscopical methods are available, and we will discuss the so-called high energy spectroscopies shown schematically in Fig. 2. In x-ray or ultraviolet photoelectron spectroscopy (XPS, UPS) an electron is excited from a core level or the valence band by a photon of known energy and the energy of the emitted electron is measured. Since an electron leaves the system a hole is left in the final state. The spectra therefore may be influenced by multielectron effects which must be taken into consideration.

In the case of O 1s core level spectroscopy for example, the 1s holes do not significantly disturb the O 2p levels near E_F [7] and hence the binding energy of O 1s electrons can be measured rather precisely. The binding energy is influenced by the neighboring atoms of the excited oxygen atom. Thus different oxygen sites show slightly different O 1s binding energies. Usually, the line width due to the finite core level life time and the resolution of the monochromator is too large to resolve these chemical shifts in XPS. On the other hand, since only electrons emitted within ~ 30 Å from the surface reach the detector, a surface contamination (O-H) was observed due to its chemical shift in many investigations.

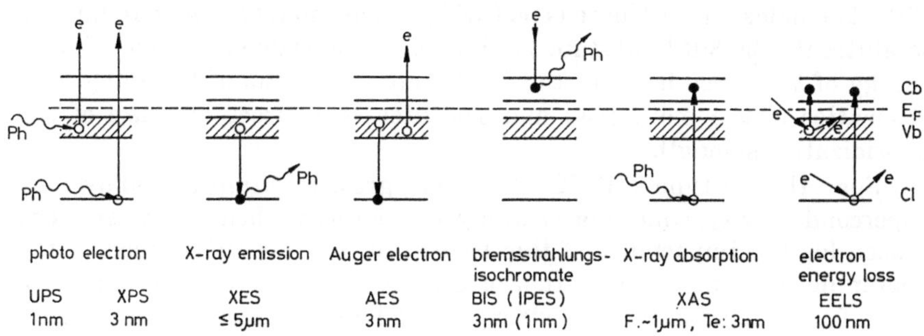

Figure 2: High energy spectroscopies and their characteristics (schematically); ph = photon, e = electron. Last line: probing depths, F = fluorescence yield, Te = total electron yield detection mode.

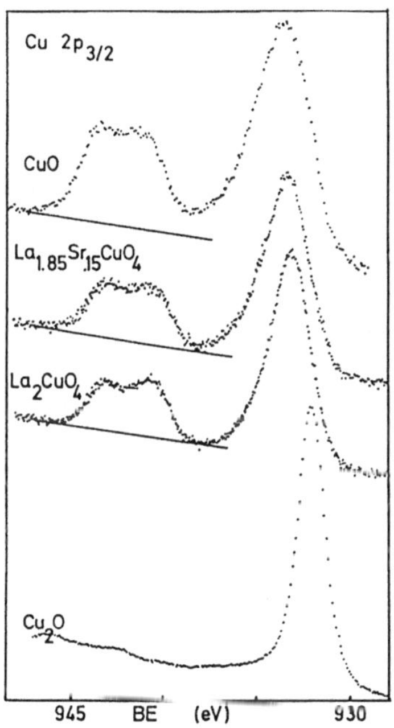

Figure 3: Cu $2p_{3/2}$ XPS peaks of CuO, $La_{2-x}Sr_xCuO_4$ ($x = 0$ and $x = 0.15$), and of Cu_2O [5].

In the case of Cu $2p_{3/2}$ XPS there is an interaction of the Cu 2p hole and the Cu 3d orbitals which clearly shows up in the spectra (Fig. 3) by a satellite structure [5]. The strong line near 933 eV seen for Cu_2O and the other compounds represents the full $3d^{10}$ shell. In the spectra of CuO and the HTSC´s this line represents the transition to Cu $\underline{2p}3d^{10}\underline{L}$ final states with a core hole in the Cu 2p state ($\underline{2p}$) and a hole in the ligand state $\underline{L}$ due to a charge transfer from the ligands (O) to the Cu ion. The satellites centered at 941 eV are due to a $\underline{2p}$ $3d^9$-like multiplet of final states [8]. The shift to higher energy is mainly caused by the Coulomb repulsion (U) of the two holes on Cu. There is no Cu^{3+}, which should show up in further satellites (Cu 2p $3d^8$ final states) at higher energies due to the correlation energy of the three holes in the final state. The ratio of the integrated intensities of the main peak and satellites is a direct measure of the hybridization strength.

For the n-type HTSC the as grown samples show no superconductivity, which appears only after a further heat treatment. The reason for the importance of this treatment was not understood in the beginning, but the investigation of Cu $2p_{3/2}$ XPS showed an enhanced ratio of main to satellite peak strength and hence an increased covalency for superconducting $Nd_{1.85}Ce_{0.15}CuO_4$ [9]. Probably, in the pristine compound the electrons are localized at the Ce sites in an impurity level. The change in the relative intensities is accompanied by a shift of the main line indicating that the structure dependent Madelung potential is an important factor controlling the hybridization strength in the ground state [10].

Recently high resolution (0.35 eV) angle resolved O 1s XPS measurements [11] were performed on $Bi_2Sr_2CaCu_2O_8$. In samples annealed in Ar two lines were observed at 528.3 eV and 529.0 eV. A new line at 530 eV became visible after annealing in 1 atm of oxygen. Annealing in 12 atm oxygen resulted in a further line at 531.2 eV. The intensity of the two lines at higher energies increase strongly at grazing incidence (10^0), which reduces the sampling depth to less than 7 Å. This indicates that extra oxygen is added, to the surface or into the BiO layer. This experiment clearly shows the problems when surface sensitive spectroscopies are used.

Applying XPS or UPS to the valence band allows to investigate the electronic structure of the valence band. The coupling of excitation and decay mechanisms at core electron photoabsorption onsets gives rise to strong resonance effects. The UPS and XPS spectra therefore are strongly energy dependent. This can be exploited to investigate site specific states by resonant photoemission spectroscopy. A direct observation of Cu^+ states in n-type HTSC was claimed to be observed for the first time with this technique [12]. The authors investigated the satellites associated with the Cu 3d ground state in the Cu 3p resonance excitation (Fig. 4). With the constant initial state technique (the photon's energy hv and the detector's energy-window E_f are changed simultaneously to investigate an initial state $E_i = hv - E_f$) a satellite structure was observed for $E_i = 17.6$ eV independent of doping. This feature, therefore, was assigned to the $3d^9$ (Cu^{2+}) ground state. Another resonance was observed at $E_i = 20$ eV for Ce doped $Nd_{2-x}Ce_xCuO_4$ only, and hence was ascribed to Cu^+.

More information on the valence band can be obtained using angle resolved ultraviolet photoelectron spectroscopy (ARUPS) on single crystalline samples. We note again that this method is highly surface sensitive and probes only the uppermost layer of the HTSC, or even only a

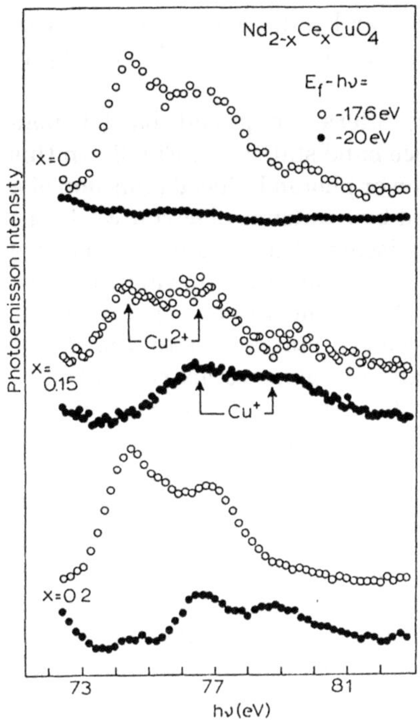

Figure 4: Constant initial state constant photoemission spectra taken at E_f-hv energies.
E_f = kinetic energy of photoelectron,
hv = energy of photon.
E_f - hv= -17.6 eV (open circles)
E_f - hv = -20 eV (solid circles) [12].

part of it. It is by no means shure that the first layer of a HTSC is superconducting at all. In the case of $Bi_2Sr_2CaCuO_8$ the bonds between the Bi-O layers are very weak, as can be seen from the easy cleavage of this compound comparable to mica. This compound therefore offers the best chance that a layer which is one lattice constant thick has nearly bulk properties. In the case of almost all the other HTSC the cleavage may drastically alter the surface unit layer and thereby prevent to get useful information on bulk properties by ARUPS. Recently, a surface specific excitation was observed for $YBa_2Cu_3O_7$ [13].

Most ARUPS investigations therefore were performed on $Bi_2Sr_2CaCu_2O_8$ cooled to low temperature in an ultrahigh vacuum and cleaved under this conditions. High energy resolution (20 meV ... 60 meV) experiments allowed the observation of the Fermi edge above T_c and the opening of a superconducting gap at lower temperatures [14]. As will be shown in the lecture by T. Takahashi (this volume) the observation of maxima changing their position in energy on changing the momentum transfer can be interpreted in terms of band dispersions [15]. The bands were found to cross the Fermi level at nearly the same places in reciprocal space as predicted by band structure calculations in the local density approximation. The dispersion was found to be less pronounced than predicted, which has been interpreted by an enhanced mass of the charge carriers. Recently, it has been discussed that the observed energy dependence of the life-time broadening near E_F increases linearly with the binding energy and is therefore not compatible with a Fermi liquid picture [16].

X-ray emission spectroscopy (XES) offers the possibility for non-surface sensitive investigations of valence band states (see Fig. 2). In this technique a core hole is formed by electron or photon induced emission of a core electron. This core hole can be filled by a valence electron while the transition energy is released by the emission of a photon. The energy spectrum of this photons gives a picture of the site specific valence band. The investigation of differnt core levels of one atomic species gives information on the character of the valence states due to the dipole selection rules for x-ray transitions. As an example, a comparison of CuK, CuL, and OK XES spectra is given in [17]. If there is more than one site for the investigated atom, the site specific spectra add up shifted with respect to each other which reflects the chemical shift at each site. More detailed information can be obtained by using a polarization sensitive x-ray detector [17, 18]. The OK_α XES spectra for a polarization of the photon in the a, b-plane of $YBa_2Cu_3O_7$ shows a maximum near 2 eV and a second peak at about 5 eV. For the polarization **E** ∥ **c** the latter peak is absent and the first peak is shifted by about 1 eV to higher energy. In comparison to UPS the energy resolution is rather poor (about 0.7 eV for O 1s XES). Due to the low efficiency of x-ray monochromators and detectors a rather high exciting power density is needed. In the experiment of Ref. 19 the electron microprobe consisted of an electron beam of 6 mW focussed on a 1μm diameter spot, and hence the power density was about 10^{11} W/cm^3. From this high dose a partial O release may result.

In Auger electron spectroscopy (AES) the recombination of a core hole state is accompagnied by the emission of an electron, thus leaving the atom in a two hole state (see Fig. 2). AES with two holes in the valence band as final state was applied to investigate the on-site correlations in HTSC. An example for CuL_3VV (Cu $2p_{3/2}$ hole (L3) filled by a valence electron (V) and a second valence electron emitted) is shown in Fig. 5. The kinetic energy of the Auger electron $E_{Kin}(CVV)$ leaving the valence band V is determined by the core level binding energy E(C), the onsite correlation energy U, and the binding energy E(V) for the two valence electrons : $E_{Kin}(CVV) = E(C) - U - 2E(V)$. This relation is, of course, over-simplified, since the involved bands are composed of multiplets. Hence U must be specified to the states involved. The determination of an effective correlation strength U_{eff} must consider that different types of experiments reach different multiplet states due to the transition matrix elements involved [20]. A rough determination for the Cu onsite correlation energy U_{dd} in HTSC can be obtained from the values of the peak energy in the CuL_3VV Auger spectrum $E_A \simeq 918$ eV, which is mainly due to the $3d^8\underline{L}$ state, the XPS binding energy $E_C \simeq 931$ eV experimentally derived for the $3d^{10}\underline{L}$ final state and the valence band

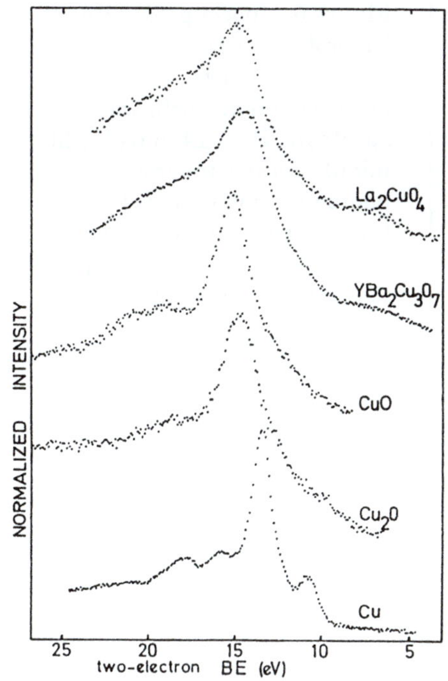

Figure 5: CuL_3VV Auger spectra of Cu compounds [19].

binding energy determined by high energy XPS (to get mainly the Cu spectrum) $E_V \simeq 4$ eV. This yields $U_{eff} \sim 5$ eV, which is at the lower end of values commonly accepted ranging up to ~9 eV (for a detailed evaluation see [21]). Similarly, the onsite correlation strength for oxygen in the HTSC has been estimated to be about 5 eV.

3. Spectroscopy of unoccupied states

For the investigation of unoccupied states (see Fig. 2), a core electron may be excited by X-ray absorption (XAS) or by electrons (EELS), or the absorption of electrons may be observed by bremsstrahlung-isochromat spectroscopy (BIS). Similar to XPS and UPS, the cross section of BIS - often named inverse photoelectron spectroscopy (IPES) at lower energies - shows strong resonance effects, which can be exploited for the detection or identification of site specific empty states. Fig. 6 shows a comparison of BIS spectra on $Bi_2Sr_2CaCu_2O_8$ for different energies of incoming electrons [22, 23] and an O 1s absorption spectrum from an EELS investigation [24]. For high energy electrons (1486.6 eV, upper curve) the cross section for Ca 3d and O 2p states are only 2 % and 3 %, respectively, of that for Cu 3d. For lower energies (intermediate curves) resonances can be observed. For an electron energy of 40 eV the Ca 3d, Cu 3d and O 2p cross sections are almost the same, while near 17 eV the Bi 6p cross section is strongest,

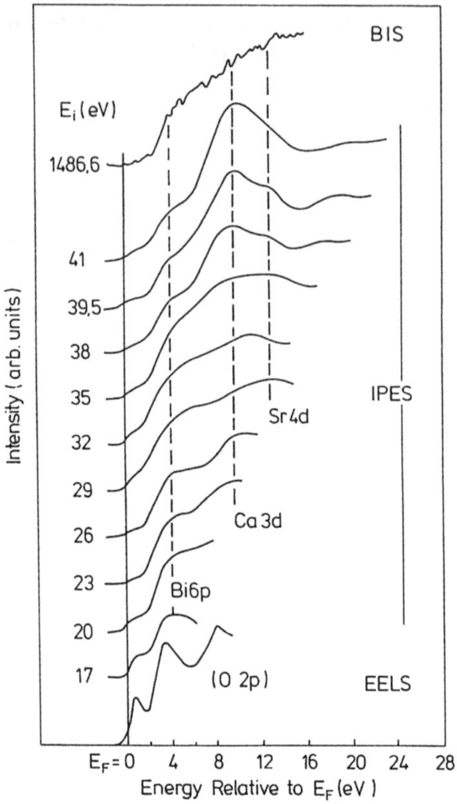

Figure 6: Unoccupied states of $Bi_2Sr_2CaCu_2O_8$. Uppermost curve: bremsstrahlen isochromat spectrum, $E_i = 1486.6$ eV [22] middle curves: inverse photoemission spectra [23] lower curve: O 1s absorption spectrum by electron energy loss spectroscopy [24].

followed by that of O 2p being 3 times lower. These spectra show that Bi 6p and O 2p states are dominant at low energies. In comparison, the O 1s absorption spectrum (lower curve) shows unoccupied oxygen 2p states only.

Since low energy electrons are strongly absorbed, IPES is rather surface sensitive as are XPS and UPS. The low efficiency of high resolution X-ray detectors demands for high electron doses, which may damage the surface of the HTSC.

Similar to ARUPS, angle resolved IPES (ARIPES) allows the investigation of the structure of unoccupied states. As an example, Fig. 7a shows results of ARIPES along the Γ-X direction of $Bi_2Sr_2CaCu_2O_8$ [25]. Fig. 7b shows band dispersions derived from these results compared to band structure calculations [26]. ARUPS and ARIPES on $Bi_2Sr_2CaCu_2O_8$ will be discussed in more detail by T. Takahashi (this volume).

XAS became very powerful after highly monochromatic, linearly polarized, and easily tunable intensive beams became available from synchrotrons [27]. The absorption of a photon is usually monitored in the (partial or total) yield method, counting all Auger electrons from the recombination of the core hole in a fixed solid angle and energy range.

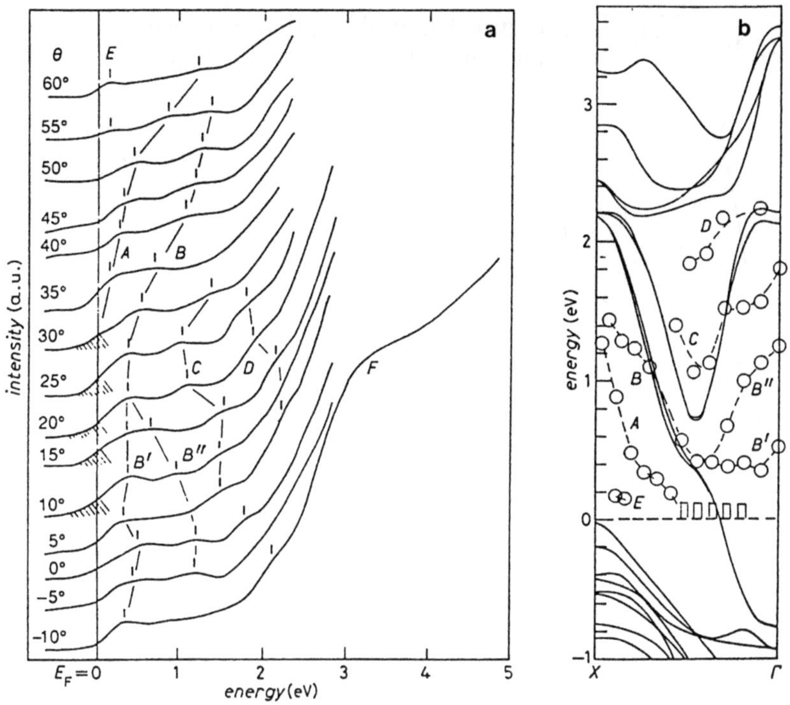

Figure 7a: Angle resolved inverse photoemission spectra along the Γ-X direction of $Bi_2Sr_2CaCu_2O_8$ (hν = 9.5 eV, Δhν = 0.35 eV) [25] (left). b: band structure derived from the result shown (Fig. 7a) compared to band structure calculation [26] (right).

Due to the limited escape depth of these electrons this method is surface sensitive. In contrary, the fluorescence yield detection mode, where the recombination photons are counted, is not surface sensitive. High energy EELS in transmission is an equivalent method. In this case, the absorption is monitored by the relative number of electrons with an energy reduced by the energy needed in this process. Sample thicknesses are about 100 nm to 200 nm

A wealth of information on HTSC's was obtained from these techniques applied to the absorption from Cu K, Cu $L_{2/3}$, O K and other core levels.

As a first example we discuss the O 1s absorption of $La_{2-x}Sr_xCuO_{4+\delta}$ performed by EELS [28] as shown in Fig. 8. Similar data have been obtained by XAS [29]. The lowest spectrum shows the absorption from an undoped sample annealed at 250^{0}C in ultra high vacuum, which is necessary to remove excess oxygen (δ). A peak (labeled C) at 530.2 eV with a threshold at 529.7 eV is observed. This peak is followed by a strong rise in intensity above 531 eV, which is assigned to O 2p states hybridized with La 5d and 5f states of the LaO layers. The peak C slightly moves to

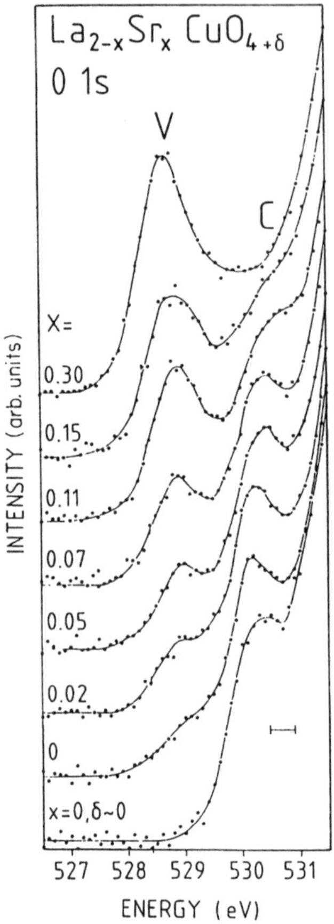

Figure 8: O 1s absorption spectra of $La_{2-x}Sr_xCuO_{4+\delta}$ ($0 \leq x \leq 0.3$) [28].

higher energies on Sr doping, loosing spectral weight and disappearing at the highest doping level ($x = 0.3$).

However, upon doping La_2CuO_4 a new peak (labeled V) emerges at the O 1s binding energy 528.4 eV, which is at E_F. The intensity of this peak V is proportional to the Sr doping and is ascribed to holes in the valence band with predominantly O 2p character. The threshold of peak C is about 1.5 eV higher in energy, a value close to the gap width (2.0 eV) observed by optical spectroscopy [30]. Therefore, the second peak is assigned to the oxygen admixture to the conduction band, which may be considered as the upper Hubbard band (see Fig. 1). These states can be observed in the O 1s absorption spectrum due to the covalency of the Cu-O bonds in the HTSC´s.

The strong reduction of the intensity of the conduction band peak C is probably not primarily due to a transition of a highly correlated system ($x < 0.15$) to a less correlated one ($x > 0.3$). The appearance of unoccupied states below the conduction band states upon doping influences the

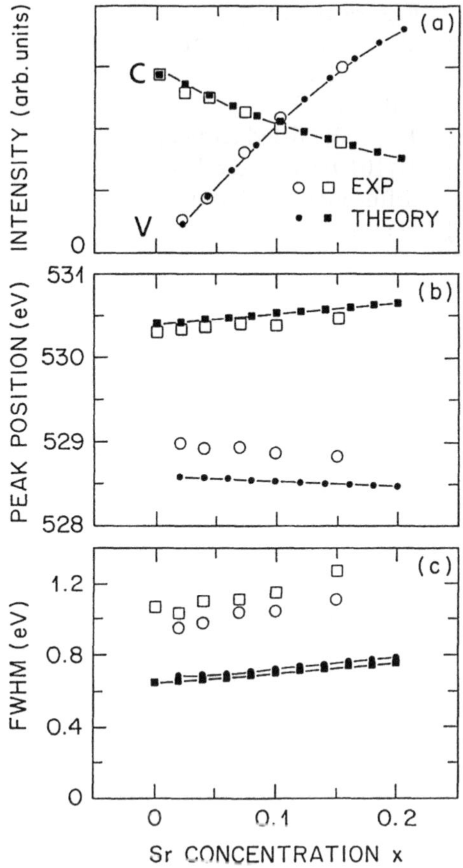

Figure 9: Comparison of experiment and cluster calculation for the conduction band (C) and the valence band (V) peak in O 1s absorption measurement of $La_{2-x}Sr_xCuO_{4+\delta}$
a: peak intensities
b: peak positions
c: peak width as a function of Sr concentration x. [29]

transition matrix elements, thus reducing the transition probability to the oxygen hole part of the Cu $3d^{10}$ $\underline{L}$ final states. This has been shown explicitely by a three band Hubbard model cluster calculation [29] which is in excellent agreement with experiment as can be seen from Fig. 9. These investigations show a smooth increase of the number of holes with no indication for the onset of the metallic behavior near x = 0.06. The observed absorption spectra can be well described by a doped charge transfer insulator. There is no evidence for midgap or impurity states within the charge transfer gap as inferred from optical measurements [30]. The corresponding peaks in these optical investigations may rather be explained by transitions from occupied to unoccupied states within the valence band.

Another example for XAS and EELS O 1s absorption measurements concerns the symmetry of O 2p holes in the $YBa_2Cu_3O_7$ and $YBa_2Cu_4O_8$ compounds [24, 31, 32]. Band structure as well as cluster calculations gave contradictory symmetries for the O 2p orbitals involved in the CuO_2 planes and the Cu-O chains. Small changes in the parameters used may

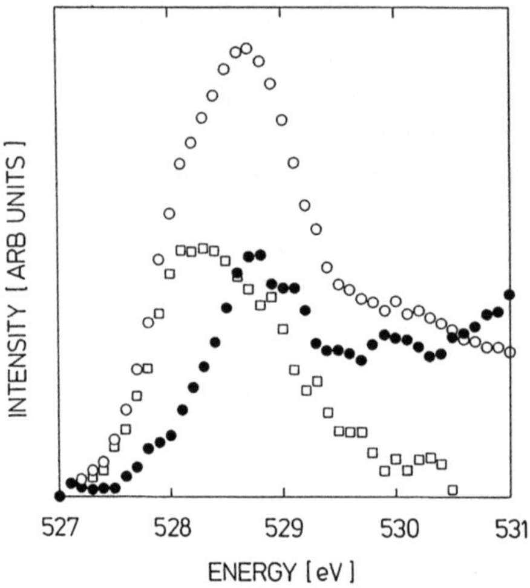

Figure 10: x-ray absorption spectrum of twinfree $YBa_2Cu_3O_7$ E ∥ [100] : filled circles, E ∥ [010] : open circles, difference spectrum: rectangles

shift the bands to or away from the Fermi surface. Unsufficient treatment of onsite correlations further enhance the uncertainty. Band structure calculations for $YBa_2Cu_3O_7$ yielded a narrow band with high density of states at E_F due to the Cu-O chains with dp π symmetry [33]. For $YBa_2Cu_4O_8$ this band should be below E_F [34].

Strong evidence against π bonded oxygen in $YBa_2Cu_3O_7$ was derived from ^{89}Y nuclear magnetic resonance [NMR] investigations [35] yielding a negative Knight shift. This result is further supported by NMR on ^{17}O [36]. An electric field gradient calculation [37] for $YBa_2Cu_3O_7$ and a comparison of NMR results yields O 2pσ orbitals as well and is not in line with O 2pπ orbitals.

The relative transition probabilities to O 2p orbitals in the three crystallographic axis were measured on oriented and twinfree samples of $YBa_2Cu_3O_7$ and $YBa_2Cu_4O_8$ by polarized absorption spectroscopy. In the polarized XAS investigation the mono-crystals were oriented with respect to the electrical field vector of the soft X-rays to single out excitations into orbitals extending along a particular crystal axis. In the EELS experiments the momentum transfer **q** was likewise chosen to be parallel to a crystallographic axis. As a typical result the absorption spectra of twinfree $YBa_2Cu_3O_7$ are shown in Fig. 10 [32] for polarisations **E** ∥ [100] and **E** ∥ [010]. The latter spectrum gives the sum of unoccupied O 2p states for the CuO_2 plane in the **b**-direction and those of the O (1) atom in the Cu-O chain. Since the states in the CuO_2 plane can be expected to be symmetric with respect to the **a** and **b** directions, the contributions of the O 2p states of the chains can be derived, from the difference of the spectra observed in the **b** and the **a** directions (Fig. 10). Comparing the relative intensities observed for polarisations in the **a**-, **b**-, and **c**-crystal axes, the number of holes for the individual oxygen orbitals (σ or π) can be derived. A hole distribution comparable to BSC results is in favour of O 2pσ

orbitals at E_F. The classification of the orbitals to the atomic sites is further supported by a comparison of observed and calculated O 1s core level binding energies.

A similar controversy concerns the copper orbitals. It is now commonly accepted that mainly Cu $3d_{x^2-y^2}$ states contribute to the CuO_2 plane density of states at E_F. It has been shown by several Cu $2p_{3/2}$ core level absorption experiments that Cu $3d_{3z^2-r^2}$ orbitals contribute at E_F to a small extent only (see for example [31]). One group of authors [38] repeatedly reported a strong increase of the admixture of $3d_{3z^2-r^2}$ states to $3d_{x^2-y^2}$ states with increasing superconducting transition temperature [T_c] and shifts of the relative peak energies assigned to these transitions. Since these experiments were performed by XAS in the total yield detection mode on a,b-surfaces (i.e. (001) surfaces), the [001] components representing the $3d_{3z^2-r^2}$ orbitals had to be observed at grazing angles. Hence, this technique was highly surface sensitive, and independent experimental evidence is needed to confirm this result. This admixture of the $3d_{3z^2-r^2}$ orbitals may be relevant to superconductivity in the HTSC´s. Two further XAS experiments yielded conflicting results concerning the energy of the two kinds of orbitals [39, 40]. No shifts were observed in EELS investigations [31].

4. Excitations

Finally, we report on results of an experiment on long wavelength collective excitations of charge carriers in $Bi_2Sr_2CaCu_2O_8$ [41]. Fig. 11 shows the loss function Im $[-1 / \varepsilon (q,\omega)]$ measured by EELS for momentum transfers $q = 0.1$ Å^{-1} and $q = 0.3$ Å^{-1}, the latter for momentum transfers parallel to the a-axis and to the [110] direction. For small q in the a, b-plane, $q \leq 0.1$ Å^{-1}, a pronounced plasmon is observed at 1 eV. With increasing momentum transfer the plasmon peak broadens and shifts to higher energy. For a given momentum transfer the plasmon frequency is highest for polarizations in the **a** or **b** direction and lowest for the [110] direction. For each polarisation a dispersion quadratic in q is observed. For the [100] and the [110] direction the size of the quadratic plasmon dispersion differs by a factor of about two. This strong in-plane anisotropy can be explained in the framework of a random phase approximation using a simple tight-binding ansatz for the Cu-O band having a quadratic Fermi surface. The calculation yields: $(\hbar\omega_q)^2 = (\hbar\omega_p)^2 + 1/6\,(ta)^2 q^2 (3/2 + 1/2 \cos 4\Phi)$, or a variation by a factor of two for the angle Φ between 0 and $\pi / 4$ (ω_q = plasma frequency for momentum q, $\omega_p = \omega_{(q=0)}$, t= transfer integral ~ 1.5 eV, a = lattice parameter, a=b assumed). This is in good agreement with experiment as shown in Fig. 12.

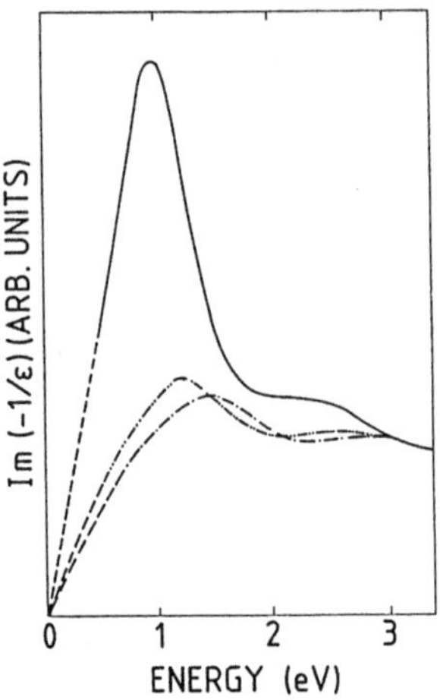

Figure 11: Loss function $\mathrm{Im}[-1/\varepsilon(q,\omega)]$ for $Bi_2Sr_2CaCu_2O_8$. Solid line: momentum transfer $q = 0.1$ Å^{-1}, **q** ∥ a, b-plane dash dotted line: $q = 0.3$ Å^{-1}, **q** ∥[100] dash double-dotted line: $q = 0.3$ Å^{-1}, **q** ∥[110] dashed line: extrapolation to zero. All curves are normalized between 3 eV and 4 eV [41].

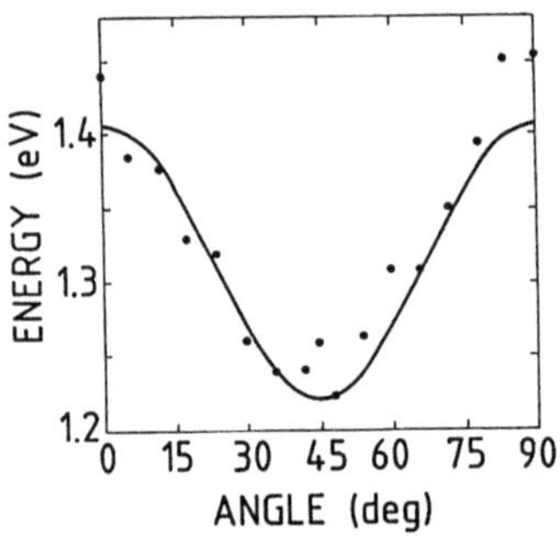

Figure 12: Variation of the energy of the charge-carrier plasmon in $Bi_2Sr_2CaCu_2O_8$ for momentum transfer $q = 0.3$ Å^{-1} as a function of the rotation angle Φ of the direction of the momentum transfer around the [001] - axis. $\Phi = 0^0$: **q** ∥[010]; $\Phi = 45^0$: **q** ∥[110]; $\Phi = 90^0$: **q** ∥[100]. Solid line: theoretical estimate [41].

5. Summary

In this lecture the ability of high energy spectroscopies in investigating the electronic structure of HTSC was discussed guided by some experiments. Only a rather limited number of investigations could be reported, the selection of which is necessarily somewhat arbitrary. Likewise, only a limited number of references could be given, and the interested reader is referred to these publications to pursue his search of the literature.

References

1. J.G. Bednorz and K.A. Müller, Z. Phys. B **64**, 189 (1986)
2. L.F. Mattheiss, Phys. Rev. Lett. **58**, 1028 (1987)
3. N.P. Ong, Z.Z. Wang, J. Clayhold, J.M. Tarascon, L.H. Greene, and W.R. Mc Kinnon, Phys. Rev. B **35**, 8807 (1987)
4. N. Nücker, J. Fink, J.C. Fuggle, P. J. Durham, and W.M. Temmerman, Phys. Rev. B **37**, 5158 (1988)
5. N. Nücker, J. Fink, B. Renker, D. Ewert, C. Politis, P.J.W. Weijs, and J.C. Fuggle, Z. Phys. B **67**, 9 (1987)
6. I. Terasaki, T. Nakahashi, S. Takebayashi, A. Maeda, and K. Uchinokura, Physica C **165**, 152 (1990)
7. J. Pflüger, J. Fink, G. Crecelius, K.-P. Bohnen, and H. Winter, Solid State Commun. **44**, 489 (1982)
8. G.v.d. Laan, C. Westra, C. Haas, and G.A. Sawatzky, Phys. Rev. B **23**, 4369 (1981)
9. S. Kohiki, J. Kawai, T. Kamada, S. Hayashi, H. Adachi, K. Setsune, and K. Wasa, Physica C **166**, 437 (1990)
10. J.M. Tranquada, S.M. Heald, W. Kunnmann, A.R. Moodenbaugh, S.L. Qiu, Y. Xu, and P.K. Davies, Preprint
11. F. Parmigiani, Z.X. Shen, D.B. Mitzi, I. Lindau, W.E. Spicer, and A. Kapitulnik, Phys. Rev. B **43**, 3085 (1991)
12. Y. Hwu, M. Marsi, A. Terrasi, D. Rioux, Y. Chang, J.T. Mc Kinley, M. Onellion, G. Margaritondo, M. Capozi, C. Quaresima, A. Campo, C. Ottaviani, P. Perfetti, N.G. Stoffel, and E. Wang, Phys. Rev. B **43**, 3678 (1991)
13. J. Fink, N. Nücker, H. Romberg, M. Alexander, P. Adelmann, R. Claessen, G. Mante, T. Buslaps, S. Harm, R. Manzke, and M. Skibowski, Proceedings Nato Science Forum´90, Biarritz, France Sept. 1990; G. Mante et al., Phys. Rev. B, in print
14. J.-M. Imer, F. Patthey, B. Dardel, W.-D. Schneider, Y. Baer, Y. Petroff, and A. Zettl, Phys. Rev. Lett. **62**, 336 (1989)
15. T. Takahashi, H. Matsuyama, H. Katayama-Yoshida, K. Seki, K. Kamiya, and H. Inokuchi, Physica C **170**, 416 (1990)
16. C.G. Olson, R. Liu, D.W. Lynch, R. S. List, A.J. Arko, B.W. Veal, Y.C. Chang, P.Z. Jiang, and A.P. Paulikas, Phys. Rev. B **42**, 381 (1990)
17. G. Dräger, F. Werfel, and J.A. Leiro, Phys. Rev. B **41**, 4050 (1990)
18. A. Kottman, F. Burgäzy, D.J. Lam, Y. Fang, P. Lamparter, and S. Steeb, submitted to Physica C
19. J.C. Fuggle, P.J.W. Weijs, R. Schoorl, G.A. Sawatzky, J. Fink, N. Nücker, P.J. Durham, and W.M. Temmerman, Phys. Rev. B **37**, 123 (1988)
20. D.v.d. Marel and G.A. Sawatzky, Phys. Rev. B **37**, 10674 (1988)
21. J. Ghijsen, L.H. Tjeng, J. v. Elp, H. Eskes, J. Westerink, G. A. Sawatzky, and M.T. Czyzyk, Phys. Rev. B **38**, 11322 (1988)

22. A.E. Bocquet, S. Ogawa, H. Eisaki, H. Takagi, S. Uchida, A. Fujimori, and S. Suga, Physica C **169**, 1 (1990).
23. T.J. Wagener, Y. Hu, Y. Gao, M.B. Jost, J.H. Weaver, N.D. Spencer, and K.C. Goretta, Phys. Rev. B **39**, 2928 (1989)
24. N. Nücker, H. Romberg, X.X. Xi, J. Fink, B. Gegenheimer, and Z.X. Zhao, Phys. Rev. B **39**, 6619 (1989)
25. H.J. Bernhoff, K. Tsushima, and J.M. Nicholls, Europhys. Lett. **13**, 537 (1990)
26. H. Krakauer, and W.E. Pickett, Phys. Rev. Lett. **60**, 1665 (1988)
27. C.T. Chen, and F. Sette, Rev. Sci. Instrum. **60**, 1616 (1989)
28. H. Romberg, M. Alexander, N. Nücker, P. Adelmann, and J. Fink, Phys. Rev. B **42**, 8768 (1990)
29. C. T. Chen, F. Sette, Y. Ma, H. S. Hybertsen, E. B. Stechel, W. M. C. Foulkes, M. Schluter, S.W. Cheong, A.S. Cooper, L.W. Rupp, Jr., B. Batlogg, Y.S. Soo, Z.H. Ming, A. Krol, and Y.H. Kao, Phys. Rev. Lett. **66**, 104 (1991)
30. S. Uchida, T. Ido, H. Takagi, T. Arima, Y. Tokura, and S. Tajima, Phys. Rev. B **43**, 7942 (1991)
31. N. Nücker, H. Romberg, M. Alexander, and J. Fink, "Studies of High T_c Supercond." ed. A. Narlikar, Nova Science Publishers Vol **6**, 145 (1990)
32. A. Krol, C.S. Lin, Z.H. Ming, Y.H. Kao, N. Nücker, G. Roth, J. Fink, G.C. Smith, K.T. Park, J. Yu, A.J. Freeman, A. Erb, G. Müller-Vogt, J. Karpinski, E. Kaldis, and K. Schönmann, to be published.
33. S. Massidda, J. Yu, A.J. Freeman, and D.D. Koelling, Phys. Lett. **A 122**, 198 and 203 (1987)
34. J. Yu, K.T. Park, and A.J. Freeman, Physica C **172**, 467 (1991)
35. M. Takigawa, A. P. Reyes, P. C. Hammel, J. D. Thompson, R. H. Heffner, Z. Fisk, and K.C. Ott, Phys. Rev. B **43**, 247 (1991)
36. H. Alloul, T. Ohno, and P. Mendels, Phys. Rev. Lett. **63**, 1700 (1989)
37. F.J. Adrian, Physica C **171**, 505 (1990)
38. A. Bianconi, P. Castrucci, A. Fabrizi, M. Pompa, A. M. Flank, P. Lagarde, H. Katayama-Yoshida, and G. Calestani, Springer Series in Solid State Sciences, **90**, 407 (1990)
39. M. Abbate, M. Sacchi, J.J. Wnuk, L.W.M. Schreurs, Y.S. Wang, R. Lof, and J.C. Fuggle: Phys. Rev. B **42**, 7914 (1990)
40. S. Suzuki, T. Takahashi, T. Kusunoki, S. Sato, and H. Katayama-Yoshida: Proc. 2nd ISSP Intern. Symp. Phys. Chem. (Tokyo, Jan 1991)
41. N.Nücker, U. Eckern, J. Fink, and P. Müller submitted to Phys. Rev.

Angle-Resolved Photoemission, Inverse Photoemission, and X-Ray Absorption Studies of Superconductive and Non-Superconductive Bi-Sr-Ca-Y-Cu-O

T. Takahashi

Department of Physics, Tohoku University, Sendai 980, Japan

Abstract. Comparative angle-resolved photoemission (ARPES), angle-resolved inverse photoemission (ARIPES), and angle-resolved x-ray absorption (ARXAS) studies of $Bi_2Sr_2Ca_{1-x}Y_xCu_2O_8$ (x=0.0-0.6) were reported. ARPES and ARXAS results show that hole-doping in an insulator (x=0.6) produces new electronic states in the charge-transfer gap and the density of the new electronic states increases with the hole-concentration. In the "over-doped" region (x=0.0 and 0.2), a slight rigid shift of the bands crossing the Fermi level was observed, suggesting that a rigid-band picture is recovered in the "over-doped" region. ARIPES result confirms that there are two bands which cross the Fermi level centered at the Γ point in the Brillouin zone. ARXAS shows that hole-doping causes a transfer of weight of the electronic states from the lowest unoccupied states (upper Hubbard band) to the mid-gap states produced by the hole-doping in the vicinity of the Fermi level.

1. Introduction

The existence of the Fermi-liquid states in the high-Tc superconductor has been established by angle-resolved photoemission spectroscopy (ARPES)[1-4]. A next step to approach the high-Tc mechanism is to elucidate the detailed structure of the Fermi-liquid states and its formation-process upon hole-doping. In this paper, we report the results of three different spectroscopic studies on superconductive and non-superconductive Bi-Sr-Ca-Y-Cu-O. One is a comparative ARPES study on a series of $Bi_2Sr_2Ca_{1-x}Y_xCu_2O_8$ (x=0.0-0.6) single crystals, where we paid a special attention to the change of the electronic structure from an insulator to a superconductor and from a superconductor to a normal metal. We also performed angle-resolved inverse photoemission spectroscopy (ARIPES) on a $Bi_2Sr_2CaCu_2O_8$ single crystal. There is a serious discrepancy among ARPES experiments [1-3] in the number of bands crossing the Fermi level. Inverse photoemission spectroscopy is a complementary technique to photoemission, giving a direct information on the unoccupied electronic structure. X-ray absorption spectroscopy also provides a crucial information on the nature of unoccupied electronic states. Here we also report the result of a high-energy-resolved angle-resolved oxygen K absorption spectroscopy on $Bi_2Sr_2Ca_{1-x}Y_xCu_2O_8$ (x=0.0 and 0.6) single crystals.

2. Experimental

Single crystals of $Bi_2Sr_2Ca_{1-x}Y_xCu_2O_8$ were grown with a floating-zone method and/or a self-flux method. The atomic ratio between Ca and Y atoms was determined by EPMA (Electron-Probe-Micro-Analysis) and x-ray diffraction analysis of the c-axis length [5]. The resistivity and magnetic susceptibility measurements showed that the crystals of x=0.0 and 0.2 have the Tc of 80K and 85K, respectively. This indicates that the non-doped crystal (x=0.0) is in the "over-doped" region close to the normal-metal region. The crystal of x=0.6 did not show superconductivity even at 4K, while that of x=0.4 showed an incomplete resistivity drop at 50K but again behaved like an insulator at the lower temperature.

Photoemission measurement was performed at UVSOR, Institute for Molecular Science of Japan. The spectrum was recorded at the energy and angular resolutions of about 0.15 eV and 2°, respectively. Inverse photoemission measurement was done with an angle-resolved inverse photoemission spectrometer constructed at Tohoku University [6]. The energy and angular resolutions were 0.35 eV and 2°, respectively. X-ray absorption measurement was carried out at Photon Factory, National Laboratory for High Energy Physics of Japan. The oxygen K absorption spectrum was recorded by the total-photo-yield method with the energy resolution of 70 meV. All the measurement were done at room temperature. Single crystals were cleaved in the spectrometer to obtain a clean flat surface for the measurement.

3. Results and Discussion

3.1 Angle-Resolved Photoemission of $Bi_2Sr_2Ca_{1-x}Y_xCu_2O_8$ (x=0.0-0.6)

Figure 1 shows comparison of ARPES spectra in the vicinity of the Fermi level of $Bi_2Sr_2Ca_{1-x}Y_xCu_2O_8$ (x=0.0-0.6). The spectra were measured at the polar angle of emitted electrons of 0°, representing the Γ point in the Brillouin zone. As shown in Fig. 1, the intensity of the spectrum at the Fermi level gradually increases with the hole-concentration, in other words, with decreasing the Y content (x). This clearly shows that the hole-doping produces new electronic states in the vicinity of the Fermi level instead of causing a rigid shift of the density of states relative to the Fermi level.

Figure 2 shows comparison of the whole valence-band structure determined from ARPES measurements for the crystals of x=0.0 (open circles), 0.4 (triangles), and 0.6 (filled circles). It is noticed again that there are no dispersive bands near the Fermi level for the insulator (x=0.6). As for x=0.4, we find small structures near the Fermi level, but they do not touch the Fermi level. It is remarked that band C is sensitively shifted by the hole-doping from 2.3 eV for x=0.6 to 1.4 eV for x=0.0. In contrast with the remarkable change in the small-binding-energy region, such a large change is not observed in the higher-binding-energy region (bands D-F), except for a slight energy shift of 0.2-0.3 eV. These results show that the change of the electronic structure caused by the hole-doping is not explained by a simple rigid shift of bands as expected from the band structure calculation.

Figure 3 shows comparison of the experimental band structure near the Fermi level between the crystals of x=0.0 (open circles) and 0.2 (filled circles). As shown in Fig. 3, both crystals have two dispersive bands crossing the Fermi level and the bands of

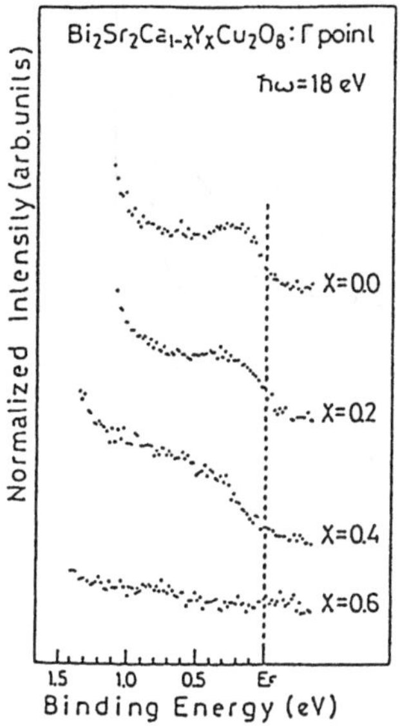

Figure 1 Comparison of angle-resolved photoemission spectra in the vicinity of the Fermi level for $Bi_2Sr_2Ca_{1-x}Y_xCu_2O_8$ (x=0.0-0.6). All spectra were measured at the polar angle of 0°, representing the Γ point in the Brillouin zone.

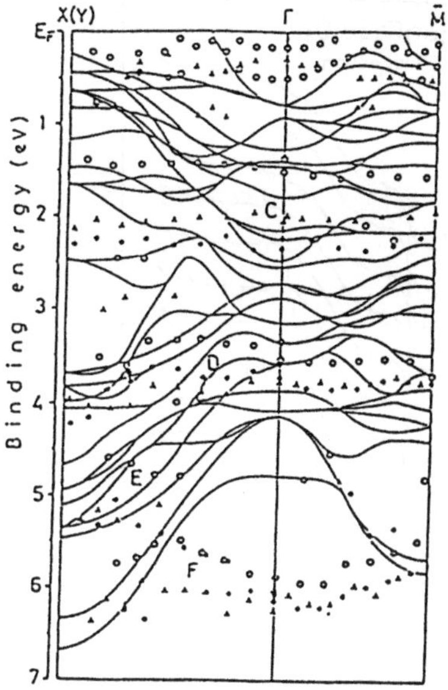

Figure 2 Comparison of whole valence-band structure among $Bi_2Sr_2Ca_{1-x}Y_xCu_2O_8$, where x=0.0 (open circles), 0.4 (triangles) and 0.6 (filled circles). A representative band structure calculation [9] is shown by thin solid lines.

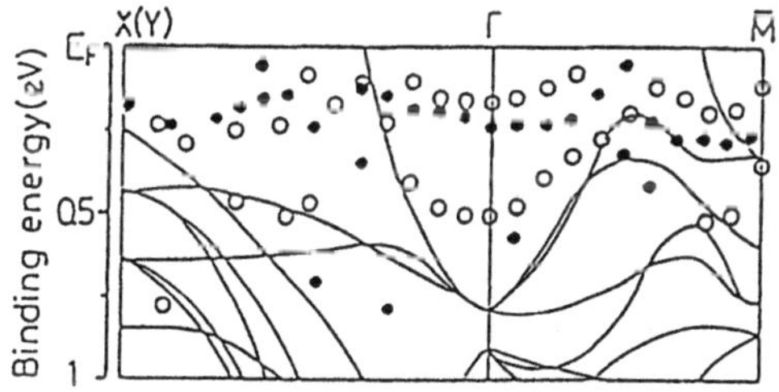

Figure 3 Comparison of the experimental band structure near the Fermi level of $Bi_2Sr_2Ca_{1-x}Y_xCu_2O_8$ [x=0.0 (open circles) and x=0.2 (filled circles)]. Note that dispersive two bands crossing the Fermi level are shifted as a whole by 0.1-0.2 eV toward high-binding-energy direction from x=0.0 to x=0.2.

x=0.2 are shifted as a whole toward the higher-binding-energy direction by 0.1-0.2 eV. This suggests that a rigid band picture may be recovered in the "over-doped" region.

3.2 Angle-Resolved Inverse Photoemission of $Bi_2Sr_2CaCu_2O_8$

Figure 4 shows angle-resolved inverse photoemission (ARIPES) spectra of $Bi_2Sr_2CaCu_2O_8$ measured for the ΓX direction in the Brillouin zone. The ARIPES spectrum at θ=0° has a narrow peak at 3eV which gradually moves toward high-energy direction as θ increases. This dispersive feature is ascribed to a surface-image-

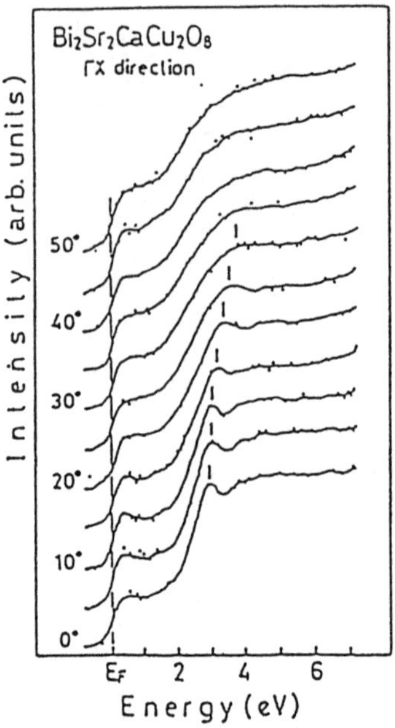

Figure 4 Angle-resolved inverse photoemission spectra of $Bi_2Sr_2CaCu_2O_8$ measured for ΓX direction. Polar angle of incident electrons referred to the surface normal (θ) is indicated on each spectrum.

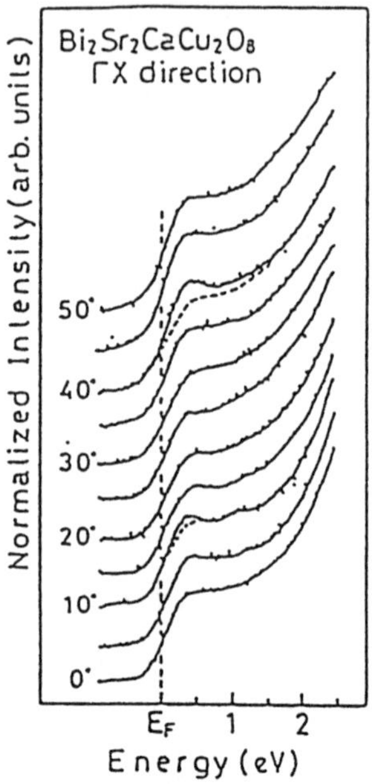

Figure 5 Angle-resolved inverse photoemission spectra of $Bi_2Sr_2CaCu_2O_8$ in the vicinity of the Fermi level measured for ΓX direction. Spectral intensity is normalized to that at the energy of 7 eV. Broken lines on 10° and 40° spectra represent 0° and 30° spectra, respectively, for comparison.

potential state. The observation of energy dispersion of a surface-image-potential state indicates that the cleaved surface is atomically flat and clean.

Figure 5 shows ARIPES spectra in the vicinity of the Fermi level. As described above, all the spectra have a clear "Fermi edge". This differs from a previous ARIPES study [7] which shows that the inverse photoemission intensity at the Fermi level (E_F) is almost zero at $\theta=0°$ and abruptly increases at $\theta=40°$. The reason for this discrepancy is not clear at present, but a recent ARIPES study by Wagener et al.[8] shows a substantial intensity at E_F even for the normal-incidence ($\theta=0°$) measurement. We tentatively ascribe this "Fermi edge" appearing in all the spectra to a phonon-assisted indirect transition which smears out k-resolved information.

In Fig. 5, we find a slight enhancement of inverse photoemission intensity near the Fermi level at $\theta=10°$ and 40°. This spectral change was reproducible. This enhancement is due to crossing of a band over the Fermi level. These points are shown in Fig. 6 as rectangles together with the ARPES result by Takahashi et al.(filled circles)[1] and a band structure calculation by Massidda et al.(thin solid liners)[9]. So far, two ARIPES studies on $Bi_2Sr_2CaCu_2O_8$ were reported [7,10]. According to Claessen et al.[7], the ARIPES intensity at E_F abruptly increase at

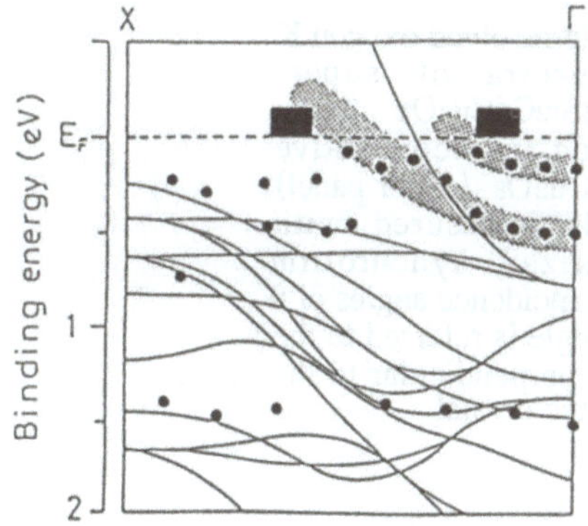

Figure 6 Comparison of the present angle-resolved inverse photoemission result (rectangles), an angle-resolved photoemission result (filled circles)[1], and a band structure calculation (thin solid lines)[9]. Shades indicate a possible connection of the photoemission and inverse photoemission results.

θ=40°, corresponding to 0.65ΓX from the Γ point. On the other hand, Wagener et al.[8] observed a band crossing at 0.25ΓX. Thus, although the previous two ARIPES studies observed either one of the two bands, the position at which a band crosses the Fermi level is almost the same between the previous studies (0.25ΓX and 0.65ΓX)[7,12] and the present one (0.2ΓX and 0.6ΓX).

Here we make a comment on the two previous ARPES experiments [2,3] which reported only one band crossing the Fermi level. A comparison between the two ARPES experiments shows that the energy position of the bands crossing the Fermi level are clearly different from each other. The result by Manzke et al.[3] shows that the band is located at about 0.2 eV below the Fermi level at Γ point while that observed by Olson et al.[2] is positioned at about 0.4 eV at the same location in the Brillouin zone. Thus it seems more natural to conclude that the two bands observed by the two ARPES experiments [2,3] are different from each other and they [2,3] observed either one of the two bands reported by Takahashi et al.[1].

3.3 Angle-Resolved Oxygen K Absorption of $Bi_2Sr_2Ca_{1-x}Y_xCu_2O_8$ (x=0.0 and 0.6)

Figure 7 shows angle-resolved oxygen K absorption spectra of superconductive $Bi_2Sr_2CaCu_2O_8$ and non-superconductive $Bi_2Sr_2Ca_{0.4}Y_{0.6}Cu_2O_8$ single crystals measured at two incidence angles of θ=0° and 75°. The incidence angle (θ) is referred to the surface normal. For the superconductor (lower panel in Fig. 7), a prominent peak appears at about 528.5 eV in the normal-incidence measurement while it is almost lost at the grazing-incidence (θ=75°) measurement. This indicates that the electronic states at the Fermi level have a dominant $O2p_{x,y}$ (in-plane) nature. For the non-superconductor (upper panel in Fig. 7), such a prominent peak is not observed, but a large shoulder appears at about 529.5 eV in the normal-incidence measurement and is also lost for the grazing-incidence measurement. This shoulder is not found in the superconductor. These experimental results indicate that (1) lowest unoccupied electronic states in a non-superconductor are located about 1 eV above the Fermi level and have a strong in-plane symmetry and (2) hole-doping causes a transfer of weight of the electronic states from the lowest unoccupied states (upper Hubbard band) to the mid-gap states produced by the hole-doping in the charge-transfer gap.

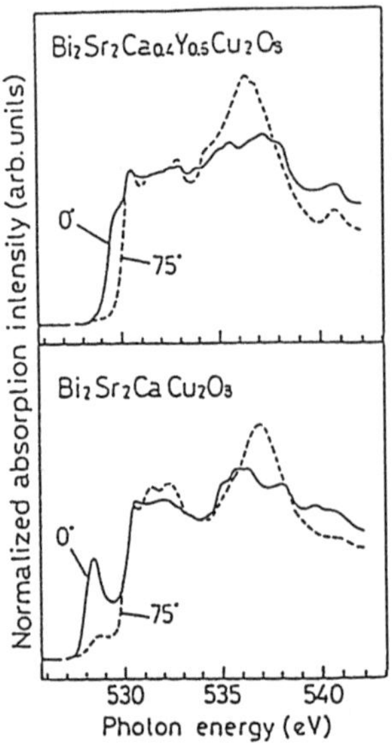

Figure 7 Angle-resolved oxygen K absorption spectra of superconductive $Bi_2Sr_2CaCu_2O_8$ (lower panel) and non-superconductive $Bi_2Sr_2Ca_{0.4}Y_{0.6}Cu_2O_8$ (upper panel) single crystals, measured with linearly polarized synchrotron radiation at two incidence angles of 0° and 75°. The angle is referred to the surface normal perpendicular to the CuO_2 plane in the crystal.

In all four spectra in Fig. 7, we find a relatively sharp peak at about 530.5 eV, whose intensity and energy-position are independent of the composition of the sample and the incidence angle of light. We ascribe this peak to the Bi6p band since the Bi6p states should be independent of the ratio between Ca and Y atoms and have an equal weight in x, y, and z directions. This assignment supports the previous scanning-tunneling-spectroscopy result [10,11] suggesting that a BiO plane is an insulator, but is different from the band structure calculation which has predicted the existence of a Bi6p electron pocket at the Brillouin-zone boundary [9].

4. Conclusion

We have performed three different kinds of spectroscopies on $Bi_2Sr_2Ca_{1-x}Y_xCu_2O_8$ (x=0.0-0.6) Angle-resolved photoemission study on a series of $Bi_2Sr_2Ca_{1-x}Y_xCu_2O_8$ shows that (1) hole-doping in an insulator (x=0.6) produces new electronic states in the charge-transfer gap and (2) a rigid-band picture is recovered in the "over-doped" region (x=0.0 and 0.2). Angle-resolved inverse photoemission indicates that there are two bands crossing the Fermi level centered at the Γ point, supporting the angle-resolved photoemission result [1]. Angle-resolved oxygen K absorption spectroscopy shows that (1) the lowest unoccupied electronic states in nonsuperconductor are located about 1 eV above the Fermi level and have a strong in-plane symmetry and (2) hole-doping causes a transfer of weight of the electronic states from the lowest unoccupied states (upper Hubbard band) to the mid-gap states produced by the hole-doping in the vicinity of the Fermi level.

Acknowledgements. The author thanks T. Kusunoki, S. Suzuki, T. Watanabe, K. Kamiya, S. Hosoya, and H. Katayama-Yoshida for their collaboration in the experiment. He also thanks A. Yamanaka, F. Minami, and S. Takekawa for providing a single crystal produced by a floating-zone method. This work was supported by grants from the Foundation for Promotion of Material Science and Technology of Japan and from the Ministry of Education, Culture and Science of Japan.

References

1. T. Takahashi, H. Matsuyama, H. Katayama-Yoshida, Y. Okabe, S. Hosoya, K. Seki, H. Fujimoto, M. Sato, and H. Inokuchi, Nature 334 (1988) 691.
2. C.G. Olson, R. Liu, A.-B. Yang, D.W. Lynch, A.J. Arko, R.S. List, B.W. Veal, Y.C. Chang, P.X. Jiang, and A.P. Paulikas, Science 245 (1989) 731.
3. T. Manzke, T. Buslaps, R. Claessen, M. Skibowski, and J. Fink, J. Euro. Phys. Lett. 9 (1989) 477.
4. J.C. Campuzano, G. Jennings, M. Faiz, L. Beaulaigue, B.W. Veal, J.Z. Liu, A.P. Paulikas, K. Vandervoort, H. Claus, R.S. List, A.J. Arko, and R.J. Bartlett, Phys. Rev. Lett. 64 (1990) 2308.
5. T. Tamegai, K. Koga, K. Suzuki, M. Ichihara, F. Sakai, and Y. Iye, Jpn. J. Appl. Phys. 28 (1989) 112.
6. H. Ohsawa, T. Takahashi, T. Kinoshita, Y. Enta, H. Ishii, and T. Sagawa, Solid State Commun. 61 (1987) 347.
7. R. Claessen, R. Manzke, H. Carstensen, B. Brundt, T. Buslaps, M. Skibowski, and J. Fink, Phys. Rev. B 39 (1989) 7316.
8. T.J. Wagener, Y.-J. Hu, M.B. Jost, J.H. Weaver, Y.F. Yan, X. Chu, and Z.X. Zhao, Phys. Rev. B 42 (1990) 6317.
9. S. Massidda, J. Yu, and A.J. Freeman, Physica C 152 (1988) 251.
10. M. Tanaka, T. Takahashi, H. Katayama-Yoshida, S. Yamazaki, M. Fujimani, Y. Okabe, W. Mizutani, M. Ono, and K. Kajimura, Nature 339 (1989) 691.
11. C.K. Shih, R.M. Feenstra, J.R. Kirtley, and G.V. Chandrashekhar, Phys. Rev. B 40 (1989) 2682.

Electronic Structure of Electron- and Hole-Doped 3d Transition-Metal Oxides

A. Fujimori

Department of Physics, University of Tokyo, 7-3-1 Hongo, Bunkyo-ku, Tokyo 113, Japan

Abstract. Many 3d transition-metal oxides can be doped with electrons or holes and can be made metallic. We investigate the electronic structure of pseudo-ternary oxides in the charge-transfer and Mott-Hubbard regimes by photoemission, inverse photoemission, and x-ray absorption spectroscopy. In most cases, mid gap-states are formed upon carrier doping whereas in some cases they are not. Both electron correlation and impurity potential play an important role in the formation of mid-gap states.

1. Introduction

The parent compounds of copper-oxide superconductors are charge-transfer insulators, in which the band gap is of a ligand p-to-metal d charge-transfer type. Doping these compounds with extra holes or electrons turns the insulators into metals and superconductivity is realized. The origin and nature of these metallic states have been the subject of much controversy. Although the elucidation of the metallic states is essential in understanding the mechanism of high-temperature superconductivity, the situation is complicated by the strongly correlated nature of d electrons and the structural disorder accompanying the chemical substitution and/or oxygen stoichiometry control necessary to introduce charged carriers. In this article, we discuss this subject from a wider scope based on the spectroscopic results of various carrier-doped 3d transition-metal compounds.

2. Band Gaps of the Parent Compounds

The atomic orbital character of doped carriers is determined by that of the band gap of the parent compound. In a Mott-Hubbard insulator, the band-gap is of a d-d type ($d^n + d^n \rightarrow d^{n+1} + d^{n-1}$), and its magnitudes is determined by the on-site d-d Coulomb interaction U. Doped electrons and holes then have primarily d character. In a charge-transfer insulator, the band gap ($d^n + d^n \rightarrow d^{n+1} + d^n\underline{L}$, where $\underline{L}$ denotes a ligand hole) is determined by a ligand-to-metal charge-transfer energy Δ. Doped electrons have d character whereas doped holes have ligand p character.

The compounds can be of either Mott-Hubbard or charge-transfer type depending on the relative size of Δ and U [1]. In Fig. 1, we plot the Δ and U values of many 3d transition-metal compounds deduced from the analysis of transition-metal core-level photoemission spectra [2]. The figure shows systematic changes in these parameters when the 3d element, ligand, or cation valence is changed. Oxides of Cu to Fe are of the charge-transfer type whereas those of Ti and V are of the Mott-Hubbard type (although they are not shown in the figure). Mn oxides are located in the intermediate region.

Figure 1 also shows that some compounds have very small or negative Δ's. (We may refer to this parameter range as negative-Δ or "valence-bond" regime.) In that case, the dominant ground-state configuration is $d^{n+1}\underline{L}$ rather than d^n, and the band gap is of a p-p type ($d^{n+1}\underline{L} + d^{n+1}\underline{L} \rightarrow d^{n+1} + d^{n+1}\underline{L}^2$) [3]. Both doped electrons and holes, therefore, have ligand p character.

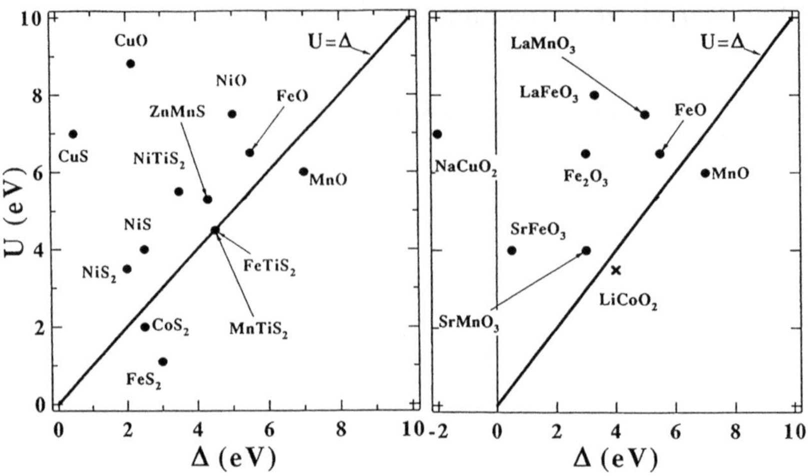

Fig. 1. Zaanen-Sawatzky-Allen Δ-U plot for divalent (left) and higher-valent (right) 3d transition-metal compounds [2]. The Δ-U plane is divided into the Mott-Hubbard (Δ >U), charge-transfer (0<Δ <U), and negative-Δ regimes.

3. Formation of Mid-Gap States

3. 1 Electron Correlation and Hybridization

The above insulators can be doped with extra electrons or holes by metal-ion substitution or oxygen stoichiometry control. Substitution is often made for rare-earth or alkaline-earth ions at the A site of ternary oxides. For the perovskite-type pseudo-ternary compounds $La_{1-x}Sr_xMO_3$, the valence of the 3d ion M can be "controlled" from 3+ to 4+. Therefore the system can be viewed either as electron-doped $SrMO_3$ or hole-doped $LaMO_3$.

Assuming that the only effect of A-site substitution is to control the valence of the 3d ion, electron or hole doping for the correlated 3d systems leads to the formation of mid-gap states. Let us first consider hole doping in a Mott-Hubbard insulator in the limit of low hole concentration. Upon doping, $d^{n-1} \rightarrow d^n$ spectral weight appears above the Fermi level (E_F) since mobile d^{n-1} sites are created in the ground state [4]. The bottom of the new $d^n \rightarrow d^{n-1}$ band (mid-gap states) touches the top of the lower Hubbard ($d^n \rightarrow d^{n-1}$) band, and the E_F is located between them schematically as shown in Fig. 2(a). This is because the excitation across E_F from the top of the Hubbard band to the lowest mid-gap state (..$d^nd^{n-1}d^n$..$d^nd^nd^n$.. →..$d^nd^nd^n$..$d^nd^{n-1}d^n$..) do not alter the ground state as long as the system has a translational symmetry.

The formation of various types of mid-gap states are described in Table 1, where one can see that mid-gap states are formed above and below E_F for hole and electron doping, respectively. If there were no p-d hybridization, hole doping in a charge-transfer insulator would simply lower the E_F into the p band. The p-d hybridization, however, induces more complex behavior. If the hybridization is strong enough, a split-off $d^n\underline{L}$ state is formed above the p-hole ($d^n\underline{L}$) continuum, invalidating the simple picture. Also, the hybridization causes a transfer of spectral weight from outside the band gap to the mid-gap states [4], analogously to a Kondo resonance in valence-fluctuating rare-earth compounds.

Table 1. Atomic-orbital origin of the single-particle density of states for the band gaps of parent compounds and that of the states induced by carrier doping. PES and BIS correspond to states below and above E_F, respectively. "*" denotes candidates for mid-gap states.

		Mott-Hubbard	Charge-transfer	Negative-Δ
Band gap	PES	$d^n \rightarrow d^{n-1}$	$d^n \rightarrow d^n\underline{L}$	$d^{n+1}\underline{L} \rightarrow d^{n+1}\underline{L}^2$
	BIS	$d^n \rightarrow d^{n+1}$	$d^n \rightarrow d^{n\mp1}$	$d^{n+1}\underline{L} \rightarrow d^{n+1}$
Electron doping	PES	$*d^{n+1} \rightarrow d^n$	$*d^{n+1} \rightarrow d^n$	$*d^{n+1} \rightarrow d^{n+1}\underline{L}$
	BIS	$d^{n+1} \rightarrow d^{n+2}$	$d^{n+1} \rightarrow d^{n+2}$	$d^{n+1} \rightarrow d^{n+2}$
Hole doping	PES	$d^{n-1} \rightarrow d^{n-2}$	$d^n\underline{L} \rightarrow d^n\underline{L}^2$	$d^{n+1}\underline{L}^2 \rightarrow d^{n+1}\underline{L}^3$
	BIS	$*d^{n-1} \rightarrow d^n$	$*d^n\underline{L} \rightarrow d^n$	$*d^{n+1}\underline{L}^2 \rightarrow d^{n+1}\underline{L}$

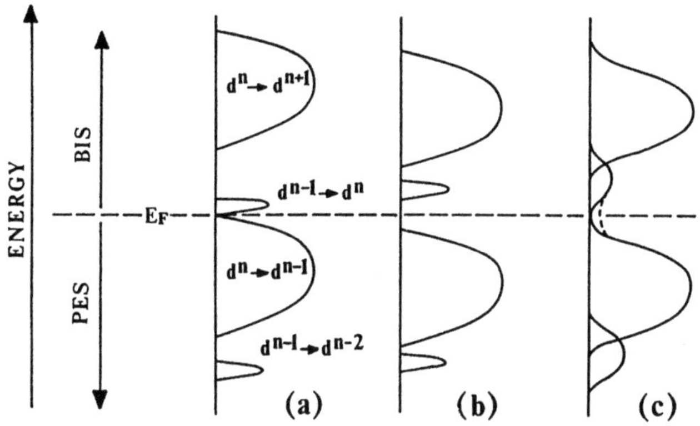

Fig. 2. Electronic structure of a hole-doped Mott insulator. (a) Low-carrier concentration limit without impurity potential; (b) with impurity potential; (c) higher doping concentration (dashed curve: metallic).

3.2 Effects of Impurity Potential

Many theoretical studies [5,6] have demonstrated that mid-gap states appears near the top of the valence band and the bottom of the conduction band for hole doping and electron doping, respectively, and that the E_F is located around there. In real materials, however, carrier doping is carried out by substituting metal ions or introducing excess oxygens or vacancies. Such defects have effectively positive and negative charges for electron and hole doping, respectively, and will therefore trap the carriers at least for low doping concentration [7,8]. In order to ionize the trapped carriers, additional energy (of order ~1 eV) is required corresponding to the binding energy of the carriers, which means that mid-gap states are shifted away from the Fermi level by this amount [Fig. 2(b)].

For higher defect concentration, each trapped carrier feels different potential from neighboring defects because of the random distribution of the defects. Thus the mid-gap states are distributed over a wider energy range [Fig. 2(c)]. The gap in a disordered semiconductor is a "soft" gap called Coulomb gap [9] as shown in Fig. 2.(c).

3.3 Spectroscopic Results

The above picture has been supported by spectroscopic studies. The position of E_F is found near the center of the band gap both for hole-doped $La_{2-x}Sr_xCuO_4$ and electron-doped $Nd_{2-x}Ce_xCuO_4$ [10], which cannot be explained without considering the effect of Sr^{2+} or Ce^{4+} impurity potential. Figure 3 shows the electronic structure of the hole-doped charge-transfer insulator $La_{2-x}Sr_xNiO_4$ [7] and the electron-doped semiconductor $La_xSr_{1-x}TiO_3$ [11] deduced form photoemission, inverse photoemission (BIS), and O 1s core-level x-ray absorption spectroscopy (XAS). The position and width of the mid-gap states suggest the importance of impurity potential, particularly for $La_xSr_{1-x}TiO_3$, in which the $d^1 \rightarrow d^0$ spectral feature would have no multiplet nor dispersional width in the low concentration limit and hence look like Fig. 2(a) if there were no La^{3+} potential. With increasing x, the "impurity band" is gradually transformed into the lower Hubbard ($d^1 \rightarrow d^0$) band of $LaTiO_3$.

Results on various pseudo-ternary 3d oxides are summarized in Table 2. (XAS results for Cu oxides are taken from [13] and [15].) As expected, mid-gap states are found in most systems predominantly above and below E_F for hole and electron doping, respectively. There are, however, a few cases in which the expected mid-gap states are not clearly observed (e.g., hole-doped $LaMnO_3$ [18] and $LaTiO_3$ [19]). It appears that this happens when the band gap of the parent compound is small or zero. Probably, the "mid-gap" states are formed out of the band gap and are obscured.

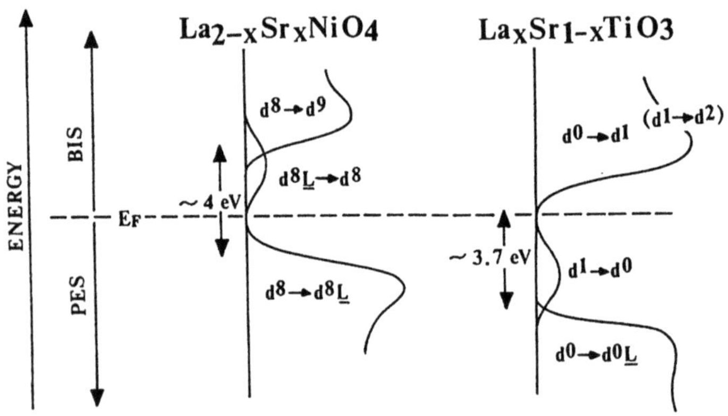

Fig. 3. Schematic electronic structure of hole-doped $La_{2-x}Sr_xNiO_4$ (0<x<0.8) [7] and electron-doped $La_xSr_{1-x}TiO_3$ (0<x<0.7) [11] in the semiconducting region.

4. Degree of Itinerancy of Doped Carriers

An important question is which factors control the degree of localization/itinerancy of doped carriers and hence the critical carrier concentration for the insulator-to-metal transition.

The atomic orbital (metal 3d versus oxygen 2p) character of the hole/electron wavefunction is certainly important considering the fact that an extra hole in $La_{2-x}Sr_xCuO_4$ is more than 90 % oxygen p-like and the system has a critical concentration as low as 5 %. However, this is not the only controlling factor since a doped electron in Nd_2CuO_4 and a doped hole in La_2NiO_4 have similar amount (60-70 %) of oxygen p character in spite of the quite different critical concentrations. The small the band gap is, the lower the critical concentration appears to be (Table 2) although La_2CuO_4 is an

Table 2. Observation of mid-gap states in pseudo-ternary 3d transition-metal oxides. The magnitude of the band gap and the critical carrier concentration for the insulator-to-metal transition are also indicated.

Parent compound		Substitution	Type of carriers	Mid-gap states		Band gap (eV)	Critical concentration
				PES	BIS/XAS		
La_2CuO_4	$Cu^{2+}(d^9)$	La→Sr	hole	weak[12]	yes [13]	2.0[20]	5 %[22]
Nd_2CuO_4	$Cu^{2+}(d^9)$	Nd→Ce	electron	yes [14]	no [15]	1.5[20]	15 %[23]
La_2NiO_4	$Ni^{2+}(d^8)$	La→Sr	hole	no [7]	yes [7]	4.0[7]	80 %[24]
$LaFeO_3$	$Fe^{3+}(d^5)$	La→Sr	hole	weak[16]	yes [18]	~3	<100 %
$SrFeO_3$	$Fe^{4+}(d^4)$	Sr→La	electron	-	no [18]	0	0
$LaMnO_3$	$Mn^{3+}(d^4)$	La→Sr	hole	-	no [18]	~<1	<10 %[25]
$SrMnO_3$	$Mn^{4+}(d^3)$	Sr→La	electron	-	no [18]	2-3	50 %[25]
ZnV_2O_4	$V^{3+}(d^2)$	Zn→Li	hole	no [17]	yes?[19]	>0	40 %[26]
$LaTiO_3$	$Ti^{3+}(d^1)$	La→Sr	hole	no [11]	no [19]	0	0
$SrTiO_3$	$Ti^{4+}(d^0)$	Sr→La	electron	yes [11]	no [19]	3.7[21]	70 %[27]

exception. The symmetry of doped carriers ($pd\sigma$ versus $pd\pi$) would also affect the insulator-to-metal transition [7].

The magnitude of the 3d-ion moment of the parent compound will also affect the mobility of doped carriers: Undoped Cu^{2+} oxides are spin-1/2 antiferromagnets, in which quantum spin fluctuation is maximum (antiferromagnetic correlation is minimum) and thus the mobility of carriers is enhanced [7]. The presence of itinerant carriers with low density as in the doped Cu oxides may be a necessary condition for the occurrence of superconductivity [14]. It is interesting to note that in $La_{1-x}Sr_xMnO_3$ the mobility is enhanced by the ferromagnetic ordering of Mn local moments [25]. Holes doped into $LaMnO_3$ may be as itinerant as those in the Cu oxides but the ferromagnetic ordering in the metallic state is unfortunately hostile to superconductivity.

Acknowledgements

I would like to thank many collaborators including H. Namatame, A. E. Bocquet, T. Mizokawa, I. Hase, M. Nakamura, Y. Fujishima, Y. Tokura, H. Eisaki, H. Takagi, S. Uchida, N. Tsuda, Y. Takeda, M. Takano, J. van Elp, P. Kuiper, G.A. Sawatzky, S. Hosoya, M. Abbate, F.M.F. de Groot and J.C. Fuggle. This work is supported by a Grant-in Aid for Scientific Research from the Ministry of Education, Science and Culture of Japan.

References

1. J. Zaanen, G.A. Sawatzky and J.W. Allen, Phys. Rev. Lett. **55**, 418.
2. A.E. Bocquet, A. Fujimori, T. Mizokawa and H. Namatame, submitted to Phys. Rev. B.
3. T. Mizokawa, H. Namatame, A. Fujimori, K. Akeyama, H. Kondoh, H. Kuroda, and N. Kosugi, submitted to Phys. Rev. Lett.
4. H. Eskes and G.A. Sawatzky, submitted to Phys. Rev. Lett.
5. H. Matsumoto. M. Sasaki, and M. Tachiki, Solid State Commun. **71**, 829 (1989).
6. W. Stephan and P. Horsch, in Dynamics of Magnetic Fluctuations in High Temperature Superconductors, edited by G. Reiter, P. Horsch, and G. Psaltakis (Plenum, New York, 1990).
7. H. Eisaki, S. Uchida, H. Namatame, A. Fujimori, J. van Elp, P. Kuiper, G. A. Sawatzky, S. Hosoya, and H. Katayama-Yoshida, submitted to Phys. Rev. B.

8. A. Fujimori, in Strong Correlation and Superconductivity, edited by H. Fukuyama, S. Maekawa, and A. P. Malozemoff (Springer, Berlin, 1989) p. 300.
9. J. H. Davies and J. R. Franz, Phys. Rev. Lett. **57**, 475 (1986).
10. H. Namatame, A. Fujimori, Y. Tokura, M. Nakamura, K. Yamaguchi, A. Misu, H. Matsubara, S. Suga, H. Eisaki, T. Ito, H. Takagi, and S. Uchida, Phys. Rev. B **41**, 7205 (1990).
11. A. Fujimori, I. Hase, H. Namatame, T. Fujishima, Y. Tokura, M. Nakamura, M. Abbate, F.M.F. de Groot, and J.C. Fuggle, unpulished.
12. H. Eisaki, T. Ido, H. Takagi, S. Uchida, T. Mizokawa, H. Namatame, A. Fujimori, Y. Tokura, J. van Elp, P. Kuiper, G. A. Sawatzky, S. Hosoya, H. Katayama-Yoshida, K. Oka, H. Unoki, Physica B **165&166**, 1225 (1990).
13. C.T. Chen, F. Sette, Y. Ma, M.S. Hybertsen, E.B. Stechel, W.M.C. Foulkes, M. Schluter, S-W. Cheong, A.S. Cooper, L.W. Rupp, Jr, B. Batlogg, Y.L. Soo, Z.H. Ming, A. Krol, and Y.H. Kao, Phys. Rev. Lett. **66**, 104 (1990).
14. A. Fujimori and H. Namatame, Physica C, to be published.
15. M. Alexander, H. Romberg, N. Nucker, P. Adelmann, J. Fink, J. T. Markert, M. B. Maple, S. Uchida, H. Takagi, Y. Tokura, A. C. W. P. James, and D. W. Murphy, Phys. Rev. B **43**, 333 (1991).
16. A. E. Bocquet, A. Fujimori, H. Namatame, S. Suga, N. Kimizuka, Y. Takeda, and M .Takano, submitted to Phys. Rev. B.
17. A. Fujimori, K. Kawakami, and N. Tsuda, Phys. Rev. B **38**, 7889 (1988).
18. M. Abbate, F.M.F. de Groot, J.C. Fuggle, A. Fujimori, M. Takano, Y. Takeda, H. Eisaki, S. Uchida, O. Strebel, F. Lopez, M. Domke, G. Kaindl, and G.A. Sawatzky, unpublished.
19. M. Abbate, F. M. F. de Groot, J. C. Fuggle, A. Fujimori, Y. Fujishima, Y. Tokura, J. van Elp, G. A. Sawatzky, O. Strebel, M. Domke, G. Kaindl, M. Sacchi, and N. Tsuda, Phys. Rev. B, in press.
20. S. Uchida, T. Ido, H. Takagi, T. Arima, Y. Tokura, and S. Tajima, Phys. Rev. B **43**, 7942 (1991).
21. M. Cardona, Phys. Rev. **140**, A651 (1965).
22. H. Takagi, T. Ido, S. Ishibashi, M. Uota, S. Uchida and Y. Tokura, Phys. Rev. B **40**, 2254 (1989).
23. Y. Tokura, H. Takagi, and S. Uchida, Nature **337**, 345 (1989).
24. Y. Takeda, R. Kanno, M. Sakano, O. Yamamoto, M. Takano, Y. Bando, H. Akinaga, K. Takita, and J.B. Goodenough, Mat. Res. Bull. **25**, 293 (1990).
25. J. H. Santen and G. H. Jonker, Physica **16**, 599 (1950).
26. K. Kawakami, Y. Sakai, and N. Tsuda, J. Phys. Soc. Jpn. **55**, 3174 (1986).
27. Y. Maeno, S. Awaji, H. Matsumoto, and T. Fujita, Physica B **165&166**, 1185 (1990).

Anisotropy of Cuprate Superconductors Under STM/STS

K. Kitazawa, T. Hasegawa, H. Ikuta, M. Nantoh, W. Wei, and N. Motohira

Department of Industrial Chemistry, University of Tokyo, 7-3-1- Hongo, Bunkyo-ku, Tokyo 113, Japan

Abstract. The atom site specified tunneling spectroscopy has been performed by scanning tunneling microscopy to characterize the properties of each of the component layers independently in $Bi_2Sr_2CaCu_2O_y$ single crystal. The origin of anisotropy of the high temperature superconductors is discussed based on the role of each layer. The potentiality of the STM/STS as well as its problems are also discussed. The overall shape of the spectrum has been examined as a function of the distance between the STM tip and the sample surface.

1. Introduction

The origin of the HTSC in cuprates, although yet being unestablished, is thought to be deeply related with the two dimensional electronic nature of the CuO_2 network layers. This, however, leads to the high anisotropy of the cuprate superconductors. In terms of the effective mass of the carriers, the anisotropy ratio reported ranges from a few tens in case of $Y_1Ba_2Cu_3O_y$[1] to more than several thousands in case of $Bi_2Sr_2CaCu_2O_y$[2]. Because of this and the high critical temperature the superconducting coherence length ξ in the cuprates becomes extremely short along the c-axis according to the equation

$$\xi=\frac{hv_F}{\pi\Delta}.$$

That the coherence length ξ is even shorter than the lattice parameter along the c-axis is causing the various technical problems which must be overcome in order for the practical application. One of the most remarkable appearances of this short ξ is the tendency towards decoupling of the superconductivity to result in superconducting weak links at interfaces and the weak pinning potential of the vortices directed along the c-direction. Also caused by the short ξ is the strong superconducting fluctuation phenomenon under a strong magnetic field applied along the c-axis[3]. Consequently, it is of vital importance to understand the nature of the high anisotropy from the view point of practical application as well.

2. Origin of Anisotropy

According to Tokura and Arima[4] the superlattice structure of the cuprates may be conveniently expressed as the pile of CuO_2 layer in which mobile carriers are created and the blocking layer which serves as the charge reservoir.

As shown in Fig.1, the most recently discovered $(Sr,Nd)CuO_2$ can be expressed as the alternative pile up of CuO_2 conductive- and (Sr,Nd) blocking- layers, whereas $Y_1Ba_2Cu_3O_7$ as CuO_2-Y-CuO_2-{BaO-CuO-BaO} layer sequences. From this view point, the higher anisotropy in $Bi_2Sr_2CaCu_2O_y$ and $(La,Sr)_2CuO_4$, which have the sequences of CuO_2-Ca-CuO_2-{SrO-BiO-BiO-SrO} or CuO_2-{(La,Sr)O-(La,Sr)O}, respectively, can be understood of the consequence of the probably more insulative nature of {SrO-BiO-BiO-SrO} or {(La,Sr)O-(La,Sr)O} blocking layer. On the other hand, {BaO-CuO-BaO} in $Y_1Ba_2Cu_3O_7$ seems to provide less anisotropy perhaps due to the conductive nature of the one dimensional chain in the {BaO-CuO-BaO} layer and to a higher electron transfer probability

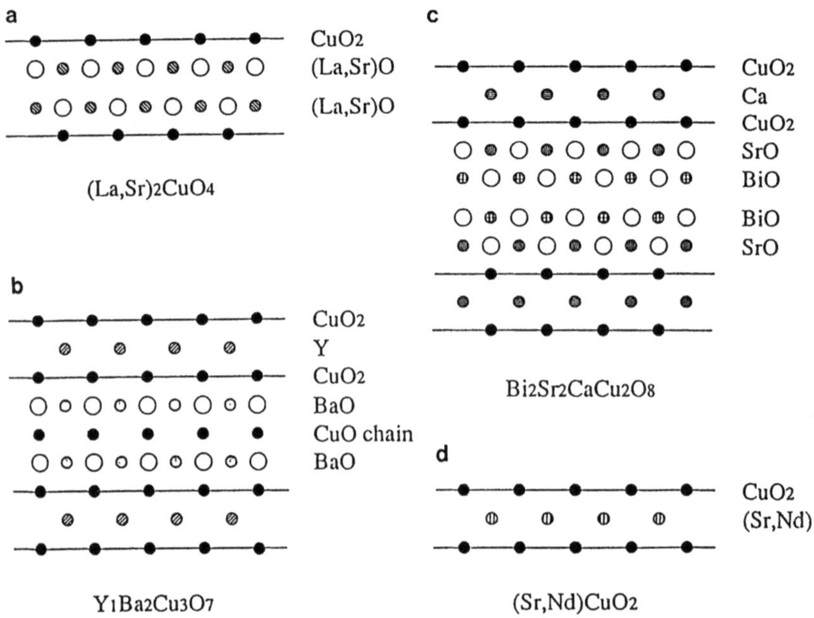

Fig. 1. Schematic representation of the structure of (a) $(La,Sr)_2CuO_4$, (b) $Y_1Ba_2Cu_3O_7$, (c) $Bi_2Sr_2CaCu_2O_8$ and (d) $(Sr,Nd)CuO_2$.

across the thin Y (or Ca) blocking layer. Quite recently it has been found that $TlBa_2Ca_2Cu_3O_y$ exhibits as high a pinning force at 77 K as 123 system, which indicates that this material is as less anisotropic as 123 and hence that the blocking layer {BaO-TlO-BaO} is as conductive as {BaO-CuO-BaO}[5].

Some photoelectron spectroscopy studies[6] as well as band calculations[7], however, have been arguing a rather conductive nature of the BiO layer in the {SrO-BiO-BiO-SrO} blocking layer. This is questionable if the extremely strong anisotropy of the resulting compounds is considered.

Therefore it is important to understand the properties of CuO_2 layer and the blocking layer independently by a local microscopic probe.

3. Electronic Structure of the Cuprate Layer

The strong electronic correlation expected on the CuO_2 layer, as first suggested by Anderson[8], has been a key controversial issue with respect to whether the electronic structure of the CuO_2 layer can be understood in the concept of Fermi liquid or not. Experimentally, there have been many reports dealing with this point[9]. This problem is vitally important because the BCS mechanism can be conceivable only on the Fermi liquid concept. The BCS mechanism discusses about the instability of the electronic states near the Fermi surface induced by the attractive forces between each pair of electrons with the aid of phonons or other possible excitations. On the other while, the non-BCS mechanism is the natural consequence if the non-Fermi liquid concept applies.

The current situation in this regard is still controversial. The main reason for this lasting ambiguity is due to the lack in sufficiently precise measurement tools to probe the electronic excitations in the vicinity of the Fermi surface. The energy associated with the occurrence of superconductivity in the cuprates is still at the most in the order of 0.01 eV. Therefore the electronic structure must be determined with the precision of the comparable energy scale or better in order to discuss the mechanism of the superconductivity. Most of the measurements, so far attempted, unfortunately lack the precision of this level.

In this context, the tunneling spectroscopy appears to be one of the most powerful tools due to its high energy resolution in probing the low-lying excitation spectrum. This method, however, has been subjected to two essential difficulties. Firstly, the reproducibility has been very poor and the results reported so far have scattered over a wide range[10]. This is attributable to the non-ideal surface state of the samples. Secondly, the spectrum obtained has inevitably been the averaged one over a macroscopic area in spite of the rather local nature of the electronic structure suspected.

Consequently, it is desired to obtain microscopic tunneling spectra on well characterized sample surfaces. The scanning tunneling spectroscopy (STS) is probably the best suited method for this purpose.

4. Technique of Scanning Tunneling Spectroscopy (STS)

The STM/STS technique has been being developed rapidly during last several years. Now it has become possible to take tunneling spectrum by interrupting the scanning process of the STM tip by fixing its position for a short period at a specified atomic site. The scanning can be re-started to complete the STM image. Then one can be sure of the surface state which he is looking at. In order to achieve this, at the first step, one has to get a stable STM image for an enough long period to allow repeated tunneling spectrum measurements at various positions during a single run of STM imaging. In this case one can be sure of the good surface state and of the atomic locations where the spectra are taken. The STS spectrum independently taken of STM imaging can not claim for this advantage.

Although there have been several reports of tunneling measurements on cuprate superconductors using STM[11], unfortunately none have been taken under the conditions stated above. This has been mainly due to the insufficient stability of the STM system and to the insufficient development of the STS technique. Comparing with STM on graphite or the other layer compound such as TaS_2, the atomic resolution STM is more difficult on cuprates according to our experience perhaps because of the lower contrast of the atomic image. The present authors recently achieved the atomic resolution STS for the first time on cuprate superconductors[12]. We believe the simultaneous STM/STS has a potentiality to open a new field in the study of the HTSC, by locally examining the excitation spectrum both in the normal and superconducting states to reveal the role of each component layer.

5. Tunneling Spectroscopy on the BiO Mono-Layer of Single-Crystalline $Bi_2Sr_2CaCu_2O_y$ at 4.2 K

Fig.2 illustrates the atomic image on the cleaved surface of single-crystalline $Bi_2Sr_2CaCu_2O_y$ at 4.2 K which is thought to be covered with a mono-layer of BiO. The modulation structure with a periodicity about 2.7 nm along the b-axis is clearly seen in Fig.2a while 2b is the scaled-up image[13]. The atoms seen in the picture are presumed to be Bi atoms.

The tunneling spectra taken at various locations, on and off the atomic position in the BiO surface layer, were essentially the same,

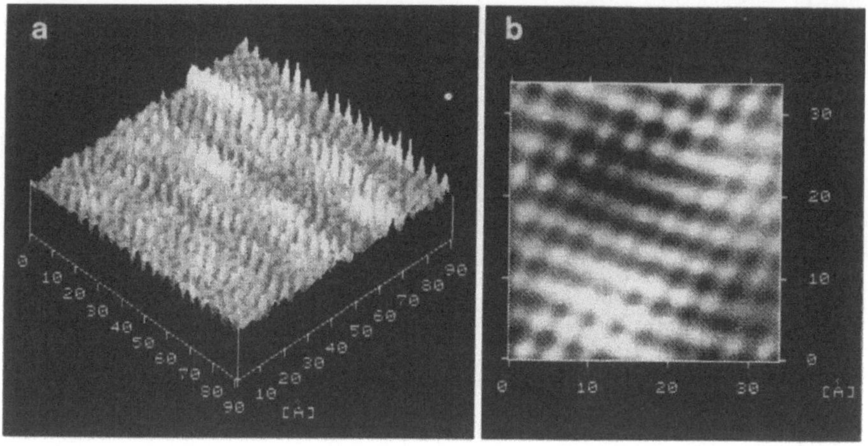

Fig. 2. Wide-scan (a) and narrow-scan (b) STM images on the cleaved surface of $Bi_2Sr_2CaCu_2O_y$ single crystal at 4.2 K. Tip bias voltage was -0.75 V, and tunneling current was 5 nA[13].

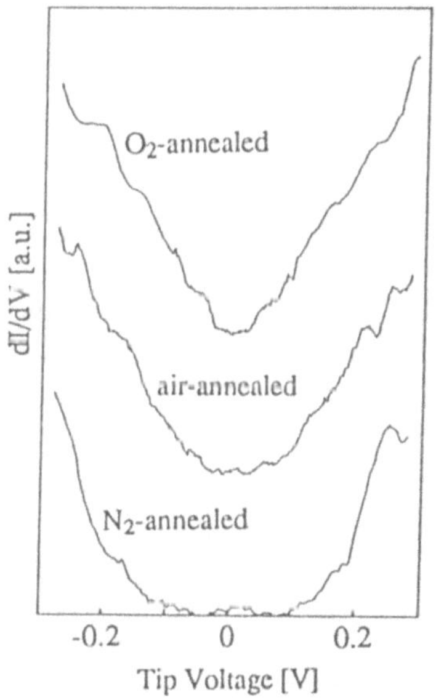

Fig. 3. Tunneling spectra on the cleaved surface of $Bi_2Sr_2CaCu_2O_y$ single crystals prepared by annealing with different conditions after crystal growth.

appearing to be semiconductive with a distinctive energy gap of about 0.1 eV on the air-annealed crystal as shown in Fig.3. But when the crystal was better oxygenated by annealing under pressurized oxygen atmosphere, the surface became more conductive. This observation contradicts with the prediction of the band calculation[7] which claims that the Bi6p-O2p nature band should cross the Fermi level to give substantial density of states at the Fermi level. The present observation

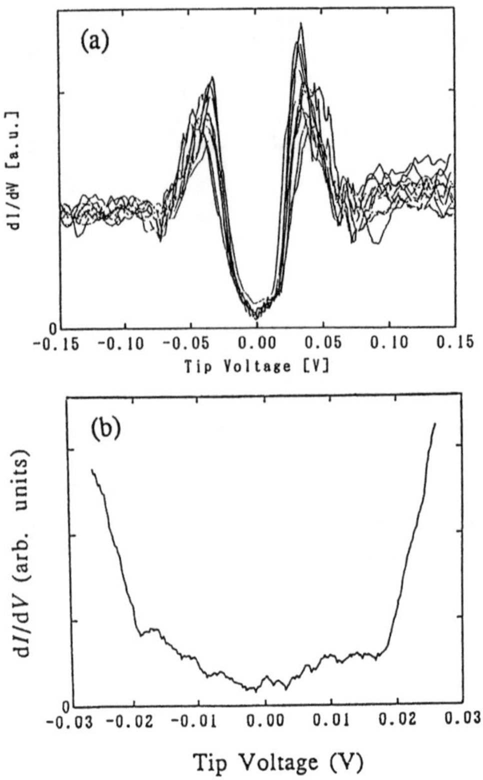

Fig. 4. Tunneling spectra on the cleaved surface of $Bi_2Sr_2CaCu_2O_y$ single crystal at 4.2 K taken with the energy resolution of 10 meV (a) and 1 meV (b). Tip bias voltage and tunneling current were -0.2 V and 20 nA, respectively. 12 spectra obtained at different measurement locations were superimposed[13].

tells us that the electronic structure of BiO layer as a function of the oxygen content should net be described by the rigid band picture.

No clear superconducting gap structures were observed so far when the high contrast atomic structure was observable. However, we frequently observed the tunneling spectra as shown in Fig.4, which seemed to be related with superconductivity. But when the superconducting gap was observed, the STM image lost the contrast in resolving the atomic image. The gap structure observed did not depend upon the position of measurements as shown in the figure. Assuming that these spectra are the signal due to the bulk superconductivity, there are several interesting features to be noted.

First of all, as seen in Fig. 4b there are no apparent inner-gap structures as have been suggested by several macroscopic tunneling measurements[14]. No zero-bias conductance peak[15] was observed,

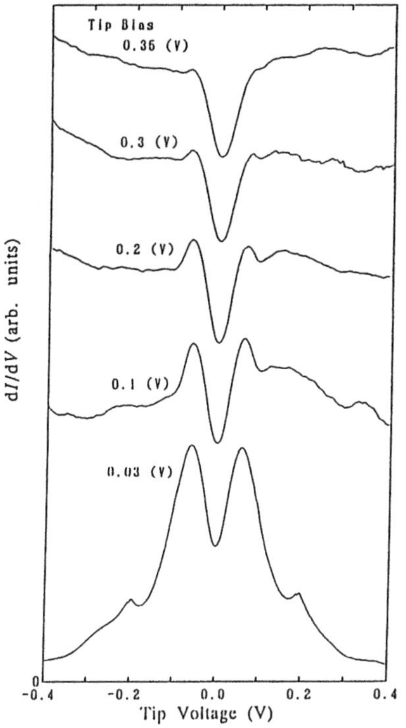

Fig. 5. Tip bias voltage dependence of the tunneling spectrum on the cleaved surface of $Bi_2Sr_2CaCu_2O_y$ single crystal at 4.2 K. Tunneling current was kept constant, 20 nA for various bias voltages. The tip-to-surface distance is shorter for the lower bias voltage. Energy resolution was 10 meV.[13]

either. The shape of the spectra, however, did not fit the ideal BCS curve even on considering the effect of short relaxation time of the pseudo-particles[16]. Therefore, it was not possible to unambiguously define the superconducting gap value 2Δ. But if we take the peak-to-peak value(first method) as 2Δ, it ranged from 70 to 90 meV which leads to an extraordinary large $2\Delta/k_BT_c$ value of 9 to 12. If adopting the definition to use the crossing points of the gap curve and the extrapolated line of the background(second method), $2\Delta/k_BT_c$ is roughly obtained as 6 to 8 which is still much larger than the BCS predicted value in the weak coupling regime, 3.52. Therefore we may at least conclude that the superconductivity in $Bi_2Sr_2CaCu_2O_y$ is not in the BCS-weak coupling regime.

Another unique feature to be noted was the fact that the shape of the tunneling spectrum changed with the STM tip-to-sample surface distance as shown in Fig. 5[13]. The distance(z) was adjusted by changing the bias voltage of the tip for the constant bias current while

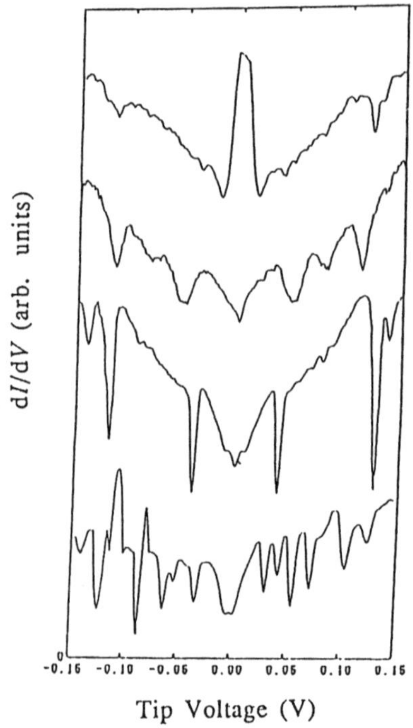

Fig. 6. Typical examples of tunneling spectra on the cleaved surface of $Bi_2Sr_2CaCu_2O_y$ single crystal at 4.2 K observed after the collision of the probe tip onto the sample surface[13].

the tip position(x-y) was fixed. The peaks became more exaggerated as the distance decreased (i.e., lower bias voltage). Through these measurements, the peak-to-peak superconducting gap value Δ_{pp} remained nearly constant while the background-extrapolated value Δ_{ext} depended significantly on the distance(z). Hence the second method to define the gap value Δ_{ext} loses its justification.

Furthermore, when the tip-to-surface distance was further shortened so that the tip directly touched onto the sample surface, the zero-bias peak did appear as shown in Fig.6. In this case, the proximity effect at the contact between the electrode in the normal state and the sample surface in the superconducting state could give the conductance peak near zero-bias. Also multi-gap structure appeared when the tip was pushed onto the sample surface as also shown in Fig.6. The multi-gap structure has often been reported in the studies of the HTSC[17]. We presume, however, that the multi-gap structures observed in the superconducting state should rather be attributed to extrinsic origins at the junction rather than to the intrinsic effects. When the tunneling region is damaged, small chipped particles may be present in the junction and give rise to the charging effect which can explain the multi-gap structure.

6. STM Image and Tunneling Spectra on the Plane Perpendicular to the Layers of $Bi_2Sr_2CaCu_2O_y$ at Room Temperature

The atomic structural STM image on the bc-plane at room temperature was as shown in Fig. 7[12]. Comparing with the TEM projection along the a-axis[18], we assigned the CuO_2 layer and $(BiO)_2$ double layer regions as indicated in the figure. The absence of the perfect periodicity is due to the modulation structure of this material along the b-axis. Image in Fig.7a was taken with the tip bias of -1.5 V, i.e., under the condition for electrons to tunnel from the tip into the sample at the electronic states 1.5 eV higher than the Fermi level. Under this condition both the CuO_2 and $(BiO)_2$ layers are of nearly equal brightness in the figure, indicating that the two layers contribute to the comparable extent to the electronic density of states at this rather high energy level.

When the tip bias voltage was reduced, the brightness of the STM image decreased drastically in the regions of the $(BiO)_2$ layer while it remained high in the CuO_2 layer regions as shown in Fig.8. This means that the density of states are locally less contributed from the $(BiO)_2$ layer region at energy levels near E_F.

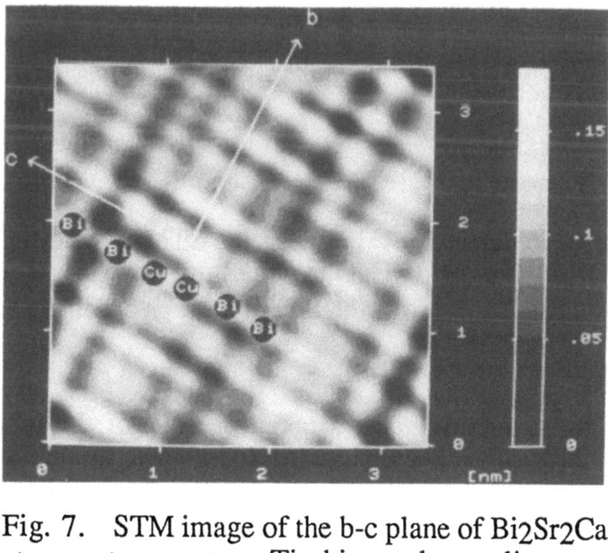

Fig. 7. STM image of the b-c plane of $Bi_2Sr_2CaCu_2O_y$ single crystal at room temperature. Tip bias and tunneling current were -1.5 V and 0.7 nA, respectively[12].

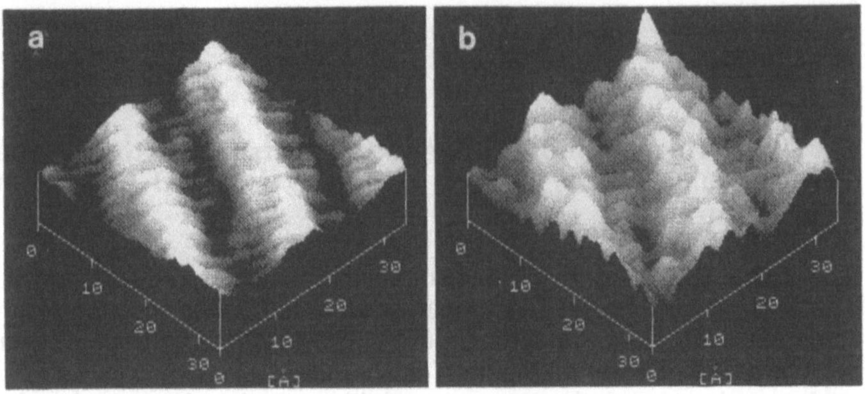

Fig. 8. STM image of the b-c plane of $Bi_2Sr_2CaCu_2O_y$ single crystal at room temperature taken at a tip bias voltage of -0.6 V and tunneling current of 0.4 nA (a), and at a tip voltage of +0.1 V and tunneling current of 0.6 nA (b)[12].

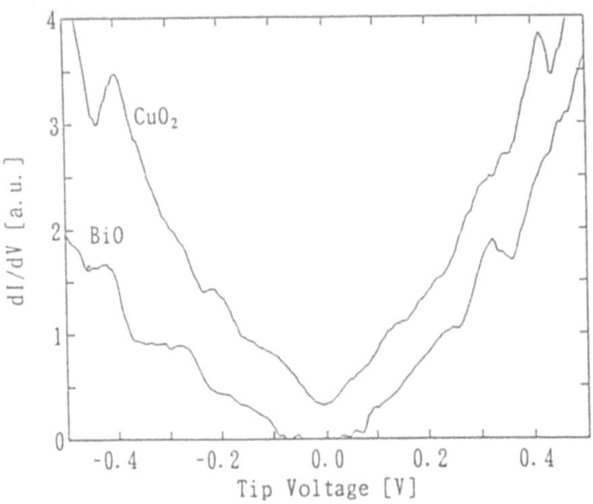

Fig. 9. Tunneling conductance *dI/dV* between the STM tip and the sample surface in $(BiO)_2$ (a) and $(CuO_2)_2$ (b) layer regions as a function of tip bias voltage. To improve the S/N ratio, each spectrum is the average of about 30 equivalent points.

Fig.9 illustrates the tunneling spectra taken on each of the two regions; $(BiO)_2$ and CuO_2 layer regions, respectively. The tunneling current signals were averaged over the equivalent 30 points distributed with equal separation on each layer region in order to improve the S/N ratio. It turned out that the spectrum on the $(BiO)_2$ layers appears to be typical as that of a semiconductor with an energy gap of 0.1 eV for the non-oxygenated specimen. Therefore, it is because of the semiconductive energy gap in the $(BiO)_2$ layers that the STM image in this region becomes dark when the bias voltage nears zero (i.e., E_F).

On the other hand, the tunneling conductance was much higher on the CuO_2 layer region even at zero bias voltage, indicating the continuous distribution of density of states at E_F and hence that the CuO_2 layer region is essentially metallic. For a typical metal, however, the density of states should be flatly distributed across the Fermi energy. But actually the spectrum in this region was not flat but it was V-shaped. That is, the apparent density of states decreases nearly linearly towards E_F on both sides of the Fermi level. This strange behavior was reported by the macroscopic tunneling measurements widely on cuprate superconductors above the critical temperature[14,19] and has been the controversial issue in modeling the electronic structure of the HTSC. The present results further specify that this behavior is solely due to the nature of the CuO_2 layer. Furthermore, the strange behavior should not be attributed to the spurious surface effect as has often been worried about, because the spectra were taken while the atoms were being observed and the vacuum tunneling conditions were secured.

As observed in Fig. 5, the V-shape experimentally was such that it changed the slopes of the two linear branches. On approaching the STM tip towards the sample surface, the V-shape gradually opened up, i.e., the slopes decreased and finally changed the sign. Therefore one must admit that the shape of the tunneling spectrum depends on the distance of the tunneling gap. This presents a very difficult but challenging problem. Conventionally this tunneling gap dependence of the overall shape of the spectrum seems to be not considered deep enough in the tunneling spectroscopy experiments. Although we do not go deep into this problem, this observation suggests that the spectrum can be subjected to some artificial parameters like the barrier height or/and the current density of the measurements. Until this problem is solved, it could be mislead to uncorrect interpretation of the density of states distribution based on the V-shaped tunneling spectra. Hence the understanding of the mechanism of tunneling itself should further be developed.

Although the observation of tunneling spectra in the superconducting state on this sample surface is highly desired we have not yet succeeded in the atomic resolution STS on this plane at the cryogenic temperature.

7. Conclusions

It has been cleared that 1) CuO_2 layer is a metal but seems to be a strange metal in terms of the V-shaped density of states spectrum around the Fermi level. To clarify this, the mechanism that the shape of the tunneling spectrum depends on the tip-sample distance must be further understood. 2) On the other hand $(BiO)_2$ layer in the bulk is semiconductive with a finite band gap energy. 3) BiO mono-layer on the cleaved surface is also semiconductive with a comparable band gap energy. 4) The band gap of the BiO mono-layer on the cleaved surface decreases as the sample contains more oxygen. 5) No superconducting gap structure has been observed so far on the BiO mono-layer as far as the atomic structure is observed. 6) Occasionally the superconducting gap structure was observable on the BiO mono-layer when the atomic image loses its contrast. 7) Based upon the superconducting gap observed on the cleave surface, the gap is much larger than that expected from the BCS theory. 8) The zero-bias peak in the superconducting gap region is due to the direct contact between the electrode and the sample. 9) The multiple superconducting gap so far often reported can be attributed to the extrinsic effects due to the damaged junction region.

So far we are still far from the complete understanding of the local nature of the material examined because of the insufficient stability of our STM system. Especially, STS observation on the edge of the layered structure is greatly desired at the low temperatures to examine the local variation of the superconducting order parameter. The limited informations so far obtained, however, are still very suggestive in regard to the origin of the anisotropy of $Bi_2Sr_2CaCu_2O_y$. The material may be viewed as the superlattice composed of the locally metallic and locally insulative layers. Further development of this technique has to be awaited.

References.

[1] D. E. Farrell, C. M. Williams, S. A. Wolf, N. P. Bansal and V. G. Kogan, Phys. Rev. Lett. **61** (1988) 2805.

[2] D. E. Farrell, S. Bonham, J. Foster, Y. C. Chang, P. Z. Jiang, K. G. Vandervoort, D. J. Lam and V. G. Kogan, Phys. Rev. Lett. **63** (1989) 782.

[3] K. Kitazawa, S. Kambe and M. Naito, "Strong Correlation and Superconductivity", ed. H. Fukuyama, S. Maekawa and A. P. Malozemoff, (Springer-Verlag, Berlin, 1989), p.361.

[4] Y. Tokura and T. Arima, Jpn. J. Appl. Phys. **29** (1990) 2388.

[5] T. Doi, M. Okada, A. Soeta, T. Yuasa, K. Aihara, T. Kamo and S. Matsuda, Physica C, submitted.

[6] B. O. Wells, Z. -X. Shen, D. S. Dessau, W. E. Spicer, C. G. Olson, D. B. Mitzi, A. Kapitulnik, R. S. List and A. Arko, Phys. Rev. Lett. **65** (1990) 3056.
[7] S. Massida, J. Yu and A. J. Freeman, Phisica C 152 (1988) 251.
[8] P. W. Anderson, Science **235** (1987) 1196.
[9] *see for example* "Proceedings of the 19th International Conference on Low Temperature Physics", ed. D. S. Betts, (North-Holland, 1990).
[10] *for the current situation of the tunneling spectroscopy measurements on high-Tc superconductors, see the review by* T. Ekino and J. Akimitsu (preprint).
[11] M. C. Gallagher, J. A. Adler, J, Jung and J. P. Franck, Phys. Rev. **B37** (1988) 7846, S. Vieira, J. G. Rodrigo, M. A. Ramos, K. V. Rao and Y. Makino, Phys. Rev. **B40** (1989) 11403, R. Wilkins, M. Amman, R. E. Solits, E. Ben-Jacob and R. C. Jaklevic, Phys. Rev. **B41** (1990) 8904, S. Vieira, M. A. Ramos, M. Vallet-Regí and J. M. Gonzalez-Calbet, Phys. Rev. **B38** (1988) 9295.
[12] T. Hasegawa and K. Kitazawa, Jpn. J. Appl. Phys. **29** (1990) L434.
[13] T. Hasegawa, M. Nantoh and K. Kitazawa, Jpn. J. Appl. Phys. **30** (1991) L276.
[14] M. Gurvitch, J. M. Valles, Jr., A. M. Cucolo, R. C. Dynes, J. P. Garno, L. F. Schneemeyer and J. V. Waszczak, Phys. Rev. Lett. **63** (1989) 1008.
[15] M. Furuyama, I. Iguchi, K. Shirai, T. Kusumori, H. Ohtake, S. Tomura and M. Nasu, Jpn. J. Appl. Phys. **29** (1990) L459.
[16] R. C. Dynes, V. Narayanamurti and J. P. Garno, Phys. Rev. Lett. **41** (1978) 1509.
[17] I. Iguchi, H. Watanabe, Y. Kasai, T. Mochiku, A. Sugishita and E. Yamaka, Jpn. J. Appl. Phys. **26** (1987) L645.
[18] Y. Matsui, H. Maeda, Y. Tanaka and S. Horiuchi, Jpn. J. Appl. Phys. **27** (1988) L372.
[19] T. Ekino and J. Akimitsu, Phys. Rev. **B40** (1989) 6902.

Unusual (Anomalous) Properties of the Normal and Superconducting State of Cu-Oxides

H.R. Ott

Laboratorium für Festkörperphysik, ETH-Hönggerberg,
CH-8093 Zürich, Switzerland

Abstract. Although the superconducting Cu-oxides are essentially metals, their normal-state properties are markedly different from those of common metals and it is argued that new theoretical concepts are needed for their description. The reason for the occurrence of superconductivity in these materials at relatively high temperatures is still unknown and debates with respect to triggering mechanisms and characteristics of the superconducting state continue. Below we briefly review experimental data that reflect anomalous behaviour, but we also mention observations that are compatible with conventional models of normal and superconducting states of metals.

1. Introduction

Listing and discussing "unusual" properties of copper-oxide high-T_c superconductors requires that some criteria are given which justify the phrasing unusual or anomalous. This is commonly done by comparing certain physical properties of the materials to be discussed here with those of other substances which have been well studied and whose behaviour is thought to be well understood. Alternatively, the experimental results are often compared with predictions of pertinent models and since the superconducting copper-oxides have to be regarded as metals (albeit of possible strange nature) the most obvious choice for such a model is that of the Fermi liquid which is usually thought to be adequate for metals with interacting conduction electrons.

With this in mind, it should be realized that hardly any group of materials of comparable chemical complexity has been investigated in such great detail and with so many experimental methods as is true for these copper-oxides. Hence it seems possible that "unusual" behaviour might also be found in materials that are thought to be well understood, given the fact that they were equally well studied in detail as these high-T_c superconductors. The synthesis of the cuprate materials is often not trivial and in many cases, multiphase substances result. Therefore, certain "anomalous" properties may frequently be traced back to the multiphase nature of these compounds.

In order to stimulate discussions, several issues concerning unusual physical properties are formulated, mostly as questions, and structured in three categories. The next section deals with structural aspects and instabilities, then we discuss electronic properties of the normal state and finally address questions related with the superconducting phase transition and characteristics of the superconducting state.

2. Structures and Instabilities

An obvious question that may arise is whether the perovskite-type unit cell or subunits thereof are of importance here. This comes to mind by recalling that for previously established oxide superconductors, the octahedral arrangement of oxygen atoms around atoms of metallic elements is a common feature. This is true for doped $SrTiO_3$ with a critical temperature T_c of less than 1 K, but also for doped $BaBiO_3$ and $LiTi_2O_4$ both of which, prior to the discovery of high-T_c cuprates, were regarded as singular oddities with surprisingly high critical temperatures between 12 and 13 K. The answer is most likely *no*, because the superconducting cuprates show a considerable tendency to anisotropic physical properties, a result of the particular stacking combinations of insulating and conducting building blocks, a feature that is absent in the above mentioned materials.

Nevertheless it seems interesting to note that both the narrow range in chemical composition favouring the appearance of superconductivity and the tendency to structural instabilities is observed in both low- and high-T_c oxide superconductors. In doped $BaBiO_3$, either with Pb for Bi or with K for Ba, superconductivity occurs in the phase with the highest symmetry of the crystal lattice (tetragonal in the first case and cubic in the second). This is not true for cuprate superconductors, where rising the symmetry from orthorhombic to tetragonal is deadly for superconductivity, as observed for Ba-doped La_2CuO_4. These observations may indicate, that the structural instabilities are, to a large extent, of electronic origin.

Assuming that the above mentioned anisotropy features are relevant, it may be argued that aspects of lower dimensionality are essential for the occurrence of superconductivity at temperatures above 30 K. This may again not be a general trend, however, and is certainly not true for $Ba_{1-x}K_xBiO_3$, where T_c's of the order of 30 K have been obtained. Another class of known oxide superconductors, the bronzes, show tendencies to 1D behaviour but for them, the critical temperatures are found at rather low temperatures. There is also no clear trend that would support a direct connection between increasing anisotropy (more 2D character) and enhancement of T_c in the copper oxides.

For the copper oxides, the structural instabilities and related metal-insulator transitions are often also connected with the occurrence of a magnetically-ordered (afm) ground state of the insulating phase. Since this is a particular feature of copper-oxides, it has been argued that the occurrence of superconductivity at elevated temperatures is intimately connected with this and a lot of work has been devoted to investigate the microscopic aspects of magnetism in these materials using various experimental techniques.

Another type of instability that might be of relevance here is related with the valence stability of the atoms providing the charge carriers in the metallic state. This is obviously true for copper, the first element after the 3d transition-metal series. One may wonder whether other elements in similar positions in the periodic table of the elements and, as constituents of compounds, embedded in a suitable chemical environment might be candidates for high-T_c superconductivity.

As is often observed, optimal conditions for superconductivity seem to be restricted to a rather narrow range of chemical composition for oxide materials. This, of course, raises the question, whether phase purity plays an important role for obtaining maximum values of T_c. Quite generally, one may also consider that these materials have an inherent tendency to phase separation, both on microscopic and macroscopic scale. Especially the first possibility would greatly hamper a conventional characterization of these substances. An instructive case for macroscopic phase separation is met in oxygen overdoped $La_2CuO_{4+\delta}$, where a separation into two distinct phases, of which one is superconducting, is observed when lowering the temperature from below the melting temperature. Extreme difficulties for synthesizing phase-pure material are also met in the n-doped Cu oxides adopting the T' crystal structure, where the optimal oxygen content is of particular importance.

To conclude this section we mention that following ideas related with lattice instabilities and hence possible enhanced electron-phonon interactions finally led to the discovery of high-T_c superconductors. Up to this point it is not clear whether mechanisms of this sort are really responsible for the much larger critical temperatures than observed in more conventional metals and alloys.

3. Electronic Properties of the Normal State

Assuming a conventional metallic state for the conducting copper-oxides results in difficulties of explaining various experimental observations on simple grounds. A prime example of such a dilemma is the frequency dependence of the conductivity $\sigma(\omega)$ as it is obtained from optical experiments. From many measurements of different workers, the decay of σ with increasing frequency ω is not consistent with the Drude-type ω^{-2} dependence as it is observed in common metals, but rather follows a ω^{-1}-behaviour. In some of the relevant materials it is possible to change the characteristics of the normal state by introducing large amounts of charge carriers. Although the transport properties, such as, e.g., $\sigma(\omega)$ approach then those of conventional metals, superconductivity at lower temperatures is unfortunately lost. The above mentioned optical data imply that the relaxation rate at low frequencies is proportional to ω, compatible, of course, with the linear temperature dependence of the dc resistivity. The latter is one of the most obvious common features of copper-oxide superconductors for which the conductivity is dominated by moving charge carriers in the Cu-O planes. Surprisingly this particular T dependence is quite robust with respect to sample imperfections.

Similar difficulties appear in the interpretation of spectroscopic experiments like Raman scattering or photoemission of electrons from near the Fermi energy. In both cases it is claimed that background contributions are observed that should not be there, if the superconducting copper oxides were merely common metals. More recent reports of analogous measurements on other more common materials, however, seem to provide results with similar characteristics.

In addition, measurements of relaxation rates T_1^{-1} related with nuclear resonances of nuclei of different atoms (Cu,O) in these compounds give results that seem impossible to be reconciled with common observations in metals. Again it is only in metallic but non-superconducting compounds where the expected Korringa-type temperature dependence is observed, at least approximately.

Finally there are various aspects in transport phenomena which are claimed to be incompatible with common behaviour. These include the large anisotropy of the conductivity, coupled with the observation of small conductivities beyond the Mott limit in the direction perpendicular to the Cu-O planes, and the temperature dependence of the Hall constant R_H for currents in the Cu-O planes. The latter experimental result is obviously not compatible with a single-band description of the charge-carrier spectrum, as it is the basis of some existing theoretical models.

All these experimental facts seem to be consistent, however, with an unusual frequency dependence of scattering processes, not characteristic for a Fermi-liquid-type model as commonly used for the description of metallic states. This has led to a variety of suggestions for new theoretical concepts for modelling the electronic properties of these compounds, some of which were also discussed during this school. They are certainly all very interesting variations from the usual perception of how metallic states in condensed matter should be described.

Taking into account some of the remarks that were made in section 2, it may also be possible that structural pecularities of these compounds may lead to electronically inhomogeneous systems. It may even be speculated that these inhomogeneities are time dependent on very short time scales. If these time scales interfere with lifetimes of quasiparticles, complications in the description of such systems may be anticipated. In connection with superconductivity (coherence length) also the spatial extent of such inhomogeneities might play an essential role.

A lot of work using different kinds of experimental techniques has been invested into investigations of how the conducting state develops out of the insulating state. Absorption-type experiments both with electromagnetic radiation and electrons are often interpreted by postulating a simple addition of new states in the energy gap of a conventional excitation spectrum upon doping the materials with suitable changes of the chemical composition. Above a certain doping level, these states are then thought to form a band in the usual sense and one is left with the obvious dilemma, how this rather conservative point of view may be reconciled with the quite unusual experimental results mentioned above.

In this context, it should be remembered that another class of exotic metals, the heavy-electron systems, show equally surprising properties, which are usually ascribed to extremely strong interactions between conduction electrons. What is clearly different in these systems is the lack of tendency towards lower-dimensionality behaviour. This is much more to be considered in certain classes of organic conductors, some of which are also superconducting. It seems natural that these classes of unusual metals may require new concepts for adequately describing all their properties.

4. Superconductivity

From a simple point of view, the most anomalous property of these superconductors is still the high critical temperature T_c. Just as for ordinary superconductors, studies of variations of T_c for a particular material, induced by external influences (change of chemical composition, external pressure or magnetic field) are made in attempting to get a clue of what might be the really determining factors for T_c. As mentioned above the superconducting phase of copper oxides typically appears close to a transition from an insulating to a conducting state. The insulating ground state is antiferromagnetically ordered and doping with charge carriers results in a gradual decrease of the Néel temperature T_N and eventually, above a critical value of doping concentration x, the ground state is superconducting.

For hole doping, the phase boundaries $T_{N,c}(x)$ between the two different types of ground state don't seem to share common points, whereas for n-doped cuprate superconductors, a narrow range of superconducting phase is adjacent to the afm-phase boundary. The latter is reminiscent of phase diagrams in the (T,p) plane of ground states adopted in some organic superconductors.

As always it would be useful to establish simple rules for the occurrence of superconductivity in these compounds. From results of various μSR studies it was suggested that the ratio of the charge-carrier concentration divided by the value of their effective mass, n/m*, is a decisive parameter for the magnitude of T_c. Although n can be influenced by varying the chemical composition, m* is rather elusive to experiment and therefore this "rule" is not very useful for tailoring new superconductors.

A remarkable early finding was the observation that in the most popular high-T_c superconductor, $Ba_2YCu_3O_7$, the trivalent Y ions could be replaced by most of the elements of the rare-earth series and in spite of their local magnetic moments, no sizeable decrease of T_c occurred. The obvious important parameter is the trivalency of the ions implanted on the Y site. Nevertheless, some of such compounds reveal long-range magnetic order among the rare-earth ionic moments at very low temperatures. In previously known magnetic superconductors like the rare-earth rhodium borides or Chevrel phases, T_c clearly varies as a function of g_J^2 J(J+1), where g_J is the ionic Landé factor and J the Hund's rule ground-state moment, thus at variance with observations mentioned above.

While in the case of YBCO superconductors, magnetic moments on sites between the Cu-O planes appear not to be pair breaking, rather small amounts of particular elements, e.g. Zn, replacing Cu in those planes, are devastating for T_c. It is not yet clear, what the cause for this drastic effect really is. More work for establishing clear trends in pair-breaking effects for Cu-O superconductors appears necessary.

In common superconductors pair formation implies the observation of so-called coherence effects which depend on the symmetry of the applied interfering interaction related with the experimental probe. The conventional Hebel-Slichter peak in $T_1^{-1}(T)$ data is notoriously absent in cuprate superconductors, as it is also in heavy-electron superconductors. Recent theoretical work suggests that for the cuprates, this doesn't necessary imply that unconventional superconductivity must be considered. Also the observation of enhanced real-part conductivities σ_1 (T) at frequencies below the gap frequency and at temperatures just below T_c are due to other causes than coherence effects.

Next we consider the formation of a gap in the excitation spectrum, its magnitude vis à vis T_c and its anisotropy, including the possibility of gap zeroes due to higher-momentum pairing. Studies of this sort are done using many experimental techniques. As a novelty, the improved resolution in photoemission experiments was for the first time sufficient to demonstrate the expected loss of intensity at E_F and a corresponding energy shift of spectral weight by cooling a cuprate superconductor to below T_c. Other spectroscopies which probe the gap formation include Knight-shift evaluations and $T_1(T)$ measurements in NMR, various forms of tunneling experiments and, again, optical experiments in the far-infrared region. All of these measurements confirm the expected gap formation but in each case, certain details do not seem to find simple explanations or are not consistent with other data. Optical data, e.g., seem to imply that the gap frequency evaluated at low temperatures doesn't change very much with increasing temperature, certainly not as predicted by the BCS theory. Knight-shift data, on the other hand, are very well approximated with the classical Yosida function, itself compatible with a BCS-type gap function below T_c. Tunneling data pose the problem that most often, non-zero conductivities in the gap region are observed, by far exceeding expected non-zero temperature effects.

A quantity which clearly only involves the superconducting condensate is the London penetration depth $\Lambda_L(T)$. The evaluation of its T dependence also gives information on the T dependence of the gap and an overwhelming part of all available data indicates a simple BCS behaviour. These and most other results of measurements with small degrees of ambiguities seem to indicate that no gap zeroes have to be considered and therefore the superconducting state is most likely not a higher-momentum state. This latter possibility can, however, not be completely ruled out either.

As a last topic we briefly touch upon the question of a gap formation above T_c in cuprate superconductors. This possibility gets some support from comparing the magnitude of the specific-heat $C_p(T)$ anomaly at T_c with the normal-state magnetic properties of respective compounds.Unusually small C_p anomalies are observed for substances where either susceptibility measurements or spectroscopic experiments (neutron scattering, NMR) indicate a loss of magnetic degrees of freedom with decreasing temperature above T_c. It is, however, certainly premature to suggest a definite link between these observations. Finally it should not be forgotten, that some of the unusual results of studies of the superconducting state of these materials may be caused by the extremely short coherence lengths.

Instead of giving a long list of original references, the interested student is recommended to consult these proceedings and the following short list of suitable publications.

References

1. Strong Correlation and Superconductivity, H. Fukuyama, S. Maekawa and A.P. Malozemoff, eds., Springer Series in Solid State Sciences, Vol. 89, (Springer, Berlin 1989

2. High-Temperature Superconductivity, Proc. of the Los Alamos Symposium 1989, K.S. Bedell, D. Coffey, D.E. Meltzer, D. Pines and J.R. Schrieffer, eds., (Addition-Wesley, Redwood City 1990)

3. Proc. of the 3rd Int. M^2S-HTSC Conference, Kanazawa, Japan, 1991.

Anisotropy in Elastic Properties of HTSC Seen by Ultrasonics

T. Fujita and T. Suzuki

Department of Physics, Faculty of Science,
Hiroshima University, Hiroshima 730, Japan

Abstract. After a brief introduction to ultrasonic experiments, elastic properties seen by ultrasonic response are presented for a single crystal of $La_{1.86}Sr_{0.14}CuO_4$. Elastic coefficients for various modes reveal not only anisotropic lattice properties in the normal state, but also anisotropic coupling between superconductivity and lattice deformation. Effect of magnetic field applied along various axes is also discussed.

1. Introduction: Attenuation and Elastic Coefficient

Let us consider the case in which pulsed ultrasound travels in a sample and is reflected at the surface. If the sample of length l has two parallel surfaces on both sides, a train of echo pulses are observed with a time interval of $2\Delta t$. The height of successive echo pulses decreases exponentially with time t: $I=I_0\exp(-\beta t)$. From this firsthand information, we obtain important physical quantities such as sound velocity $v=l/\Delta t$, attenuation coefficient $\alpha=\beta/v$ and elastic coefficient $c=\rho v^2$, where ρ is the mass density. These coefficients provide information not only about lattice properties but also about electron–phonon interaction and electronic properties.

In a normal state of pure metals, the attenuation coefficient α_n is proportional to relaxation time of electrons or electrical conductivity. In the superconducting state which has an isotropic gap Δ of the BCS type, the coefficient α_s for a longitudinal mode is simply proportional to the Fermi distribution: $\alpha_s/\alpha_n=2/\{\exp(\Delta/kT)+1\}$, whereas α_s for a transverse mode exhibits an abrupt decrease immediately below the superconducting transition temperature T_c due to the shorting out effect by supercurrent. If the superconductor has an anisotropic gap which vanishes at the equator of the Fermi surface or at the pole, α_s/α_n for a longitudinal mode has a T or T^2 dependence on temperature [1], respectively.

Thermodynamically the elastic coefficient c is defined as the second derivative of free energy with respect to the strain ε. Consequently the coefficient c is a second rank tensor which has various components sensitive to anisotropy in crystallized materials. For a tetragonal lattice, there are five components with pure symmetry: $(c_{11}-c_{12})/2, c_{44}$, c_{66}, c_{33} and $(c_{11}+c_{12})/2$. The first three are components for transverse modes corresponding to the shear strains $\varepsilon_{xx}-\varepsilon_{yy}$, $(\varepsilon_{yz}, \varepsilon_{zx})$ and ε_{xy}. In case of c_{33} both propagation vector and displacement are along the c axis. The last one, which corresponds to a non–propagating mode, can be calculated from $(c_{11}-c_{12})/2$ and c_{11}. For conventional isotropic superconductors, the longitudinal elastic coefficient is expected to exhibit a discontinuous drop across T_c in contrast to the transverse component showing no discontinuity [2].

In this article, we report our ultrasonic experiment on a single crystal of $La_{1.86}Sr_{0.14}CuO_4$ and discuss anisotropic properties both in the normal and superconducting states.

2. Structural Transition of $La_{2-x}Sr_xCuO_4$

A high–T_c superconductor $La_{2-x}Sr_xCuO_4$ with x=0.14 undergoes a structural transition from a tetragonal phase with the space group I4/mmm to an orthorhombic phase with Bmab around T_d=200 K, which is triggered by tilting the CuO_6 octahedron toward the <110> direction [3]. For simplicity, we use the crystalline coordinates of the high–temperature tetragonal phase even for the low–temperature phase. In the orthorhombic phase, oxygen atoms of the CuO_2 layer displace slightly along the <001> direction so that Cu and O atoms form a zigzag plane.

Figure 1 shows the anisotropic response of a single crystal of $La_{1.86}Sr_{0.14}CuO_4$ to various modes of ultrasound. Two components c_{11} and c_{66}, both of which are the elastic constants for longitudinal and transverse modes in the basal CuO_2 plane, exhibit pronounced softening around T_d, reflecting the lattice instability. We find a substantial decrease especially in c_{66}, which is the response to the symmetry breaking strain ε_{xy} with an irreducible representation B_{2g}. Below T_d, c_{66} was not observed presumably because the twin boundaries

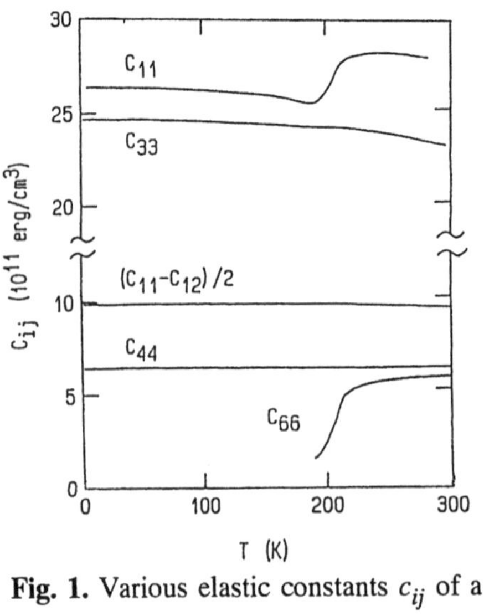

Fig. 1. Various elastic constants c_{ij} of a single crystal $La_{1.86}Sr_{0.14}CuO_4$.

produced by the transition heavily scatter the ultrasound. The elastic anomalies associated with the structural transition at T_d are so weak in the longitudinal c_{33} as well as in the transverse $(c_{11}-c_{12})/2$ and c_{44} that any change can not be displayed in the scale of Fig. 1.

Preliminary measurement performed on a polycrystalline sample of $La_{1.875}Ba_{0.125}CuO_4$ shows a clear increase both in attenuation and elastic coefficient below 60 K for the longitudinal ultrasound, corresponding to another structural transition to the low temperature tetragonal phase with the space group $P4_2/ncm$ [4,5].

3. Elastic Properties in Superconducting State

Figures 2(a) and (b) are close-up illustrations of the elastic constants of $La_{1.86}Sr_{0.14}CuO_4$ at low temperatures. All elastic constants except c_{44} exhibit

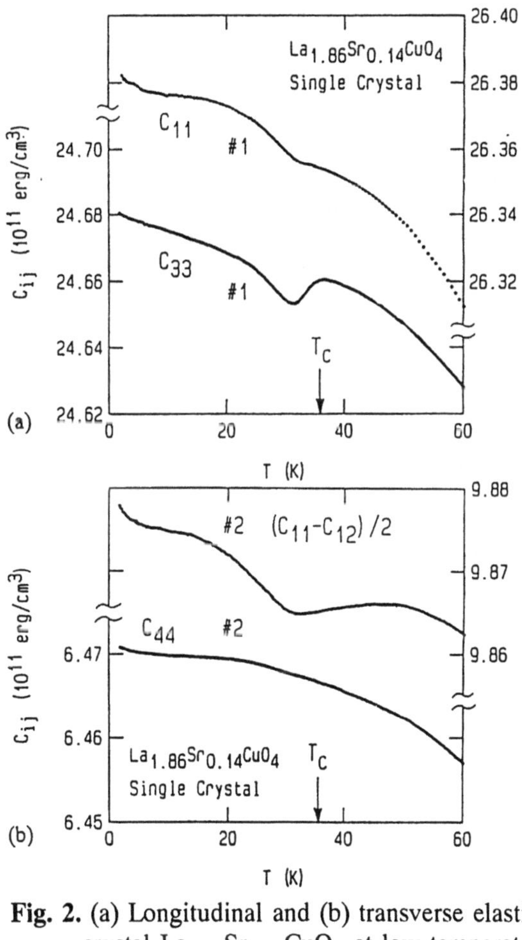

Fig. 2. (a) Longitudinal and (b) transverse elastic coefficients of a single crystal $La_{1.86}Sr_{0.14}CuO_4$ at low temperatures around T_c.

a clear anomaly around T_c. c_{33} shows abrupt softening at T_c which is followed by gradual hardening. This behavior is similar to that expected for conventional superconductors [2]. According to the Ehrnfest relation [6], the relative value of softening $\Delta c_{33}/c_{33}$=-6.7×10^{-4} and the reported jump of specific heat [7] gives an estimation of $dT_c/d\sigma_z \approx 0.49$ K/kbar. This stress dependence of T_c is consistent with the reported pressure dependence of T_c [8]. The longitudinal c_{11} only shows hardening below T_c. It is surprising that transverse $(c_{11}-c_{12})/2$ remarkably softens below about 50 K, which is ~15 K higher than T_c. This softening might be ascribed to some other origin rather than the superconducting transition. The very weak response of c_{44} to the superconducting transition can be understood because the relevant strain ε_{zx} with E_g acts to slide stacks of CuO_2 layers parallel to <100> or <110> rather than to deform the CuO_2 lattice in the plane.

To explain the discontinuous softening observed in c_{33} on the basis of a phenomenological Ginzburg–Landau theory, we need to include in the free energy a coupling term between the superconducting order parameter Ψ and strain ε in the form of $\gamma|\Psi|^2\varepsilon$ in addition to the original term of the elastic energy $c^0\varepsilon^2/2$. On the other hand, the lowest term of the coupling should have a form of $\eta|\Psi|^2\varepsilon^2$ to describe a change in the slope of the elastic constants against temperature at T_c. We believe that this anisotropic coupling between superconducting order parameter and strain implies an essential character of superconductivity in the high–T_c oxide superconductors.

When magnetic field H is applied parallel to the a–b plane, the longitudinal c_{33} is enhanced below T_c as shown in Fig. 3 (a). Since the enhancement appears to be proportional to H^2, this hardening can be explained by the contribution from the stiffness of the flux line lattice trapped in the pinning potential. In the field along the c axis, a direct coupling is not expected between H and c_{33}. However, the discontinuous softening in c_{33} just below T_c is smeared out with increasing field. This effect should be ascribed to the same origin as the broadening in the resistive transition in the applied fields. The H dependence of c_{33} at low temperatures could be attributed to the suppression of the superconducting order parameter and recovery toward the behavior in the normal state.

The field dependence of the in–plane mode c_{11} is shown in Fig. 3 (b). When the field is applied along the same direction <100> as the ultrasonic displacement vector, the elastic constant below T_c decreases with increasing field. This may be explained also in terms of the suppression of the superconducting order parameter. In the field parallel to the a–b plane but perpendicular to the ultrasonic displacement, we find an additional hardening, which is probably due to the contribution from the stiffness of the flux line lattice developed below T_c. The most mysterious behavior is seen for c_{11} in magnetic fields along the c axis. Especially the origin of the low–temperature hardening remains to be unknown.

All of the ultrasonic attenuation coefficients α_{ij}'s show an strong anisotropy with a large remnant value well below T_c. It is noted that α_{ij}'s do not show the BCS–like exponetial decrease with decreasing temperature below T_c. The non–

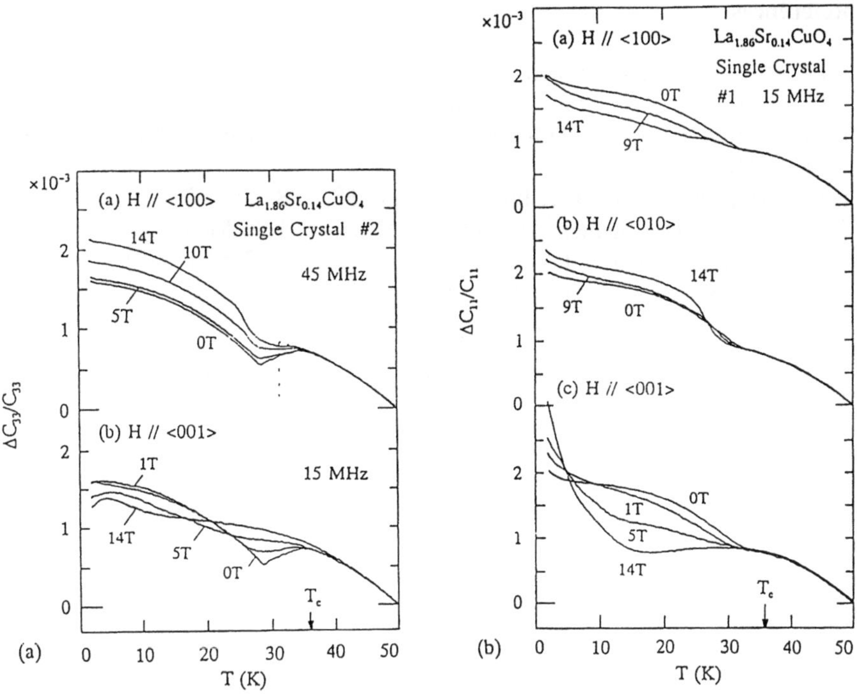

Fig. 3. Magnetic–field dependence of (a) c_{33}, and (b) c_{11} in the superconducting state.

exponential decrease is likely to suggest the anisotropic wave–vector dependence of the superconducting energy gap.

4. Conclusion

In summary, we reported ultrasonic experiments for various strain modes in $La_{1.86}Sr_{0.14}CuO_4$ and found the symmetry–dependent anisotropic coupling between the superconductivity and the lattice. The attenuation coefficients do not exhibit the exponential decrease below T_c. These anisotropic response to the ultrasound serves as a critical checkpoint to discriminate the mechanism of high–T_c superconductivity from a variety of proposals.

We would like to express our sincere thanks to Prof. H. Kojima and Dr. I. Tanaka for supplying us with single crystals of $La_{1.86}Sr_{0.14}CuO_4$ with high quality. We also thank Mr. M. Nohara and Mr. T. Tomita for their collaboration in the ultrasonic experiments.

References

[1] K.Miyake, J. Magn. Magn. Mater. **63** & **64**, 411 (1987).
[2] L.R.Testardi, Phys. Rev. B **12**, 3849 (1975).
[3] B.H.Toby, T.Egami, J.D.Jorgensen and M.A.Subramanian, Phys. Rev. Lett. **64**, 2414 (1990).
[4] T.Suzuki and T.Fujita, J. Phys. Soc. Jpn. **58**, 1883 (1989).
[5] J.D.Axe, A.H.Moudden, D.Hohlwein, D.E.Cox, K.M.Mohanty, A.R.Moodenbaugh and Y.Xu, Phys. Rev. Lett. **62**, 2751 (1989).
[6] A.J.Millis and K.M.Rabe, Phys. Rev. B **38**, 8908 (1988).
[7] A.Amato, R.A.Fisher, N.E.Phillips and J.B.Torrance, Physica B **165** & **166**, 1337 (1990).
[8] C.Murayama, N.Mori, S.Yomo, H.Takagi, S.Uchida and Y.Tokura, Nature, **339**, 293 (1989).

Some Consequences of Anisotropy

T.H. Geballe

Stanford University, Department of Applied Physics,
Stanford, CA 94305, USA

Anisotropy can affect superconductivity in structures which are three- dimensional as well as quasi one and two-dimensional. Consideration of existing data shows that quasi two-dimensional features have particular advantages for superconductivity; 3 special systems are discussed. It seems likely that there are many more new superconductors awaiting discovery.

Anisotropy in 1, 2, and 3 dimensions

The superconducting state is the most common ground state of non magnetic metals. However, almost all the known superconductors have transitions (T_c's) which occur below 15K. The distribution in critical temperature of these superconductors can be understood in terms of parameters of the Bardeen Cooper Schrieffer BCS theory such as the Debye temperature, the electron-phonon coupling constant, and the retarded Coulomb pseudopotential. These superconductors are essentially three dimensional with very little anisotropy. In contrast, structural anisotropy does become important in the compounds with the cubic A15 structure where nonintersecting chains of metal atoms along the cubic directions give rise to planar Fermi surface features. Nb_3X compounds (X = Sn, Al, Ga, Ge and Si), to the extent they can be prepared as stoichiometric compounds, have many unusual properties in both the normal and superconducting state including T_c's of up to 23K, [1]. These were the highest known superconductors before the discovery of Bednorz and Mueller in 1986. A weak structural (martensitic) transition at temperatures somewhat above T_c can occur which is due to an instability resulting from the nesting of the planar sections of the Fermi surface. This nesting eliminates a small portion of the Fermi surface and consequently lowers T_c by a small amount.

In general anisotropy of the Fermi surface which results from quasi one-dimensional structural features gives rise to spin and charge wave instabilities which compete with superconductivity. The competition is too strong in one dimensional structures such as organic charge transfer salts where spin density waves inevitably are found. It is only under high pressures or when anions are used which introduce quasi two-dimensional bonding that superconductivity is found in these structures [2]. The inorganic polymer $(SN)_x$ which becomes superconducting at very low temperatures also has two dimensional features [3]. A charge density wave occurs in organic one-dimensional structures such as the platinum chain compounds [4]. It is when the quasi-one-dimensional feature are confined to only portions of the Fermi surface as in the A15 compounds, that a superconducting ground state is found.

The situation in so far as superconductivity is concerned is more favorable in the case of quasi two-dimensional structures. Studies of the transition metal dichalcogenides have been particularly revealing. These are layered structures which consist of sandwiches of XMX-XMX where X is the chalcogen, M is a transition metal and the two-dimensional characteristic is introduced by the weak Van der Waals bonds which occur between adjacent chalcogen planes. The group V compounds such as TaS_2 have a conduction band which is constructed from one d electron per Ta. At high temperatures where the equilibrium phase has an octahedral disposition of sulfur atoms (which can easily be quenched) the instabilities are strong and semiconducting behavior is observed at low temperatures because the Fermi surface is wiped out by the formation of commensurate charge density waves. In the room temperature stable polymorph, the Ta or Nb has a trigonal prismatic coordination and the charge density wave instability affects only part of the Fermi surface. For $NbSe_2$, T_c =7.2 K is raised to over 8 K when the weak incommensurate charge density wave is suppressed [5]. In TaS_2 the weak Van der Waals bonding between successive sulfur layers make it possible to intercalate a wide variety of organic and inorganic molecules. The intercalate prevents the charge density wave instability and raises T_c. It can easily be seen why the intercalation prevents the charge density wave from forming. Namely, due to the ever present electron-phonon interaction there must be a lattice distortion (commensurate or incommensurate with the periodicity of the lattice). The distortion costs elastic energy which increases in the presence of the intercalate. Gamble et al [6] studied the changes in T_c for a series of aliphatic amine intercalates as a function of the length of the amine. Initially the amine lie parallel to the layers (constant interlayer spacing) and the reduced charge transfer from the amine to the conduction band per unit area results in a monotonic decrease in T_c. For the longer chains (C>8) the amines stack perpendicularly to the layers, the layer spacing increases with chain length while the bond density and hence charge transfer per unit area remains constant, as does the T_c, all the way to spacings of 57 angstroms. The theory of Lawrence and Doniach [7] in which the layers are ultimately connected by Josephson coupling was formulated to explain the superconducting transition found in such quasi two-dimensional structures.

The parallelism of the intercalated dichalcognides with the high temperature superconducting cuprates discovered by Bednorz and Mueller is clear. In both cases the energy required to deform the intercalate, or in the cuprates what has come to be called the "block" layers between the CuO_2 planes, tends to suppress competing charge wave instabilities. In both cases the intrinsic anisotropy is enormous and as a result is difficult to quantify experimentally. On the one hand cracks or increased separations of the weakly bonded Van der Waals layers can increase the apparent resistivity perpendicular to the layers, and, on the other hand shorts due to various kinds of stacking faults and other imperfections will decrease the measured anisotropy. In the case of the dichalcogenides even though the measured resistivity anisotropy was very large, the temperature dependence of the resistivity in the perpendicular direction could be mapped on to that in the planar direction simply by a geometric scale factor. This could be understood in terms of the screw dislocations which were found in the crystals. The helicity of the screw provides a path for perpendicular conduction which is always in the high conductivity planes. Factors of 10^6 in the intrinsic anisotropy can be shorted by dislocation densities less than 10^5 per square cm of the dichalcogenides. In the cuprates screw dislocations are also found under some conditions of growth [8] and almost certainly will provide a shorting mechanism. Such shorting is evidently not dominant in single crystals which are grown from the melt (as distinct from vapor phase growth) because the resistivity perpendicular to the conduction planes has a negative, semiconductor-like temperature dependence at low temperatures.

Enhanced Anisotropy via superlattices

Superlattices provide a controlled means for studying two-dimensional behavior. Superlattices are called artificial structures when they are grown from the vapor phase by sequential deposition of layers of differing composition and do not occur naturally. It is instructive to consider three specific examples.

MoGe/Ge sequences form a model system in which the Josephson coupling can easily be controlled. The mass anisotropy can be extended from less than, to more than, that occurring in the high T_c cuprates [9]. White et al have found that the coupling between superconducting MoGe layers which are 60 angstroms thick through layers of insulating Ge layers which are 35 angstroms gives a mass ratio about the same as in pure $YBa_2Cu_3O_7$. Resistively-measured transitions as a function of magnetic field are found to have similar shapes and activation energies as found in $YBa_2Cu_3O_7$. It is evident that many more studies can be achieved by proper doping of the germanium layers.

A number of informative studies of superlattice cuprate structures have been carried out, particularly in the $YBa_2Cu_3O_7/PrBa_2Cu_3O_7$ structure. An excellent discussion of the behavior has been given in the lecture by Prof. Fischer at this school. I would like to call attention to the recent work of Bozovic, Eckstein, and co-workers at Varian concerning BiSrCaCuO superlattices with the 2201/2212 sequence [10]. The end members are pure 2201 (T_c = 10 K) and pure 2212 (T_c = 85 K). This series is probably the most perfect series of high T_c superlattices studies to date. They are made by layer by layer sequential deposition in which the formation of each layer is monitored by its RHEED pattern. The only difference between the two end members is than an extra Ca and CuO_2 layers is inserted every half unit cell in the 2212 structure. Thus, for example in the superlattice consisting of 5 one-half unit cells of 2201 followed by a single one-half unit cell of 2212, the periodicity requires only that an extra Ca and CuO_2 layer by inserted in the 2201 sequence after every twenty-five layers. The results are striking in so far as T_c is concerned because the T_c (and temperature dependent resistivity when normalized) is almost the same as the pure 2212 phase. There is no appreciable decrease in the resistive T_c as the separation between successive double pyramidal CuO_2 planes in the 2212 structure (which themselves are separated by 3 angstroms) is increased to 74 angstroms. The 85 K transition is evidently a property of the double pyramidal sequence. This is in contrast to the $YBa_2Cu_3O_7/PrBa_2Cu_3O_7$ system where T_c falls gradually as the external separation between the double pyramidal sequences of the $YBa_2Cu_3O_7$ units is increased. As a consequence of the Varian results it seems possible that the fall off in the Y-based superlattices is extrinsic, due to the increased strain, and/or interfusion of the cations during growth. One can further conclude that single CuO_2 layers Josephson-coupled through 12 angstroms of intervening layers (in the 2201 structure at least) have a T_c of 10 K which is well within the range of normal BCS superconductivity as mentioned in the opening paragraph.

As a final example I would like to call attention to the superlattices which can be formed when the CuO_2 planes are perpendicular to the substrate, that is in the a-axis orientation as distinct from the c-axis orientation discussed in the previous paragraph [11]. In the a-axis superlattices it is possible to divide the CuO_2 planes into alternating superconducting and insulating strips by sequentially depositing $YBa_2Cu_3O_7$ and $PrBa_2Cu_3O_7$ units. There is an advantage over c-axis

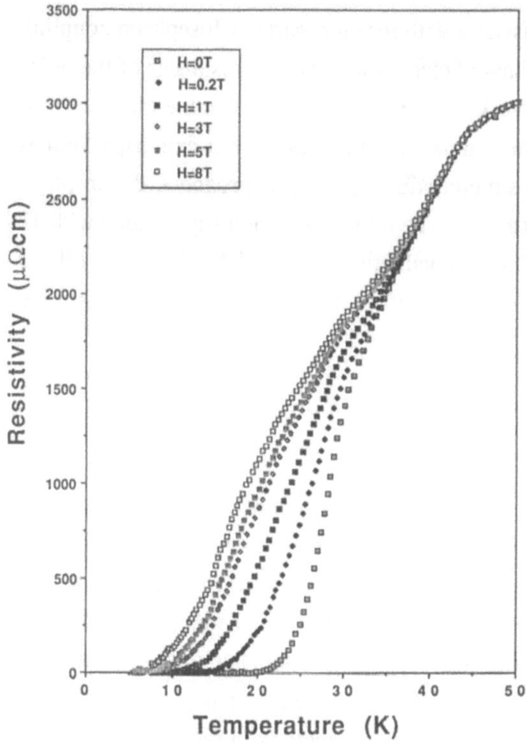

Figure 1. Resistivity versus temperature for 24Å/24Å a-axis $YBa_2Cu_3O_7/PrBa_2Cu_3O_7$ superlattice and for different magnetic fields between 0 and 8T. These curves illustrate a crossover at a characteristic temperature T^*: above T^* the resistivity is insensitive to magnetic fields up to 8T, below T^* some flux flow related broadening occurs. The origin of this effect is discussed in the text.

depositions when growing from a composite target, as is usually done in sputtering or laser ablation, because there are sites for all the cations on each surface layer during growth. Fig 1 shows for a 24Å/24Å YBCO/PBCO a-axis that an interesting crossover occurs in the resistive transition as a function of field. A field-independent regime (up to 8 T, the maximum available) is evident in the higher temperature regime where thermal fluctuations can suppress coupling and prevent the formation of vortices. As can be seen vortices must form at the low temperature end of the transition where a thermally activated flux flow regime is found. The crossover between the two regime is designated by T^* in the figure. A rough estimate indicates that a separation between the superconducting strips of 100 angstroms will cause complete decoupling over the measured temperature range. This estimate seems inconsistent with recent reports of Josephson coupling across $PrBa_2Cu_3O_7$ barriers in trilayer heterostructures over distances of the order of 1000 angstroms [12]. Further work will be required to resolve the difference. It is possible to make quasi one-dimensional structures in these a-axis superlattices when the in-plane width of the superconducting strip approaches the in-plane coherence length of 15 to 20 angstroms. The initial study of a superlattice with 12 angstrom spacing does not show superconductivity above 4 K. Extrinsic effects such as diffusion must be sorted out before the behavior can be studied quantitatively.

Concluding Remarks

In conclusion it is possible to make a few general remarks. The first is that it seems evident that quasi two-dimensional structures have a natural advantage over quasi one-dimensional structures in creating conditions favorable for superconductivity. In the former case intervening layers can be incorporated which serve the dual purpose of doping the conducting layers to some optimum value in-so-far as superconductivity is concerned and at the same time introducing a stiffness which suppresses the other instabilities which compete with the superconducting transition. No one has yet developed a good strategy suppressing the competing instabilities in one-dimensional structures. In three dimensional structures it is possible for anisotropy to enhance the superconductivity, although in studies so far the effects are weaker than in quasi two-dimensional structures. It is possible to control the doping in three-dimensional ternary structures such as $BaKBiO_3$ by chemical substitution on sites which are not involved with the conduction band. In such systems the maximum T_c frequently occurs near or at a phase transition to a non superconductor. In the related but lower T_c superconductor $Ba(BiPb)O_3$, where the doping takes place on the same sites from which the conduction band is formed it has been found [13] that charge density waves persist in the superconducting state.

Finally it should be emphasized that there is plenty of reason to believe that many more superconductors will be found in the coming years. One need only to mention the recent discovery of superconductivity in the C_{60} alkali metal intercalates [14]. It should be noted in the context of the present discussion that the C_{60} structural units (so called "buckyballs"), are smaller than the superconducting coherence length and thus the superconducting transition is quasi 0--dimensional. In this sense there may be other classes of cluster or mesoscopic systems in which the relevant length scales can produce high T_c superconductivity without requiring anisotropic interactions. It is easy to believe that as more knowledge is accumulated that superconductors with even higher T_c's than are known at present will be synthesized.

References

[1] R. M. White, R. H. Geballe, Long Range Order in Solids, Academic Press NY (1979) pages 234-243.

[2] N. M. Williams, et al, Science **232** 1501, (1991), and D. Jerome ibid, 1509 (1991).

[3] R. L. Greene and G. B. Sheet Chemistry and Physics of One Dimensional Structures p. 167, Plenum Press, NY (1977) H. S. Keller, Ed.

[4] R. Comes, M. Lambert, H. Launois and H. R. Zeller, Phys Rev B 8, 571 (1973).

[5] Ref. 1, p. 197, R. M. White, T. H. Geballe, Long Range Order in Solids, Academic Press, NY (1979).

[6] F. R. Gamble, J. H. Osiecki, M. Cais, R. Pisharody, F. J. DiSalvo and T. H. Geballe, Science, 174 493 (1971).

[7] W. E. Lawrence and S. Doniach, Proc. 12th International Conference on Low Temperature Physics, E. Kande, Ed. Tokyo, (1970).

[8] C. Gerber, D. Auselmetti, J. G. Bednorz, J. Manhart and D. G. Schlom, Nature, 350 279 (1991); M. Hawley, I. D. Raistrack, J. G. Berry and R. J. Houlton, Science 251 1587 (1991).

[9] W. R. White, A. Kapitulnik and M. R. Beasley, PRL **66** 2826 (1991).

[10] I. Bozovic, J. Eckstein, J. of Superconductivity (Plenum Press) in press.

[11] C. B. Eom, J. M. Triscone, Y. Suzuki and T. H. Geballe, to be published in the proceedings of the Conference on the Materials and Mechanisms of Superconductivity - High Temperature Superconductors, Kanazawa, Japan (1991); J. M. Triscone, C. B. Eom, J. Suzuki and T. H. Geballe, J. Less Com Metals (in press).

[12] J. B. Barner, C. T. Rogers, A. Inam, R. Ramesh, and S. Bersey, submitted for publication.

[13] J. M. Boyce, F. G. Bridges, T. Claeson, Phys Rev B 44 6961 (1991).

[14] A. F. Hebard, M. J. Rosseinsky, R. C. Haddon, D. W. Murphy,S. H. Glarum, T. T. M. Palstra, A. P. Ramirez, and A. R. Kortan, Nature 350 600 (1991).

Anisotropic Properties of Layered Superconductors

Y. Iye

Institute for Solid State Physics, University of Tokyo,
Roppongi, Tokyo 106, Japan

Abstract. Quasi-two-dimensional anisotropy and extremely short coherence length are the two principal keywords characterizing the superconducting properties of cuprates. Similarities and differences are contrasted between the high T_c cuprates and some prototype layered (conventional) superconductors. The dynamical properties of the flux lines and transport dissipation in the mixed state are also strongly influenced by the layered structure.

1. Introduction

All the members of hitherto discovered cuprate-based high temperature superconductors have layered crystal structures[1]. Quasi-two-dimensional character of the electronic state has a rather profound influence both on the basic mechanism of superconductivity and on the superconducting properties. To illustrate some aspects of the relevance of quasi-two-dimensionality to superconductivity, we pick up a few representatives of layered conventional superconductors and compare them with high T_c cuprates. The statics and dynamics of flux lines in the mixed state are also strongly affected by the layered structure.

2. Layered Superconductors

There exist numerous metallic systems with layered crystal structures, some of which exhibit superconductivity at low temperatures. Figure 1 depicts a few representative layered superconductors including the high T_c cuprates. Among the conventional layered superconductors, perhaps the most extensively studied are the transition metal dichalcogenides, such as $NbSe_2$, and their intercalation derivatives[2]. Graphite intercalation compounds constitute a unique group of layered superconductors with their well-ordered stage structure[3]. The superconductivity in graphite intercalation compounds may attract renewed interest in relation with the recently discovered super-

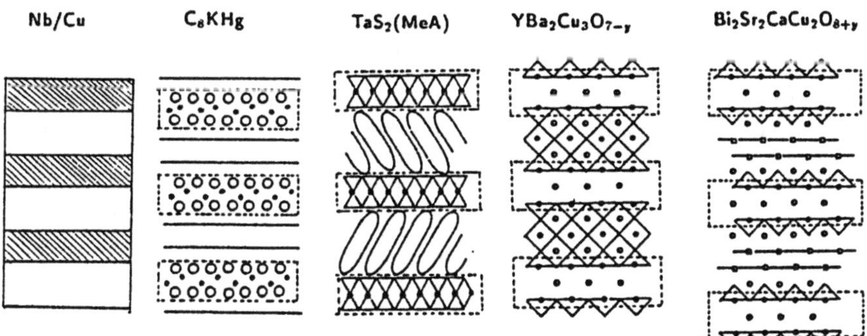

Fig. 1 Crystal structures of representative layered superconductors.

conductivity in doped fullerene, $C_{60}K_x$[4]. Many of organic superconductors also have quasi-two-dimensional structures[5]. Another important class of layered superconductors is formed by artificial metallic multilayers composed of various combinations of superconducting and other materials[6]. The cuprate high temperature superconductors have many features in common with those conventional layered superconductors, while exhibiting, undoubtedly, many novel properties.

Many of these layered superconductors can be, to a first approximation, viewed as consisting of alternate stacking of superconducting layer (S-layer) and non- (or weakly-) superconducting layer (N-layer). The anisotropy (or quasi-two-dimensionality) is determined by the strength of coupling between adjacent S-layers, and depends strongly on the nature of the intervening N-layer. In this respect, the artificial metallic multilayer structure offers a unique oppotunity to tailor the interlayer coupling strength.

The cuprate high temperature superconductors can be also viewed in this way, as pointed out in many literatures. It is well established that the CuO_2 plane is the main stage for the superconductivity. The other layers, such as CuO chain layer in $YBa_2Cu_3O_{7-y}$ (YBCO) and Bi_2O_2 layer in $Bi_2Sr_2CaCu_2O_{8+y}$ (BSCCO), essentially act as charge reservoir and supply mobile carriers to the CuO_2 layers by charge transfer. The nature of the N-layer in these two materials are considerably different as will be shown below.

The influence of quasi-two-dimensionality on the mechanism of superconductivity has many facets, of which we pick up a few examples: (1)Quasi-two-dimensional electronic structure leads to narrow bands and hence a large density of states at the Fermi level, which is favorable to occurence of superconductivity. The density of states at the Fermi level may be further enhanced if the Fermi level coincides with a van Hove singularity. (2)An important problem frequently encountered in low dimensional electron systems is the competition of superconductivity with other types of instability including CDW, SDW *etc.*. In some cases, the balance is delicate and can be tipped by external perturbations like pressure, or by introduction of pair-breakers that act differently to the competitors. (3)It has been established that doping is necessary to make the cuprates metallic and superconducting. In a three dimensional bulk structure, doping inevitably leads to simultaneous introduction of random potential in the conducting path, which may be detrimental to superconductivity. For a layered structure, disorder in the conducting path can be minimized by doping only into the "other" layers; an idea not unlike the so-called modulation doping in semiconductor superlattices.

3. Phenomenology of Anisotropic Superconductivity

Figure 2 compares the transition temperature, T_c and the anisotropic coherence lengths, ξ_a and ξ_c among various layered superconductors. The transition temperatures of the materials included in thie figure span a wide range from $T_c \sim 0.1\ K$ of C_8K to $\sim 100\ K$ of the cuprates. The geometrical average of the coherence length roughly scales inversely with T_c, as expected from the BCS relation $\xi = \frac{\hbar v_F}{\pi\Delta}$. The magnitude of anisotropy shows considerable diversity among these materials, depending on the nature of the "N-layer". It is the extremely short ξ_c of high T_c cuprates that characterizes the superconducting properties of cuprates, such as fluctuation and unusual behavior of flux line system, to be discussed in the next section.

The anisotropic nature of superconductivity is manifest in the angular dependence of the critical fields. The degree of anisotropy is governed by the interlayer coupling. If the coupling is not too weak, the anisotropy can be incorporated by an anisotropic effective mass tensor and the angular dependence of the upper critical field can be expressed by the anisotropic Ginzburg Landau model, sometimes called effective mass model. When the interlayer coupling is made very weak, one approaches the two-dimensional limit. A signature of the two-dimensional behavior is a cusp of $H_{c2}(\theta)$ at $\theta = 0$, as observed in the case of a thin film superconductor.

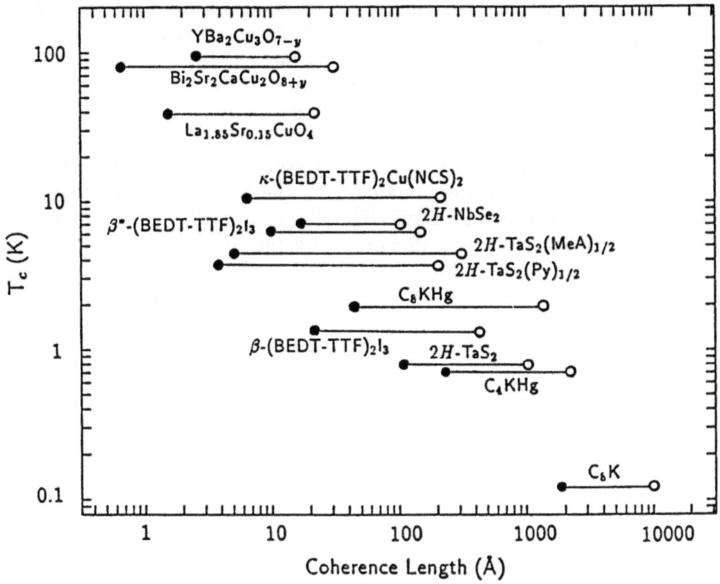

Fig. 2 Comparison of T_c and anisotropic coherence lengths among layered superconductors. The open and solid symbols represent coherence length within the basal plane and that along the c-direction, respectively.

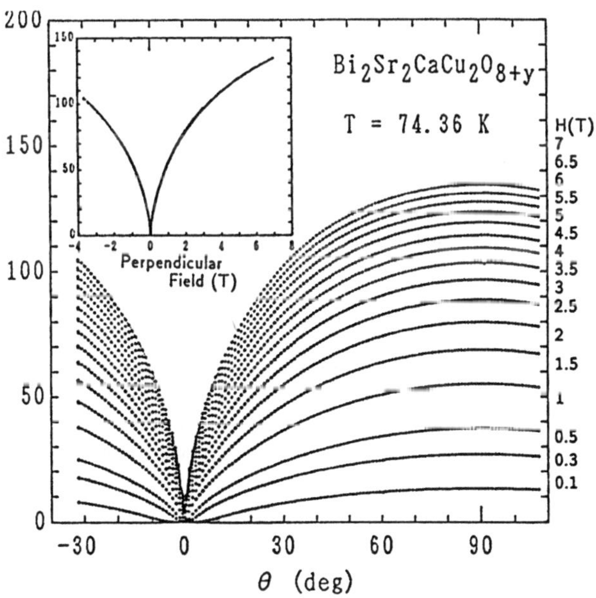

Fig. 3 Resistivity in the mixed state of BSCCO as a function of magnetic field angle relative to the basal plane. The inset shows a replot of the data as a function of the normal component of the field.

The main panel of Fig. 3 shows the dependence of resistivity on the field angle θ in BSCCO. The extremely sharp feature at $\theta \approx 0$ reflects the large anisotropy in this material very close to the two-dimensional limit. YBCO is the least anisotropic material among the high T_c cuprates. The anisotropy in a fully oxygenated YBCO, which is the least anisotropic material among the high T_c cuprates, can be fitted with the effective mass formula. It becomes more two-dimensional when it is prepared in an ultrathin film, or made somewhat oxygen deficient. That the interlayer coupling in YBCO is very sensitive to the oxygen deficiency is closely related with the diversity of the reported temperature dependences of the c-axis resistivity[7]. The estimated values for the coherence lengths are, $\xi_a \approx 15$Åand $\xi_c \approx$ 2-3Åfor fully oxygeneted YBCO. For BSCCO, $\xi_a \approx 30$Åand ξ_c may be shorter than ≈ 1Å.

4. Anisotropy and Flux Dynamics

Transport energy dissipation in the mixed state of a type II superconductor is caused by movement of flux lines in response to the Lorentz force exerted by the externally applied transport current. Usually, various types of defects act as pinning centers for the flux lines. It is in fact this pinning that makes a type II superconductor a material of practical use. In high temperature superconductors, because of their high operating temperatures, thermally activated motion of flux lines plays an important role. The rate of thermally activated hopping out of a potential minimum is proportional to e^{-U/k_BT}, where U is the barrier height depending on various parameters. An interesting possibility currently persued is that even at very low temperatures where thermal activation rate becomes negligibly small, flux lines might still move via quantum mechanical tunneling.

Difference in the magnitude of anisotropy between YBCO and BSCCO is reflected in the flux dynamics and dissipation. A flux line piercing through a layered superconductor with very weak interlayer coupling may be viewed as a stack of two-dimensional vortex discs, or "pancake" vortices. The flux line can wander away from a straight line configration. The positional correlation between two discs in adjacent layers is characterized by a correlation length ζ_c depending on the interlayer coupling strength. When the external field is tilted from the normal direction, and is set at an angle θ from the basal plane, the areal density of the "pancakes" in each layer is given by the normal component of the field, $H sin\, \theta$. In the two-dimensional limit, the vortex discs in adjacent layers are completely decoupled, so that their motion becomes uncorrelated. The dissipation is then solely determined by the number of the "pancakes" in each layer[8]. The inset of Fig. 3 is a replot of the data in the main panel with the abscissa scaled by the normal component of the applied field. It is seen that such s scaling makes all the data collapse into a single curve, which indicates that BSCCO comes very close to a realization of the two-dimensional limit model. If, however, one looks more closely at the behavior in the immediate vicinity of $\theta = 0$ in a logarithmic scale, one decerns deviation from the two dimensional model[9].

Dissipation process for field directions parallel and nearly parallel to the layer deserves further investigations. In the case of BSCCO, dissipation shows an unusual feature that it is independent of the relative angle between $\vec{H}$ and $\vec{J}$ when they are both in the ab-plane[10]. In the case of YBCO, dissipation does depend on the relative angle between $\vec{H}$ and $\vec{J}$, but there also exists an angular independent contribution to dissipation. For YBCO, a feature associated with pinning of flux lines by twin planes are also observed in the angular dependence. The apparent Lorentz force independence of dissipation in the $\vec{H} \parallel ab$-plane configuration is presumably associated with a strong pinning potential modulation arising from the layered structure.

A flux line parallel to the basal plane direction find themselves most confortable when its core lies in the "N-layer". Potential as a function of the position of the flux line therefore has a modulation with the priodicity of the layer repeat distance. The modulation amplitude is larger for larger anisotropy and shorter ξ_c, and give rise to

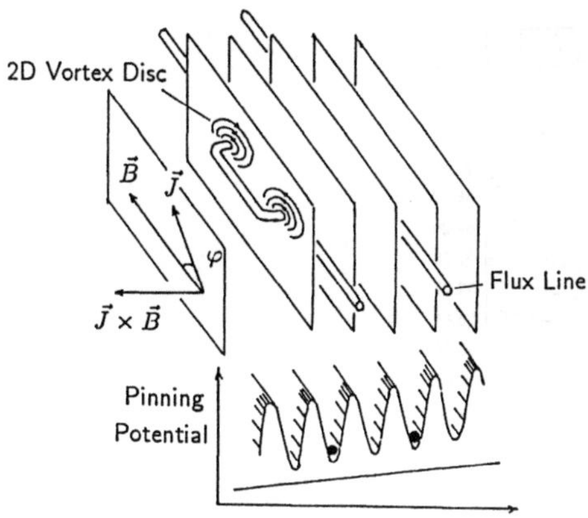

Fig. 4 Schematic drawing of the intrinsic pinning potential for a parallel flux line and a thermally nucleated kink-anitkink pair ("pancake" and anti-"pancake"). Dessociation of the pair by the Lorentz force leads to a net motion of the flux line along the c-direction.

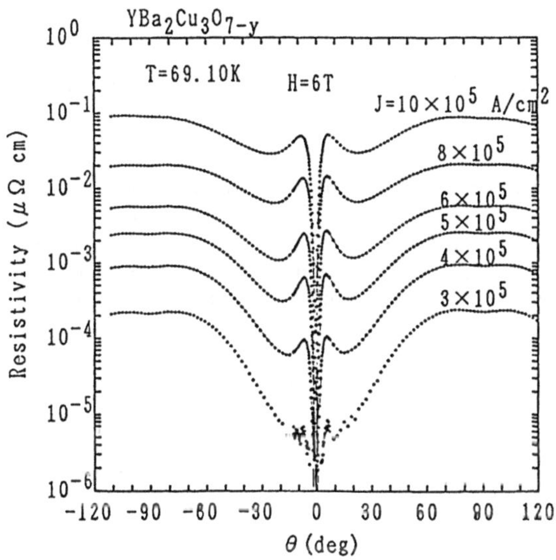

Fig. 5 Anomalous angular dependence of dissipation in YBCO at elevated transport current densities.

intrinsic pinning of flux lines[11]. This is analogous to the Peierls-Nabbaro potential encountered in the motion of dislocations in a crystal lattice. In the presence of large pinning potential, a flux line trapped in one of the potential minima cannot move in the c-direction as a rigid line object. An energetically less costly excitation in such a circumstance can be created by a local hop of a short segment of the flux line, which nucleate a pair of kink and antikink[12], as illustrated in Fig. 4. For a hop in the 'right' direction, the Lorentz force drives the kink-antikink pair to dissociate and yields a net motion of the original flux line by one unit along the c-direction. As the flux line can wander relatively freely within the potential well, it is likely that even in the case of $\vec{H} \parallel \vec{J}$, local Lorentz force is always at work, leading to dissipation comparable in magnitude as the case of $\vec{H} \perp \vec{J}$.

At elevated current densities, a new feature emerges in the angular dependence near $\theta = 0$, as shown in Figs. 5 and 6 for YBCO and BSCCO, respectively[9,13]. As seen

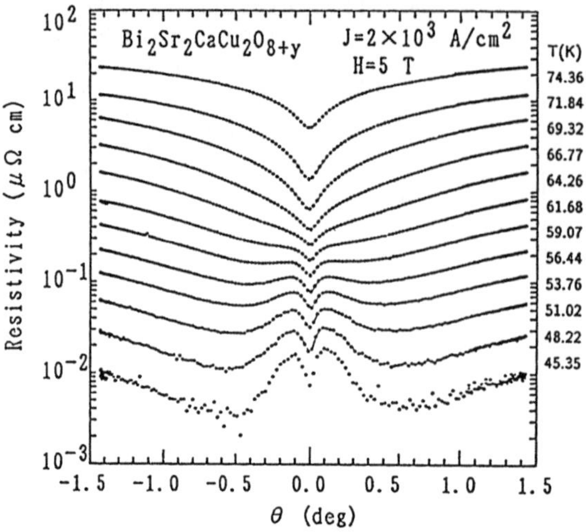

Fig. 6 Anomalous angular dependence of dissipation in BSCCO in a narrow range of angle near $\theta = 0$.

in these figures, there occurs a pronounced peak of dissipation at a certain angle of field from the parallel direction. The characteristic angle for the anomalous peak in dissipation is $\sim 5°$ for YBCO and $\sim 0.1°$ for BSCCO. In Fig. 5 it is seen that the peak resistivity can become even higher than that for $\vec{H} \parallel c$-axis. The origin of these anomalous enhancement of disspation is attributed to contribution of the vertical motion of the lateral segment of the flux lines caused by a strong Lorentz force in the c-direction[14].

5. Conclusion

The layered crystal structure and quasi-two-dimensional electronic state in high T_c cuprates have significant implications for the mechanism and the phenomenology of high temperature superconductivity. The parallel field configurations offers particularly interesting problems of dynamics of a flux line system in a strong periodic potential.

References

[1] See for example, Y. Tokura's lecture note in this volume.
[2] "*Intercalated Materials*", ed. by F.A. Levy (Reidel, Dordrecht, Holland, 1979).
[3] Y. Iye and S. Tanuma, Synthetic Metals 5 (1983)257.
[4] A.F. Hebbard *et al.*, Nature 350 (1991) 600.
[5] "Physics and Chemistry of Organic Superconductors", *ed.* G. Saito and S. Kagoshima, (Springer, Heidelberg, 1990).
[6] B.Y. Jin and J.B. Ketterson, Adv. in Phys. 38 (1989) 189.
[7] S.W. Tozer *et al.*, Phys. Rev. Lett. 59 (1987) 1768: S.J. Hagen *et al.*, Phys. Rev. B37 (1988) 7928: Y. Iye *et al.*, Physica C153-155 (1988) 26: T. Itoh *et al.*, Nature
[8] P.H. Kes *et al.*, Phys. Rev. Lett. 64 (1990) 1063.
[9] Y.Iye *et al.*, Physica C174 (1991) 227;
[10] Y. Iye *et al*, Physica C159 (1989) 433.
[11] M. Tachiki and S. Takahashi, Solid State Commun. 70 (1989) 291.
[12] S. Chakravarty *et al.*, Phys. Rev. Lett. 64 (1990) 3187: Y. Iye *et al.*, Physica C169 (1990) 62: C. Dasgupta and T.V. Ramakrishnan, *Proc. Int. Conf. on Superconductivity* ed. by S.K. Joshi *et. al.*, (World Sci. Pub., Singapoire, 1991) 432.
[13] Y. Iye *et al.*, Physica C177 (1991) 393.
[14] Y. Iye *et al.*, Proc. HTSC-M^{2}S (Kanazawa, July 1991), Physica B, to be published.

Thin Films and Superlattices of High-T_c Superconductors

Ø. Fischer[1], *J.M. Triscone*[1], *L. Antognazza*[1], *O. Brunner*[1], *L. Miéville*[1], *I. Maggio-Aprile*[1], *A.D. Kent*[1], and *O. Eibl*[2]

[1]Départment de Physique de la Matière Condensée, Université de Genève, 24, Quai E.-Ansermet, CH-1211 Genève 4, Switzerland

[2]Siemens Research Laboratories, Otto Hahn Ring 6, W-8000 Munich 83, Fed. Rep. of Germany

Abstract. This paper reviews recent work on the properties of $YBa_2Cu_3O_7$/ $PrBa_2Cu_3O_7$ superlattices. The study of these materials allows one to obtain important information about the role played by the interlayer coupling in the high temperature superconductors. Furthermore, investigations of the resistive transition in a magnetic field has shed new lights on the physical process leading to the dissipation in the superconducting state.

1. Introduction.

The starting point of this lecture is the well known fact that the high temperature superconducting cuprates all have a layered crystalline structure which causes them to have strongly anisotropic properties. Another common feature of all these compounds is the presence of CuO_2 planes and that superconductivity is thought to be due to holes or electrons, introduced for instance by doping, moving essentially in these planes. The actual superconducting properties depend very much on the stacking of these CuO_2 planes with other metal, or metal oxide, planes in the various compounds so far investigated. For this reason it is of great interest to be able to investigate materials with different sequences of the various layers. In particular it is a remarkable challenge to be able to alter this stacking by artificial means, i. e. to produce artificial superlattices of various types. Today several different types of superlattices made from these materials have already been fabricated and the goal of this lecture is to review some of the work done and to point to some of the open questions and challenges.

The first superlattices based on these materials were made from $YBa_2Cu_3O_7$ and other $REBa_2Cu_3O_7$ (RE=rare earth) compounds. In particular $YBa_2Cu_3O_7/DyBa_2Cu_3O_7$ (YBCO/DyBCO) [1] and $YBa_2Cu_3O_7/PrBa_2Cu_3O_7$ (YBCO/PrBCO) [2, 3, 4] superlattices have been investigated. Superlattices have also been grown with YBCO and $Nd_{2-x}Ce_xCuO_4$ [5] and more recently the stacking sequence of the CuO_2 planes in the BiSrCaCuO system has successfully been modified [6,7,8]. In this lecture we shall focus especially on the work carried out on the YBCO/PrBCO multilayers. These systems are interesting for various reasons. First, the lattice mismatch between the different REBCO is small so

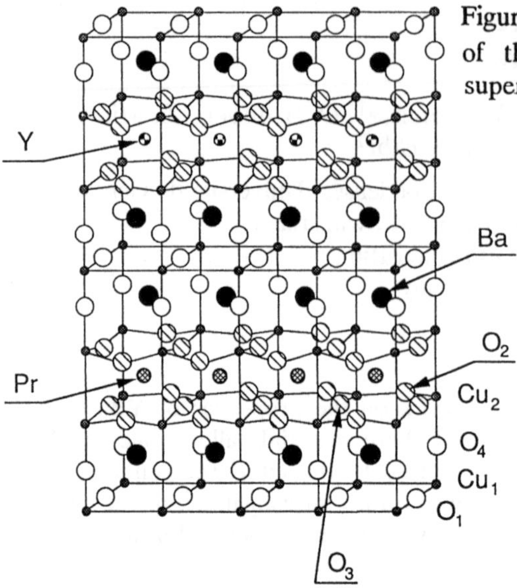

Figure 1: Schematic representation of the structure of a 12Å/12Å superlattice.

that the different compounds can grow epitaxially on each other. Second, it is only necessary to get these materials to grow one unit cell by one unit cell, i. e. in deposition techniques like sputtering or laser ablation one can use targets made of the two compounds and it is not necessary to control the growth atomic layer by atomic layer. Third, PrBCO is an insulator (with a hopping type conduction) and thus intercalating PrBCO layers between the YBCO layers allows to change the anisotropy of the material in a controlled manner.

In Fig. 1 we show schematically a superlattice composed of alternating layers of YBCO and PrBCO, with each a unit cell thickness (11.62 Å, for YBCO and 11.76 Å for PrBCO) in the c-direction. Since PrBCO is insulating the two CuO_2 planes on each side of the Pr-plane are expected to be insulating and thus the superconducting CuO_2-planes on each side of the Y-plane are now separated by more (20Å) than a unit cell dimension from the next plane. Consequently, we expect that this material is much more anisotropic than YBCO. Such superlattices have been made and indeed one finds that they are considerably more anisotropic.

The fact that the HTS compounds are very anisotropic have led many theorists to view these materials as essentially two dimensional (2D) and to take a two dimensional model based on the CuO_2 plane as a starting point in their efforts to explain the properties of these materials. Furthermore, most of the properties, like for instance the remarkable broadening of the resistive transition in a magnetic field, are clearly strongly influenced by the anisotropy. For these and other reasons it is of central interest to be able to control and vary the anisotropy and to study the modification of the properties as one goes from a 3 dimensional anisotropic to a 2 dimensional material.

Such studies have already been started with the YBCO/PrBCO system, and its properties have been studied for various thicknesses of the individual layers of the two compounds. This is certainly one important motivation for making multilayers, but it is by far not the only. For instance, making new materials by modifying the stacking sequence within one unit cell is clearly a rich field of possible investigations. First results in this direction have been reported for modifications of the $Bi_2Sr_2CaCu_2O_8$ system [6,7]. In view of the large number of compounds in the perovskite class of materials we expect that there will be a large number possible and interesting superlattices to make. Let us finally mention the central interest of this kind of approach for future applications. Fabrications of junctions, of metal or insulating overlayers without perturbing the properties of the other layers will be essential for obtaining Josephson junctions and weak links for microelectronics applications. Furthermore, the physics of the interfaces, which can be studied with the multilayers, is also central to future applications.

2. Growth of thin films and multilayers.

Superlattices have actually been made by different deposition techniques: sputtering [1,2,6,9], laser ablation [3,4,5], e-beam deposition [10] and molecular beam [7,8] epitaxy. To the extent that a comparison can be made, it appears that at this point none of the techniques are significantly superior to the others; any differences are probably more related to optimalisation problems of the individual processes. Nevertheless, the synthesis of superlattices put stringent conditions on the nature of the growth of the films. First, the films should be atomically flat over reasonably large surfaces. Second, the films should grow, at least locally, in a layer by layer mode, where the individual layers may be as thick as the unit-cell dimension. The quality of the superlattices already obtained show that this is largely the case under the conditions used to produce the multilayers, a fact that has recently been experimentally verified by RHEED studies [11].

In what follows we shall first describe the fabrication method used in Geneva and then describe some of the characterization methods used.

The superlattices made in our laboratory were made using d.c. magnetron sputtering [12]. Figure 2 shows a schematic view of our deposition system. We use two sputtering heads, containing each a composite target of YBCO and PrBCO, respectively. The substrate is fixed on a heated substrate holder and can be made to face either one of the two sputtering guns by means of a computer controlled step motor. The deposition rates, which can be kept stable over long time intervals, were calibrated and the substrate kept in front of a given sputtering head for the time necessary to produce the wanted number of unit cells. The superlattice was then produced simply by alternately moving the substrate in front of the two sputtering guns. Although superlattices have been made on both MgO and $SrTiO_3$ substrates we shall report here mainly results obtained on MgO. In this case the substrate was heated to about 700°C. Before starting the deposition, the sputtering guns were run for about half an hour and

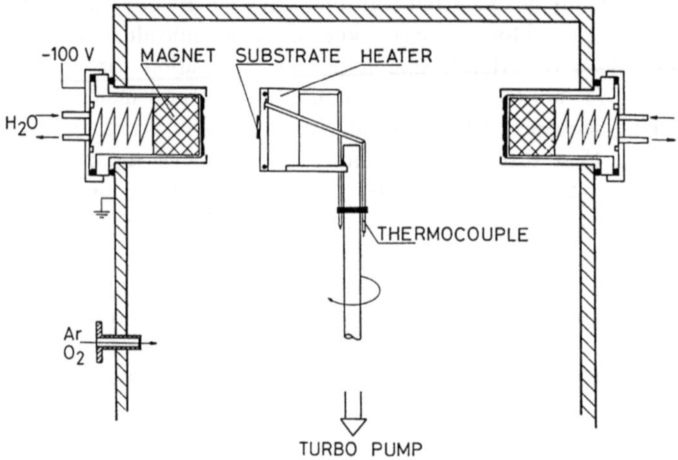

Figure 2: A magnetron sputtering system used for the fabrication of superlattices [19]

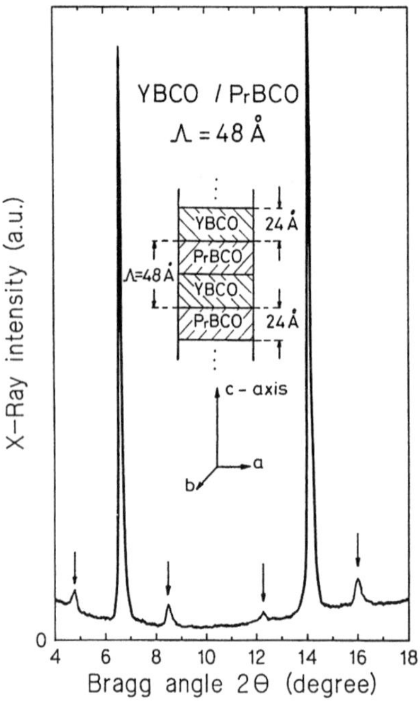

Figure 3: X-ray diffractogram for a 24Å/24Å multilayer [2].

the substrate was brought to the nominal temperature for the same period. The sputtering gas was a mixture of Argon and Oxygen and the pressure was kept at 300 mTorr, with a partial pressure of 20mTorr O_2. After deposition the oxygen pressure was increased to 1 Torr and the temperature was slowly reduced until it reached 450°C, where the sample was kept in a O_2 pressure of 20Torr for about 30 min.

Superlattices with different thicknesses of the individual layers have been produced this way. In what follows we shall mainly consider c-axis films (i.e. with the c-axis perpendicular to the film plane) and for simplicity we shall denote the superlattices as if the c-axis unit-cell was 12Å. Thus for instance a superlattice with individual YBCO layers with a thickness of two unit cells and individual PrBCO layers with 4 unit-cell thicknesses is denoted a 24Å/48Å superlattice. In the actual calibration we used the exact values of the unit cells. The first check that a superlattice is obtained, is to verify that the corresponding satellite peaks appear in the X-ray diffractogram. Figure 3 shows the satellite peaks of the (001) and the (002) peaks of a 24Å/24Å superlattice [2]. Two satellite peaks appears corresponding to the new 48Å period. In fact, the presence of these peaks allows a rather precise calibration of the sputtering guns and is systematically used to control the thickness of the individual layers.

Although these X-ray diffractograms are essential to characterize the films, they only give an information on the average layering. However, off-axis studies have confirmed the in plane crystalline order of the films. Further information, especially about the local crystalline order have been obtained from TEM studies by various groups [5, 13]. The results confirm that a high degree of perfection is obtained and furthermore that practically no interdiffusion between the Y-layers and the Pr-layers take place. An important question for the growth of superlattices is about the importance of the substrate and its surface quality. It turns out that imperfections of the substrate can partly be healed by the YBCO film itself. One finds in fact that the first few layers of a film are strongly perturbed whereas a higher quality is obtained in the following layers [13]. This is illustrated in Fig 4 where the interface of a $SrTiO_3$ substrate and the first YBCO layers is shown. One can easily distinguish the surface of the substrate which contains a few one unit cell steps. These steps clearly perturb the growth of the 123 compound and disordered areas develop around the steps. However, these areas disappear after a few unit cells have been grown, and one can clearly see in Fig. 4 that the YBCO film grows with the a,b plane having a small angle with respect to the substrate lattice. This angle reflects the average angle of the substrate surface due to the presence of the steps. A result of this observation is that imperfections of the substrate are eliminated after a few monolayers of YBCO and so YBCO produces itself a high quality substrate for the remaining part of the film. A consequence of this is that whereas it has so far proven impossible to grow superconducting unit-cell thick films on any substrate, this can be made in YBCO/PrBCO multilayers. By growing a few unit-cells of PrBCO one obtains a nearly perfect substrate onto which unit-cell thick superconducting films can be grown [10].

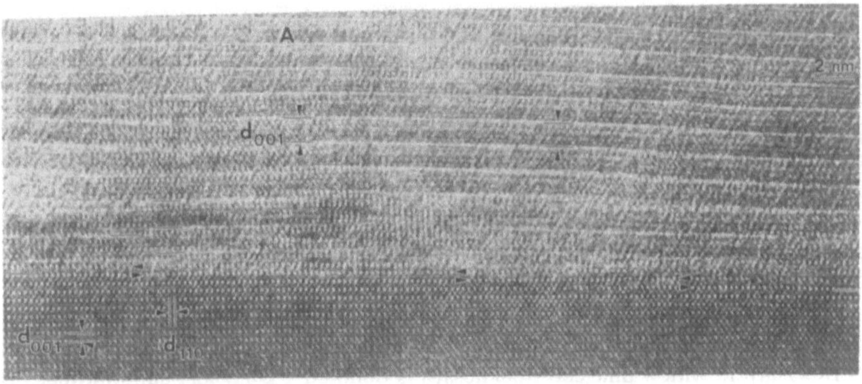

Figure 4: TEM cross-section of the interface $SrTiO_3$ YBCO [13].

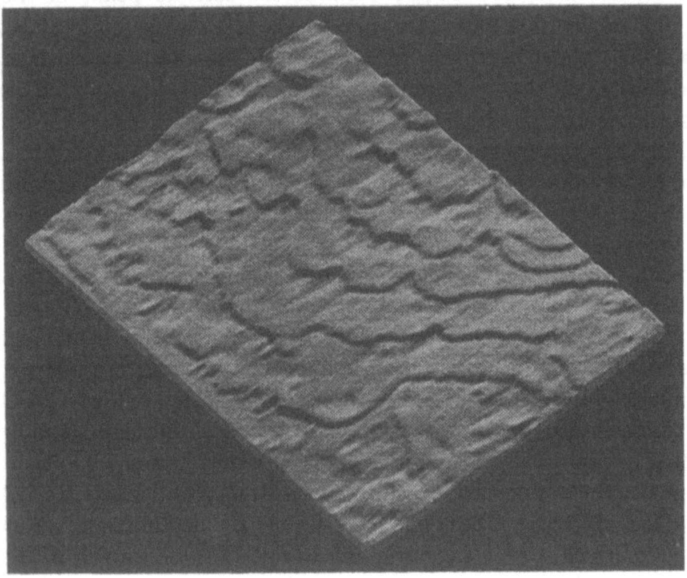

Figure 5: STM image of the surface topography of a sputtered film. The area shown is 5000 x 5000Å^2 [16].

New information about the growth of such films has recently been obtained from scanning tunneling microscopy. In particular it has been reported that thin films fabricated by sputtering may show domains with a spiral growth around screw dislocations [14,15]. These spirals were found to have various sizes ranging from typically a few hundred to a few thousand Å in diameter. We have also observed such spiral growth in our laboratory [16]. We find that the appearance of such growth spirals depend very much on the exact sputtering conditions. For instance, another sample made under somewhat different deposition conditions (lower substrate temperature) did not reveal any spiral growth, but showed instead broad terraces (typically 1000Å by 4000Å) with steps of the order of a few unit cells (Figure 5). Obviously, for the fabrication of multilayers and heterostructures one wants to have as broad terraces as possible. Thus, a fine tuning of the deposition conditions is necessary. These conditions are at this point specific to each deposition system and no clear picture has emerged yet as to the exact criteria for a large surface layer by layer growth. However, we have found that the films grown on MgO under the same conditions as the multilayers tend to be very smooth, even more so than shown in Fig. 5 for a film grown on $SrTiO_3$. To which extent the alternate growth of two different compounds in a multilayer further stabilizes a layer by layer growth is the subject of current investigations.

Since the b and c lattice parameters fit very well to the $SrTiO_3$ substrate the natural epitaxial growth on $SrTiO_3$ is with the a-axis perpendicular to the film. By lowering the substrate temperature the number of a-axis domains increases and several groups have produced essentially pure a-axis films [17]. The a-axis films tend to grow with smooth surfaces on the micron scale. However, when looked at with a STM a grainy structure appears, presumably corresponding to grains, about 100Å high, and typically 1000Å by 400Å in surface, having different orientations of the b and c axes. Surprisingly, multilayers can also be grown for a-axis films [18]. In this case the modulation is within each Y/Pr layer, demonstrating that the interdiffusion even within the Y/Pr layer is negligible at the deposition temperature. These materials have thus the very interesting property that they consist of thin ribbons of superconducting material, coming close to a one dimensional version of the high temperature superconductors.

3. The critical temperature of YBCO/PrBCO superlattices.

Rather extensive studies of the critical temperature of the YBCO/PrBCO multilayers have been carried out by different groups [2,3,4,12,19]. A characteristic result is shown in Fig. 6. It shows the resistive transition of a 24Å/96Å and a 96Å/96Å superlattice together with a 1500Å thick YBCO film. With 96Å PrBCO separation between the YBCO layers it is found that the interlayer coupling can safely be neglected. The observed behaviour is therefore characteristic of individual YBCO layers sandwiched between PrBCO layers. This has recently been confirmed by the investigation of PrBCO/YBCO/PrBCO trilayers [10]. As the thickness of the YBCO layers becomes smaller we observe that the onset T_{co} shifts to lower temperature. Furthermore we find that the transitions becomes gradually broader so that the "zero resistance" T_c decreases faster than T_{co}. However, a 24Å, i. e. a two unit cell thick YBCO layer still has a T_c of about 60K and it is found that even a one unit cell thick layer is superconducting with a zero resistance "transition temperature" of the order of 10K to 30K. The decrease of T_{co} reflects a change in the "mean field" critical temperature and must be related either to a kind of proximity effect due to the presence of the PrBCO layers or to a reduction of the effective electron-electron interaction due to the suppression of the interlayer coupling at the interface.

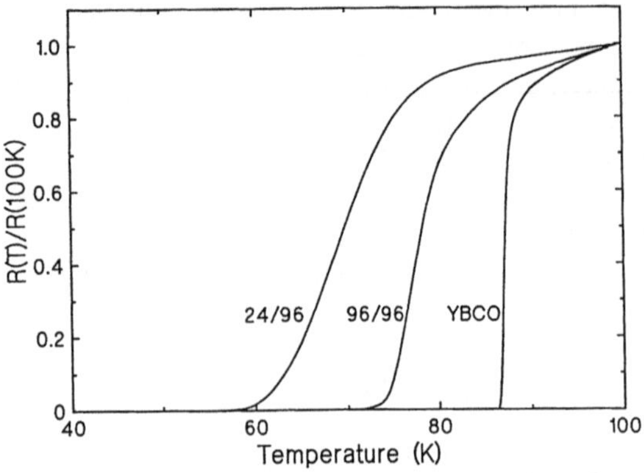

Figure 6. Resistive transitions for a 24Å/96Å and a 96Å/96Å superlattice compared to the transition of a pure YBCO film.

The latter case would imply that a significant contribution of the electron-electron interaction would come from an interlayer coupling.

The increased broadening of the transition is characteristic of increased 2D like fluctuations and in an ideal system the zero resistance T_c should be identified with the Berezinskii-Kosterlitz-Thouless transition. In a bulk YBCO crystal T_c will be very near to T_{co} and the systematic study of a possible BKT transition is difficult. Using multilayers or ultrathin superconducting layers for the study of the BKT transition in the oxides allows to increase the temperature region of interest and thus to study the systematic variation as the thickness is modified. Indeed, the universal relation

$$k_B T_c = (\phi_o)^2 d / 32\, \pi^2\, \lambda^2(T_c)$$

where d is the thickness of the individual superconducting layers, show that T_{co}-T_c does gradually increase as d is decreased. Several recent investigations have analyzed the experimental data in more detail and find a behaviour that is consistent with a Berezinskii-Kosterlitz-Thouless type behaviour [20, 21]. On the other hand several theoretical papers have recently investigated the T_c variation with the thickness of YBCO and PrBCO [22,23,24]. The central problem here is how to explain the strong decrease of T_c as the layers become gradually less coupled and thinner. From the experimental point of view, several questions still remains open, as for instance the role of imperfections in the films. Part of the broadening of the transitions might be due to inhomogeneities due to the island type growth discussed above. From the point of view of comparisons with the theory, it is certainly important that one observes the different characteristic signatures of the BKT transition consistently in the same samples. Summarizing this part, the YBCO/PrBCO multilayers provide ideal model systems to study the presence of Berezinskii-Kosterlitz-Thouless behaviour in the high temperature superconductors and to clarify certain open questions.

When the PrBCO layers are made very thin, say 12Å or 24Å, the YBCO layers seem to couple across and thus in superlattices with thin PrBCO one can presumably study how the properties changes when one goes from decoupled layers to the bulk material. It has been observed that T_c of multilayers having a given number of unit cells of YBCO decreases gradually when the PrBCO thickness is increased but saturates when the latter is larger than 60Å [2,3,4,12,19]. This illustrates the importance of the coupling between the layers. However, the nature of this coupling is not yet understood, in particular the reason for the gradual change up to a PrBCO thickness as large as 60Å. So far rather different theoretical explanations have been proposed [22,23], but experiments are at this point not able to discriminate between the various models.

4. Investigations of the thermally activated flux flow in perpendicular magnetic fields.

In this section we shall discuss how the superlattices can be used to study properties related to the fluxline lattice. A well known, but yet remarkable property of the high temperature superconductors, is the broadening that occurs in the resistive transition when a magnetic field is applied. It turns out that the shift that is usually observed in low temperature superconductors is very small in the oxides since their upper critical field H_{c2} is very high. The broadening is therefore not related to H_{c2} but is understood as dissipation that occurs in the superconducting state due to thermal motion of the flux lines.

Whereas the "visible" part of the resistive transition is found to correspond to a flux flow, the resistive tail for say $\rho < 10^{-2}\ \rho_n$ (ρ_n is the normal state resistivity) has been shown to have an activated Arrhenius behaviour[25]:

$$\rho(T,H) = \rho_0 \exp(-U(T,H)/k_B T)$$

corresponding to a thermally activated flux motion (TAFF) [26]. The temperature and field dependent activation energy corresponds to the minimum energy necessary to create a flux motion. We can write formally this energy as

$$U = p\,(\mu_0 H_c^2/2) V_c\ ,$$

where V_c is the correlation volume involved in the activated process. We further write $V_c = R_c^2 L_c$ where R_c^2 is the correlation area perpendicular to the field and L_c the correlation length in the field direction. L_c is a quantity that depends on the tilt modulus of the flux line lattice and therefore on the anisotropy. It is a measure of the stiffness of the fluxlines. Palstra et al. [27] noticed that the activation energy of bulk YBCO is much larger than for bulk BiSCCO and suggested that the difference can be explained as a result of the higher anisotropy of the latter compound.

We have used our superlattices to test these ideas [28,29,30]. In Fig. 7 is shown the Arrhenius plots for different fields for a 96Å/96Å superlattice. When comparing this with the plots for bulk YBCO [25] one finds that the slope and

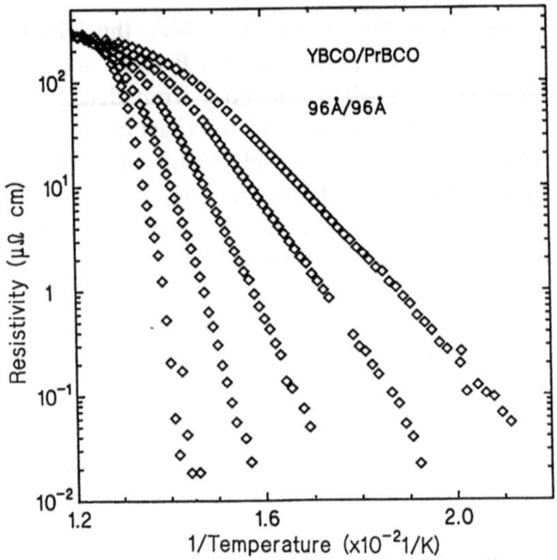

Figure 7: Arrhenius plots for the 96Å/96Å superlattice and magnetic fields of 0,1,3,6,9 Tesla.

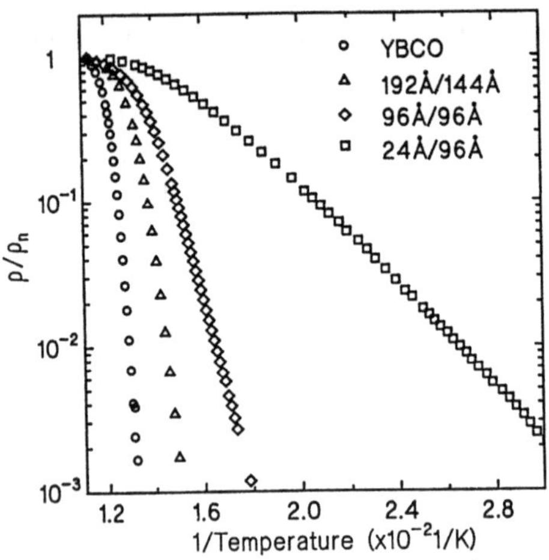

Figure 8: Arrhenius plots for a 1500Å film and a 24Å/96Å, a 96Å/96Å and a 192Å superlattices. The magnetic field is 3 Tesla and is applied perpendicular to the a,b plane.

thus the activation energy of the multilayer is considerably smaller. Figure 8 shows the Arrhenius plots for 3 Tesla, for a 1500Å YBCO film and for a 24Å/96Å, a 96Å/96Å and a 192Å/144Å superlattice. Since we have verified that the activation energies are independent of the PrBCO thickness when the latter is larger than 48Å, we interprete the behaviour shown in Fig. 8 as a dependence of U on the thickness of the YBCO layers. We have determined the activation

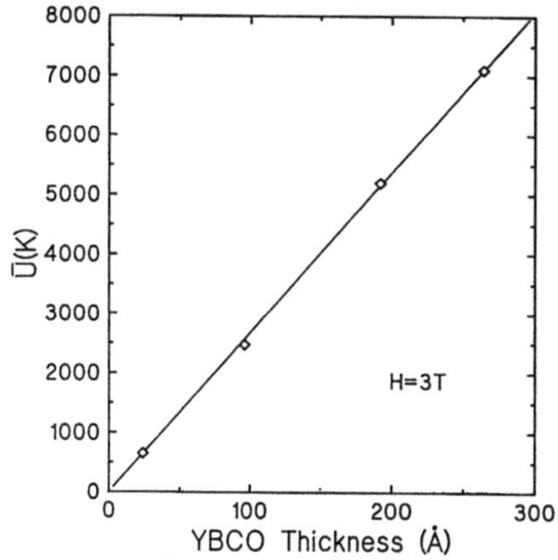

Figure 9: $\bar{U}$ versus the thickness of the individual YBCO layers.

energy from the average slope in the resistivity range $10^{-2}<\rho(t)/\rho_n<10^{-4}$ and in Fig. 9 we plot the obtained values as a function of the YBCO thickness and obtain a remarkable straight line. Thus the activation energy determined in the above manner, denoted hence forth $\bar{U}$, is proportional to the YBCO thickness. Clearly, when the thickness gets large enough the value for $\bar{U}$ must saturate. The value of the YBCO thickness where the crossover from thin film behaviour to bulk behaviour occurs must be of the order of L_c. This shows that L_c for bulk YBCO must be several hundred Å. When comparing with the 1500Å YBCO film, which is not necessarily representative for bulk YBCO, we estimate that L_c must be of the order of 450Å or more in bulk YBCO. We note at this point that contrary to measurements of bulk YBCO our Arrhenius plots cover a large temperature range and that for the thinnest YBCO layers (typically <100Å) we find that $\bar{U}$ represents to a very good precision the zero temperature activation energy. This point is discussed in more detail in ref.[28].

The fact that $\bar{U}$ increases proportionally to the YBCO thickness shows on the one hand that the Josephson coupling between the CuO_2 planes in pure YBCO is strong enough to produce relatively rigid vortices. On the other hand we interprete the proportionality as a demonstration that the vortex lattice is 2D in nature in this range of thickness, that is, they behave as (thick) pancakes vortices [31]. For such a 2D fluxline lattice it is expected that the thermally activated flux flow comes from the creation of edge dislocation pairs in the vortex lattice [32,33]. In the collective pinning regime one expects an activation energy of the order:

$$U = (\phi_o^2 d/16\mu_o\pi^2\lambda^2)\ln(a/\xi_o),$$

where a is the fluxline lattice parameter and ξ_o the BCS coherence length.

This predicts a magnetic field dependence of the form $U/d = -\alpha\ln B + \beta$, with $\alpha = 3.7$ K/Å, $\beta = 23$ K/Å and B in Tesla. In Fig. 10 we plot $\bar{U}$ versus logB for the 24Å/96Å and the 96Å/96Å superlattice. We find indeed a logarithmic

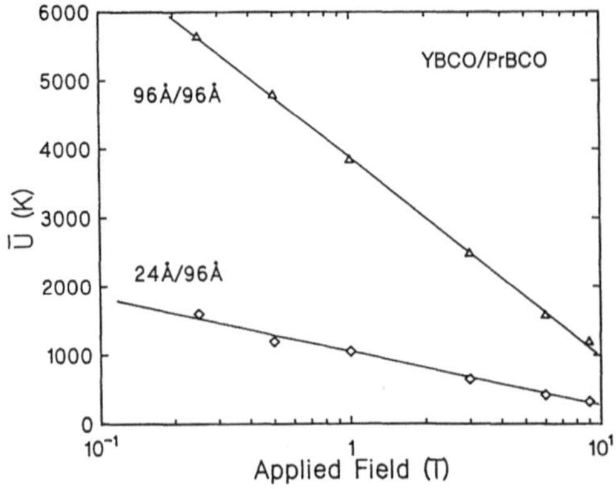

Figure 10: The activation energy as a function of logB for a 24Å/96Å and a 96Å/96Å superlattice.

field dependence. Taking an average over several multilayers we find $\bar{U}/d = -13\ln B + 45K/Å$. Taking into account the approximate nature of the theoretical expression this must be considered as a very good agreement. Thus we find evidence for the existence of thermally activated edge dislocation pairs in our multilayers. Of course the theory is based on the assumption of collective pinning and thus at the present stage we cannot completely rule out the possibility of thermal depinning of individual vortices. In this case our results would imply a high density of very strong pinning centers, of the order of one per unit cell. Very interesting results are also obtained when the field is applied parallel to the a,b plane. For a discussion, we refer to refs. 30 and 34.

5. Concluding remarks.

We have shown that superlattices can be grown with different high temperature superconducting cuprates. The detailed investigations already carried out on YBCO/PrBCO superlattices show how these materials can be used to study basic properties of the high temperature superconductors.

In view of the fact that these cuprates superconductors are members of the large class of perovskite-like compounds, we believe that numerous other types of interesting superlattices can be grown and that a wealth of new and exiting properties, both for basic science and for applications, may be found.

References.

1. J.-M. Triscone, M. G. Karkut, L. Antognazza, O. Brunner, and Ø. Fischer, Phys. Rev. Lett. 63, 1016, (1989)
2. J.-M. Triscone, Ø. Fischer, O. Brunner, L. Antognazza, A. D. Kent, and M. G. Karkut. Phys. Rev. Lett. 64, 804, (1990]
3. Q. Li, X. X. Xi, X. D. Wu, A. Inam, S. Vadlamannati, W. L. McLean, T. Venkatesan, R. Ramesh, D. M. Hwang, J. A. Martinez, and L. Nazar. Phys. Rev. Lett. 64, 3086, (1990).
4. D. H. Lowndes, D. P. Norton, and J. D. Budai. Phys. Rev. Lett. 65, 1160, (1990).
5. A. Gupta, R. Gross, E. Olsson, A. Segmüller, G. Koren, and C. C. Tsuei. Phys. Rev. Lett. 64, 3191, (1990).
6. T. Matsushima, Y. Ichikawa, H. Adachi, K. Setsune, and K. Wasa. Solid State Comm. 76, 1201, (1990).
7. M. Kanai, T. Kawai, and S. Kawai, Appl. Phys. Lett. 57, 198, (1990).
8. I. Bozovic, J. N. Eckstein, M. E. Klausmeier-Brown, and G. Virshup. Preprint.
9. G. Jacob, P. Przyslupski, C. Stölzel, C. Tomé-Rosa, A. Walkenhorst, M. Schmitt, and H. Adrian, Preprint.
10. Y. Bando, T. Terashima, K. Shimura, T. Sato, Y. Matsuda, S. Komiyama, K. Kamigaki, and H. Terauchi. Proceedings of the International Workshop on HTSC Thin Films; Rome, Italy, April 1991.
11. T. Terashima, Y. Bando, K. Iijima, K. Yamamoto, K. Hirata, K. Hayashi, K. Kamigaki, and H. Terauchi. Phys. Rev. Lett. 65, 2684, (1990).
12. J.-M. Triscone, Ø. Fischer, L. Antognazza, O. Brunner, A. D. Kent, L. Miéville, and M. G. Karkut in Science and Technology of thin Film Superconductors 2. Edited by R.D.McConnell and R.Noufi . Plenum Press, New-York (1991), p.13.
13. O. Eibl, H. E. Hoenig, J. M. Triscone, Ø. Fischer, L. Antognazza, and O. Brunner. Physica C 172, 365, (1990).
14. Ch. Gerber, D. Anselmetti, J. G. Bednorz, J. Mannhart, and D. G. Schlom., preprint submited to Nature.
15. M. Hawley, I. D. Raistrick, J. G. Beery, and R. J. Houlton. Science 1587, (1991).
16. I. Maggio-Aprile, A. D. Kent, Ph. Niedermann, L. Antognazza, L. Miéville, O. Brunner, J.-M. Triscone and Ø. Fischer. To be published in the proceedings of the International Conference on Scanning Tunneling Microscopy, Interlaken, Switzerland, august 1991.
17. C. B. Eom, A. F. Marshall, S. S. Ladermann, R. D. Jacowitz, T. H. Geballe, Science 249, 1549, (1990).
18. C. B. Eom, A. F. Marshall, J. M. Triscone, B. Wilkens, S. S. Ladermann, T. H. Geballe. Science 251, 780, (1991).

19. Ø. Fischer, J.-M. Triscone, L. Antognazza, O. Brunner, A. D. Kent, L. Miéville, and M. G. Karkut. J. Less Comm. Metals 164&165, 257, (1990).
20. D. H. Lowndes and D. P. Norton, Proceedings of the University of Miami Workshop on "Electronic Structure and Mechanisms for High Temperature Superconductivity" Miami, January 1991.
21. S. Vadlamannati, Q. Li, T. Venkatesan, W. L. McLean and P. Lindenfeld. Preprint.
22. D. Ariosa and H. Beck Phys. Rev. B. 43, 344, (1991)
23. T. Schneider, Z. Gedic, and S. Ciraci. Europhys. Lett. 14, 261, (1991).
24. M. Rasolt, T. Edis, and Zlatko Tesanovic. Phys. Rev. Lett. 66, 2927, (1991).
25. T.T.M. Palstra, B. Batlogg, L. F. Schneemeyer, and J. V. Waszczak. Phys. Rev. Lett. 61, 1662, (1988).
26. P. H. Kes and J. van den Berg in Studies of High Temperature Superconductors, ed. A. Narlikar,NOVA Science Publishers, New York, Vol. 5, p. 83.
27. T. T. M. Palstra, B. Batlogg, R. B. van Dover, L. F. Schneemeyer, and J. V. Waszczak. Phys. Rev. B 41, 6621, (1990).
28. O. Brunner, L. Antognazza, J. M. Triscone, L. Miéville and Ø. Fischer. To be published.
29. Ø. Fischer, J.-M. Triscone, O. Brunner, L. Antognazza, M. Affronte and L. Miéville. To be published in the proceedings of the NATO Science Forum: "Highlights of the eighties and future prospects in condensed matter physics", Biarritz, France, September 1990.
30. J.-M. Triscone, L. Antognazza, O. Brunner, L. Miéville, M. G. Karkut, O. Eibl, P. van der Linden, J. A. A. J. Perenboom and Ø. Fischer. To be published in the proceedings of the third international Conference on Materials and Mechanisms of Superconductivity-High Temperature Superconductivity, Kanazawa, Japon, July, 1991.
31. J. R. Clem, Phys. Rev. B. 43, 7837 (1991).
32. M. V. Feigelmann, V. B. Geshkenbein and A. I. Larkin, Physica C 167, 177, (1990).
33. V. M. Vinokur, P. H. Kes, and A. E. Koshelev. Physica C 168, 29, (1990).
34. O. Brunner, M. G. Karkut, L. Antognazza, L. Miéville, P. van der Linden, J. A. A. J. Perenboom , J.-M. Triscone, and Ø. Fischer. To be published in the proceedings of the third international Conference on Materials and Mechanisms of Superconductivity-High Temperature Superconductivity, Kanazawa, Japon, July, 1991.

Critical Currents in High-T_c Superconductors

J. Mannhart

IBM Research Division, Zurich Research Laboratory,
Säumerstr. 4, CH-8803 Rüschlikon, Switzerland

Abstract. An introduction to critical currents in high-T_c superconductors is given which includes

- Measurement techniques
- Limiting mechanisms
- Overview of the state of the art, strategies to raise J_c.

1. Introduction

A simple circumstance has prevented us from using high-T_c superconductors for such tempting applications as motors, generators or power transmission lines: To date, no wire or tape made of these materials has been able to carry a sufficiently high electric current without becoming resistive. Typically required current densities are shown in Fig. 1 and compared to values of high-quality samples of high-T_c superconductors [1-8]. Whereas the critical current density, i.e. the density of the maximum supercurrent, of thin films of high-T_c superconductors is already high enough for many intriguing applications, polycrystalline materials, which are required for making wires or tapes, are characterized by lower current densities, especially in high magnetic fields.

In this lecture, an overview of critical currents in high-T_c superconductors shall be given in order to provide an introduction to the field on the graduate student's level. The fundamental mechanisms that control critical currents as well as techniques to measure critical current densities will be discussed briefly. Exciting improvements of the critical current density achieved in the past few years will be presented and an overview of the technologies which have helped to raise it to the present level (May '91) will be provided.

There are some ambiguities and caveats associated with the definition of the critical current I_c and of the critical current density J_c to be addressed in the beginning:

There is no system in nature with resistivity exactly zero for a finite electric current ($T > 0$ K) [9].

Springer Series in Solid-State Sciences, Vol. 106 **Physics of High-Temperature Superconductors**
Eds.: S. Maekawa and M. Sato

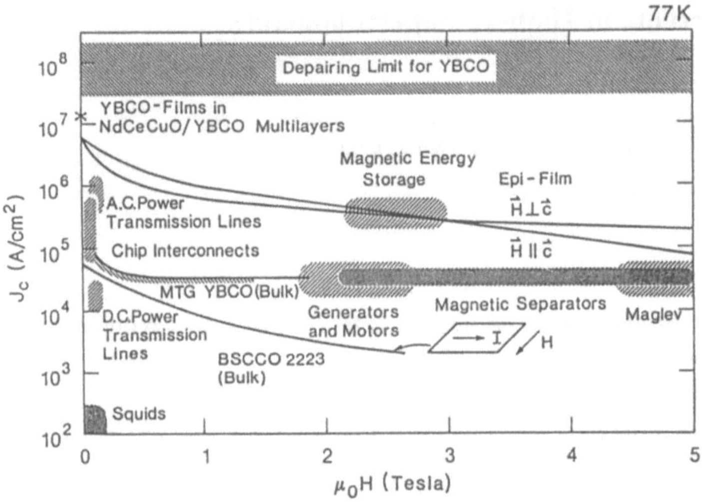

Fig. 1 Critical current densities required for various applications compared to transport J_c's achieved today as a function of applied magnetic field at 77 K (from [1]). The following abbreviations are used: YBCO = $YBa_2Cu_3O_{7-\delta}$, BSCCO 2223 = $Bi_2Sr_2Ca_2Cu_3O_{10}$, MTG = Melt Textured Growth, Maglev = Magnetically Levitated Train. The data have been compiled from the following sources: Chip interconnects [2], other applications [1], epi-film [3], BSCCO 2223 [4], MTG YBCO [5-8]. The MTG YBCO data refer to J_c in the *a,b*-plane, $\vec{H} \perp \vec{J}_c$.

For this reason the term 'supercurrent' has to be used for a current flowing with a resistance that is negligible in a specific context. This implies that for any practical purpose [10] I_c and J_c have to be defined with some arbitrariness as the current or the current density which give rise to an electric field, to a resistivity, or to power dissipation that equals some threshold criterion (Fig. 2a). The value of this threshold (typically 1 μV/cm) may reflect the requirements of an application or experiment.

Therefore, the measured critical current density depends on the threshold criterion selected, especially in samples with a small normal state resistivity. Only samples with a linear current-voltage characteristic ($I(V)$ curve) offer a way out of this dilemma: the current density at which the extrapolated resistive part of the $I(V)$ curve intersects the $V=0$ axis can be used for the definition of J_c (Fig. 2b) [11].

The average critical current density $\bar{J}_c = I_c/A$, where A is the sample cross section, is not only a function of material properties, but also depends on the geometry of the sample.

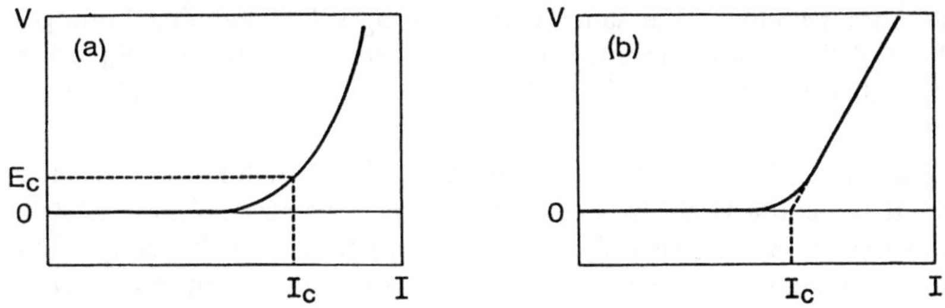

Fig. 2 Sketch of two $I(V)$ characteristics for the definition of J_c.

The reasons for this are clear: in large samples, $\bar{J}_c$ is decreased because the self-field (i.e. the magnet field generated by the transport current) lowers J_c at the surface. In addition, surface effects such as surface pinning, which in general enhance $\bar{J}_c$, are less helpful. For example, in granular high-T_c materials, $\bar{J}_c$ may be changed by more than one order of magnitude just by altering the sample cross section. For this reason, in addition to $\bar{J}_c$, the critical current I_c is an important quantity for sample characterization.

In any transport measurement, $\bar{J}_c$ will be controlled by the cross section of the sample with the smallest $\bar{J}_c$.

For instance, in thin films $\bar{J}_c$ may simply reflect the properties of the cross section grown on the most deeply scratched area of the substrate. Such effects, which have been studied in a spatially resolved manner by low-temperature scanning electron microscopy (LTSEM) [12], have severe implications for the analysis of $I(V)$ curves: Increasing the sample current beyond I_c leads to a growth of the resistive area simply because of the higher current load and the heat generated. Hence, $I(V)$ curves reflect not only intrinsic properties of the superconducting material, but also an increase of resistance due to these spatial effects.

These remarks show that one should take J_c data with a grain of salt and be open-minded when comparing samples of different quality and shape. For this reason the J_c values in these notes are merely given as approximate numbers.

2. Techniques Used to Measure J_c

Two techniques – transport and magnetic measurements – are widely used to measure critical current densities.

2.1 Transport Measurements

For this type of J_c measurement the sample is biased with a so-called transport current and the voltage along the sample is measured with a four-point technique. J_c is usually obtained by dividing the critical current by the smallest cross section of the sample implying $J_c \simeq \bar{J}_c$, assuming homogeneous flow of the critical current. The main advantage of this technique lies in the fact that J_c is measured in a straightforward way, and the $I(V)$ curve needed for most applications is obtained directly.

A drawback of transport measurements is the need for low resistance contacts. If the contacts are bad, the bias current generates excessive heat in the contact pads, which not only shifts the temperature of the sample but also causes temperature gradients at the contacts and in the superconductor. Resulting thermoelectric and thermomagnetic effects can easily generate several microvolts. Heating can be reduced by pulsing the current, by using large contact pads, and by placing all contacts far away from each other and from the part of the sample to be measured. A careful overview of problems associated with transport measurements can be found in Ref. 13, a detailed discussion of contact effects is given in Ref. 14, and the fabrication of high-quality contacts is described in Ref. 15.

2.2 Magnetic Measurements

It is convenient to have an alternative technique to measure J_c which requires neither contacts nor patterning: In magnetic measurements critical currents are induced into the sample by an external magnetic field H; the critical current density is inferred from the magnetic moment generated by the currents. The basic principle goes back to C.P. Bean [16] and is readily understood: A superconductor tries to shield itself from H by compensating H with a magnetization M. M is generated by ring currents as shown in Fig. 3. The Bean model, as well as other critical state models, is based on the notion that it is energetically favorable for the superconductor to shield itself with the highest current density possible, which of course is J_c (in this case the sample is in the 'critical state'): For small H, the fraction of the sample in which the current flows (and which therefore is unshielded) is minimized; if H is so large ($> H^*$ see Fig. 4, [17]) that currents flow in the entire sample, M is maximized. Therefore one can infer $J_c(H)$ from the magnetization loop $M(H)$.

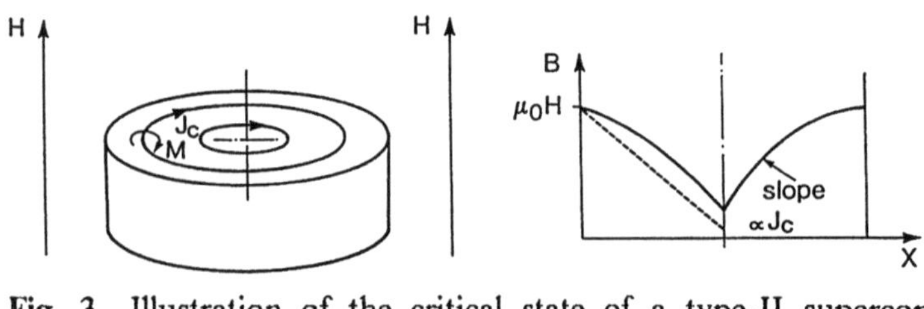

Fig. 3 Illustration of the critical state of a type-II superconductor in a magnetic field H. H induces shielding currents of density J_c, generating a magnetization M. The figure at right shows the flux density B inside the superconductor for $J_c(H)$ = constant (broken line) and for a field-dependent J_c (solid line).

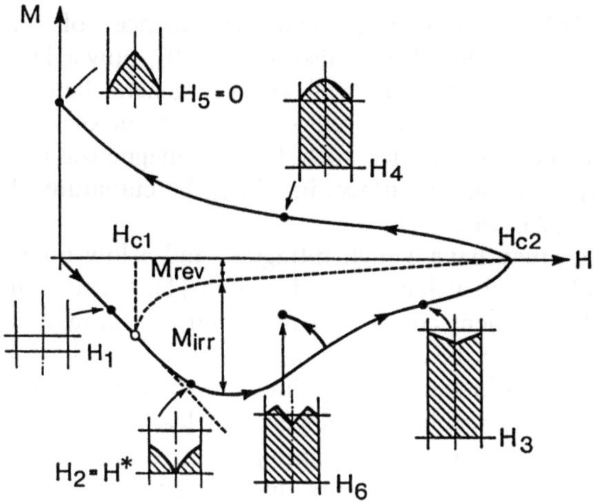

Fig. 4 Sketch of the magnetization process in a type-II superconductor. For various points of the magnetization curve the flux density profile inside the superconductor is shown (areas penetrated by the field are hatched). From [17].

2.2.1 Bean model

In the Bean model [16], a relation between J_c and $M(H)$ is derived by using the following assumptions and approximations:

- the density of the shielding current is either 0 or J_c,
- the sample is homogeneous and isotropic,
- the demagnetization factor $N = 0$ (e.g. cylinder with axis parallel to H),
- surface effects are negligible and $H_{c1} = 0$,
- the maximum value of H in the $M(H)$ loop is greater than $2H^*$,
- J_c is independent of H.

With these approximations, one obtains for cylindrical samples (SI units, M in Tesla) [18]:

$$J_c(H) = \frac{3}{2\mu_0}\frac{M^+(H) - M^-(H)}{R}, \tag{1}$$

where R is the sample radius. The values of the magnetization measured while the field is ramped to higher and lower values, M^- and M^+, (and not merely M) are used in (1) because M is the sum of the reversible and irreversible magnetizations, and $J_c = 0$ if the latter is 0.

2.2.2 Extensions of the Bean model

The Bean model has been extended in many ways to circumvent the approximations mentioned above. Some examples of this are:

- In the Kim model [19] a magnetic field dependence of J_c $J_c(H) = J_c(0)/(1 + B/B_0)$ was introduced, but also linear [20], power law [21] and exponential dependences [20] as well as more general behaviors [22] were employed to model $J_c(H)$ in high-T_c superconductors.
- Models have been developed for samples with finite demagnetization, such as for plates, H perpendicular to surface, in which the curvature of the flux lines is important [23, 24].
- In granular samples, which may have a high intragrain and a low intergrain J_c, it is questionable which value to use for R in (1), because in such samples the magnetic moment consists of intergrain and intragrain moments. A similar scenario may be given for inhomogeneous epitaxial films. In awkward cases, the ratio between the intergrain and intragrain moments is temperature and magnetic field-dependent, leading to confusing $J_c(T,H)$ dependences. To avoid such ambiguities, a number of groups [25-29] are using ring-shaped samples, in which the magnetic moment of the ring dominates the intragrain moments.
- Effects arising from $H_{c1} > 0$ and from surface barrier effects are considered in various models [30, 31] as well as implications of anisotropy [32, 33].

With all this in mind, it is no surprise that only some groups report satisfactory agreement between transport and magnetic measurements of J_c's [34-37], whereas others do not [38-40]. Differences in the results obtained by the two techniques can be attributed to some extent to differences in the time scales involved (fractions of a second compared to seconds or minutes), to the electric field criteria (typically μV/cm compared to nV/cm or even pV/cm) and to the fact that these techniques may sense different areas of different orientations as the weakest cross sections.

In addition to direct measurements of the $M(H)$ loops, measurement techniques have been developed which use other tools to analyze the critical state, for example torque measurements, studies of the flux profile inside the superconductor [41], from the response of the sample to small ac-signals superimposed on H [42], or magnetic imaging. For more extensive discussions of techniques to measure J_c, the reader is referred to Refs. 43-46.

3. Mechanisms which Limit J_c

There are three mechanisms which control critical currents in superconductors: Depairing, depinning, and decoupling, the definitions of which are given below. Depinning and decoupling are usually the active mechanisms for practical materials, with the depairing limit being reached only in extreme cases.

3.1 Depairing

In the London model, the depairing critical current density J_c^{dp} can be calculated as the current density for which the kinetic energy of the charge carriers equals the condensation energy: $J_c^{dp}(T) = B_c(T) / (\mu_0 \lambda(T))$, where $B_c(T)$ is the thermodynamic critical field and $\lambda(T)$ the London penetration depth. According to the BCS theory, J_c^{dp} is only a few percent higher than $\tilde{J}_c^{dp}$, the current density for which the energy gap goes to zero for a clean superconductor, which is given by $\tilde{J}_c^{dp} = n^*(v)e^*\Delta(0)/\hbar k_F = n^*(v)e^*\hbar/\pi m\xi_0$ ($T = 0$ K) where, n^*, e^*, and v are the density, charge and velocity of the Cooper pairs, Δ is the gap energy, $\hbar k_F$ is the momentum of the quasiparticles at the Fermi edge and m their mass [47]. For high-T_c superconductors the short coherence length ξ leads to quite a large depairing critical current density of $\simeq 10^9$ A/cm^2 at low temperatures [48].

3.2 Depinning

Above $H_{c1}/(1-N)$, magnetic fields enter type-II superconductors as flux lines which carry single flux quanta Φ_0. The fields may be supplied by external sources or by the self-field of the transport current. The transport current exerts a so-called Lorentz force $\vec{F}_L = \vec{J} \times \Phi_0 \vec{B}/B$ per unit length on each flux line. If $\vec{F}_L$ accelerates a flux line to a velocity $\vec{v}$, an electric field $\vec{E} = \vec{B} \times \vec{v}$ is induced in the superconductor, which then becomes resistive. If no field is applied externally, loops of flux lines generated by the self-field enter the sample and, if they are not pinned, shrink and desist at the sample axis. In the case of a thin isotropic film, flux lines of opposite signs enter at the edges of the film, move to the center and nullify each other there.

To achieve reasonable critical currents the movement of the flux lines has to be blocked. This blocking can be done by defects: If the normal core of a flux line is located at a defect that has a condensation energy reduced by an amount ΔE_{con}, a pinning force $F_P = \partial(\Delta E_{con})/\partial x$ results. Here, x is the relative position of the flux line with respect to the defect. The reduction of the defect's condensation energy can be caused by a lower critical temperature or by variations of the Ginzburg-Landau parameter κ at the defect. Whereas κ depends for instance on the electron mean free path, T_c may be reduced due to the stochiometry of the defect, by lower electron-phonon coupling, or due to changes in the density or elasticity of the material [49]. In the latter case the defect interacts with the vortex via elastic interactions (in the vortex core the density of a superconductor is slightly higher ($\simeq 10^{-7}$) and the material is a bit stiffer). Aside from this type of pinning, called core pinning, magnetic pinning may hinder the flux line movement. In magnetic pinning, the pinning force is due to magnetic interactions of the flux line with defects or with the sample surface. Whereas for core pinning the most efficient pinning sites depress the order parameter of the superconductor to zero within about one coherence length, the length scale for magnetic pinning is the magnetic penetration depth.

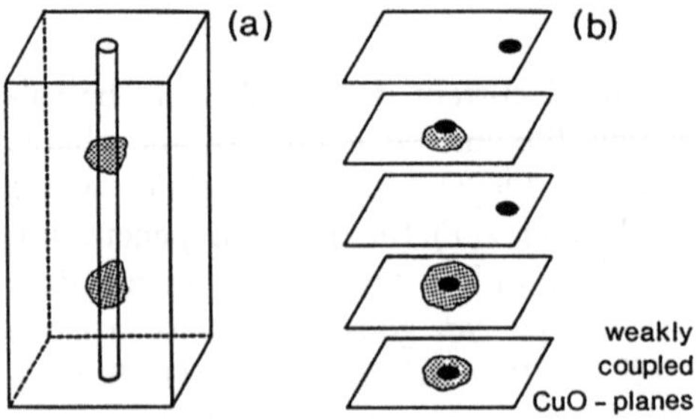

Fig. 5 Sketches of flux lines in (a) isotropic and (b) very anisotropic superconductors. The shaded areas depict pinning sites.

A flux line can also be constrained by interactions with the neighboring flux lines (e.g. for the B,T regime in which a flux line lattice exists). Calculating the pinning force that acts on the entire flux line arrangement (the global pinning force) on the basis of the elementary pinning forces of the individual flux lines is not an easy task (the "summation problem"). But it is understandable that due to randomly distributed pinning sites, some softness of the flux line lattice is required to obtain a large global pinning force. The critical current density is proportional to the pinning force and is given by the balance of $\vec{F}_L = \vec{F}_P$, which defines the critical state and J_c. It is noteworthy that if the pinned flux lines lead to magnetic field gradients as described in critical state models (compare with Fig. 4), I_c is enhanced. Whereas in the Meissner state the current flows only within surface layers, the magnetic field gradients lead to a more uniform supercurrent distribution [50].

In the case of superconductors with pronounced anisotropy, like in the Bi-based high-T_c cuprates, pinning becomes a more challenging problem [51]. This is illustrated in Fig. 5, which is a sketch of a flux line in a superconductor so anisotropic that it consists of a stack of planes coupled only by the Josephson effect, so that the flux line can be visualized as a pile of flux disks [52]. Such flux disks can move almost independently of each other [52] because only a small amount of magnetic coupling energy, but no condensation energy, is required for one disk to slide out of the pile. The Lorentz force and the thermal energy kT act on every disk. Therefore, until every disk is pinned individually, the superconductor will have a low J_c [53].

For detailed discussions of the various pinning mechanisms the reader is referred to Refs. 43-45 and 50.

3.3 Decoupling of Weak Links

J_c can be reduced rather drastically even by thin layers of non-superconducting material blocking the flow of the supercurrent. If these layers are insulators, the supercurrent has to tunnel, if they are semiconductors or normal metals, weak superconductivity can be induced into them by the proximity effect. In the following, insulating barriers are considered. Two superconductors can still be coupled by a thin insulator: the energy of the system is reduced by a coupling energy [54] which is proportional to $\cos \gamma$, where γ is the difference between the phases of the two superconductors. Furthermore, a supercurrent tunnels through the junction, which in the case of small junctions, theoretically has a density of

$$J(T) = \frac{\pi \Delta_i(T)}{2eR_N} \tanh\left(\frac{\Delta_i(T)}{2kT} \right) \sin \gamma \ , \tag{2}$$

where R_N is the junction resistance in the normal state, and $\Delta_i(T)$ is the gap energy at the interface between superconductor and insulator. Equation (2) implies that on increasing the phase difference γ from 0 to $\pi/2$ the supercurrent density increases to J_c. At $J_c(T)$ the coupling energy vanishes and the phases of the two superconductors decouple. Since the value Δ_i of the gap at the *interface* enters the Ambegaokar-Baratoff [55] equation (2), $J_c(T)$ is indirectly dependent on the coherence length ξ. While in superconductors with large ξ the gap value Δ_i corresponds to the gap value deep inside the material, in superconductors, where ξ is comparable to the lattice spacing, Δ_i can be depressed easily by structural and chemical variations of the superconductor close to the interface. Furthermore, as pointed out by Deutscher and Müller [56], the Ginzburg-Landau model shows that even for perfect superconductors, Δ_i is lowered if ξ is small, owing to the boundary conditions on $\Delta(r)$ at the interface. This implies that in high-T_c superconductors, boundary layers reduce J_c rather effectively.

In contrast to critical current densities limited by depinning or by depairing, current densities limited by decoupling are usually highly sensitive to magnetic fields. Magnetic flux densities of a few Gauss are typically large enough to insert a flux quantum into the barrier, which enforces a variation of the phase difference γ along the junction, so that J_c is reduced.

In some cases several of the mechanisms described above work together to control J_c. For example, in Superconductor-Weak Superconductor-Superconductor type weak links (SS′S junctions) the transport current may reduce the gap in S close to the SS′ interfaces (onset of depairing), so that flux lines in S′ become depinned more easily (depinning). Other examples are given by inhomogeneous tunnel junctions which are so extensive that Josephson vortices (i.e. flux lines without a normal core) can exist in the barrier. I_c is then controlled by the depinning of such vortices, the pinning energy being correlated with variations of the coupling energy of the junction.

For high-T_c superconductors, it is of particular importance that for $0 < J$ the system is not in thermodynamic equilibrium: thermodynamic fluctuations can decouple Josephson junctions and depin flux lines, giving rise to thermally activated phase slip [57-59] and to thermally activated flux flow and flux creep [60-63]. The frequency scale of these processes is proportional to $\exp(-U(T)/kT)$, the term $U(T)$ being the relevant coupling or pinning energy. Owing to the small pinning energies involved, thermally activated depinning is very marked in the BiSrCaCuO compounds.

Although thermally activated processes are more important at higher temperatures, they do not rule out in principle the existence of room temperature superconductors with high J_c's because, taking $YBa_2Cu_3O_{7-\delta}$ as a standard, U would simply have to be increased fourfold to compensate for the higher temperature.

4. Overview of the State of the Art – Strategies to Increase J_c

In this chapter critical currents of $YBa_2Cu_3O_{7-\delta}$ and BSCCO superconductors will be discussed. Due to space reasons we have to restrict ourselves to these systems, but they show many features which are characteristic for other compounds too.

4.1 $YBa_2Cu_3O_{7-\delta}$

4.1.1 Single crystals

The critical current density of $YBa_2Cu_3O_{7-\delta}$ single crystals lies in the range of 10^4 A/cm^2 (in the *a*,*b*-plane at 77 K, 1 T parallel to c). As shown by many experiments in which single crystals have been irradiated, for example by neutron or proton beams, their critical current density is limited by depinning. Up to a maximum irradiation dosage, J_c increases as a function of irradiation, (Fig. 6) because the irradiation creates defects which act as pinning sites. In some cases an enhancement of J_c by two orders of magnitude has been reported [64, 65]. At higher doses the abundance of defects destroys the order parameter and J_c deteriorates again. It is not practical to use irradiation to fabricate samples for large-scale applications; the main

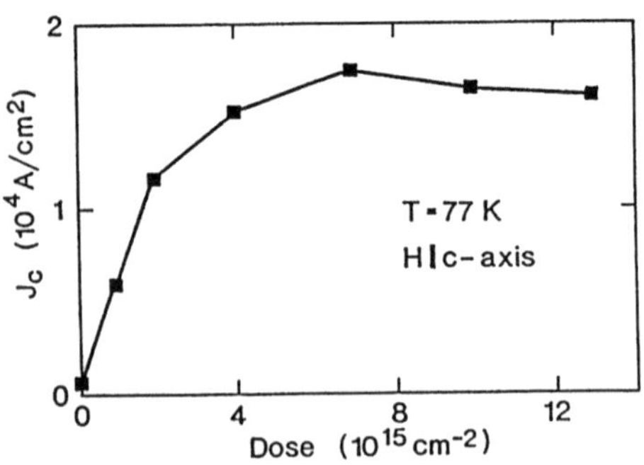

Fig. 6 Critical current density of a $YBa_2Cu_3O_{7-\delta}$ single crystal at 77 K and 1 T as a function of irradiation dose (3 MeV protons). J_c, measured magnetically, is oriented parallel to the *a*,*b*-planes. From [64].

objective of such experiments is to improve the understanding of the pinning. Interestingly, flux profile measurements of powdered single crystals as a function of particle size indicate the presence of intragrain weak links [41] in these samples, which have not yet been correlated with structural properties of the crystals.

High-quality single crystals have a relatively low defect density and because the density of their twin boundaries can be varied by thermal treatment they are a convenient system to study pinning at twin boundaries. Decoration experiments and magnetic relaxation studies clearly indicate that if the flux lines are aligned parallel to the twins, the boundaries contribute to the pinning at least for temperatures above 60 K [32, 66-69].

Additional pinning is provided by the layered structure of the high-T_c cuprates: because the coherence length in the c-direction is shorter than the lattice spacing, the order parameter can vary along the c-direction, so that the condensation energy density is highest at the CuO planes. Therefore, if the magnetic field is aligned parallel to the CuO planes, the flux lines tend to sit between the CuO planes and are blocked from movement in the c-direction (intrinsic pinning) [70, 71].

4.1.2 Epitaxial films

The highest critical current densities of high-T_c superconductors have been achieved with epitaxial $YBa_2Cu_3O_{7-\delta}$ films. Elementary requisites for high current densities are uncracked films of well-superconducting material with little mosaic spread [72]. For these reasons, high-J_c films were first grown on substrates like $SrTiO_3$ {100} which pose no diffusion problems and are closely matched in lattice size to $YBa_2Cu_3O_{7-\delta}$ {100}. In the meantime, current densities above 10^6 A/cm^2 at 77 K (self-field) have been reported for quite a number of other substrates too, such as MgO {100} [35], $NdGaO_3$ {100} [73], $La_{0.95}Gd_{0.05}GaO_3$ [73], $LaAlO_3$ {100} [74] and $Zr(Y)O_2$ {100} [75]. Even for Si {100} and sapphire {$\bar{1}012$}, current densities of 4.6×10^6 A/cm^2 (77 K, self-field) [76] and of 5×10^6 A/cm^2 (4.2 K, self-field) [77] have been obtained by using $Zr(Y)O_2$ buffer layers to block the diffusion of Si and low growth temperatures to minimize the reaction of Al with the $YBa_2Cu_3O_{7-\delta}$ films, respectively.

Transport studies [78], measurements of the magnetic relaxation [79] and irradiation experiments [80] show that in high-quality films, J_c is limited by depinning, which raises questions as to the nature of the pinning sites. Measurements of the angular dependence of $J_c(B)$ find a maximum for J_c if the magnetic field is applied parallel to the a,b-planes, [3], giving evidence of intrinsic pinning. Similar experiments indicate pinning by the film/substrate interface [3] by the film surface if the flux lines are aligned parallel to those planes and by the film edges [81]. But these pinning mechanisms cannot explain the high current densities seen when the magnetic field is applied parallel to the c-axis.

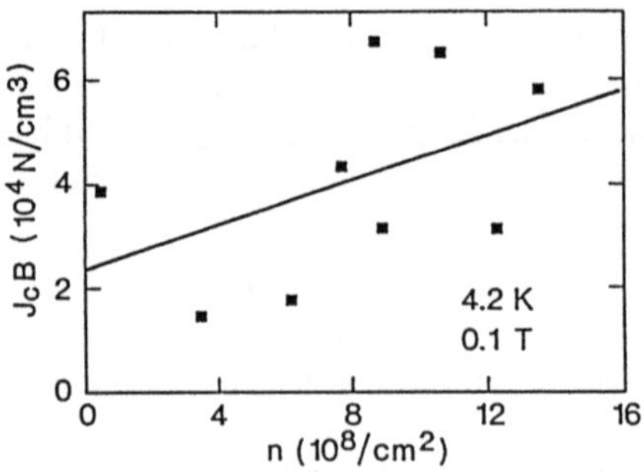

Fig. 7 Pinning force per unit film volume as a function of screw dislocation density in sputtered *c*-axis oriented $YBa_2Cu_3O_{7-\delta}$ films at 4.2 K for $H = 0.1$, $T // \vec{c}$. The screw dislocation density was measured using STM as described in [84].

Twin planes are pinning sites as shown by the single crystal experiments. But if they were the main pinning source in epitaxial $YBa_2Cu_3O_{7-\delta}$ films, one would expect no significant difference between the current densities of single crystals and epitaxial films. So there are mainly three types of defects left as candidates for pinning sites: point defects, dislocations and precipitates. Figure 7 indicates that at least for sputtered films most of the pinning is indeed caused by defects. The pinning force per unit volume of these films is correlated with the density of screw dislocations and it is reasonable to assume that the total defect density (and not an intrinsic pinning mechanism) is monitored by the dislocation density.

It is important to note that the flux lines have to be pinned along a good fraction of their length [82]: In the critical state the pinning force per unit length (for $J_c = 2\text{-}5 \times 10^7$ A/cm^2, 4.2 K) equals $F_P = F_L = J_c{}^*\Phi_0 \simeq (0.5\text{-}1) \times 10^{-3}$ N/m. On the other hand, the maximum pinning force per unit length for $\vec{B} // \vec{c}$ can be estimated roughly from the condensation energy density U_{con}: $F_p = U_{\text{con}}/\xi \simeq B_c^2\xi/2\mu_0 \simeq \Phi_0^2\xi_{\text{ab}}/(16\pi\mu_0\lambda_{\text{ab}}^2)$, which yields for $YBa_2Cu_3O_{7-\delta}$ ($\lambda_{\text{ab}} \simeq 1400$ Å, $\xi_{\text{ab}} \simeq 15$ Å), $F_p^{\max} \simeq 2 \times 10^{-3}$ N/m. Hence the flux lines have to be pinned along $\simeq 1/4\text{-}1/2$ of their total length! If this pinning were due to point defects alone, a rather high point defect density $> 4 \times 10^{19}$ cm^{-3} would be required [82].

Similar arguments apply for precipitates or misoriented grains, which in good films do not seem to be numerous and small enough to provide the necessary pinning force by their own [83].

If the pinning were due to dislocations, dislocation densities $n \simeq m/4$ are necessary, where $m = B/\Phi_0$ is the density of flux lines. In typical self-field conditions, $m \simeq 10^9$ cm^{-2} for good films, which requires excessively high dislocation densities. Surprisingly, in recent studies of sputtered films using scanning tunneling microscopy [84, 85], screw dislocations with densities of $n \simeq 10^9$ have indeed been observed (see Fig. 8). Edge dislocations are expected to be even more numerous. For example, 10^{11} edge dislo-

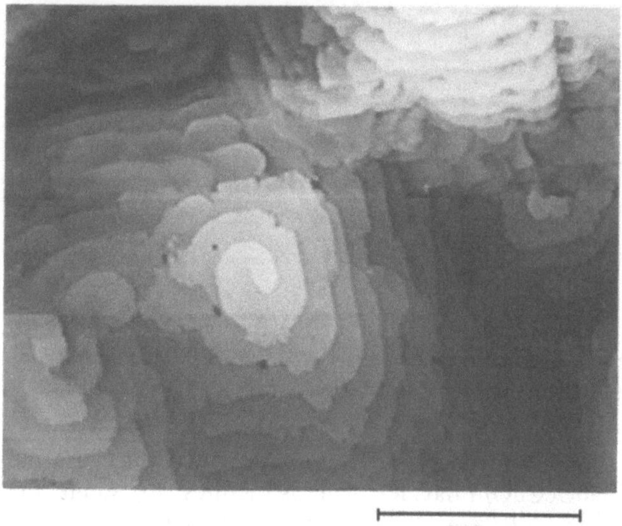

Fig. 8 Screw dislocation in a sputtered, c-axis oriented $YBa_2Cu_3O_{7-\delta}$ film $\simeq$120 nm thick. The step height of the terraces is $\simeq$12 Å.

cations per cm^2 were observed resulting from coalescence of $YBa_2Cu_3O_{7-\delta}$ nuclei in ultrathin sputtered films [86]. Therefore dislocation pinning could account for J_c at low fields (see also [87]). As a function of applied field, J_c
their total length! If this pinning were due to point defects alone, a rather high point defect density $> 4 \times 10^{19}$ cm^{-3} would be required [82].

Similar arguments apply for precipitates or misoriented grains, which in good films do not seem to be numerous and small enough to provide the necessary pinning force by their own [83].

If the pinning were due to dislocations, dislocation densities $n \simeq m/4$ are necessary, where $m = B/\Phi_0$ is the density of flux lines. In typical self-field conditions, $m \sim 10^9$ cm^{-2} for good films, which requires excessively high dislocation densities. Surprisingly, in recent studies of sputtered films using scanning tunneling microscopy [84, 85], screw dislocations with densities of $n \simeq 10^9$ have indeed been observed (see Fig. 8). Edge dislocations are expected to be even more numerous. For example, 10^{11} edge dislocations per cm^2 were observed resulting from coalescence of $YBa_2Cu_3O_{7-\delta}$ nuclei in ultrathin sputtered films [86]. Therefore dislocation pinning could account for J_c at low fields (see also [87]). As a function of applied field, J_c drops rather slowly as shown in Fig. 9, demonstrating that besides the screw dislocations, additional pinning sites have to be active to provide the pinning force in the field range above a few hundred Gauss, where $m \gg n$ [88]. Candidates for these pinning centers, which may well be weaker because they do not have to manage $J_c(H = 0)$, are edge dislocations, point defects and twin planes.

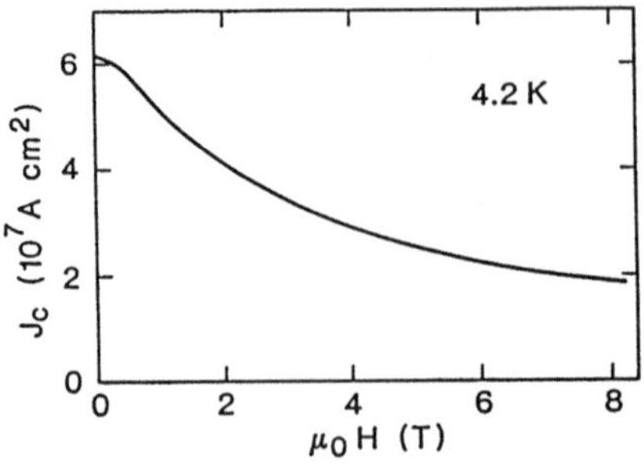

Fig. 9 Magnetic field dependence of a sputtered $YBa_2Cu_3O_{7-\delta}$ film with $\simeq 13.5 \times 10^8$ screw dislocations per cm^2 at 4.2 K (100 μV/cm criterion, $I_c \simeq 0.5$ A). From [88].

Pinning by dislocations is also suggested by experiments which use the lattice mismatch of NdCeCuO/YBaCuO heterostructures to strain the $YBa_2Cu_3O_{7-\delta}$ layers. If the $YBa_2Cu_3O_{7-\delta}$ layers are strained to the extent at which the formation of dislocations is expected, a maximum of J_c of 1.1×10^7 A/cm^2 (77 K, self-field) for the $YBa_2Cu_3O_{7-\delta}$ layers is observed [89].

The highest value of J_c in epitaxial films ($\simeq 1.3 \times 10^9$ A/cm^2, 77 K, self-field, resistivity criterion of $\simeq 3 \times 10^{-14}$ Ω cm, $I_c \simeq 100$ mA) was measured recently in a 500 Å wide line of $YBa_2Cu_3O_{7-\delta}$ [90], by taking advantage of the small self-field and large surface pinning in such samples.

In conclusion, microscopic studies combined with transport measurements indicate a variety of pinning sources of different strengths and densities active in epitaxial $YBa_2Cu_3O_{7-\delta}$ films. This agrees with magnetic relaxation measurements of polycrystalline and single crystalline samples which give evidence for a spectrum of pinning energies [79]. The relative weight of these pinning mechanisms may be temperature-dependent and may vary from sample to sample. For further discussions of pinning in high-T_c superconductors the reader is referred to Refs. 36, 82, 83, 91-93.

4.1.3 Polycrystalline $YBa_2Cu_3O_{7-\delta}$

A variety of effects is responsible for the low J_c of polycrystalline high-T_c superconductors. Certainly, J_c in polycrystals is lowered by voids, cracks and second phases. Furthermore, if the grains are not oriented, the current has to flow in some grains along the c-direction which has a low J_c [94]. Finally, grain boundaries in high-T_c superconductors reduce J_c [95, 96], in contrast to certain low-T_c superconductors (e.g. A-15 compounds) in which grain boundaries enhance J_c by acting as pinning sites. In the following, the properties of individual grain boundaries will be presented and the behavior of polycrystals shall be discussed on that basis.

Single grain boundaries. Critical current densities of individual grain boundaries and of their adjacent grains have been measured directly in

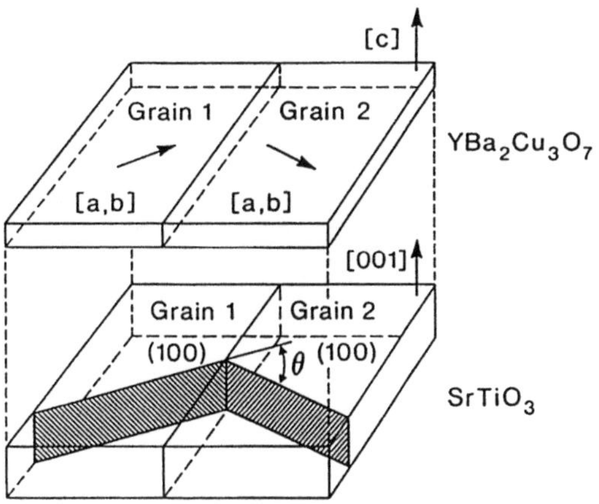

Fig. 10 Schematic diagram of a bicrystalline sample. The *c*-axis oriented $YBa_2Cu_3O_{7-\delta}$ bicrystals are grown epitaxially on $SrTiO_3$ bicrystals. From [97].

bicrystalline films of $YBa_2Cu_3O_{7-\delta}$ grown on $SrTiO_3$ bicrystals, as shown in Fig. 10. This technique allows the fabrication of millimeter-sized films with single, well-defined grain boundaries [97]. With a laser ablation technique [98] three superconducting bridges were patterned into each film, two bridges located inside the grains and one bridge straddling the grain boundary, allowing the transport properties of the grain boundary to be compared directly with the behavior of the adjoining grains. The main results of these measurements are as follows:

- The critical current densities of all bicrystals were limited by the grain boundaries. Grain boundary J_c's are reduced by a factor of 2-500 relative to the grain values.
- The grain boundary J_c decreases rapidly with grain misorientation. As shown in Fig. 11, there is one master curve for J_c as a function of grain misorientation for < 100 > tilt, < 001 > tilt and < 100 > twist boundaries. The grain boundary J_c scales with the J_c of the grains.
- Grain boundaries with an misorientation angle > 5-10° behave like Josephson junctions. Their critical current densities are highly sensitive to magnetic fields.

A more complete presentation of the properties of individual grain boundaries may be found in [99], complimentary experiments in [100], discussions of the mechanisms controlling the grain boundary J_c in [101-104] and investigations of their microscopic structure in [105-108]. Transport measurements of polycrystalline $YBa_2Cu_3O_{7-\delta}$ films which had a high portion of low-energy grain boundaries revealed a critical current density of such films of 5×10^5 A/cm^2 (77 K, self-field), suggesting that certain crystallographically special, high-angle grain boundaries can support a high J_c,

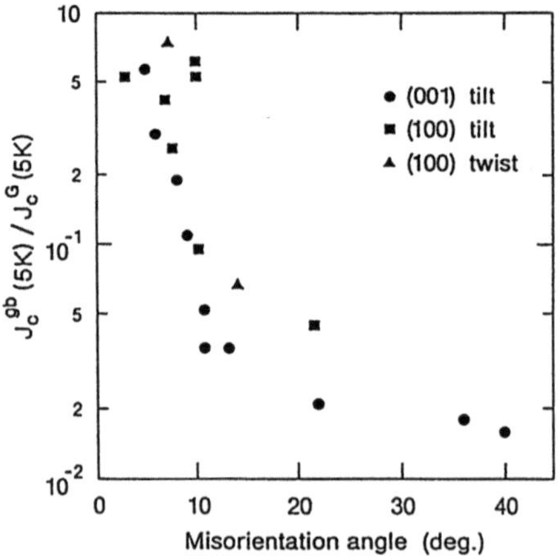

Fig. 11 Ratio of the grain boundary critical current density to the average value in the adjacent grains at 4.2 K as a function of misorientation angle. The different symbols distinguish the primary component of the misorientation. From [99].

at least for $H = 0$ [109]. This would be nice, but the conclusion is not compelling since grain boundary networks can have a J_c which is higher than that of the individual grain boundaries.

Studies of single grain boundaries have been performed with bulk bicrystals, too. Whereas a grain boundary limited J_c was also found for the Bi compounds [110], the grain boundary J_c of a bulk YBCO sample was reported to be no lower than the intragrain value [111], this may be attributed to the very low intragrain J_c's of these samples ($\simeq 3 \times 10^3$ A/cm^2 at 77 K in self-field). Studies of single grain boundaries of $BaPb_{1-x}Bi_xO_3$ which show similar behavior as the grain boundaries described above are described in [112].

Grain boundary networks. Owing to a number of factors, the critical current density of a polycrystalline high-T_c superconductor may differ from that of its individual grain boundaries, being higher or lower than the average grain boundary J_c [113]. It is difficult to model supercurrent flow in polycrystals quantitatively, because the grain boundaries form disordered, three-dimensional networks of Josephson junctions, of which the elements have a distribution of J_c's. This implies that the J_c of one grain boundary may depend on the current load of all its neighbors. For example, the critical current of the grain boundaries at the sample surface will be depressed by the self-field of the entire system. Nevertheless, some models have been developed to calculate the critical current density of polycrystalline high-T_c superconductors [26, 113-118]. Also, the current distribution in polycrystals has been studied to some extent by LTSEM [119].

Despite the difficulties involved in quantitatively understanding the J_c of high-T_c polycrystals, it is obvious that J_c can be increased by

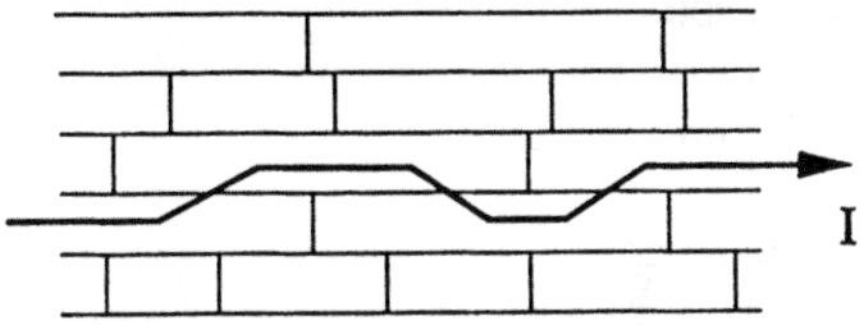

Fig. 12 Sketch of a polycrystal with a large effective grain boundary area.

- avoiding weak links
- maximizing the grain boundary J_c
- maximizing the grain boundary area.

The strategy used to achieve these goals is based on the grain boundary properties: The first two objectives can be achieved by aligning the grains biaxially to within a few degrees. In addition one may try doping the grain boundaries (e.g. with Ag) to increase their coupling strength. The effective grain boundary area can be increased by optimizing the aspect ratio and the packaging of the grains as suggested by Fig. 12 [113].

Grain alignment. There are basically three ways known to orient grains of the high-T_c cuprates: compactation, magnetic orientation and directional solidification as employed by the zone melting and melt texture techniques. In magnetic orienting, powdered samples are suspended in organic solvents or in epoxy and are oriented with a magnetic field of several Teslas taking advantage of the paramagnetic anisotropy of these compounds [120-124]. After the solvent is evaporated, the field can be switched off for sintering. Uniaxial alignment with an angular spread of $\lesssim 10°$ [122] and studies of biaxial alignment [124] have been performed, yielding increases of J_c of factors of 5 to $\simeq$2000 A/cm^2 (77 K, self-field).

Doping. Many groups have studied the effects of increasing the inter-grain and intergrain critical current density of polycrystals by adding dopants such as Ag, Au, Fe [93], Ni [93], Co [125] or Bi [126] to the material. Silver, depending on the processing technique, may either form precipitates at the grain boundaries, or substitute Cu inside the grains. If the Ag is placed at the grain boundaries, 3-10% of silver can increase the critical current density of the polycrystals by a factor of two to some 10^4 A/cm^2 (77 K, self-field), depending somewhat on the quality of the undoped sample [127]. This effect seems to result from a closer packing of the grains and due to wetting of the grain boundaries by Ag, increasing cohesion (Fig. 13) [128]. In addition, Ag may reduce cracking. For silver placed inside the grains, an increase of the intragrain pinning energy by a factor of 5 has been reported, which may arise from Ag in the Cu chain sites of $YBa_2Cu_3O_{7-\delta}$ or from small Ag precipitates [129]. An overview of the effects on silver doping can be found in [129, 130].

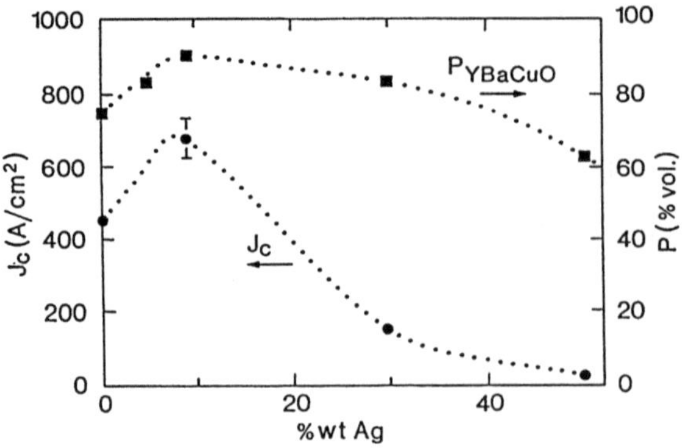

Fig. 13 J_c and $YBa_2Cu_3O_{7-\delta}$ volume fraction as a function of silver doping in polycrystalline $YBa_2Cu_3O_{7-\delta}$. From [128].

Melt texturing. Melt texturing, introduced by Jin and co-workers [131], has led to a significant increase of the reported values of J_c in YBCO polycrystals/bulk samples. Melt texturing allows the growth of well-aligned samples with large grain boundary areas. In melt texturing, the samples are heated above their peritectic decomposition temperature ($\simeq$1010°C in air), where $YBa_2Cu_3O_{7-\delta}$ melts incongruently into a liquid phase and Y_2BaCuO_5, and is then cooled so slowly that large grains grow (up to a centimeter in length). Related processes have been developed [5, 6, 39, 132] in some of them temperature gradients are employed to enhance grain alignment. Critical current densities of such samples are approaching 10^5 A/cm² in self-field, for $H \simeq 1.8$ T values of 4×10^4 A/cm² [7] ($I_c > 220$ A, transport data at 77 K with criteria of 1-10 μV/cm) have been found.

The high J_c values in large fields demonstrate that these samples are not dominated by weak links, but by depinning. Twin boundaries, stacking faults, point defects, dislocations, grain boundaries (!) and precipitates, in particular micron-sized particles of Y_2BaCuO_5 (211) and their stress fields have been suggested as pinning sites. The impact of the 211 particles on J_c has been studied [132-135] but the assessment of their pinning force is difficult because their presence is correlated to other microstructural properties of $YBa_2Cu_3O_{7-\delta}$: Samples with many 211 particles have less BaCuO, thinner grains and less cracks [134].

However, these outstanding *transport* data have only been reported for centimeter sized samples in which the current had to pass but a few grain boundaries. It is challenging to extend the melt texturing technique to the fabrication of long wires. Longer samples require cladding to constrain the superconductor, but cladding is problematic for $YBa_2Cu_3O_{7-\delta}$. At the high processing temperatures involved the cladding material diffuses into the cuprate, thereby usually degrading superconductivity. Nevertheless, using Ni cladding, 25-cm-long, partially melt-processed and Ag-doped $YBa_2Cu_3O_{7-\delta}$

wires have been fabricated which support transport currents of more than 600 A with a J_c of almost 10^5 A/cm^2 at 4.2 K and 8 T ($\vec{H} \parallel \vec{I}$) [136]. Long wires with much lower J_c's, however, have been fabricated by more conventional techniques. Multifilamentary $YBa_2Cu_3O_{7-\delta}$ wires, 20 m long, with $J_c \simeq$ 4000-6000 A/cm^2 ($I_c \simeq$ 10 A, 50 K, self-field) are commercially available [137].

4.2 The BSCCO compounds

The melting point of the Bi-based high-T_c cuprates is lower than that of $YBa_2Cu_3O_{7-\delta}$ ($\simeq$880°C compared to $\simeq$1010°C). The lower sintering temperature allows the use of Ag cladding. For Ag sheathed wires, tapes and ribbons of the Bi-(Pb)-(Sb)-Sr-Ca-Cu-O system, fantastic critical current densities have been found at low temperatures in large magnetic fields (Fig. 14) [138-144]. Above 15 T their critical current density exceeds that of the best low-T_c superconductors, suggesting the use of the Bi compounds in high-field magnets. Whereas low intergrain current densities are reported for standard Bi materials, these high-J_c wires are obviously not plagued by weak links. This is attributed to the huge areas of the grain boundaries (Fig. 15), which in some samples are combined with pronounced texturing (the necessity of texturing is pointed out in [145]). Unfortunately, above 30 K, disappointing $J_c(H)$ characteristics are observed (Fig. 14). These low J_c values result from the low pinning energy of the very anisotropic Bi compounds [138, 146], which is already indicated by the lack of hysteresis seen in Fig. 14.

Higher J_c's at 77 K have been achieved in silver sheathed $Bi_2Sr_2Ca_2Cu_3O_{10}$ wires, characterized by extensive texturing and small-angle grain boundaries [4, 147, 148]. Taking advantage of finely dispersed non-superconducting phases, current densities of $\simeq 10^4$ A/cm^2 were achieved at 77 K in such wires or tapes with 1 T applied parallel to the tape surface

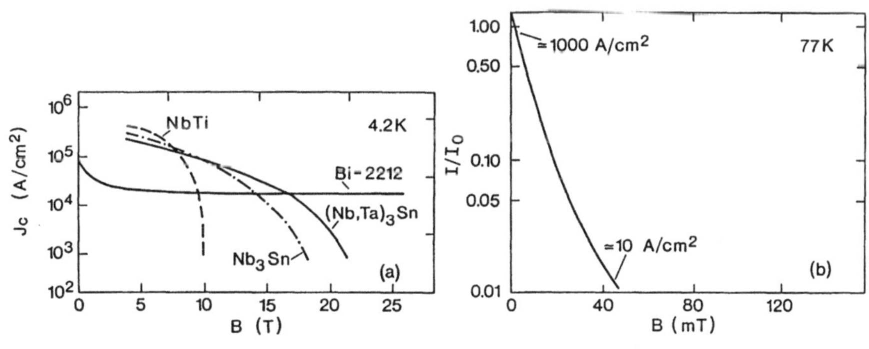

Fig. 14 J_c of a $Bi_2Sr_2Ca_1Cu_2O_8$/Ag wire at 4.2 K and at 77 K as a function of applied magnetic field compared to commercial low-T_c superconductors (1 μV/cm criterion, outer diameter of the wire 1.5 mm). From [138].

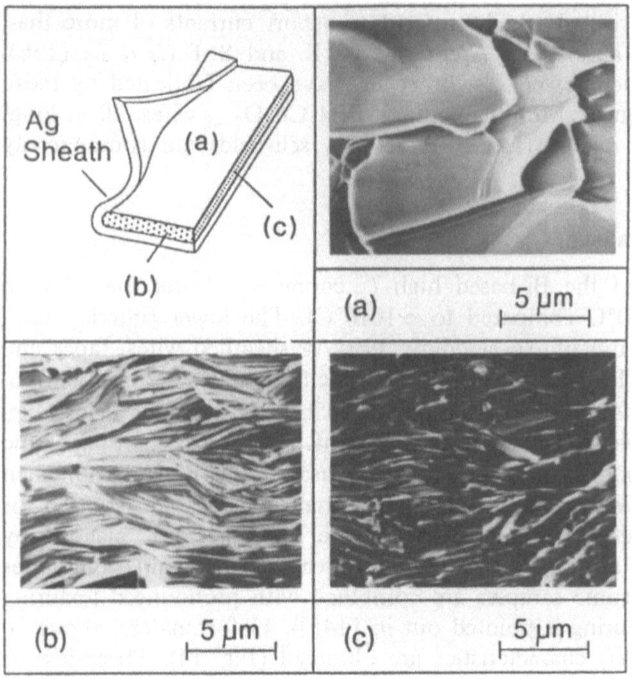

Fig. 15 SEM micrographs of a silver sheated BiPbSrCaCuO tape featuring large effective grain boundary areas (see Fig. 12) between well-aligned, platelet shaped grains. At 4.2 K such tapes have a J_c of 1.5×10^5 A/cm^2 in self-field and 5×10^4 A/cm^2 at 10 T ($\vec{I} \perp \vec{H}$, $\vec{H} \perp \vec{c}$, 0.2 μV/cm criterion). From [139].

(1 μV/cm criterion, $I_c \simeq 10$ A) [4]. These results are very encouraging, in particular as the wires are 50 cm long and even show a certain degree of mechanical flexibility.

Higher J_c's at 77 K have been achieved in silver sheathed $Bi_2Sr_2Ca_2Cu_3O_{10}$ wires, characterized by extensive texturing and small-angle grain boundaries [4, 147, 148]. Taking advantage of finely dispersed non-superconducting phases, current densities of $\simeq 10^4$ A/cm^2 were achieved at 77 K in such wires or tapes with 1 T applied parallel to the tape surface (1 μV/cm criterion, $I_c \simeq 10$ A) [4]. These results are very encouraging, in particular as the wires are 50 cm long and even show a certain degree of mechanical flexibility.

5. Outlook

During the past few years considerable insight has been gained into the mechanisms that limit critical current densities in high-T_c superconductors. The problems encountered in increasing their critical current densities have their roots in the small coherence length, the complex chemistry and the high operation temperature of these materials. Impressive progress has been achieved in the fabrication of high-J_c cuprates. Thin films with $J_c \simeq 3\text{-}5 \times 10^6$ A/cm^2 at 77 K are routinely fabricated; the current density of high-J_c Bi compound wires exceeds at low temperatures and high magnetic fields that of the best low T_c superconductors and has reached more than 10^4 A/cm^2 at 77 K in 1 T for some magnetic field orientations.

The critical current densities of epitaxial films are sufficient for most applications and are already approaching the depairing limit. Therefore, as far as films are concerned, interest will be shifting from improving record J_c values to growing high J_c films at modest growth temperatures over large areas on substrates which have attractive electronic, mechanical and structural properties.

In the realm of polycrystals, three objectives have to be achieved: high J_c ($> 10^5$ A/cm^2)

- at high temperatures (≥ 77 K)
- in large magnetic fields (several T) of any orientation
- in long (hundreds of meters), flexible samples.

To date, no sample has been fabricated which meets all three requirements, but specimens are at hand that basically fulfill any two of them, pointing out two potential routes to practical high-T_c, high-J_c superconductors: Extending melt texturing to the fabrication of sufficiently long samples with high transport J_c, or increasing the operation temperature of the high-J_c Bi materials by further enhancing pinning.

It is encouraging to see that the increase of the critical current densities of high-T_c superconductors is basically a technological quest. Although there are still challenges ahead, there is no fundamental physical problem blocking the way to intriguing applications.

The author gratefully acknowledges valuable discussions with J.G. Bednorz, K.A. Müller, D.G. Schlom, A. Baratoff, E.H. Brandt, M. Ferrari, P. Martinoli and S. Shinde, and thanks Sumitomo Heavy Industries Ltd. for permission to reproduce Fig. 15.

References

[1] A.M. Wolsky, R.F. Giese, E.J. Daniels: Sci. Am., Feb. 1989, pp.45
[2] S. Shinde, private communication
[3] B. Roas, L. Schultz, G. Saemann-Ischenko: Phys. Rev. Lett. 64, 479 (1990)
[4] K. Sato, T. Hitaka, H. Mukai, M. Ueyama, N. Shibuta, T. Kato, T. Masuda, M. Nagata, K. Iwata, T. Mitsui: IEEE Trans. Mag. 27, 1231 (1991)
[5] P. McGinn, W. Chen, N. Zuh, M. Lanagan, U. Balachandran: Appl. Phys. Lett. 57, 1455 (1990)

[6] K. Salama, V. Selvamanickam, L. Gao, K. Sun: Appl. Phys. Lett. **54**, 2352 (1989)
[7] D. Shi, M.M. Fang, J. Akujieze, M. Xu, J.G. Chen, C. Segre: Appl. Phys. Lett. 57, 2606 (1990)
[8] M. Murakami, S. Gotoh, N. Koshizuka, S. Tanaka, T. Matsushita, S. Kambe, K. Kitazawa: Cryogenics **30**, 390 (1990)
[9] Since any system with a ground state of $R=0$ has an exited state with $R>0$ which is separated by a finite energy barrier, thermal fluctuations always lead to dissipative processes resulting in $R>0$ for the system. But of course, R may be quite small.
[10] There are non-arbitrary definitions of J_c, which, however, are more of theoretical than of practical importance: For example, one may calculate $\lim(T\rightarrow 0)\, J_c(T)$ to avoid the problems associated with thermal activation. Or one may model thermally activated processes and use deviations from the expected behavior for a definition of J_c. The term J_c may also be defined as the current density at which the smallest barrier to flux motion or to decoupling is reduced to zero. See M. Tinkham: "Considerations Limiting Critical Currents in High Temperature Superconductors," proceedings of the ISTEC Workshop on Critical Currents in High-T_c Superconductors, Feb. 1989
[11] J.W. Ekin: Appl. Phys. Lett. **55**, 905 (1989)
[12] R. Gross, J. Mannhart, K. Hipler, R.P. Huebener, P. Chaudhari, D. Dimos, C.C. Tsuei, J. Schubert, U. Poppe: Physica C **162-164**, 1603 (1989)
[13] L.F. Goodrich, S.L. Bray: Cryogenics **30**, 667 (1990)
[14] H. Jones, L. Cowey, D. Dew-Hughes: Cryogenics **29**, 795 (1989)
[15] J.W. Ekin, T.M. Larson, N.F. Bergen, A.J. Nelson, A.B. Swartzlander, L.L. Kazmerski, A.J. Panson, B.A. Blankenship: Appl. Phys. Lett. **52**, 1819 (1988)
[16] C.P. Bean: Phys. Rev. Lett. **8**, 250 (1962); C.P. Bean: Rev. Mod. Phys. **36**, 31 (1964)
[17] H.W. Weber, O. Hittmayer: "Supraleitung," Thiemig Taschenbücher (Verlag Karl Thiemig, Munich, 1979)
[18] W.A. Fietz, M.R. Beasley, J. Silcox, W.W. Webb: Phys. Rev. **136**, A 335 (1964)
[19] Y.B. Kim, C.F. Hempstead, A.R. Strnad: Phys. Rev. Lett. **9**, 306 (1962); Y.B. Kim, C.F. Hempstead, A.R. Strnad: Phys. Rev. **129**, 528 (1963)
[20] A. Sanchez, D.X. Chen, J.S. Munoz, Y.-Z. Li: Physica C **175**, 33 (1991)
[21] F. Irie, K. Yamafuji: J. Phys. Soc. Jpn. **23**, 255 (1967)
[22] M. Xu, D. Shi, R.F. Fox: Phys. Rev. B **42**, 10773 (1990)
[23] D.J. Frankel: J. Appl. Phys. **50**, 5402 (1979)
[24] M. Däumling, D.C. Larbalestier: Phys. Rev. B **40**, 9350 (1989)
[25] M. Polak, V. Windte, W. Schauer, J. Reiner, A. Gurevich, J. W Wühl: Physica C **174**, 14 (1991)
[26] W. Paul, Th. Baumann: Physica C **175**, 102 (1991)
[27] C. Mee, A.I.M. Rae, W.F. Vinen, C.E. Gough: "Flux Creep in High-T_c Superconductors," preprint
[28] K. Bernhardt, R. Gross, M. Hartmann, R.P. Huebener, F. Kober, D. Koelle, T. Sermet: Physica C **161**, 468 (1989)
[29] J.S. Satchell, R.G. Humphreys, P. Sidhu, N.G. Chew, S.W. Goodyear, J.A. Edwards: Physica C **162-164**, 625 (1989)
[30] A.K. Grover, C. Radhakrishnamurty, P. Chaddah, G. Ravi Kumar, G. V. Subba Rao: Pramana J. Phys. **30**, 569 (1988)

[31] M.W. McElfresh, Y. Yeshurun, A.P. Malozemoff, F. Holtzberg: Physica A **168**, 308 (1990)
[32] F.M. Sauerzopf, H.P. Wiesinger, H.W. Weber: Cryogenics **30**, 650 (1990)
[33] E.M. Gyorgy, R.B. van Dover, K.A. Jackson, L.F. Schneemeyer, J.V. Waszczak: Appl. Phys. Lett. **55**, 283 (1989)
[34] T.R. McGuire, A. Gupta, G. Koren, R.B. Laibowitz, D.Dimos: Physica C **162-164**, 131 (1989)
[35] M.J. Scharen, A.H. Cardona, J.Z. Sun, L.C. Bourne, J.R. Schrieffer: Jpn. J. Appl. Phys. **30**, L15 (1991)
[36] C.B. Eom, J.Z. Sun, B.M. Lairson, S.K. Streiffer, A.F. Marshall, K. Yamamoto, S. Anlage, J.C. Bravman, T.H. Geballe: Physica C **171**, 354 (1990)
[37] J.S. Satchell, R.G. Humphreys, P. Sidhu, N.G. Chew, S.W. Goodyear, J.A. Edwards: Physica C **162-164**, 625 (1989)
[38] S.E. Male, J. Chilton, A.D. Caplin, C.N. Guy, S.B. Newcomb: Supercond. Sci. Technol. **2**, 9 (1989)
[39] L. Gao, R.L. Meng, Y.Y. Xue, P.H. Hor, C.W. Chu: Appl. Phys. Lett. **58**, 92 (1991)
[40] J. Kase, J. Shimoyama, E. Yanagisawa, S. Kondoh, T. Matsubara, T. Morimoto, M. Suzuki: Jpn. J. Appl. Phys. **29**, L277 (1990)
[41] H. Küpfer, I. Apfelstedt, R. Flükiger, C. Keller, R. Meier- Hirmer, B. Runtsch, A. Turowski, U. Wiech, T. Wolf: Cryogenics **29**, 268 (1989)
[42] K.-H. Müller, A.J. Pauza: Physica C **161**, 319 (1989); K.H. Müller, J.C. MacFralane, R. Driver: Physica C **158**, 69 (1989)
[43] A.M. Campbell, J.E. Evetts: Adv. Phys. **21**, 199 (1972)
[44] R.P. Huebener: "Magnetic Flux Structures in Superconductors," Springer Series Solid-State Science, Vol. 6 (Springer, Berlin, Heidelberg, 1979)
[45] H. Ullmayer: "Irreversible Properties of Type II Superconductors" Springer Tracts in Modern Physics, Vol. 76 (Springer, Berlin, Heidelberg, New York, 1975)
[46] D.M. Kroeger, C.C. Koch, J.P. Charlesworth: J. Low Temp. Phys. **19**, 493 (1975)
[47] M. Tinkham: "Introduction to Superconductivity," (McGraw-Hill, New York, 1975)
[48] C.C. Tsuei, J. Mannhart, D. Dimos: Proceedings of Topical Conference on High-Tc Superconducting Thin Films, Devices and Applications, AIP Conference Proceedings No. 182 (ed. by G. Mararitondo, R. Joynt, and M. Onellion, 1989), pp. 194-207
[49] P.H. Kes: "Flux Pinning and the Critical Current" in "Concise Encyclopedia of Magnetic and Superconducting Materials" ed. by J.E. Evetts (Pergamon, Oxford)
[50] E.H. Brandt: Z. Phys. B **80**, 167 (1990)
[51] T.T.M. Palstra, B. Batlogg, L.F. Schneemeyer, J.V. Waszczak: Phys. Rev. B **43**, 3756 (1991)
[52] J. R. Clem: Phys. Rev. B **43**, 7837 (1991)
[53] G. Deutscher, A. Kapitulnik: Physica A **168**, 338 (1990)
[54] B.D. Josephson: Phys. Lett. **1**, 251 (1962)
[55] V. Ambegaokar, A. Baratoff: Phys. Rev. Lett. **10**, 486 (1963) and (errata) Phys. Rev. Lett. **11**, 104 (1963)
[56] G. Deutscher, K.A. Mueller: Phys. Rev. Lett. **59**, 1745 (1987)
[57] V. Ambegaokar, B.I. Halperin: Phys. Rev. Lett. **22**, 1364 (1969)
[58] T. Tsuneto: J. Phys. Soc. Jpn. **57**, 3499 (1989)

[59] R. Gross, P. Chaudhari, D. Dimos, A. Gupta, G. Koren: Phys. Rev. Lett. **64**, 228 (1990)
[60] P.W. Anderson: Phys. Rev. Lett. **9**, 309 (1962)
[61] D. Dew Hughes: Cryogenics **28**, 674 (1988)
[62] Y. Yeshurun, A.P. Malozemoff: Phys. Rev. Lett. **60**, 2202(1988)
[63] M. Tinkham: Helv. Phys. Acta 61, 443 (1988)
[64] L. Civale, A.D. Marwick, M.W. McElfresh, T.K. Worthington, A.P. Malozemoff, F.H. Holtzberg, J.R. Thompson, M.A. Kirk: Phys. Rev. Lett. **65**, 1164 (1990)
[65] R.B. van Dover, E.M. Gyorgy, L.F. Schneemeyer, J.W. Mitchell, K.V. Rao, R. Puzniak, J.V. Waszczak: Nature **342**, 55 (1989)
[66] L.J. Swartzendruber, D.L. Kaiser, F.W. Gayle, L.H. Bennett, A. Roytburd: Appl. Phys. Lett. **58**, 1556 (1991); L.J. Swartzendruber, A. Roytburd, D.L. Kaiser, F.W. Gayle, L.H. Bennett: Phys. Rev. Lett. **64**, 483 (1990)
[67] J.Z. Liu, M.D. Lan, P. Klavins, R.N. Shelton: Phys. Lett. A **144**, 265 (1990)
[68] W.K. Kwok, U. Welp, K.G. Vandervoort, Y. Fang, and G.W. Crabtree: Appl. Phys. Lett. **57**, 1268 (1990); W.K. Kwok, U. Welp, G.W. Crabtree, K.G. Vandervoort, R. Hulscher, J.Z. Liu: Phys. Rev. Lett. **64**, 966 (1990)
[69] G.J. Dolan, G.V. Chandrashekhar, T.R. Dinger, C. Field, and F. Holtzberg: Phys. Rev. Lett. **62**, 827 (1989)
[70] M. Tachiki and S. Takahashi: Solid State Commun. **70**, 291 (1989)
[71] D. Feinberg, C. Villard: Phys. Rev. Lett. **64**, 919 (1990)
[72] W. Schauer, X.X. Xi, V. Windte, O. Meyer, G. Linker, Q. Li, J. Geerk: Cryogenics **30**, 586 (1990)
[73] G. Koren, A. Gupta, E.A. Giess, A. Segmüller, R.B. Laibowitz: Appl. Phys. Lett. **54**, 1054 (1989)
[74] A. Inam, X.D. Wu, L. Nazar, M.S. Hedge, C.T. Rogers, T. Venkatesan, R.W. Simon, K. Daly, H. Padamsee, J. Kirchgessner, D. Moffat, D. Rubin, Q.S. Shu, D. Kalokitis, A. Fathy, V. Pendrick, R. Brown, B. Brycki, E. Belohoubek, L. Drabeck, G. Gruner, R. Hammond, F. Gamble, B.M. Lairson, J.C. Bravman: Appl. Phys. Lett. **56**, 1178 (1990)
[75] J. Geerk, G. Linker, O. Meyer: "Epitaxial Growth and Properties of YBCO Thin Films" Manuscript KfK4601 Kernforschungszentrum Karlsruhe, ISSN 0303-4003
[76] D.K. Fork, F.A. Ponce, J.C. Tramontana, N. Newman, J.M. Phillips, T.H. Geballe: Appl. Phys. Lett. **58**, 2432 (1991)
[77] K. Char, D.K. Fork, T.H. Geballe, S.S. Laderman, R.C. Taber, R.D. Jacowitz, F. Bridges, G.A.N. Connell, J.B. Boyce: Appl. Phys. Lett. **56**, 785 (1990)
[78] J. Mannhart, P. Chaudhari, D. Dimos, T.R. McGuire: Phys. Rev. Lett. 61, 2476 (1988)
[79] C.W. Hagen, R. Griessen: Phys. Rev. Lett. 62, 2857 (1989)
[80] B. Roas, B. Hensel, G. Saemann-Ischenko, L. Schultz: Appl. Phys. Lett. **54**, 1051 (1989)
[81] S. Tahara, S.M. Anlage, C.B. Eom, D.K. Fork, T.H. Geballe, M.R. Beasley: Physica C **162-164**, 1175 (1989)
[82] T.L. Hylton, M.R. Beasley: Phys. Rev. B **41**, 11669 (1990)
[83] P. Chaudhari, D. Dimos, J. Mannhart: "Critical Currents in Single-Crystal and Bicrystal-Films," in "Earlier and Recent Aspects of Superconductivity," ed. by J.G. Bednorz and K.A. Müller (Springer, Berlin, Heidelberg, New York, 1990)

[84] Ch. Gerber, D. Anselmetti, J.G. Bednorz, J. Mannhart, D.G. Schlom: Nature **350**, 279 (1991)
[85] M. Hawley, I.D. Raistrick, J.G. Beery, R.J. Houlton: Science **251**, 1587 (1991)
[86] S.K. Streiffer, B.M. Lairson, C.B. Eom, B.M. Clemens, J.C. Bravman, T.H. Geballe: Phys. Rev. B **43**, 13007 (1991)
[87] V.M. Pan, E.A. Pashitsky, C.G. Tretiatchenko: "The Electrostatic Mechanism of Abrikosov Vortex Pinning at Charged Pont Defects and Edge Dislocations in High-Tc Superconductors," preprint
[88] D.G. Schlom, D. Anselmetti, J.G. Bednorz, R. Broom, A. Catana, T. Frey, Ch. Gerber, H.J. Güntherodt, H.P. Lang, J.Mannhart, K.A. Müller: "Screw Dislocation Mediated Growth of Sputtered and Laser Ablated YBCO-Films," preprint, Z. Phys. B
[89] R. Gross, A. Gupta, E. Olsson, A. Segmüller, G. Koren: Phys. Rev. Lett. **57**, 203 (1990)
[90] H. Jiang, Y. Huang, H. How, S. Zhang, C. Vittoria, A. Widom, D.B. Chrisey, J.S. Horwitz, R. Lee: Phys. Rev. Lett. **66**, 1785 (1991)
[91] T. Matsushita: Cryogenics **30**, 817 (1990)
[92] A.M. Campbell: IEEE Trans. Mag. **27**, 1660 (1991)
[93] R. Wördenweber, K. Heinemanmn, G.V.S. Sastry, H.C. Freyhardt: Cryogenics **29**, 458 (1990)
[94] T.R. Dinger, T.K. Worthington, W.J. Gallagher, R.L. Sandstrom: Phys. Rev. Lett. **58**, 2687 (1987)
[95] P. Chaudhari, J. Mannhart, D. Dimos, C.C. Tsuei, C.C. Chi, M.M. Oprysko, M. Scheuermann: Phys. Rev. Lett. **60**, 1653 (1988)
[96] J.W. Ekin, A.I. Braginski, A.J. Panson, M.A. Janocko, D.W. Capone II, N.J. Zaluzec, B. Flandermeyer, O.F. de Lima, M. Hong, J. Kwo, S.H. Liou: J. Appl. Phys. **62**, 4821 (1987)
[97] D. Dimos, P. Chaudhari, J. Mannhart, F.K. LeGoues: Phys. Rev. Lett. **61**, 219 (1988)
[98] J. Mannhart, M. Scheuermann, C.C. Tsuei, M.M. Oprysko, C.C. Chi, C.P. Umbach, R.H. Koch, C. Miller: Appl. Phys. Lett. **52**, 1271 (1988)
[99] D. Dimos, P. Chaudhari, J. Mannhart: Phys. Rev. B **41**, 4038 (1990)
[100] S.E. Russek, D.K. Lathrop, B.H. Moeckly, R.A. Buhrman, D.H. Shin, J. Silcox: Appl. Phys. Lett. **57**, 1155 (1989)
[101] A.M. Campbell: Physica C **162-164**, 273 (1989)
[102] A.A. Golubov, M.Yu. Kuprijanov Theoretical Investigation of SNINS and SNIS Josephson Junctions with Finite Transparency of NS Boundaries, Sov. Phys. JETP (1989)
[103] D. Dimos, D. Clarke: "The Role of Grain Boundaries and Interfaces on Superconductivity" in NATO-ASI Series "Surfaces and Interfaces in Ceramic Materials" 1989.
[104] V.G. Kogan: Phys. Rev. Lett. **62**, 3001 (1989)
[105] M.F. Chrisholm, D.A. Smith: Phil. Mag. **59**, 181 (1989)
[106] H.W. Zandbergen, W.T. Fu, L.J. de Jong, G. van Tendeloo: Cryogenics **30**, 628 (1990)
[107] Y. Gao, K.L. Merkle, G. Bai, H.L.M. Chang, D.J. Lam: Physica C **174**, 1 (1991)
[108] S.E. Babcock, D.C. Larbalestier: Appl. Phys. Lett. **55**, 393 (1989)
[109] T.S. Ravi, D.M. Hwang, R. Ramesh, S.W. Chan, L. Nazar, C.Y. Chen, A. Inam, T. Venkatesan: Phys. Rev.B **42**, 10141 (1990)
[110] N. Tomita, Y. Takahashi, Y. Ishida: Jpn. J. Appl. Phys. **29**, L30 (1990)

[111] S.E. Babcock, X.Y. Cai, D.L. Kaiser, D.C. Larbalestier: Nature 347, 167 (1990)
[112] V.N. Stepankin, E.A. Protasov, A.V. Kuznetsov, S.V. Zaitsev-Zotov: Pis'ma Zh. Eksp. Teor. Fiz. 41, 23 (1985)
[113] J. Mannhart, C.C. Tsuei: Z. Phys. B 77, 53 (1989)
[114] R.L. Peterson, J.W. Ekin: Phys. Rev. B 37, 9848 (1988); R.L. Peterson, J.W. Ekin: Physica C 157, 325 (1989)
[115] T. Matsushita, B. Ni, K. Yamafuji: Cryogenics 29, 384 (1989)
[116] J. Rhyner, G. Blatter: Phys. Rev. B 40, 829 (1989)
[117] J.R. Clem: Physica C 153-155, 50 (1988)
[118] C.S. Nichols, D.R. Clarke: "Critical Currents in Inhomogeneous Triangular Josephson Arrays: A Model for Polycrystalline Superconductors," preprint
[119] J. Mannhart, R.P. Huebener, F. Kober, D. Koelle, P. Chaudhari, D. Dimos, R. Gross, A. Gupta, G. Koren, C.C. Tsuei: Physica A 168, 345 (1990)
[120] D.E. Farrell, B.S. Chandrasekhar, M.R. DeGuire, M.M. Fangh, V.G. Kogan, J.R. Clem, D.K. Finnemore: Phys. Rev. B 36, 4025 (1987)
[121] J. D. Livingston, H.R. Hart Jr., W.P. Wolf: J. Appl. Phys. 64, 5806 (1988)
[122] K. Chen, B. Maheswaran, Y.P. Liu, B.C. Giessen, C. Chan, R.S. Markiewicz: Appl. Phys. Lett. 55, 289 (1989)
[123] J.W. Ekin, H.R. Hart Jr., A.R. Gaddipati: J. Appl. Phys. 68, 2285 (1990)
[124] F. Chen, B. Zhang, R.S. Markiewicz, B.C. Giessen: Appl. Phys. Lett., 58, 531 (1990)
[125] A.M. Campbell: Cryogenics 30, 809 (1990)
[126] C.C. Tsuei, J. Mannhart, D. Dimos; unpublished.
[127] R. R. Schulz, K.W. Blazey: "Resistivity, Critical Current and Modulated Microwave Absorption of Ag_2O Doped $YBa_2Cu_3O_{7-\delta}$ Ceramics" to be published in the Proceedings of the EMRS, Strassburg, 1991
[128] B. Dwir, M. Affronte, D. Pavuna: Appl. Phys. Lett. 55, 399 (1989)
[129] S.L. Shinde, T.M Shaw, "Considerations for Improved Polycrystalline Cuprate Superconductors," in Superconductivity and Ceramic Superconductors, ed. by K.M. Nair, and E.A Giess, (The American Ceramic Soc., 1990) 579
[130] S.L. Shinde, J. Morrill, D. Goland, D.A. Chance, T. McGuire: Phys. Rev. B 41, 8838 (1990)
[131] S. Jin, T.H. Tiefel, R.C. Sherwood, M.E. Davis, R.B. van Dover, G.W. Kammlot, R.A. Fastnacht, H.D. Keith: Appl. Phys. Lett. 52, 2074 (1988)
[132] M. Murakami, M. Morita, N. Koyama: Jpn. J. Appl. Phys. 28, L1125 (1989); M. Morita, M. Murakami, K. Miyamoto, K. Sawano, S. Matsuda: Physica C 162-164, 1217 (1989)
[133] T. Matsushita, E.S. Otabe, B. Ni, K. Kimura, M. Morita, M. Tanaka, M. Kimura, K. Miyamoto, K. Sawano: Jpn. J. Appl. Phys. 30, L342 (1991)
[134] P. McGinn, N. Zuh, W. Chen, S. Sengupta, T. Li: Physica C 176, 203 (1991)
[135] S. Jin, T.H. Tiefel, G.W. Kammlot: "The Effect of Y_2BaCuO_5 Inclusions on Flux Pinning in $YBa_2Cu_3O_{7-\delta}$," preprint, submitted to Appl. Phys. Lett.
[136] G. Kozlowski, C.E. Oberly: IEEE Trans. Mag. 27, 901 (1991)
[137] L.D. Woolf, W.A. Raggio, F.E. Elsner, M.V. Fisher, R.B. Stephens, T.L. Figueroa, C.H. Shearer, J.D. Rose, K.M. Schaubel, R.A. Olstad, T. Ohkawa, D.M. Duggan, M. DiMartino, R.L. Fagaly: Appl. Phys. Lett., 58, 534 (1991)

[138] K. Heine, J. Tenbrink, M. Thoener: Appl. Phys. Lett., 55, 2441 (1989)
[139] Y. Yamada, K. Jikahara, T. Hasebe, T. Yanagiya, S. Yasuhara, M. Ishihara, T. Asano, Y. Tanaka: Jpn. J. Appl. Phys. 29, L456 (1990)
[140] N. Enomoto, H. Kikuchi, N. Uno, H. Kumakura, T. Kazumasa, K. Watanabe: Jpn. J. Appl. Phys. 29 , L447 (1990)
[141] K. Sato, T. Hikata, Y. Iwasa: Appl. Phys. Lett. 57 , 1928 (1990)
[142] S. Jin, R.B. van Dover, T.H. Tiefel, J.E. Graebner, N.D. Spencer: Appl. Phys. Lett. **58**, 868 (1991)
[143] H. Sekine, K. Itoh, K. Inoue, H. Maeda, K. Numata: Appl. Phys. Lett. 55, 2772 (1989)
[144] J. Kase, N. Irisawa, T. Morimoto, K. Togano, H. Kumakura, D.R. Dietderich, H. Maeda: Appl. Phys. Lett. 56, 970 (1990)
[145] S. Jin, T.H. Tiefel, R.B. Van Dover, J.E. Graebner, T. Siegrist, G.W. Kammlot, R.A. Fastnacht: "Field Dependence of Critical Currents in Polycrystalline Bi-Pb-Sr-Ca-Cu-O Superconductors," preprint, submitted to Appl. Phys. Lett.
[146] H.W. Neumueller, G. Ries, J. Bock, E. Preisler: Cryogenics **30**, 639 (1990)
[147] R. Flükiger, T. Graf, M. Decroux, C. Groth, Y. Yamada: IEEE Trans. Mag. 27, 1258 (1991)
[148] J. Tenbrink, M. Wilhelm, K. Heine, H. Krauth: IEEE Trans. Mag. 27, 1239 (1991)

Anisotropy of Flux Lattice Structures in the Low Field Mixed State of the High-T_c Superconductor BSCCO (2212)

P.L. Gammel

AT&T Bell Laboratories, 600 Mountain Ave.,
Murray Hill, NJ 07974, USA

Abstract. High resolution Bitter patterns have been analyzed to reveal the static structure of the flux lattice in the low field mixed state of BSCCO (2212). For field cooled experiments with $B \| \hat{c}$, the in-plane effective mass anisotropy distorts the lattice. Measured distortions in the range of 5-10% are well below that expected from resistance anisotropy and observed in YBCO (123) decorations. At low fields, the pattern is otherwise amorphous. Increasing the field above ~20G results in a sudden transition into an hexatic phase. The signature of the hexatic is long range orientational order, with the positional order remaining short ranged. The crossover field, lowered to ~8G upon annealing the sample for 24 hours in 1 atm O_2, is, by inference, strongly correlated with pinning strength. Rotating the applied field away from the $\hat{c}$ axis results in a dramatic new structure-a periodic array of vortex chains with an intervening hexagonal lattice. In extremely anisotropic systems an attractive vortex-vortex interaction perpendicular to the rotation axis stabilizes a chain structure.

1. Mixed State Structures

Almost coincident with the discovery of the oxide superconductors was the realization of their extreme $\hat{c}$-axis anisotropy. Less dramatic, but significant, anisotropy is now known to exist in the $\hat{a}-\hat{b}$ plane as well. Whether or not this provides the elusive clue to understanding the microscopics of superconductivity in these compounds, the consequences of anisotropy are both manifest and numerous. The mixed state of an anisotropic type-II superconductor, permeated by a dense array of quantized vortices, has generally been studied with macroscopically averaged probes such as transport or magnetization. Here, we present direct images of the low field vortex lattice, and discuss the consequences of anisotropy.

Theories of the mixed state have evolved along two rather disjoint paths. In one instance, Ginzburg-Laundau (G-L) theory has been extended to include both anisotropies and non localities[1]. Simplified anisotropic London theory leads to predictions of interesting distortions and orientational effects of the lattice[2], both with field normal to the planes and at other angles are nontrivial and compared with experiments here. The dynamics of vortices, which lead to finite dissipation in the mixed state have also been the subject of intense debate. The short coherence lengths, high temperatures and large anisotropy make thermal fluctuations important. These fluctuations compete with pinning and vortex-vortex interactions.

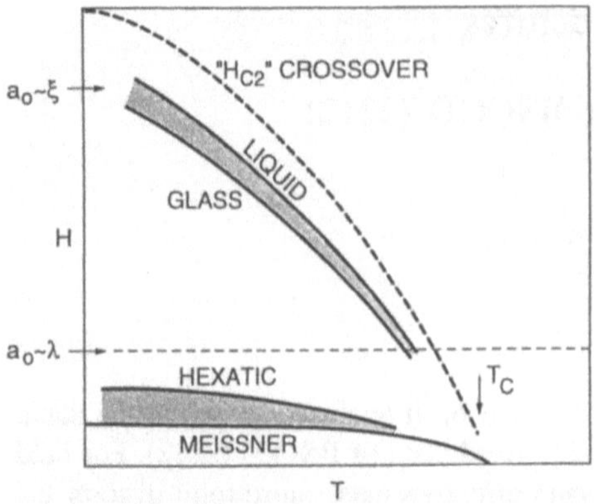

Fig. 1. Phase diagram of the mixed state of the high T_c superconductors. The solid lines, which represent phase boundaries, are not continued into regions where there is no experimental data. H_{c2} is no longer a phase transition in the presence of strong fluctuations. The vortex lattice constant is a_o.

While it has long been known that d=4 is the lower critical dimension for the vortex lattice in the presense of disorder[3], new thermodynamic phases have been proposed in d=2 and d=3[4]. Possible transitions separating the phases, similar to melting of the vortex lattice in the clean limit are under investigation[5]. A type of generic possible phase diagram is shown in fig. 1. Images as a function of field presented here show a striking and prominent order-disorder transition[6]. The ordered phase is orientational only-a candidate hexatic, at intermediate fields ~20G, but develops essentially crystalline order at 100G. This probe shows the low field part of the phase diagram, and its sensitive dependence on pinning.

2. The Bitter Pattern Technique

The experiments discussed here consist entirely of images extracted from high resolution Bitter patterns on BSCCO (2212) crystals prepared by directional solidification as described in other work[7]. The superconducting transition is sharp and the pinning energy low. In our implementation of this technique a freshly cleaved sample is cooled to 4.2K in a fixed applied field. The cleaved surface is clean and smooth in SEM images. Some studies have indicated that the Bi double layer serves as the atomic cleavage location[8]. A ferromagnetic material is then evaporated onto the sample. Typically 0.2mg Fe produced good pictures with an 2cm sample to boat spacing. A background pressure (0.17mbar) of He gas causes the formation of ~50Å thermalized particles. Particles which drift to the surface preferentially decorate the regions of high magnetic field. The sample is warmed to room temperature and the Fe particles, held immobile by van der Waals forces, are imaged in an SEM. A picture of this type is shown in fig. 4. The positions of the vortices are determined from the center of mass of the piles of particles in a digitized image. Regions of digitization were typically limited to 40x40 vortices due to the

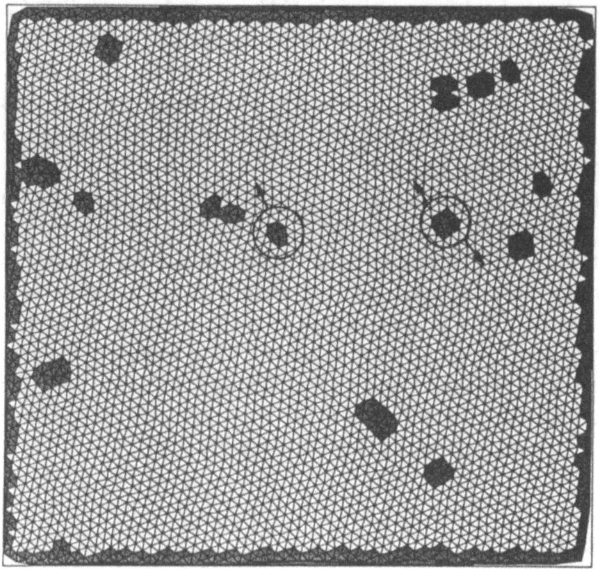

Fig. 2. Delaunay triangulation of field cooled decoration at 23G. Non sixfold coordinated sites are shaded. The arrows indicate the Burgers vectors for free (a) and paired (b) dislocations. For this picture both the orientational and translational correlation lengths are determined by the density of free dislocations.

number of pixels available and the requirements that the vortex locations be ~10 pixels apart.

Of paramount concern is the effective temperature represented by the decoration. Consider the evolution of a quenched sample. At some time, the sample passes through T_c, and the vortices appear. Then, at a slightly lower temperature T_{irr}, the long wavelength modes of the vortex lattice fall out of equilibrium, and the magnetization shows the irreversibility transition. At some lower temperature T_f, the topology of the vortex lattice becomes fixed as the short wavelength modes also fall out of equilibrium. T_f, which is unknown as yet, is the effective temperature imaged by the decoration. It has generally be assumed that $T_f \sim T_{irr} > 0.95 T_c$ for the field range under study[9]. If this is true, then the penetration depth where the picture is frozen in may be $\lambda_f \sim 3\lambda(0) \sim 0.5 \mu m$ using the two fluid form for the temperature dependence of λ. This is crucial in understanding, for example, the role of nonlocal interactions.

The analysis of the digitized images begins with the unique Delaunay triangulation[10], an example of which is shown in fig. 2. Each vortex core is represented by a vertex, with bonds drawn to all nearest neighbor vortices. Triangles boardering non six-fold coordinated topological defects such as disclinations and edge dislocations are shaded. These defects are responsible for both the positional and orientational correlation lengths, as discussed below. The orientational order is sufficient in these pictures to determine the positions of the first reciprocal lattice vectors to within 5%. The anisotropy in the magnitudes $|\vec{G}|$ ranges from 3% to 9%.

Effective mass anisotropy, determined from resistivity[11], would have predicted a much more distorted lattice, approximately 20%. In YBCO, anisotropy determined from decoration and effective mass agree well[12]. The difference here is unclear. One possibility is that stacking faults cause $\hat{a}$ and $\hat{b}$ to alternate through the thickness of the crystal, yielding a lower averaged anisotropy. Resistivity would only probe the surface layer, due to the large $\hat{c}$ anisotropy.

Equally important is the observation that the reciprocal lattice vectors show no alignment either with the crystal axes, determined by X-ray analysis, or with sample topography. Within G-L theory, all orientations of the vortex lattice are degenerate for this case[2]. In previous observations, however, observed orientations were found to be pinned to the crystal[13]. While, in general, it is believed that the effects of the crystal lattice become less important as κ increases, these represent the first case where the orientation of a vortex lattice arises spontaneously. In such a case, the broken symmetry is unambiguous evidence of a phase transition, which we have identified as the boundary of the amorphous and hexatic glass phases[6].

3. Field Dependence for $B \| \hat{c}$

As the magnitude of the applied field is changed, there is a dramatic change in the density of defects observed in the vortex lattice[6]. The fraction of disclination sites f_d varies from 0.26 at 8G to .07 at 11G to 0.00 at 69G. Our experiments only probed fields up to 100G. Analysis begins with the radially averaged Fourier transform. Apart from the anisotropy in the nearest neighbor spacing described above important information is contained in the limit $k \to 0$. The power spectral density $S(k) \sim k/n^{1/2}$, where n is the vortex density. This implies the spectrum is representative of the vortex lattice in the bulk. While there is some remain debate about the proportionality constant, the data are certainly within a factor of 10 of expectation[14]. It had been suggested that surface interactions should dominate, since the vortices emerging from the crystal surface can be mapped to a charge $1/\alpha = 137.04$ Coulomb gas[15]. In this case, $S(k) \sim k^2$, with a proportionality 10^4 different from observation.

To elucidate the nature of the long range order, we have calculated the bond orientational and translational correlation lengths. The translational correlation function $g_G(r)$, is calculated from the order parameter $\psi_G(r_j) = \exp(iG \cdot r_j)$ for each site j, where G is a first reciprocal lattice vector. We have fit the correlation functions to a simple exponential decay $g \sim \exp(-r/\xi)$. The correlation lengths range from $1a_o$ to $20a_o$ as a function of field as shown in fig. 3. Our imaging system was accurate for correlation lengths up to $50a_o$. The translational order increases with magnetic field with no sign of saturation. The measured translational correlation length is in all cases quite close to the intrinsic free dislocation density average separation. Hence, our correlations are determined by this concentration.

The six fold bond orientational order parameter $\psi_6(r) = n_j^{-1} \Sigma \exp(i6\theta_{jk})$ is summed over the set n_j of nearest neighbors with θ the bond angle. The correlation function for this order parameter $g_6(r)$ is fit to an algebraic form $g_6(r) \sim r^{-\eta}$. These fits are also shown in fig. 3. Simulations[16] in systems of 2D arrays of hard spheres

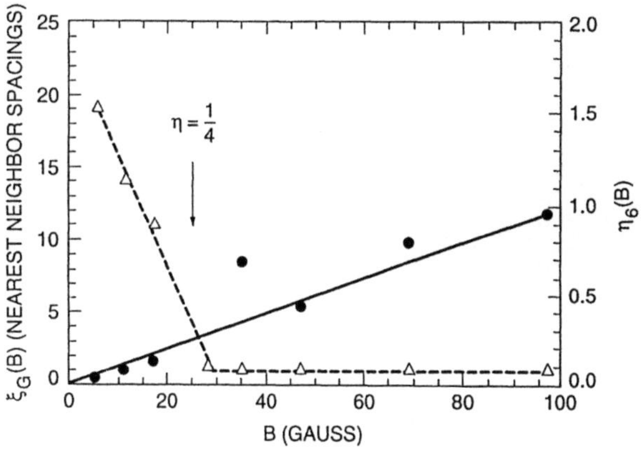

Fig. 3. Translational and orientational correlation lengths vs. applied field. The location of the Hexatic transition $\eta \sim 1/4$ and a fit to the positional correlation length using local elastic theory are included.

with trapped random dislocations find $\eta = 9c/\pi$ where c is the fraction of dislocation cores. If, in our case, we take c the fractional area of Voronoi polyhedra in which there is a net Burgers vector this gives values close to the observed results. Dislocations are responsible for both the short range positional order as well as the quasi long range orientational order. Theories of the vortex lattice involving elastic properties explicitly do not include dislocations and are irrelevant for the description of these data.

The phase with long range orientational order has been associated with an Hexatic. Nelson and Halperin's theory[17] for 2D hexatics with thermal disorder indicate that $\eta \sim 1/4$ at the hexatic to fluid phase boundary, about 20G for the as made samples, indicated by the arrow in fig. 3. In addition, the Fourier transform lineshape should go from Gaussian to Lorentzian. Our lineshapes are not sufficient to resolve this difference, as least in part due to in plane anisotropy. The hexatic phase described here has been variously interpreted as a vestige of a high temperature hexatic liquid or the true low temperature ground state produced by the competition between vortex interactions and disorder.

4. Angle Dependence: the Vortex Chain State

Decoration experiments have also been performed[18] with the applied field at an angle with respect to $\hat{c}$, with the geometry of the inset to fig. 4. The samples were mounted at fixed angle and field cooled, with all decorations at 4K. In this case, the experiments probe the large anisotropy direction, so no effort was made to align the $\hat{a}$ axis. The first result arises from the average density n of vortices on the cleaved face of the crystal. For all angles examined up to 85° the density is $n = (B/\Phi_o)\cos\theta$. The vortex lattice is induced by the component of B parallel to $\hat{c}$. This had been previously been inferred by Kes on indirect grounds[19].

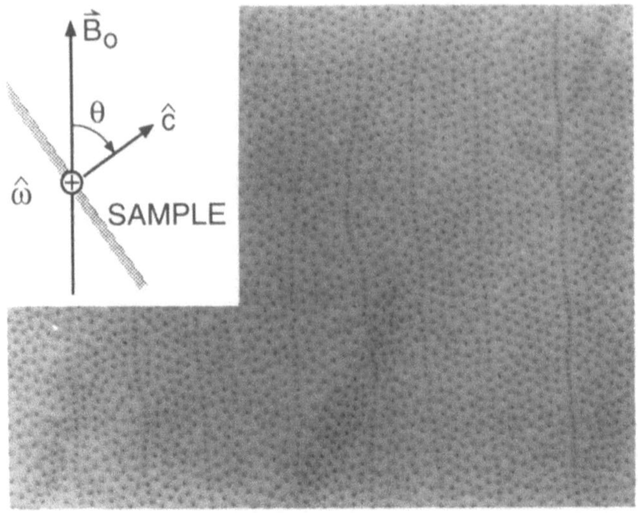

Fig. 4. Vortex chain state observed for $\theta=60°$. The applied field is 35G. Shown in the inset is the sample geometry. The chains lie in the plane defined by $\vec{B}$ and $\hat{c}$. The Abrikosov lattice between the chains is oriented with one lattice vector parallel to the chain.

The analysis of the structure is best divided into to regimes, $\theta<60°$ and $\theta>60°$. For the smaller angles, the vortex lattice is isotropic, within the 5-10% distortions discussed previously. In addition, there remains no apparent preferred orientation, either with respect to the crystallographic axes or the tilt axis. Interestingly enough, this measured isotropy already implies extreme anisotropy in the effective mass. To clarify this point, consider the case of an isotropic system. Absent surface effects, an hexagonal vortex lattice forms perpendicular to the applied field. The decorated surface is at an angle θ to this lattice. Hence, the pattern on the surface perpendicular to $\hat{c}$ will be distorted, with a distortion $\rho=a/b=1/\cos\theta$, simply due to projection. In the limit $\Gamma\rightarrow\infty$, the distortion predicted by G-L theory[2] is likewise $1/\cos\theta$. This exactly cancels the projection term, and the observed vortex lattice is isotropic. Assuming a maximum distortion of 5% at 60°, the effective mass ratio is $\Gamma^2>8$. While this is certainly consistent with other measurements, it is insensitive to Γ when the anisotropy is large.

For angles $\theta>60°$, a dramatic new structure is observed as pictured in fig. 4. The striking chains of vortices lie in the plane defined by $\vec{B}$ and $\hat{c}$. The chains have an increased density with respect to the background lattice as discussed below. The chains also orient the intervening vortex lattice so that one of the lattice vectors lies parallel to the chain axis. This is independent of the orientation of the $\hat{a}$ axis. This is the preferred orientation predicted by anisotropic G-L theory, although without mention of the prominent chain structure. Note, however, that this orientation was not defined prior to the chain formation. It has been widely reported that neutron scattering experiments actually observe a vortex lattice rotated 90° from this. This report is based on experiments on Tc,YBCO and UPt_3[20]. In all cases, the field orientation corresponded to $\theta=90°$, in which case the two orientations are degenerate within G-L theory. This experiment, therefore, is first true test of this prediction.

In the plane of the photograph, take the spacing along the chain to be D and that between chains to be C. Although C has large variations, as can be seen from the picture, both D and C are found to scale as $B^{-1/2}$ at fixed angle. Since this is the same scaling as the vortex lattice constant, the picture at fixed angle is field independent in the sense that the field only sets the overall magnification. It is also found that for all angles and fields $DC=\Phi_0/B$. Viewing the chains as a superlattice modulation, this is equivalent to one extra flux quantum per superlattice unit cell. To accommodate this extra vortex, the chains acquire a necklace of dislocation pairs with zero net Burger's vector, as determined from triangulation.

As the angle increases, the number of vortices between chains decreases. Due to the large variation in C, the exact form of this variation is difficult to pin down. One form, generated by considering what distortions of an isotropic lattice can produce chains, which agrees with the data is $D=0.75[\Phi_0/B\cos(\theta)]^{-1/2}$. Experimentally, the number of vortex lattice constants between chains varied from 11 at 60° to 2 at 85°.

Chain structures were anticipated theoretically, although not exactly matching our results[21]. Anisotropic G-L calculations near H_{c1} showed the existence of an attractive well in the vortex-vortex interaction along the chain direction. This is a remnant of the simple observation that two dipoles attract when their axes are both parallel and along he line connecting them. Further estimates showed the state near H_{c1} to consist entirely of chains. The spacing along the chain was determined to be $\sim\Lambda=(\lambda_a\lambda_b\lambda_c)^{1/3}\sim0.8\mu m$, and field independent. Recently, the finite thickness of the sample has been included[22]. The lowest energy state for intermediate sample thicknesses appropriate to this experiment was modeled with two types of vortices; chain vortices parallel to B and a hexagonal lattice parallel to $\hat{c}$. While the ratios of vortices for the lowest energy seemed to match our results, there was again no field dependence. It may be that λ appropriate to our pictures is large, as the state is frozen in near T_c. Further simulations exploring this, and other, possibilities are in progress. The role of nonlocality is also being investigated.

5. Conclusions

The vortex lattice which evolves in the mixed state of an anisotropic superconductor embraces several unusual properties. Distortions due to effect mass were calculated on the basis of G-L theory, and generally agree with observation, although the exact magnitude in 2212 is not fully understood. This includes distortions for the field at an angle from the $\hat{c}$ axis. Orientations, studied for the first time in the angle dependence of this work, also seem to follow G-L theory in rough outline. BSCCO is the first example where crystal symmetry does not seem to effect the vortex lattice orientation. For large angle tilt, the newly described vortex chain state appears. Again, the jist of the attractive interaction which stabilizes the structure are derived from an anisotropic G-L theory.

The order, both orientational and translational, is determined by the free density of dislocations. This points to the flaws inherent in using /elastic theory in describing the vortex lattice. The existence of an amorphous to Hexatic transition as the vortex density is increased is probably favored by the extreme anisotropy in this system, although the details are not yet worked out.

Clearly, anisotropies play a major role in both the static and dynamic response of the mixed state. Understanding them is essential to characterization and control of the high-T_c superconductors.

References

[1] V.G. Kogan, *Phys. Rev.* **B24,** 1572 (1981). A. Houghton, R.A. Pelcovits and A. Sudbo, preprint.

[2] L.J. Campbell, M.M. Doria and V.G. Kogan, *Phys. Rev.* **B38,** 2439 (1988).

[3] A.I. Larkin and Yu.N. Ovchinnikov, *J. Low Temp. Phys.* **34,** 409 (1979).

[4] M.P.A. Fisher, *Phys. Rev. Lett.* **62,** 1415 (1989). D.S. Fisher, M.P.A. Fisher and D.A. Huse, *Phys. Rev.* **B43,** 130 (1991). D.S. Fisher, *Phys. Rev.* **B22,** 1190 (1980).

[5] P.L. Gammel, L.F. Schneemeyer and D.J. Bishop, *Phys. Rev. Lett.* **66,** 953 (1991). R.H. Koch, V. Foglietti and M.P.A. Fisher, *Phys. Rev. Lett.* **64,** 2586 (1990).

[6] D.G. Grier, C.A. Murray, C.A. Bolle, P.L. Gammel and D.J. Bishop, *Phys. Rev. Lett.* **66,** 2270 (1991).

[7] D.B. Mitzi, L.W. Lombardo, A. Kapitulnik, S.S. Laderman and R.D. Jacowitz, *Phys. Rev.* **B41,** 6564 (1990).

[8] C.H. Chen, private communication. M.D. Kirk, J.Nogami, A.A. Baski, D.B. Mitzi, A. Kapitulnik, T.H. Geballe and C.F. Quate, *Science* **242,** 1673 (1988).

[9] L. Lombardo, M. Ng, J.S. Urbach, D.B. Mitzi and A. Kapitulnik, preprint

[10] F.F. Preparata and M.L. Shamos, **Computational Geometry, an Introduction,** (Springer-Verlag, New York, 1985). S. Fortune, *Algorithmica,* **2,** 153 (1987).

[11] S. Martin, A.T. Fiory, R.M. Fleming, L.F. Schneemeyer and J.V. Waszczak, *Phys. Rev. Lett.* **60,** 2194 (1989).

[12] G.J. Dolan, F. Holtzberg, C. Field and T.R. Dinger, *Phys. Rev. Lett.* **62,** 2184 (1989). L.Ya. Yinnikov, I.V. Grigoriera, L.A. Gurevich and Yu.A. Ossipyan in **High Temperature Superconductivity from Russia,** (World Scientific, London, 1989), eds. A.I. Larkin and N.V. Zavaritsky.

[13] R.P. Huebner, **Magnetic Flux Structures in Superconductors** (Springer-Verlag, New York, 1979). See also ref. 12.

[14] D.R. Nelson and D.S. Fisher, private communication.

[15] D.A. Huse, preprint.

[16] D.R. Nelson, M. Rubinstein and F. Spaepen, *Phil. Mag.* **A46,** 105 (1982).

[17] B.I. Halperin and D.R. Nelson, *Phys. Rev. Lett.* **41,** 121 (1978). D.R. Nelson and B.I. Halperin, *Phys. Rev.* **B19,** 2457 (1979).

[18] C.A. Bolle, P.L. Gammel, D.G. Grier, C.A. Murray, D.J. Bishop, D.B. Mitzi and A. Kapitulnik, *Phys. Rev. Lett.* **66,** 112 (1991).

[19] P.H. Kes, J. Aarts, V.M. Vinokur and C.J. van der Beek, *Phys. Rev. Lett.* **64,** 1063 (1990).

[20] For UPt_3, R.N. Kleiman, private communication. For YBCO, E.M. Forgan, private communication. For Tc, J. Schelten, H. Ullmaier, W. Schmatz, *Phys. Status Solidi* **B48,** 619 (1971). J. Schelten, G. Lippmann and H. Ullmaier, *J. Low Temp. Phys.* **14,** 213 (1974).
[21] A.I. Buzdin and A.Yu. Simonov, *JETP Lett.* **51,** 191 (1990). V.G. Kogan, N. Nakagawa and S.L. Thieman, *Phys. Rev.* **B42,** 2631 (1990).
[22] L. Daemon and L.J. Campbell, private communication.

Some Implications of a Short Coherence Length in High-Temperature Superconductors

P. Chaudhari

IBM T.J. Watson Research Center, P.O. Box 218, Yorktown Heights, NY 10598, USA

Abstract. The copper oxide based layered superconductors have both anisotropic and short coherence lengths. These have broad implications in determining a variety of physical properties such as critical currents or in the construction of devices such as Josephson junctions. In this talk a brief review of the role of short and anisotropic coherence lengths is presented with particular emphasis on their influence on the properties of grain boundaries.

There are two coherence lengths used in superconductivity. There is the Pippard coherence length, ξ_0, which was introduced by him to take into account the nonlocal electrodynamic character of superconductivity. Using the uncertainty principle, Pippard showed that

$$\xi_0 \approx \hbar V_f / kT_c \tag{1}$$

where $\hbar$ is the Planck's constant, V_f the Fermi velocity, k the Boltzmann constant, and T_c the critical temperature. For the high temperature superconductors, ξ_0 is of the order of 15-30Å in the plane of the copper oxide layers.

The second coherence length, $\xi(T)$, was introduced by Ginzburg and Landau in their phenomenological theory of superconductivity. This length is a measure of the variation of the wavefunction $\psi(r)$. For example, if a superconductor is brought into contact with another material, the Ginzburg-Landau, G-L, wavefunction is modified at the interface and recovers its value in the bulk over a distance $\xi(T)$. The G-L theory is valid near T_c and under conditions where the order parameter does not vary too rapidly.

The two coherence lengths are related, and for a clean metal $\ell > \xi_0$, where ℓ is the mean free path for elastic scattering,

$$\xi(T) = 0.74\xi_0(1-t)^{-1/2} \tag{2}$$

and for $\ell < \xi_0$

$$\xi(T) = 0.855(\xi_0\ell)^{1/2}(1-t)^{-1/2} \tag{3}$$

where $t \equiv T/T_c$.

The value of $\xi(T)$ can be determined directly from the upper critical field, H_{c2}, of a bulk superconductor

$$H_{c2} = \Phi_0/2\pi\xi^2(T) \tag{3}$$

where Φ_0 is the flux quantum ($\Phi_0 = 2.07 \times 10^{-7}$gauss-cm^2). Measurements of H_{c2} in different directions provide $\xi(T)$ along the principal axes. In the high temperature superconductors $\xi(T)$ in the copper oxide plane is about an order of magnitude larger than perpendicular to it. Hence $\xi_a(T)\approx\xi_b(T)\approx10\xi_c(T)\approx$15-30Å when extrapolated to 0K, where a, b, and c are the principle axes. The coherence length in the c-direction is less than the dimensions of the unit cell of the high temperature superconductors. It is in fact comparable to the thickness of copper oxide layers. As we discuss below, this introduces a richness to the possible structure of vortices and their interaction.

At temperatures close to T_c, $\xi(T) > \xi_0$ and $\xi(T) > a_c$ where a_c is the unit cell dimension along the c-axis; in this temperature range, close to T_c, the anisotropic Ginzburg-Landau expressions can be used. However, as the temperature is reduced, the coherence length approaches the thickness of the $Cu\text{-}O_2$ layers. As these conducting layers in the high temperature superconductors are separated by insulating layers, the behavior of the vortex can be quite complex.[1,2] For example, if the magnetic field is applied along the b-axis, the core of the vortex is no longer that of an Abrikosov vortex but rather that of a Josephson vortex. When the magnetic field is along the c-axis, the structure of the vortex may be described by an array of pancake vortices connected by Josephson strings.[2] Hence we see that the short and anisotropic coherence lengths can have a dramatic influence on the structure of a vortex. I shall not discuss this further here for it will presumably be the subject matter of the session on anisotropy.

Deutscher and Müller[3] were amongst the first to point out that the very short coherence length of the high temperature superconductors would drastically modify the value of the order parameter at a superconductor-non superconductor interface. They used de Gennes[4] expression for the boundary condition

$$\left(\frac{\hbar}{i}\nabla - \frac{2e}{c}A\right)\psi\Big|_{x=0} = \frac{i}{b}\psi \tag{4}$$

where A is the vector potential and b a real constant, whose value is given by

$$1/b = \frac{2}{L}\int_{-\alpha}^{+\alpha}\frac{\Delta(x)}{\Delta_0}\left(1 - \frac{N(x)}{N(0)}\right)dx \tag{5}$$

where $L\simeq\xi_0^2$, Δ_0 and N(O) are the bulk value of the order parameter and the normal density of states, respectively. At a superconductor-insulator interface de Gennes showed that $b\approx\xi_0^2/a$, where a is a distance from the interface of the order of the lattice parameter. If $\xi_0\approx$3Å, b is small and $\Delta(x=0)/\Delta_0$ approaches $b/\xi(T)\approx\xi_0/\xi(T)$. The order parameter at the interface is then severely depressed as one approaches T_c.[3] However, for the a- and b-axis the situation is markedly improved (except very close to T_c) because the coherence length is an order of magnitude larger. It is for this reason that a number of investigators have chosen to develop tunnel junctions in which the insulators are perpendicular to the copper oxide layers. However, even for this geometry there can be difficulties as we discuss next.

The first experimental demonstration of high critical currents in the high temperature superconductors was based on the assumption that extrinsic factors, such as grain boundaries, limit current densities.[5] Since that experiment, measurements on single grain boundaries have provided direct evidence that indeed the boundaries are weak links.[6] When the misorientation angle between the boundaries is small, the weak link behavior is dominated by flux flow. However, above a critical misorientation angle, in the neighborhood of 10° in the $Y_1Ba_2Cu_3O_{7-\delta}$ superconductor, the behavior changes from flux flow to Josephson coupling. Experimental evidence also shows that the value of the gap, as deduced from critical currents, is suppressed at the boundary and the magnitude of the depression correlates (scales) with the critical current density[7] or junction resistance.[8] Let us see how far we can carry the argument of a short coherence length to explain these two sets of data.

We shall assume, and justify later, that the grain boundary can be described as a sheath of material in which the order parameter is depressed and that the extent of depression is a function of the dislocation density. The larger the misorientation between grains, the higher the dislocation density and greater the depression of the order parameter. The two bulk superconductors have their order parameter depressed at the interfaces between the sheath and the superconductors. The sheath, in our model, is a dirty metal.

The critical current density, at low temperatures, for a Josephson junction is given by the Ambegoakar and Baratoff equation

$$J_{c,j} = \frac{\pi \Delta_j}{2e\rho_n \ell_j} \tanh \frac{\Delta_j}{2kT} \approx \frac{\pi \Delta_j}{2e\rho_n \ell_j} \tag{6}$$

where Δ_j is the order parameter at the interface, ρ_n and ℓ_j are the resistivity and width of the sheath, respectively. If the sheath has a finite order parameter, Δ_b, the critical current it supports is

$$J_{c,b} = en_s v_s = en_s \Delta_b / \hbar k_f \tag{7}$$

where Δ_b, n_s, k_f are the order parameter, density of superconducting electrons, and the Fermi wave vector, respectively, of the sheath.

Josephson coupling dominates the critical current when

$$\frac{\pi \Delta_j}{2e\rho_n \ell_j} > \frac{en_s \Delta_b}{\hbar k_f} \quad . \tag{8}$$

Using the free electron theory, this simplifies to

$$\frac{\pi\Delta_j}{\ell_j} > \frac{\Delta_b}{\ell_b} \tag{9}$$

where ℓ_b is the mean free path of the carriers in the sheath. It is not unreasonable to assume that $\ell_j \simeq \pi\ell_b$ for low angle boundaries. The value of ℓ_j is of the order of a few unit cells and ℓ_b at the boundary is determined by dislocation or impurity scattering. These distances are also likely to be of the order of a few unit cells when the inequality favoring Josephson coupling sets in. Flux flow is observed when $\pi\Delta_j/\ell_j \lesssim \Delta_b/\ell_b$ and Josephson coupling if the inequality is reversed. Starting with a low angle boundary, when the dislocation placing is large compared to the coherence length and the mean free path, we expect the boundary to have a value of Δ_b essentially that of bulk. As the misorientation angle is increased, both Δ_j and Δ_b decrease but Δ_b decreases faster than Δ_j and eventually Josephson coupling dominates the critical current.

On the basis of the G-L theory, de Gennes[4] has predicted that the critical current across a metallic boundary produces a $(1 - t)^2$ dependence near T_c where $t=T/T_c$. In contrast, an insulating junction produces a (1 - t) dependence except very close to T_c.[3] Our assumption that the sheath is metallic is based, in part, on the experimental observations that at temperature close to T_c, ($\Delta T \sim 10K$) the critical current across larger angle boundaries in $Y_1Ba_2Cu_3O_{7-\delta}$ bicrystals shows a $(1 - t)^2$ dependence.[10] The element of uncertainty in our conclusion comes from the observation that in these short coherence length materials the range of temperatures over which an insulating interface might show a $(1 - t)^2$ dependence can be quite extended.[3,9] Although values of ξ_a used to fit the temperature dependence of the critical current are somewhat smaller than generally accepted they are nevertheless not out of the range of values quoted in the literature. Hence we need additional evidence to convince us that the grain boundary is metallic rather than an insulator. This comes from the dependence of the I_cR_n product on the resistance of the junction, as we discuss next.

The I_cR_n product, where I_c is the critical current and R_n the resistance of the junction, has been found to be in proportion to $J_c^{0.6}$ for a very large number of bicrystal grain boundary samples.[7] A similar relation, where I_cR_n is proportional to R_n, has been observed when the samples are annealed under varying conditions of ambient gases such as O_2 or Ar.[8] These observations can be explained by using the preceding arguments. It can be shown that the gap at the interface between a superconductor and a metal is given by[11]

$$\Delta_j \propto \xi_n^2 / \ell_j \quad . \tag{10}$$

In the dirty limit $\xi_n^2 \propto \rho_n^{-1}$ and at low temperatures, the I_cR_n product is proportional to Δ_j. Substituting for Δ_j and ξ_n^2 we obtain

$$I_cR_n \propto (\rho_n \ell_j)^{-1} \tag{11}$$

which leads immediately to the relation

$$I_cR_n \propto (Jc)^{0.5} \tag{12}$$

We note this is close to the experimental value. If the cross-section of the junction is constant, then $I_cR_n \propto R_n^{-1}$, as observed in the annealing experiments.

Equations (10-12) are only valid at low temperatures where $l_j < \xi_n$. In the high temperature superconductors this is unlikely to be a good approximation at temperature close to T_c, the domain where the G-L approximation becomes increasingly accurate. The value of ξ_n is given by $\xi_n = (\hbar D_n/2\pi kT)^{1/2}$ where D_n is the diffusivity of the carriers in the metallic sheath. As $D_n \alpha V_f \ell_b$, ξ_n will decrease with increasing temperatures both due to a reduction in ℓ_b and an increase in T. Now $\xi_n < \xi_0$ both because $V_{fn} < V_{fs}$ and $\ell \lesssim \ell_j$. The value of ξ_0 is of the order of 15-30Å. Hence ξ_n<15-30Å. At low temperatures, this inequality can change, for a reduction in temperature increase ξ_n. The width of the junction, ℓ_j, cannot be reduced below 1.5a for geometrical reasons. In fact, recent noise measurements across grain boundary junctions suggest that a part of the grain boundary resistance is connected with oxygen deficiency[12] in the $Y_1Ba_2Cu_3O_{7-\delta}$ superconductor. This would imply that the effective width of the boundary may be larger than the geometrical width. All of these reasons would suggest that as $T \rightarrow T_c$, $\ell_j/\xi_n > 1$. Following de Gennes[4] the I_cR_n product can be written as follows

$$I_cR_n \propto (N_s / N_n)^2 (\xi_o / \ell_j)^2 (T_c^3 / T^2) (1 - t)^2 \tag{13}$$

and

$$I_cR_n \propto (N_s / N_n)^2 (\xi_o / \ell_j)^2 (T_c^2 / T) (\alpha^3 \exp - \alpha) (1 - t)^2 \quad . \tag{14}$$

Equation (13) is valid when $\alpha \equiv \ell_j / \xi_n << 1$ and Equation (14) when $\alpha >> 1$. Both equations predict that the I_cR_n product decreases a ξ_0^2. Hence a short coherence has a deleterious effect on the I_cR_n product. Secondly, for α>3, Eq. (14) predicts that the I_cR_n product is a function of α: decreasing with increasing α. As α is proportional to $\rho_n^{1/2}$, we find, as in the low temperature regime (Eq. 10-12) that the I_cR_n product will scale with resistivity or critical current density. This is in agreement with experimental observations.

In summary we note that the order parameter in the high temperature superconductors is depressed at the grain boundaries. The depression is large relative to the expected value of the order parameter in the bulk. The depressed value and many of the experimental features of grain boundary junctions can be explained by assuming that the boundary is metallic with a very short mean free path and short coherence length.

References

1. S. Tachiki and M. Takahashi, Solid State Comm. **70**, 291 (1989).
2. J.R. Clem, Phys. Rev. B **43**, 7837 (1991) and references therein.
3. G. Deutscher and K.A. Müller, Phys. Rev. Lett. **59**, 1745 (1987).
4. P.G. de Gennes, Superconductivity of Metals and Alloys, Benjamin, New York, 1966; p. 220ff; see also Rev. Mod. Physics **36**, 225 (1964).
5. P. Chaudhari, R.H. Koch, R.B. Laibowitz, T.R. McGuire, and R.J. Gambino, Phys. Rev. Lett. **58**, 2684 (1987).
6. D. Dimos, P. Chaudhari, and J. Mannhart, Phys. Rev. B **41**, 4038 (1990).
7. R. Gross, P. Chaudhari, M. Kawasaki, and A. Gupta, Phys. Rev. B **42**, 10735 (1990).
8. S.E. Russek, D.K. Lathrop, R.A. Buhrmann, and D.H. Shin, Appl. Phys. Lett., **57** 1155 (1990).
9. G. Deutscher, IBM J. Res. Develop. **33**, 293 (1989).
10. R Gross, P. Chaudhari, D. Dimos, A. Gupta, and G. Koren, Phys. Rev. Lett. **64**, 228 (1990).
11. G. Deutscher and P. Chaudhari, Phys. Rev. B (1991), in press.
12. M. Kawasaki, P. Chaudhari, and A. Gupta, submitted for publication.

Flux Quantisation in Superconducting Rings and the Mixed State

C.E. Gough

Superconductivity Research Group, University of Birmingham,
Birmingham B15 2TT, UK

Abstract. This lecture introduces the concept of the macroscopic superconducting wave function and describes how this accounts for the Meissner Effect and flux quantisation. The experimental confirmation of flux quantisation in LTC, HTC and hybrid LTC/HTC rings will be briefly discussed. Flux quantisation in the superconducting mixed state leads to the formation of the Abrikosov flux line lattice. Experiments confirming the flux line lattice in HTC superconductors by magnetic decoration and recent low angle neutron diffraction will be described. An extension of the mixed state model to granular superconductors with giant flux lines will also be briefly considered.

1. The superconducting macroscopic wave function.

London[1] was the first to propose that the electrodynamic properties of a superconductor could be explained if it was assumed that the microscopic quantum mechanical interactions leading to superconductivity resulted in a macroscopic quantum mechanical wave function of the form $|\Psi(r)| \exp[i\theta(r)]$, where $|\Psi(r)|^2$ was identified as the number density of superconducting electrons $n_s(T)$.

If the macroscopic wave function has the usual quantum mechanical properties, we can express the supercurrent density as

$$J(r) = e^*/m^* \; \{-i\hbar/2[\Psi^* \nabla\Psi - \Psi \nabla\Psi^*] - e^* A(r) |\Psi|^2\}, \quad (1.)$$

or more simply in terms of the amplitude and phase of the wave function as

$$J(r) = n_s e^*/m^* \{ \hbar \nabla\theta(r) - e^* A(r) \}, \quad (2.)$$

where e^* and m^* are the effective charge and mass of the charge carriers in the superconducting state (e.g. 2e and 2m for the Cooper pairs of BCS theory). A(r) is the magnetic vector potential, which is related to the local magnetic induction field by B(r)= curl A(r) . The term involving the vector potential account for the

Springer Series in Solid-State Sciences, Vol. 106 Physics of High-Temperature Superconductors
Eds.: S. Maekawa and M. Sato

electric fields induced by changing magnetic flux , $d\Phi/dt = \frac{d}{dt}\int B.\delta A$. The electric field is given by $E= -\text{grad}\phi - dA/dt$, where the second term can be related to induced emfs from changing magnetic fields by integration, emf = $\int E.dl = -\int A.dl = - d\Phi/dt$, which is Faraday's familiar law of electromagnetic induction.

As we show in the next section, this quantum mechanical description of a superconductor automatically accounts the exclusion of magnetic flux from the bulk of a superconductor (the Meissner effect), which implies persistent supercurrents flowing on the surface to screen out any external field. It also implies the quantisation of flux in units of h/e* within a superconducting ring or cylinder (as first recognised by London in his book on *Superfluids* where it is simply included only as a casual passing footnote!).

The above ideas were subsequently incorporated in the Ginsburg-Landau (GL) theory of superconducting phase transitions[2]. $|\Psi(r)|^2$ was interpreted as the superconducting order-parameter and the free energy density difference between the superconducting and normal states was written as

$$F_s - F_n = \alpha(T)\,|\Psi(r)|^2 + \beta\,|\Psi(r)|^4 + 1/2m^*|\, -i\hbar\nabla\Psi(r) - e^*A(r)\Psi(r)\,{}^2 . \quad (3.)$$

Minimising this expression with respect to Ψ and A in the London gauge leads to equ.1.

Subsequently Gorkov [3] and others derived the GL expansion from microscopic BCS theory, though only for restricted regions of the superconducting phase diagram - close to T_c in small fields and close to H_{c2}, the upper critical field H_{c2}, at all temperatures. For most practical purposes, however, the GL equations are generally assumed to provide a good approximation to the properties of conventional LTC type-2 superconductors at all temperatures and fields.

The derivation of the GL expansion from microscopic BCS theory proved that $e^*=2e$ and $m^*=2m$, consistent with the pairing of electrons in the superconducting state. There are two characteristic lengths associated with the GL- expansion - the superconducting coherence length $\xi = [\hbar^2/m^*\alpha(T)]^{1/2}$, which is the length scale over which spatial variations of the order parameter take place, and the London penetration length $\lambda = [\, m^*/\mu_0 e^2 n_s(T)]^{1/2}$, over which spatial variations in the local magnetic field and associated supercurrents occur.

The above formulation of the GL free energy expansion assumes isotropic superconducting properties, as expected for the s-state pairing of the original BCS model. However, HTC superconductivity introduces a number of potentially significant complications. Firstly, the cuprate HTC superconductors have a very strong 2-dimensional character despite their bulk 3-dimensional superconductivity. The superconducting wave functions reflect the anisotropy of the layered structures. In particular, the marked anisotropy of the effective mass m^* in the G-L expansion leads to significant differences in the superconducting coherence lengths ξ_{ab} and ξ_c parallel and perpendicular to the planes[4] and also in the corresponding magnetic penetration depths. Secondly, near T_c, the free energy difference, $\xi_{ab}^2\xi_c\mu_0H_c^2/2$, for a typical coherence volume, becomes comparable with kT. Thermodynamic fluctuations of the amplitude and phase of the order parameter can then dominate the behaviour near T_c, so that it may no longer be appropriate to use the standard Ginsburg Landau theory of the superconducting second-order phase transition. In particular, H_{c2} may not be a well-defined parameter, so that it is no longer posible to determine ξ via the relation $2\pi\xi^2H_{c2}= \Phi_0$. This may partly account for the large variations in reported values of ξ_{ab} and ξ_c. Additionally, HTC superconductivity may involve electrons paired in non-zero orbital momentum states resulting in a macroscopic superconducting state with a net orbital moment. The order-parameter would then be a vector-rather than a scalar-quantity, as in the A-phase helium-3, requiring a generalised vector-version of the GL free energy expansion (see ref 5). At an even deeper level, HTC superconductivity may involve exotic coupling mechanisms, such as Anderson's original RVB[6] or anyon models[7], which might lead to even more significant departures from the standard GL expansion. Such models could, for example, result in values of e* different from 2e. They could also result in departures from the zero magnetic flux Meissner state and to non-integral flux quantisation within a ring. Such complications are indeed implied by anyonic models of superconductivity. (An anion is a charged quasi-particle in an essentially 2-dimensional system with statistics intermediate between those of fermions and bosons. Whether such a model could be applied to the layered HTC depends on the extent to which coupling between the CuO planes destroys the essential 2-dimensional character required by the anyon model.)

Experimental verification of the predictions of GL-theory for HTCs is therefore important as any deviation from the expected behaviour may provide information of a fundamental nature on the symmetry of the underlying microscopic wave functions.

2. The Meissner effect and flux quantisation.

We now show how London's original ideas account for the Meissner effect and the quantisation of magnetic flux. The two geometries considered are illustrated in

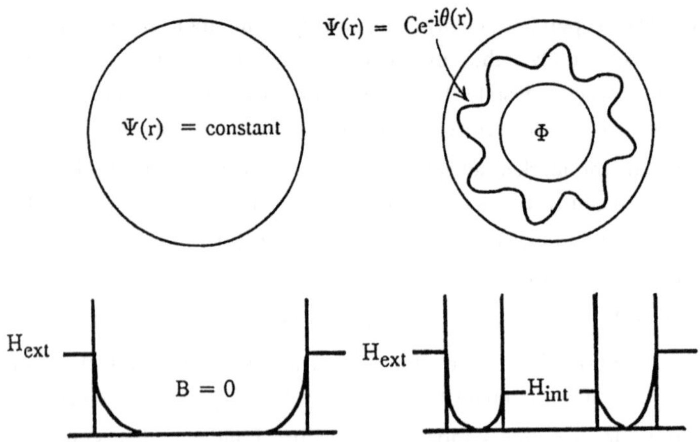

Fig.1 Examples of singly-connected and multiply- connected superconductors and the screening of magnetic fields from the bulk.

Fig.1a and 1b. In the first, an infinitely long superconducting cylinder or slab is aligned parallel to an external magnetic field H_{ext}. As the superconductor is *singly connected,* the topology dictates that the only possible solution for the superconducting wave function is $\Psi(r) =$ constant. Equation 1 therefore reduces to

$$J(r) = - n_s e^2/m\, A(r) \quad , \qquad (4.)$$

where we have replaced the effective charge and mass by the single electron values. (Strictly speaking this local relationship between the current density a vector potential is only true for a dirty (mean free path $l < \xi$) superconductor; the essential physics is unchanged for a clean superconductor and we will ignore such complications). In contrast to a superconductor, the magnetically induced acceleration of completely free, non-interacting, electrons would result in a current given by the time derivative of the above equation. Whereas a hypothetical perfect conductor would screen out any changes in magnetic field, by inducing a resistanceless screening current on its surface, a superconductor screens out even a static magnetic field, as we now show.

We first combine the above result with the Maxwell equation curl $H = J + dD/dt$. Ignoring the displacement current term (only relevant at extremely high frequencies) and taking the Coulomb gauge $\mathrm{div}A=0$, it is straightforward to show that

$$\frac{d^2B}{dz^2} = \frac{\mu_o n_s e^2}{m} B = \frac{1}{\lambda^2} B \quad , \qquad (5.)$$

where $\lambda(T) = [m/n_s\mu_o e^2]^{1/2}$ is the London penetration depth. Solutions are of the form $B(z) = \mu_o H_{ext} e^{-z/\lambda}$, where z is the distance from the surface. Any external field is therefore screened from the bulk by supercurrents flowing within a characteristic distance λ from the surface. The magnetic induction goes to zero in the bulk provided the size of the superconductor is significantly greater than λ, as illustrated in Fig 1a. The exclusion of field from the bulk of a superconductor implies a well-defined thermodynamic state in the presence of a magnetic field, independent of magnetic or thermal history. This is in contrast to our hypothetical perfect conductor, where the magnetic state depends on past thermal and magnetic history.

We now consider a hollow cylinder or ring in a longitudinal magnetic field, Fig.1b. As in the previous example, superconducting shielding currents flow within a distance λ of the outer and inner surfaces to screen out both the external and any trapped fields. Provided the thickness of the cylinder is $>> \lambda(T)$, the magnetic field and associated screening currents drop to zero well within the bulk , so that equation 3 may be written as

$$0 = e/m \{-i\hbar/2[\Psi^* \nabla\Psi - \Psi \nabla\Psi^*] - e^* A(r) |\Psi|^2\} \qquad (6.)$$

or more simply as

$$0 = n_s e/m \{\hbar \nabla\theta(r) - e^* A(r)\} . \qquad (7.)$$

For $\theta(r)$ to be single-valued, the phase of the superconducting order parameter must increase by an integral number of 2π on going once round the superconducting circuit, as illustrated in figure 1b. This implies $\int \theta(r). dl= 2n\pi$, where n is an integer. But the line integral $\int A(r).dl$ is simply the magnetic flux Φ enclosed by the integration path ($\mu_0 H_{int} A$ in the above example). Therefore, $\Phi = nh/e^* = n\Phi_0$, where $\Phi_0 = h/2e$ for Cooper pairs in a conventional BCS superconductor. The flux quantum is equal to 2.07×10^{-15} Tm^2 , somewhat less than 10^{-6} of the earth's magnetic field passing through through a wedding ring.

3. Experimental confirmation of flux quantum

The quantisation of flux in a superconducting ring was first reported in 1961 in two consecutive papers in Phys.Rev. Letters [8 ,9] both confirming quantisation in units of h/2e and hence the pairing of electrons in the superconducting state. Both involved measurement of integral quanta of flux trapped within hollow superconducting cylinders evaporated onto the surface of very fine fibres.

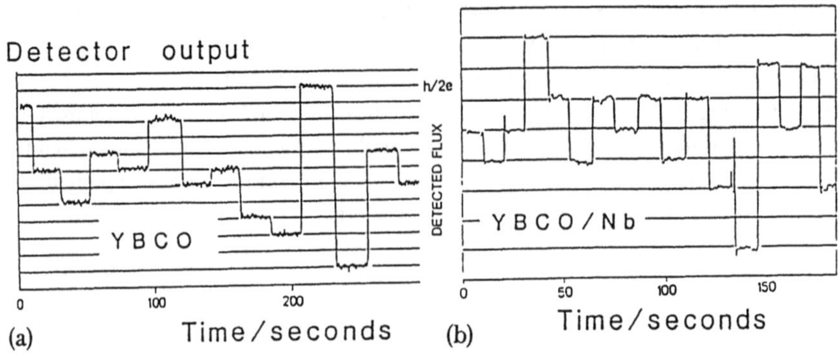

Fig2. Flux quantisation in (a) an YBCO ring sample and (b) a hybrid YBCO/Nb ring.

Subsequently, Little and Parks[10] demonstrated quantum periodicity in the field dependence of T_c determined from the resistance of a superconductor-coated quartz fibre.

Flux quantisation in HTC superconductors was first confirmed by the Birmingham Group[11] in a multi-phase granular ceramic YBCO ring using a conventional liquid helium cooled SQUID (Superconducting QUantum Interference Device) to measure the trapped flux. Unlike a solid ring of LTC material, the granular HTC material was only weakly superconducing. It was therefore relatively easy to excite the ring between its metastable quantum states, differing in trapped flux by integral multiples of the flux quantum. This implies a weakly superconducting path across the section of the ring, which allowed flux to move as flux quanta in and out of the ring. Figure 2a shows flux transitions in the YBCO ring excited by periodic bursts of electromagnetic noise. In addition to demonstrating the quantisation of flux in units of h/2e and hence the pairing of electrons in HTC superconductors, this experiment also demonstrated long-range coherence of the superconducting wave function, despite the granular nature of the sintered ceramic material. These early measurements were performed at liquid helium temperatures. Later accurate measurements of the periodicity in a rf-SQUID fabricated from single-phase 123-YBCO confirmed flux quantisation at 77K[12].

4. Phase coherence across a HTC/LTC superconductor boundary.

A subsequent variant of the above experiment[13], with a niobium superconductor bridging a slit across an YBCO ring, again confirmed flux quantisation in units of h/2e, as illustrated in Fig. 2b. This experiment demonstrated the coherence of the macroscopic superconducting wave function across the HTC/LTC interface, even though the pairing mechanisms of LTC and HTC superconductors may be totally different. As the macroscopic wave function reflects the symmetry of the paired microscopic states, the phase coherence across the HTC/LTC interface might be thought to imply the same symmetry of paired states in both superconductors. However, loss of symmetry on reflection at a boundary and complications from spin-orbit coupling [14] result in coupling between s- and higher orbital coupling state superconductors at an interface. The above experiment therefore does not prove that the pairing in a HTC superconductor is necessarily s-wave. However, it would be more difficult to account for the observed phase coherence, if HTC superconductivity involved an exotic mechanism such as anyonic superconductivity(7).

5. Flux quantisation and the mixed state.

Starting from the GL expansion for the free energy, Abrikosov[15] showed that for type-2 superconductors, for which $\lambda > \sqrt{2}\ \xi$, magnetic flux penetrates the bulk as flux vortices or flux lines, each carrying a quantum of flux. This *mixed state* is nucleated above the lower critical field $H_{c1} \sim \Phi_o/\lambda^2$, when it becomes energetically favourable for a single flux line to penetrate the bulk. It persists up to the upper critical field $H_{c2} = \Phi_0/2\pi\xi^2$, at which field interactions betwen the flux lines reduce the order parameter everywhere to zero. Because the flux lines are quantised, the density of flux lines is simply B/Φ_0.

In an ideal type-2 superconductor the flux lines form a regular triangular lattice as illustrated by the experimentally determined magnetic induction field B(r) shown in Fig. 3[16]. HTC superconductors are all extreme type-2 superconductors with

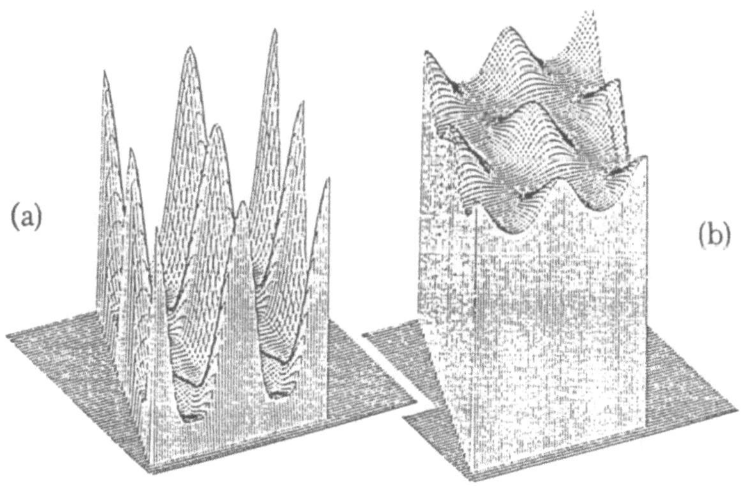

Fig.3 Variation of local magnetic flux density in the mixed state at low fields ~H_{c1} and higher fields >> H_{c1}, from neutron diffraction measurements on niobium(16)

$\lambda >> \xi$. They therefore have rather small lower critical fields H_{c1} and very high upper critical fields H_{c2} , typically 100T or larger. One of the most important unanswered questions about HTC superconductors concerns the existence of the flux line lattice at high temperatures. It is widely believed that thermodynamic fluctuations of the order parameter at high temperatures will lead to a melting of the quantised flux line lattice[17] and several experiments claim to support this view (for example ref 18), but the interpretation of such measurements remains contentious (see for example 19).

The flux line lattice can be oserved directly by allowing a fog of ultra-fine magnetic powder to fall on the surface of a magnetised superconductor. The powder falls preferentially in the high field regions at the centre of flux lines, as was first demonstrated by Essmann and Trauble[20]. The finite size of the powder limits such measurements to a few mT. Decoration measurements by Dolan et al [21] have confirmed the existence of the flux line lattice for HTC superconductors and provide strong evidence for strong interactions between flux lines and defects at these low values of flux density. Other groups have observed liquid crystal-like flux line configurations[22], which lack the long-range order of the perfect lattice.

The flux line lattice in the highly anisotropic type-2 superconductor $NbSe_2$ has recently also been observed by scanning tunnelling microscopy[23]. In this superconductor the coherence length and surface flatness is sufficient to allow measurement of the local density of superconducting electron states both within and external to the flux lines cores with almost atomic resolution.

The only currently available method for studying the flux line lattice in type-II superconductors at fields of practical importance is by low angle neutron diffraction, as in the original measurements on niobium by Schelten et Al [24]. Recently such measurements have been extended to single crystals of YBCO by Forgan and Paul for fields up to 0.6T [25, and private communication]. Well above H_{c1}, the spatial variation in local magnetic field within the mixed state, δH, is approximately independent of magnetic field and of order Φ_0/λ^2. The diffracted neutron intensity depends on $(\delta H)^2$ and is therefore proportional to $\lambda(T)^{-4}$. Because $\lambda(0)$ for the cuprate HTC superconductors is considerably larger than for most LTC superconductors ($\lambda_{YBCO} \sim$ 140nm compared with $\lambda_{Nb} \sim$40nm), the diffracted neutron intensity is rather small by comparison. The difficulty of performing such experiments is further compounded by the difficulty in growing fully oxygenated single crystals of any size.

Figs. 4a-c show recent low-angle neutron diffraction measurements (private communication) of the flux line lattice in niobium and YBCO single crystals. For

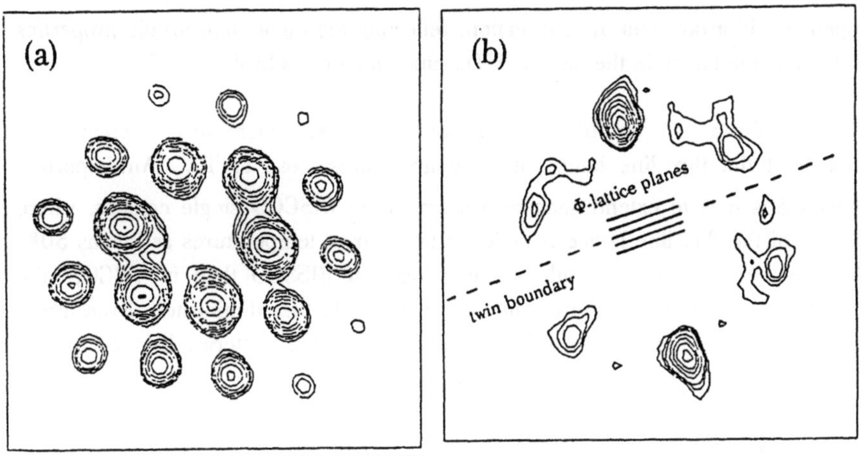

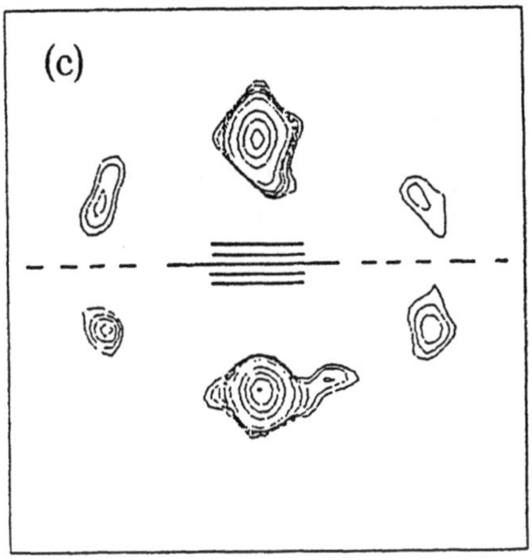

Fig.4a-c. Neutron diffraction from the flux line lattice in single crystals of (a) niobium, (b) YBCO for B parallel to the c-axis and f (c) for B at 45°to the c-axis.

the cm-sized single crystal of niobium the diffracted intensity is sufficient to resolve not only the first order diffraction reflections but also many higher order reflections. The spatial variation of field between flux lines can therefore be deduced rather precisely by an inverse spatial transform. For the much smaller YBCO single crystal with field parallel to the c-axis (perpendicular to the CuO planes), only the first order diffraction spots can be resolved. Furthermore, although vestigial evidence for a hexagonal diffraction pattern can be discerned, only diffraction spots corresponding to planes of flux lines parallel to the [110] twin boundary planes are strong. This implies a strong interaction between the flux lines and the twin boundaries at these fields. The flux lines appear to line up parallel to the twin-boundary planes, with well define parallel planes of flux lines giving the strong spots. However, the spacing of flux lines within the planes appears to rather non uniform, resulting in the relatively weak diffraction spots in the other directions. Fig. 4c illustrates the pronounced asymmetry of the flux line lattice when the field is tipped ~45^{0} away from the c-direction, reflecting the highly anisotropic properties of the flux line lattice in the layered HTC superconductors [26].

In none of the neutron diffraction measurements was there any evidence for melting of the flux line lattice up to temperatures close to T_C. An important challenge is now to extend such measurements to BISCCO single crystals, where melting of the flux line lattice is believed to occur at temperatures as low as 50K. Because the penetration length is even longer in BISSCO than in YBCO, the expected intensity of the neutron diffraction will be further diminished. Even larger single crystals are required if diffraction from low angle scattering by the flux line lattice is to be distinguished from scattering by defects.

6. Giant flux lines in the granular mixed state.

Finally we briefly consider the granular model for superconductivity, as this has been widely discussed as a model for non-crystalline HTC superconductors. The superconducting properties are then partly determined by the superconducting single crystal grains, which are assumed to behave as bulk type-II superconductors, and partly by the relatively weak superconducting junctions between such grains, which determine the bulk transport supercurrent. A schematic representation of a granular superconductor is shown in Fig.5 below.

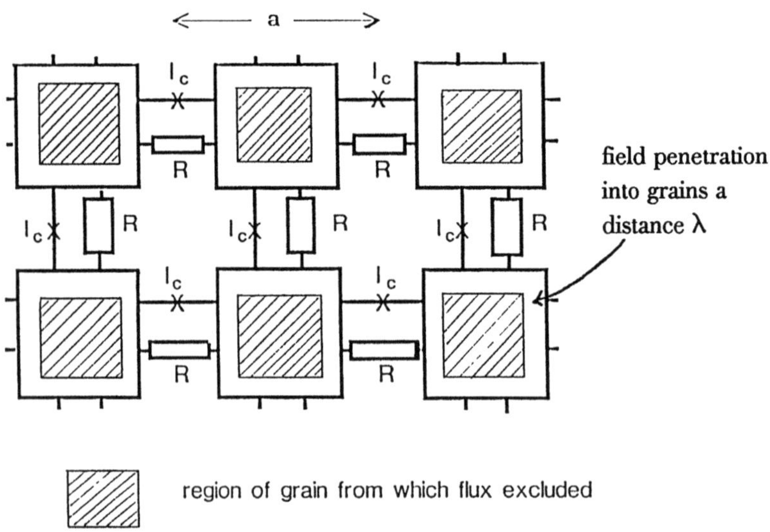

Fig.5 Granular model of a high T_C superconductor. Individual superconducting grains are assumed to be weakly coupled together by Josephson junctions with parallel resistances to represent the intergranular resistance in the normal state. The field penetrates each grain to a distance λ, as indicated.

The properties of granular superconductors have been discussed by Clem [27]. The weak links are represented as Josephson Junctions satisfying the usual Josephson current phase relation
discussed in the next lecture and a parallel resistance is included as this plays an important role in the electrodynamics. It is relatively easy to show that, within an effective medium or continuum approximation, the magnetic properties of the granular material are closely analogous to a normal type-2 superconductor with an effective magnetic penetration length $\lambda_{gran} = (\frac{\Phi_0 a}{2\pi\mu\mu_0 I_c})^{1/2}$ and an effective coherence length of order a, the intergrain spacing.

$I_c(T)$ is the critical current between grains $\sim J_c(T)\, a^2$, where J_c is the bulk critical current density. For fields greater than $\sim \Phi_0/\lambda_{gran}^2$ - of order 0.1mT or less for a typical HTC ceramic with $J_c \sim 100\text{-}1000\ Am^{-2}$ and grain size 3-10 micron - flux is predicted to penetrate the granular material as giant quantised flux-lines, the vortex supercurrents extending over a distance $\lambda_{gran} >> \lambda_{London}$. However, in practice, the relevance of this model to HTC materials is limited to only poor quality ceramic samples or to ceramics very close to T_c, where $I_c(T)$ is sufficiently small that $\lambda_{gran} >>$ the grain spacing. For typical critical current densities the magnetic penetrations depth λ_{gran} is usually of the same order as the grain size, so that the continuum model approximation ($\lambda >> a$) breaks down. For most granular HTC material it is probably more realistic to consider flux entering the material in the form of quantised flux lines threading between the grains.

References.

1. F.London, Proc Roy.Soc. A152,24 (1935): *Superfluids-Macroscopic Theory of Superconductivity*, Dover Publications, New York (1961)
2. V.L.Ginsburg and L.D.Landau, Zh. Eksperim. i Teor. Fiz. **20** 1064 (1950), see also M.Tinkham, *Introduction to Superconductivity*, McGraw Hill, New York, 1975 for a general background.
3. L.P.Gorkov, Soviet Physics-JETP 9 1364 (1959)
4. E.M.Forgan, Nature 329, 483 (1987)
5. D.R. Tilley and J. Tilley, *Superfluidity and Superconductivity* (2nd edition), Adam Hilger, Bristol, 1985
6. P.W. Anderson, Science 235, 1196 (1987)
7. F.Wilczek, *Anyons in Scientific American, May 1991*: B.G.Levi, *Physics Today, Feb.1991, p 17.*

8. R. Doll and M. Nabauer, Phys. Rev. Lett. 7 , 51 (1959)
9. B.S. Deaver and W.M. Fairbank, Phys. Rev. Lett. 7 , 43 (1961)
10. W.A.Little and R.D.Parks, Phys. Rev.,133A, 97 (1964)
11. C.E. Gough, M.S. Colclough, E.M. Forgan, R.G. Jordan, M.Keene, C.M. Muirhead, A.I.M. Rae, N. Thomas, J.S. Abell, S. Sutton, Nature 326, 855 (1987)
12. S. Harrop, C.M. Muirhead, M.S. Colclough, C.E. Gough, Physica 153-155 , 1411 (1988)
13. M.N.Keene, T.J.Jackson and C.E.Gough, Nature 40, 210 (1989)
14. A.J.Millis, Physica B135, 69 (1985)
15. A.A.Abrikosov, Soviet Physics-JETP 9, 1364 (1959)
16. J.Schelten, H.Ullmaier and G.Lippmann, Z.Phys 253, 219 (1972)
17. D.Nelson and H.S. Seung, Phys.Rev. B39, 9174 (1989)
18. P.L.Gammel, L.F.Schneemeyer, J.V.Waszsak and D.B.Bishop, Phys.Rev.Lett. 61, 1666 (1988)
19. E.H.Brandt, Physica C169, 91 (1991)
20. U. Essmann and H. Trauble, Phys. Letts. A24 , 526 (1967)
21. G.J. Dolan, F. Holtzberg, C. Feild and T.R. Dinger, Phys. Rev. Lett. 62 , 218 (1989)
22. D.J.Bishop, P.L.Gammel, C.A.Murray, D.B.Mitzi and A.Kapitulnik, Physica B 169, 72 (1991)
23. H.F.Hess, R.B.Robinson and J.V.Waszsak, Physica B169, 422 (1991)
24. J.Schelten, G.Lippman and H.Ullmaier. J.Low Temp.Phys. 14, 213 (1974))
25. E.M.Forgan, D.McK. Paul, H.A.Mook, P.A.Timmins, H.Keller, S.Sutton and J.S.Abell, Nature 343, 735 (1990)
26. A.J.Buzdin and A.Yu.Simonov, Physica B165/166, 1101 (1991)
27. J. Clem, Physica C153-155, 50 (1988)

Flux Quantisation and Its Application to rf and dc SQUIDs

C.E. Gough

Superconductivity Research Group, University of Birmingham,
Birmingham B15 2TT, UK

Abstract. The Josephson effect is briefly described and is shown to result in properties of superconducting loops containing one or more weak links which are periodic in magnetic flux in units of the flux quantum. The modes of operation of rf and dc SQUIDs are briefly described and an assessment is made of the current performance of liquid nitrogen cooled HTC SQUIDs.

1. The Josephson Effect.

In the previous lecture it was shown that the existence of a single-valued macroscopic wave function describing the superconducting state leads to the Meissner effect in a singly connected superconductor and to flux quantisation in a superconducting ring. Here we consider the consequences of phase coherence around superconducting rings intersected by one or more Josephson junctions. A more detailed analysis of quantum interference and its application to SQUIDs (Superconducting QUantum Interference Devices) is given by Barone and Patermo[1] and van Duzer and Turner[2].

An ideal Josephson junction is formed between two superconductors having an insulating layer separating them sufficiently small for there to be a finite probability of quantum mechanical tunnelling of electrons between them. In 1962, Josephson [3] showed that the dc and ac supercurrents passing through such a junction were determined by the phase difference of the superconducting wave function across the junction. He argued that mathematical terms earlier discarded as unphysical by earlier theorists implied the existence of a genuine supercurrent at zero voltage across the insulating layer, given by

$$\boxed{I = I_0 \sin(\theta_2 - \theta_1) \qquad \text{the dc- Josephson relation}} \quad , \qquad (1.)$$

where I_0 is the maximum supercurrent that can flow (determined by the tunnelling matrix for single particle tunnelling across the junction) and $(\theta_2 - \theta_1) = \Delta\theta$ is the difference in phase of the macroscopic wave function across the junction, as illustrated in Fig. 1a below. This is known as the dc- Josephson relation or

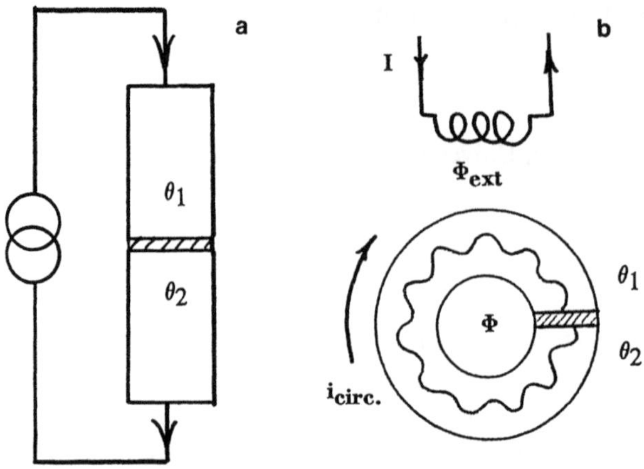

Fig. 1 A Josephson Junction biased by a current source and in a superconducting loop biased with an external flux.

sinusoidal current- phase relationship. A particularly illuminating derivation of this relationship is given by Feynman[4]. The zero-voltage Josephson current relationship was experimentally confirmed by Anderson and Rowell[5].

We now consider a Josephson junction intersecting a superconducting loop, as illustrated in Fig.1b above. Provided we ignore the small amount of flux penetrating the junction region itself, we can use the arguments of the lecture on "Flux quantisation in Superconducting Rings and the Mixed State" in this book to show that the phase difference along a line integral well within the superconducting ring (assumed to have a thickness $>> \lambda$) is simply given by

$$\Delta\theta = 2\pi \frac{\Phi}{\Phi_0}, \qquad (2.)$$

where Φ is the flux within the ring . The Josephson current, which flows with zero voltage across the junction, is thus periodic in the flux within the ring having the periodicity of the flux quantum, $\Phi_0 = h/2e$.

Josephson also considered the situation for a finite voltage V across a junction, as would for example be generated by a changing magnetic flux within the ring. The voltage induced, $V = d\Phi/dt$, which through the relationship between flux and phase across the junction is equivalent to

$$V = \frac{\Phi o}{2\pi}\frac{d\,(\Delta\theta)}{dt}. \qquad (3.)$$

Although this has been argued for a Josephson junction in a superconducting ring, it is equally valid for a Josephson junction in any superconducting circuit. The

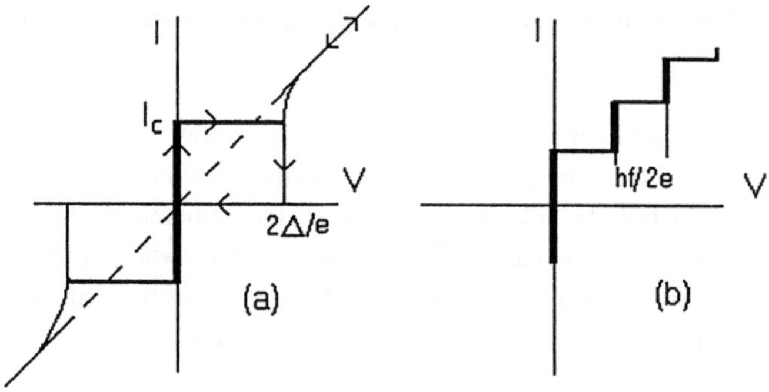

Fig.2 (a) Schematic I/V characteristics for a Josephson Junction, (b) the effect of irradiating the junction with microwaves.

current through a Josephson junction biased by a constant voltage V is therefore determined by a phase difference across the Junction(eq.2), which increases linearly with time as $\frac{2\pi V t}{\Phi_0}$. The current therefore oscillates as

$$\boxed{I = I_0 \sin \omega_J t \qquad \text{the ac Josephson effect}} \tag{4.}$$

where $\omega_J = \frac{2\pi}{\Phi_0} V$ is the Josephson frequency = 484 MHz/μV.

This may be identified with the emission of a quantum of electromagnetic energy $h\omega = 2eV$, every time an electron pair crosses the biased junction. In non-ideal Josephson junctions, the tunnelling of electron pairs may be correlated leading to higher frequency components emitted and departures from the sinusoidal current-phase relationship of critical current, which nevertheless remains periodic in the flux quantum.

For an ideal Josephson junction, Ambegaokar[6] showed that $I_c(T)$ varied as

$$I_c(T) = \frac{\pi\Delta(T)}{2eR} \tanh \frac{\Delta(T)}{2kT}, \tag{5.}$$

where $\Delta(T)$ is the temperature dependent energy gap. The I_cR product at low temperatures approaches $\frac{\pi}{2} \frac{\Delta(0)}{e}$, where R is the resistance of the junction well above T_c. Note that $2\Delta(0)/e$ is the voltage bias for single particle excitation across the gap at low temperatures, first measured by Giaever[7]. The zero-temperature current through a perfect Josephson junction, including the single particle tunnelling at finite voltage, is shown schematically in Fig 2.

The measured characteristics depend on the nature of the biasing (current or voltage bias) and also on the resistive damping of the device (see later). The irreversible characteristics illustrated in Fig.2a are for current bias and weak damping.

In practice any "weak-link" between two superconductors can support a Josephson supercurrent that is determined by the phase difference across the junction. Other types of weak-link can include a microbridge, where the superconductor is purposely thinned to restrict the supercurrent flow; a point-contact, where a sharp point is pressed into the surface to form a junction, which may be partly insulating and partly micro-bridge in character; a vacuum proximity junction with the space between the superconductors controlled using scanning tunnelling techniques; and the ubiquitous intergranular weak-links that characterise HTC ceramics. In many of these examples the temperature dependence of I_c is similar to that predicted by Ambegaokar but the I_cR product is often substantially smaller.

2. Energetics and dynamics of Josephson Junctions.

Consider in more detail the two experimental configurations illustrated in Fig.1. For consistency with the sinusoidal Josephson phase relationship, the energy for the constant current-biased junction must be given by

$$E(\Delta\theta,I) = -\frac{\Phi_0}{2\pi} I_c \cos\Delta\theta - \frac{\Phi_0}{2\pi} I\,\Delta\theta \quad , \qquad (6.)$$

where the second term represents work done by the external source

$(\int IVdt = I \int \frac{\Phi_0}{2\pi}\frac{d\theta}{dt} dt = I\frac{\Phi_0}{2\pi}\Delta\theta$). Minimising the energy with respect to $\Delta\theta$,

we recover the required sinusoidal current-phase relationship. We can express the energy associated with the supercurrent flow through the junction as $-\varepsilon_J \cos\theta$, were $\varepsilon_J = \frac{I_c\phi_0}{2\pi}$ is the Josephson junction coupling energy.

The above potential is known as the "washboard potential" shown schematically in Fig.3. For a given current bias $< I_c$, there is a multiplicity of possible stable energy states of the system differing in phase across the junction by multiples of 2π. Such states represent supercurrent flow through the junction at zero voltage. However, transitions between such states can be induced by thermal activation[8], when kT is a significant fraction of the barrier energy separating the metastable local minima (typically $kT \sim \varepsilon_J/40$), as illustrated in Fig 3a. The consequent "phase slippage" results in a voltage across the junction given by the voltage-phase relationship (eq. 3). Thermally activated phase-slippage results in thermally

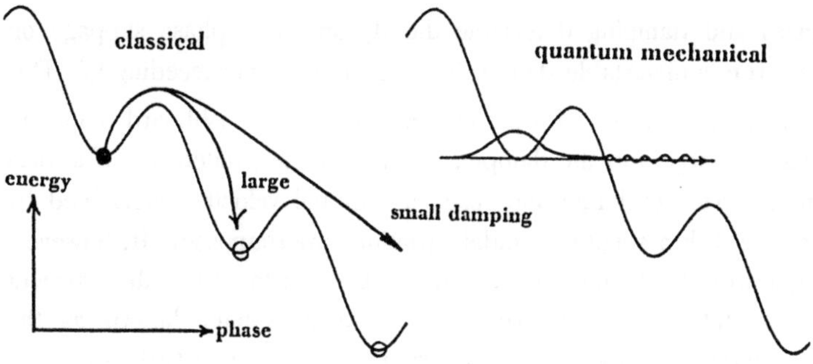

a) Thermal activation b) Macroscopic Quantum Tunnelling

Fig.3. The washboard potential for a current biased Josephson Junction illustrating (a) thermal activation and (b) macroscopic quantum tunnelling.

activated flux flow (TAFF) in ceramic HTC superconductors, where the weak links are formed between the individual grains or domains of superconducting material.

As first pointed out by Anderson, the charge and phase difference across a Josephson junction should be treated as quantum mechanical variables with a corresponding uncertainty relation - essentially the number-phase uncertainty principle. There will therefore be a finite (but usually infinitesimal) probability of the phase tunnelling across the potential barrier - as illustrated schematically in Fig.3b. This was first considered by Lhikarev[9] and Widom and Clark[10] and has been discussed in considerable detail by Legett[11]. Treating the phase (or flux within a ring) as a quantum mechanical variable conjugate with the charge across the junction capacitance allows the possibility of important experiments in the quantum theory of measurement including practical investigations of the Schrodinger Cat Paradox[12]. The existence of "macroscopic quantum tunnelling" has been confirmed in a number of elegant experiments on superconducting junctions [13,14,15]. It has been suggested that apparent magnetic flux creep in HTC superconductors at rather low temperatures [16] may be a manifestation of such tunnelling.

In determining the probability for macroscopic quantum tunnelling and the dynamics of phase slippage, it is necessary to introduce two other parameters for the junction, its capacitance C, which is equivalent to the "effective mass" in the equivalent single particle tunnelling model, and the parallel resistance, R, which accounts for damping. To observe macroscopic quantum mechanical tunnelling, the capacitance has to be very small (typically $<<10^{-12}$ F). Resistive damping inhibits macroscopic tunnelling for values of $R < h/e^2$, the quantum unit of resistance.

The effective mass and damping determine the dynamics of phase slippage on thermal excitation from a metastable state or for critical currents exceeding I_c. This is most easily considered in terms of a particle moving in the washboard potential. If the particle mass is large and the damping is small, once excited out of a local potential minimum it will accelerate and reach a terminal velocity determined by the resistive damping with a small sinusoidal variation superimposed. If, however, the particle is light and the damping is strong, it will come to rest in the adjacent potential minimum state. We therefore introduce a parameter known as the McCumber parameter[17] to characterise the damping given by $\omega_c C/R$, where $\omega_c = 2eI_cR/\hbar$ is the Josephson frequency corresponding to the I_cR product ($\sim 2\Delta/e$ for an ideal Josephson junction).

The current voltage characteristics of a junction are hysteretic or non-hysteretic depending on whether the McCumber parameter is greater or smaller than unity. For optimum performance in practical devices it is often necessary to include a parallel resistance across the junction to control the device dynamics.

We now consider a Josephson junction in a superconducting loop, biased with an external flux Φ_{ext}. The energy can then be written as

$$E = \frac{(\Phi-\Phi_{ext})^2}{2L} - \varepsilon_J \cos 2\pi\frac{\Phi}{\Phi_0}, \qquad (7.)$$

where Φ is the total flux within the loop (external flux minus less flux induced by circulating screening currents i_s) and L is the inductance of the ring. The first term corresponds to the stored magnetic energy($Li_s^2/2$), which provides a parabolic background centred on Φ_{ext}, while the second is our familiar cosine Josephson coupling energy with minima at multiples of Φ_0.

Minimising eq.7 with respect to Φ, we obtain the following expression for the external flux in terms of the flux within the ring

$$\Phi_{ext} = \Phi + L I_c \sin 2\pi\frac{\Phi}{\Phi_0} , \qquad (8.)$$

which is drawn in Fig.4a. The magnetic behaviour of such a ring depends critically on the value of $\beta = 2\pi L I_c/\Phi_0$, as illustrated by the examples drawn for β greater and less than unity. For $\beta < 1$ the circulating supercurrents provide a gentle modulation of the total flux within the ring about Φ_{ext} and the resulting magnetic properties are non-hysteretic. However, for $\beta > 1$ the circulating currents are large and tend to shield changes of flux within the ring. The magnetic

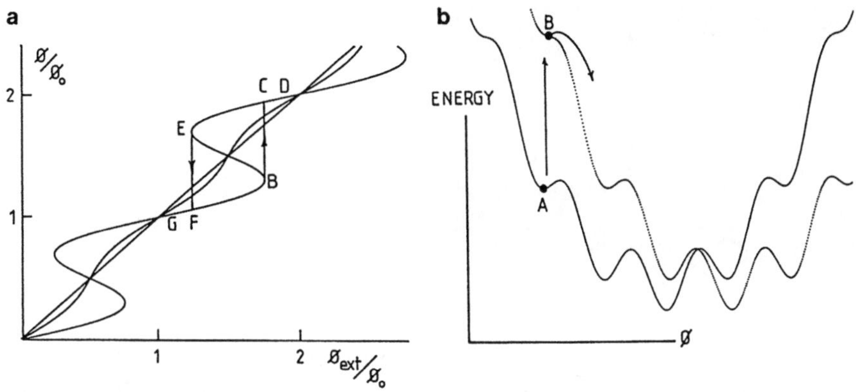

Fig.4. (a) Magnetic properties of a weak-link superconducting ring for small and large β; (b) energy of system as function of internal flux for 2 values of Φ_{ext}.

properties then become hysteretic as illustrated by the magnetic hysteresis loop on increasing Φ_{ext} from G to D and back again in Fig.4 .

The hysteretic behaviour can be understood in terms of flux transitions between the local energy minima illustrated in Fig. 4b . On increasing the external flux the background parabolic energy curve moves to the right raising the local minima from A to B, at which point it becomes unstable leading to a transition of flux in or out of the ring. The McCumber damping parameter again determines the subsequent dynamics. If the damping is small, the final state of the system is uncertain, whereas large damping limits transitions to adjacent stable states.

As before, thermal activation can cause early transitions[18] and even for $\beta > 1$ can lead to quasi-reversible magnetic behaviour, when the thermal energy is sufficient to cause many thermally activated transitions between adjacent metastable states within the time of a measurement.

3. Tho rf SQUID

The magnetic behaviour described in the previous section is the basis of operation of the rf SQUID. Full details of the operation of such devices are given in references 1 and 2. Here, we simply note that the magnetic properties of a superconducting loop intersected by a Josephson junction are periodic in the external flux. Thus, if we couple such a loop inductively to an rf coil providing the external flux, the resonant frequency and loss will vary periodically with external static bias field with the periodicity of the flux quantum. For $\beta < 1$, the magnetic properties are non-hysteretic and the effect of the coupled superconducting circuit

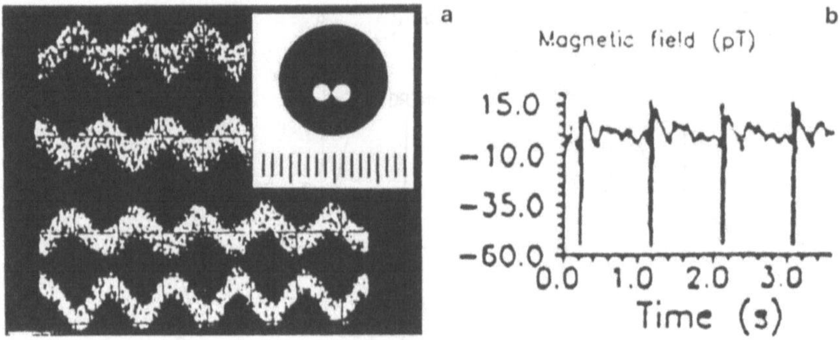

Fig.5. (a) Triangular characteristics of a HTC SQUID operating at 77K. (b) a magneto-cardiogram recorded by Vasiliev (22) using a HTC SQUID at 77K.

is to change the effective inductance and hence resonant frequency of the rf coil. This is known as the non-hysteretic operation of an rf SQUID. The more normal operation is to adjust I_C so that $\beta > 1$ and to include sufficient damping that the hysteretic behaviour is well defined. The rf level is then set so that the magnetisation is taken around one or more hysteresis loops such as that illustrated in Fig. 4 . If the rf circuit is biased with a constant rf current, the rf voltage across the coil varies in a characteristic triangular way with external flux, as illustrated in Fig.5a for a rf HTC SQUID operating at liquid nitrogen temperatures[19]. This SQUID uses the naturally occurring intergranular links between grains in a thinned down section of YBCO to provide a "weak link" of the desired strength.

RF SQUID operation was first demonstrated by Zimmerman [20] using a niobium ring with a niobium point contact as weak link . Commercial liquid helium cooled rf SQUIDs based on niobium rings with point contact or thin film junctions have been widely used for over two decades as ultrasensitive magnetometers and for measuring ultra-small currents and voltages. Zimmerman was also the first to observe rf -SQUID behaviour in a bulk HTC ring [21]. He used a "break-junction" as weak link, where the ceramic was first broken under liquid helium or nitrogen and then pressed back together to form a weak link of the desired strength. Subsequently a number of authors, including ourselves, as in Fig.5a above, developed rf SQUIDs using the naturally occurring weak-links between grains. Such devices convincingly demonstrated the potential for HTC SQUID operation, including cardiographic measurements by Vasiliev[22] (shown in Fig. 5b) and laboratory magnetometers using liquid nitrogen cooled SQUIDs [23]. However, the lack of long-term reproducibility and the difficulty of reliably fabricating weak-link junctions of the required strength prevents such devices from being much more than interesting laboratory demonstrations.

4. The dc SQUID

We now describe the operation of the dc SQUID, which is the preferred device for sensitive magnetometry using either LTC or HTC superconductors as it is ideally suited to thin film fabrication and the associated electronics also tends to give a better noise performance. The principles of operation of the dc SQUID can be illustrated in terms of the schematic representation below (from DeGennes[24]).

Consider first the current I flowing through the two Josephson junctions in parallel (assumed identical)

$$I = I_c \left[\sin(\theta_{2A} - \theta_{1A}) + \sin(\theta_{2B} - \theta_{1B}) \right] \quad . \qquad (9.)$$

The phases across the junctions are connected via the flux enclosed by the superconducting circuit. Consider first the current path 1 well within the superconducting ring along which J_S is zero. Along this path $\hbar\Delta\phi = 2eA$, so that

$$\theta_{1B} - \theta_{1A} = \frac{2e}{\hbar} \int_{C_1} A.dl \quad , \qquad (10.)$$

and
$$\theta_{2A} - \theta_{2B} = \frac{2e}{\hbar} \int_{C_2} A.dl \quad , \qquad (11.)$$

so that
$$(\theta_{2A} - \theta_{1A}) - (\theta_{2B} - \theta_{2B}) = \frac{2e}{\hbar} \int A.dl = 2\pi \frac{\Phi}{\Phi_0} . \qquad (12.)$$

Recognising that the absolute value of the phases are arbitrary and that only phase differences are significant , it is straightforward to show that

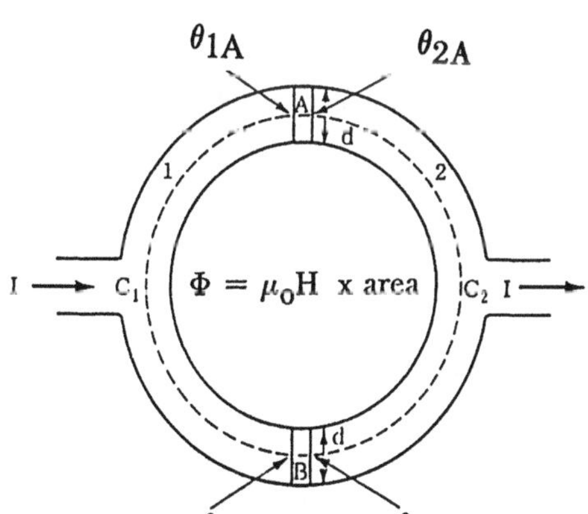

Fig.6. Schematic representation of a dc SQUID- or quantum interferometer. A and B are Josephson junctions.

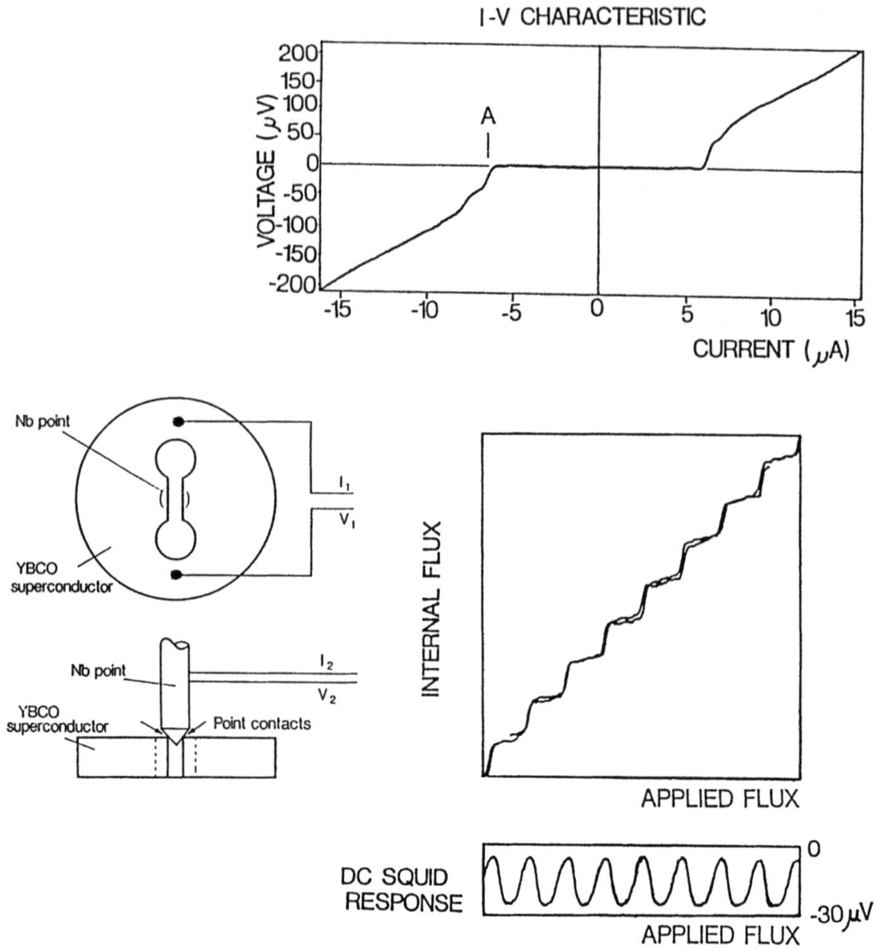

Fig. 7 Measurements of the electrical and magnetic characteristics of a hybrid 2-hole dc-SQUID formed from a bulk YBCO sample with a niobium double point contact.

$$I = 2 I_C \left|\cos\left(\pi \frac{\Phi}{\Phi_0}\right)\right| . \qquad (13.)$$

The maximum supercurrent that can flow through the device is therefore a periodic function of the flux within the ring . The situation is closely analogous to the double-slit interference pattern in optics, in the superconducting case the phase changes being determined by the flux within the ring. These predictions were strikingly confirmed by Jaklevik et al. 25] in measurements confirming the predicted cosine interference pattern modulated by a single-slit diffraction pattern from interference effects from flux penetration into the junction regions themselves.

Fig.7 illustrates the periodicity of both the electrical characteristics and the magnetic properties of a dc SQUID ring. These measurements[26] were made on a 2-hole dc SQUID fabricated from a solid block of sintered YBCO ceramic with a superconducting weak-link barrier inhibiting the transfer of flux between the holes provided by a niobium point making two junctions to the YBCO block, as illustrated in Fig.7b. The uppermost curves are the V/I characteristics for current flowing from the niobium to the YBCO across the two weak-link junctions, which together with the Nb and YBCO form a dc-SQUID. The middle curve is the flux inside one of the holes as a function of bias flux applied to the other and the lower curves shows the modulation of the V/I characteristics at the bias current A for the same bias flux and Nb point settings. This device could be operated in both dc-SQUID (as illustrated) and rf SQUID configurations. Such measurements confirmed that the measured noise was intrinsic to he sample and not to the mode of operation, though the external electronic noise was significantly less for the dc SQUID.

Various electronic circuits can be used to determine the modulation in critical current through the double-junction dc -SQUID. Practical magnetometers based on rf or dc SQUIDs generally use a feedback loop to "lock " the system on to one of the minimas of the observed periodicity. Any change in external flux is then compensated by a feedback signal, so that the device has a linear response to external field changes provided it stays "locked on" to a minima.

The sensitivity of a SQUID can be expressed in terms of its sensitivity to changes in the biasing field δH_{ext} or $\delta\Phi_{ext}$ within the superconducting loop or to the corresponding changes in magnetic energy $(\delta\Phi_{ext})^2/2L$. All measurements are made against a background of intrinsic and extrinsic noise that is wide-band in character, though 1/f noise usually dominates at very low frequencies (for high quality LTC SQUIDs at frequencies typically <1Hz) . The sensitivity of a SQUID is therefore always quoted for a given measurements bandwidth either in terms of its field sensitivity $\delta H_{ext}/\sqrt{Hz}$, flux sensitivity $\delta\Phi_{ext}/\sqrt{Hz}$, or energy sensitivity $[(\delta\Phi)^2/2L]/Hz$. Fig.8 illustrates the performances of selected commercial liquid-helium cooled rf and dc SQUIDs, laboratory LTC SQUIDs at 4.2K and state of the art HTC SQUIDs at 77K.

At low frequencies the noise performance is dominated by 1/f device-noise, either from metastably trapped flux being thermally activated in and out of the SQUID ring or from variations in the critical currents of the junctions, possibly associated with random occupation of localised electron trapping states within the junctions. The best laboratory LTC SQUIDs approach the quantum limit of energy resolution of h/Hz[27], where h is Plank's constant.

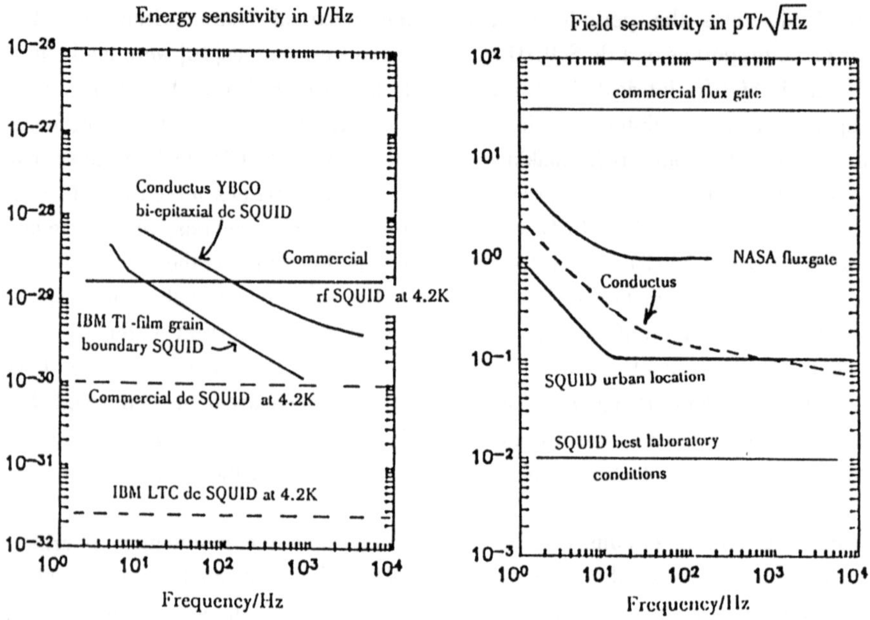

Fig. 8 Field and energy sensitivities of commercial and laboratory LTC and HTC SQUIDs.

In practical SQUIDs, the stored magnetic energy in the ring is usually designed to be of the same order as the Josephson coupling energy, $\Phi_0{}^2/2L(\text{squid ring}) \sim I_c\Phi_0/2\pi$, so that $\beta \sim 2\pi LI_c/\Phi_0$ is of order unity or somewhat greater. The Junctions have also to be shorted with a damping resistor to ensure well-behaved dynamical behaviour, as explained above. Thermal fluctuations give variations in magnetic flux $\delta\Phi$ within the SQUID ring, where $(\delta\Phi)^2/2L \sim kT$. Initially there was some concern that the performance of liquid nitrogen temperature SQUIDS would be limited by such fluctuations[28], though it is now clear that provided the inductance is sufficiently small this should not be a serious limitation. However, small inductance implies small scale devices. There is then a major problem in magnetically coupling to the external field to be measured. This requires the use of a superconducting flux-transformer, which has to be an integral part of the design of any practical dc SQUID.

Progress in the development of practical HTC dc-SQUIDS has been extremely encouraging though many formidable materials problems had to be overcome in order to produce optimally designed, reproducible devices. The first major problem was to fabricate reproducible weak-link junctions with the appropriate critical currents for liquid nitrogen temperature operation. Initially, the naturally occurring weak-links between grains in YBCO and thallium *ex-situ* processed thin films were used. The performance of such a SQUID fabricated by Koch et al. [29], using the natural grain boundaries in a layered thallium cuprate thin film, when operated at 77K was shown to be comparable with commercially available helium cooled niobium rf SQUIDs above around 100Hz, as illustrated in Fig. 8. At low frequencies the performance was dominated by a large 1/f-noise component, which from noise measurements on YBCO thin films[30] appeared likely to be magnetic in origin, probably arising from thermally activated transitions of flux between localised flux pinning states in the thin film.

More recently another IBM group [31] have succeeded in fabricating dc-SQUIDs using high quality *in situ grown* YBCO thin films with controlled strength weak-link junctions formed across the interface of bicrystal substrates. The noise performance of these devices (see Fig.8) appeared limited by junction- rather than flux-noise. Most recently Conductus[32] in association with Berkeley and Stanford Groups have reported a dc SQUID, which is claimed to have the best field sensitivity yet reported for a HTC device (also shown in Fig.8). It uses the weak-links formed at a 45° switch in thin film orientation produced by growing part of the film on a buffer layer on which the film grows at 45° to that on the rest of the substrate. The same group have succeeded in developing a multilayer thin film deposition technology, which allows the fabrication of the superconducting flux transformers necessary to couple such SQUIDs to the fields to be measured. Currently, the flux transformer has to be deposited on a second substrate and placed face downwards on the SQUID to achieve the necessary coupling.
Progress in fabrication of multilayer structures, which preserve both the epitaxy and superconducting properties of the YBCO films, is such that a fully integrated dc SQUID with integrated flux transformer looks increasing likely within the next year or so.

References.

1. A.Barone and G.Paterno, *The Physics and Applications of the Josephson Effect,* John Wiley, New York, 1982
2. T.van Duzer and C.W.Turner, *Principles of Superconducting Devices and Circuits,* Elsevier, New York, 1981
3. B.D.Josephson, Physics. Letts. **1**, 251 (1962)
4. R.P.Feynman Lectures on Physics vol 3 ch 21 (1965), Addison-Wesley, New York.
5. P.W.Anderson and J.M.Rowell, Phys.Rev.Lett.**10**, 230 (1963)
6. V.Ambegaokar and A.Baratoff, *Phys.Rev.Lett.* 10 (1963) 486
7. I.Giaever, Phys.Rev.Lett.**5**, 147 (1960)

8. V.Ambegaokar and B.I.Halperin, Phys.Rev.Lett.**24**, 876 (1961)
9. K.K.Lhikarev, Rev.Mod.Phys. **51**, 101 (1979)
10. A.Widom and T.D.Clark, Nuovo Cimento Lett.**28**, 186 (1980)
11. see, for example, A.J.Leggett, Phys.Rev.B **30**, 1208 (1984)
12. see for example A.I.M.Rae,*Quantum Mechanics : Illusion or Reality,* Cambridge University Press (1986).
13. M.H.Devoret, J.M.Martinis and J.Clarke, Phys.Rev.Lett .**55**, 1908 (1985)
14. D.B.Schwartz, B.Sen, C.N.Archie and J.E.Lukens, Phys.Rev.Lett. **55**, 1547 (1985)
15. S.Washburn, R.A.Webb, R.F.Voss and S.M.Farris, Phys.Rev.Lett.**54**, 2712 (1985)
16. A.Hamzic, L.Fruchter and I.A.Campbell, Nature **345**, 515 (1990)
17. D.E.McCumber, J.Appl.Phys.**39**,297 (1967)
18. J.Kurkijarvi, Phys. Rev. B**6**, 832 (1972)
19. S.Harrop, C.E.Gough, M.N.Keene and C.M.Muirhead, Supercond.Sci.Technol. **1**,68 (1988)
20. A.H.Silver and J.E.Zimmermann, Phs.Rev.Lett. **15**,888 (1965)
21. J.E.Zimmerman, J.A.Beall, M.W.Cromar and R.H.Ono, Appl. Phys. Lett. **51**, 617 (1987)
22. B.V.Vasiliev, preprints (DUBNA) and private communications
23. V.N.Zavaritski and N.V. Zavaritski, Physica B **165&166B**, 1447 (1990)
24. De Gennes, *Superconductivity of Metals and Alloy,* Benjamin, New York, 1966
25. R.C.Jaklevic, J.Lambe, A.H.Silver and J.E.Mercereau, Phys.Rev.Lett.**12**, 159 (1964): R.C.Jaklevic, J.Lambe, J.E.Mercereau and A.H.Silver. Phys.Rev.**140**, A1628 (1965)
26. M.N.Keene, S.P.Harrop, T.J.Jackson, C.M.Muirhead and C.E.Gough, Supercond. Sci Technol. **3**, 263 (1990)
27. J.Clarke , Nature **333**, 29 (1988)
28. C .M.Pegrum and G.B.Donaldson, *EEC Workshop on HTSC and Potential Applications,* Genoa, p125 (1987)
29. R.H.Koch, W.J.Gallagher, B.Bumble and W.Y.Lee, Appl.Phys.Lett.**54**, 951 (1989)
30. M.J.Ferrari. M.Johnson, F.C.Wellstood, J.Clarke, P.A.Tosenthal, R.H.Hammond and M.R.Beasley, Appl.Phys.Lett.**53**, 695 (1988)
31. R.Gross, P.Chaudhari, M.Kawasaki, M.B.Ketchen and A.Gupta, Appl.Phys.Lett.**57**, 727 (1990)
32. A.H.Miklich, J.J.Kinston, F.C.Wellstood, J.Clarke, M.S.Colclough, K.Char, and G.Zaharchuk to be published in Appl.Phys.Letts.

Index of Contributors

Springer Series in Solid-State Sciences

Editors: M. Cardona P. Fulde K. von Klitzing H.-J. Queisser

1 **Principles of Magnetic Resonance** 3rd Edition By C. P. Slichter

2 **Introduction to Solid-State Theory** 2nd Printing By O. Madelung

3 **Dynamical Scattering of X-Rays in Crystals** By Z. G. Pinsker

4 **Inelastic Electron Tunneling Spectroscopy** Editor: T. Wolfram

5 **Fundamentals of Crystal Growth I** Macroscopic Equilibrium and Transport Concepts. 2nd Printing By F. Rosenberger

6 **Magnetic Flux Structures in Superconductors** By R. P. Huebener

7 **Green's Functions in Quantum Physics** 2nd Edition By E. N. Economou

8 **Solitons and Condensed Matter Physics** 2nd Printing Editors: A. R. Bishop and T. Schneider

9 **Photoferroelectrics** By V. M. Fridkin

10 **Phonon Dispersion Relations in Insulators** By H. Bilz and W. Kress

11 **Electron Transport in Compound Semiconductors** By B. R. Nag

12 **The Physics of Elementary Excitations** By S. Nakajima, Y. Toyozawa, and R. Abe

13 **The Physics of Selenium and Tellurium** Editors: E. Gerlach and P. Grosse

14 **Magnetic Bubble Technology** 2nd Edition By A. H. Eschenfelder

15 **Modern Crystallography I** Symmetry of Crystals. Methods of Structural Crystallography By B. K. Vainshtein

16 **Organic Molecular Crystals** Their Electronic States. By E. A. Silinsh

17 **The Theory of Magnetism I** Statics and Dynamics. 2nd Printing By D. C. Mattis

18 **Relaxation of Elementary Excitations** Editors: R. Kubo and E. Hanamura

19 **Solitons.** Mathematical Methods for Physicists. 2nd Printing By G. Eilenberger

20 **Theory of Nonlinear Lattices** 2nd Edition By M. Toda

21 **Modern Crystallography II** Structure of Crystals By B. K. Vainshtein, V. M. Fridkin, and V. L. Indenbom

22 **Point Defects in Semiconductors I** Theoretical Aspects By M. Lannoo and J. Bourgoin

23 **Physics in One Dimension** Editors: J. Bernasconi, T. Schneider

24 **Physics in High Magnetic Fields** Editors: S. Chikazumi and N. Miura

25 **Fundamental Physics of Amorphous Semiconductors** Editor: F. Yonezawa

26 **Elastic Media with Microstructure I** One-Dimensional Models. By I. A. Kunin

27 **Superconductivity of Transition Metals** Their Alloys and Compounds By S. V. Vonsovsky, Yu. A. Izyumov, and E. Z. Kurmaev

28 **The Structure and Properties of Matter** Editor: T. Matsubara

29 **Electron Correlation and Magnetism in Narrow-Band Systems** Editor: T. Moriya

30 **Statistical Physics I** Equilibrium Statistical Mechanics By M. Toda, R. Kubo, N. Saito

31 **Statistical Physics II** Nonequilibrium Statistical Mechanics By R. Kubo, M. Toda, N. Hashitsume

32 **Quantum Theory of Magnetism** 2nd Edition By R. M. White

33 **Mixed Crystals** By A. I. Kitaigorodsky

34 **Phonons: Theory and Experiments I** Lattice Dynamics and Models of Interatomic Forces. By P. Brüesch

35 **Point Defects in Semiconductors II** Experimental Aspects By J. Bourgoin and M. Lannoo

36 **Modern Crystallography III** Crystal Growth 2nd Edition By A. A. Chernov

37 **Modern Crystallography IV** Physical Properties of Crystals Editor: L. A. Shuvalov

38 **Physics of Intercalation Compounds** Editors: L. Pietronero and E. Tosatti

39 **Anderson Localization** Editors: Y. Nagaoka and H. Fukuyama

40 **Semiconductor Physics** An Introduction 4th Edition By K. Seeger

41 **The LMTO Method** Muffin-Tin Orbitals and Electronic Structure By H. L. Skriver

42 **Crystal Optics with Spatial Dispersion, and Excitons** 2nd Edition By V. M. Agranovich and V. L. Ginzburg

43 **Structure Analysis of Point Defects in Solids by Multiple Magnetic Resonance Spectroscopy** By J.-M. Spaeth, J. R. Niklas, and R. H. Bartram

44 **Elastic Media with Microstructure II** Three-Dimensional Models By I. A. Kunin

45 **Electronic Properties of Doped Semiconductors** By B. I. Shklovskii and A. L. Efros

46 **Topological Disorder in Condensed Matter** Editors: F. Yonezawa and T. Ninomiya

47 **Statics and Dynamics of Nonlinear Systems** Editors: G. Benedek, H. Bilz, and R. Zeyher

48 **Magnetic Phase Transitions** Editors: M. Ausloos and R. J. Elliott

49 **Organic Molecular Aggregates** Electronic Excitation and Interaction Processes Editors: P. Reineker, H. Haken, and H. C. Wolf

MIX
Papier aus verantwortungsvollen Quellen
Paper from responsible sources
FSC® C105338

If you have any concerns about our products,
you can contact us on
ProductSafety@springernature.com

In case Publisher is established outside the EU,
the EU authorized representative is:
Springer Nature Customer Service Center GmbH
Europaplatz 3, 69115 Heidelberg, Germany

Printed by Libri Plureos GmbH
in Hamburg, Germany